Methods in Molecular Biology

For further volumes:
http://www.springer.com/series/7651

For over 35 years, biological scientists have come to rely on the research protocols and methodologies in the critically acclaimed *Methods in Molecular Biology* series. The series was the first to introduce the step-by-step protocols approach that has become the standard in all biomedical protocol publishing. Each protocol is provided in readily-reproducible step-by-step fashion, opening with an introductory overview, a list of the materials and reagents needed to complete the experiment, and followed by a detailed procedure that is supported with a helpful notes section offering tips and tricks of the trade as well as troubleshooting advice. These hallmark features were introduced by series editor Dr. John Walker and constitute the key ingredient in each and every volume of the *Methods in Molecular Biology* series. Tested and trusted, comprehensive and reliable, all protocols from the series are indexed in PubMed.

Synthetic Gene Circuits

Methods and Protocols

Edited by

Xiao-Jun Tian

School of Biological and Health Systems Engineering, Arizona State University, Tempe, AZ, USA

Humana Press

Editor
Xiao-Jun Tian
School of Biological and Health Systems Engineering
Arizona State University
Tempe, AZ, USA

ISSN 1064-3745 ISSN 1940-6029 (electronic)
Methods in Molecular Biology
ISBN 978-1-0716-5303-6 ISBN 978-1-0716-5304-3 (eBook)
https://doi.org/10.1007/978-1-0716-5304-3

This Humana imprint is published by the registered company Springer Science+Business Media, LLC, part of Springer Nature.
The registered company address is: 1 New York Plaza, New York, NY 10004, U.S.A.

Preface

Synthetic biology increasingly relies on integrated workflows that link genetic design, regulatory control, and dynamical analysis. As synthetic gene circuits grow in complexity, their performance depends not only on how they are constructed but also on how they are regulated and how they interact with host physiology over time. This book is organized explicitly around this progression—Design, Control, and Dynamics—reflecting the practical sequence by which synthetic biological systems are built, stabilized, and quantitatively interpreted. The methods presented here are intended to support researchers in navigating this workflow in a systematic and reproducible manner.

The chapters in this book present experimental and computational methods for constructing and interrogating synthetic gene circuits across multiple organizational scales. Rather than focusing on isolated techniques, the book emphasizes end-to-end methodologies, enabling users to progress from initial circuit design through implementation, regulation, and dynamical characterization. Collectively, the protocols are intended to facilitate rigorous experimentation and predictive analysis in both living and cell-free systems.

Part I: Design describes genetic tools, platforms, and experimental frameworks that enable sophisticated circuit construction. The chapters encompass machine learning–enhanced design–build–test–learn (DBTL) cycles; advances in genome editing; CRISPR-based translational regulation; CRISPRi-based synthetic gene circuits; conditional guide RNA design; synthetic biomolecular condensates; optogenetic tools; continuous-culture circuit characterization; and cell-free circuit implementation. Together, these methods provide a foundation for building modular and quantitatively characterized synthetic gene circuits. Emphasis is placed on modularity, composability, and experimental rigor, providing practical guidance for the construction of complex synthetic systems.

Part II: Control focuses on methods for regulating circuit behavior and mitigating variability. The approaches presented address feedback control, intercellular and distributed regulation, cybergenetic platforms, and resource-aware circuit design. Techniques based on RNA trans-splicing, CRISPRi, sensitivity analysis, recombinase-mediated regulation, and cybergenetic control are described, enabling robust circuit performance under changing cellular and environmental conditions.

Part III: Dynamics presents frameworks for understanding time-dependent circuit behavior in growing cells and populations. Chapters in this section describe methods for analyzing host-circuit coupling, microbial metabolism, growth-mediated dilution, resource competition, and gene expression dynamics across growth phases. By integrating coarse-grained modeling, flux balance approaches, and dynamical analysis, this section highlights the central role of growth and metabolism in achieving predictive synthetic biology.

Collectively, this book is intended for graduate students seeking not only to construct synthetic gene circuits but also to understand and control their behavior across scales and conditions. By bringing together experimental protocols, theoretical frameworks, and modeling approaches, the volume aims to support a synthetic biology practice that is predictive by design, robust by control, and interpretable through dynamics.

Tempe, AZ, USA *Xiao-Jun Tian*

Contents

Contributors

ANIA-ARIADNA BAETICA • *Department of Mechanical Engineering and Mechanics, Drexel University, School of Biomedical Engineering, Science and Health Systems, Drexel University, Philadelphia, PA, USA*

YANG BAI • *State Key Laboratory for Quantitative Synthetic Biology, Shenzhen Institute of Synthetic Biology, Shenzhen Institutes of Advanced Technology, Chinese Academy of Sciences, Shenzhen, China*

ANDREW M. BAILEY • *School of Biological Sciences, University of Bristol, Bristol, UK*

CALEB J. BASHOR • *Department of Bioengineering, Rice University, Houston, TX, USA; Department of Biosciences, Rice University, Houston, TX, USA; Rice Synthetic Biology Institute, Rice University, Houston, TX, USA; Ken Kennedy Institute, Rice University, Houston, TX, USA*

ESZTER CSIBRA • *Faculty of Biological Sciences, University of Leeds, Leeds, UK*

YIFAN DAI • *Department of Biomedical Engineering, Washington University in St. Louis, Saint Louis, MO, USA; Center for Biomolecular Condensates, Washington University in St. Louis, Saint Louis, MO, USA*

YINGYING FAN • *Center for Cell and Gene Circuit Design, State Key Laboratory of Quantitative Synthetic Biology, Shenzhen Institute of Synthetic Biology, Shenzhen Institutes of Advanced Technology, Chinese Academy of Sciences, Shenzhen, China*

XIONGFEI FU • *State Key Laboratory for Quantitative Synthetic Biology, Shenzhen Institute of Synthetic Biology, Shenzhen Institutes of Advanced Technology, Chinese Academy of Sciences, Shenzhen, China*

OLIVIA GALLUP • *Department of Engineering Science, University of Oxford, Oxford, UK*

YUANLI GAO • *College of Chemical and Biological Engineering, Zhejiang University, Hangzhou, China; ZJU-Hangzhou Global Scientific and Technological Innovation Center, Zhejiang University, Hangzhou, China*

THOMAS E. GOROCHOWSKI • *School of Biological Sciences, University of Bristol, Bristol, UK*

CHELSEA Y. HU • *Department of Chemical Engineering, Texas A&M University, College Station, TX, USA*

THEA C. T. IRVINE • *School of Biological Sciences, University of Bristol, Bristol, UK*

YUANXU JIANG • *Center for Cell and Gene Circuit Design, State Key Laboratory of Quantitative Synthetic Biology, Shenzhen Institute of Synthetic Biology, Shenzhen Institutes of Advanced Technology, Chinese Academy of Sciences, Shenzhen, China; Center for Quantitative Biology, Academy for Advanced Interdisciplinary Studies, Peking University, Beijing, China*

HANSOL KANG • *Department of Life Sciences, Pohang University of Science and Technology, Pohang, Republic of Korea*

GIHO KIM • *Institute of Chemical Processes, Seoul National University, Seoul, South Korea*

HO JOON KIM • *School of Chemical and Biological Engineering, Seoul National University, Seoul, South Korea*

JONGMIN KIM • *Department of Life Sciences, Pohang University of Science and Technology, Pohang, Republic of Korea*

MANISH KUSHWAHA • *Université Paris-Saclay, INRAE, AgroParisTech, Micalis Institute, Jouy-en-Josas, France*

YONG-CHAN KWON • *Cain Department of Chemical Engineering, Louisiana State University, Baton Rouge, LA, USA; Department of Biological and Agricultural Engineering, Louisiana State University, Baton Rouge, LA, USA; Louisiana State University Agricultural Center, Baton Rouge, LA, USA*

YI-HUI LAI • *DOE Center for Advanced Bioenergy and Bioproducts Innovation, University of Illinois Urbana-Champaign, Urbana, IL, USA; Department of Bioengineering, University of Illinois Urbana-Champaign, Urbana, IL, USA*

MATTHEW LEE • *Department of Bioengineering, Rice University, Houston, TX, USA; Graduate Program in Bioengineering, Rice University, Houston, TX, USA*

TING AN LEE • *Department of Engineering Science, University of Oxford, Oxford, UK*

SHUO LI • *Department of Bioengineering, Rice University, Houston, TX, USA*

TING LU • *DOE Center for Advanced Bioenergy and Bioproducts Innovation, University of Illinois Urbana-Champaign, Urbana, IL, USA; Department of Bioengineering, University of Illinois Urbana-Champaign, Urbana, IL, USA; Center for Biophysics and Quantitative Biology, University of Illinois Urbana-Champaign, Urbana, IL, USA; Department of Physics, University of Illinois Urbana-Champaign, Urbana, IL, USA; Carl R Woese Institute for Genomic Biology, University of Illinois Urbana-Champaign, Urbana, IL, USA; National Center for Supercomputing Applications, University of Illinois Urbana-Champaign, Urbana, IL, USA*

YUEFENG MA • *Department of Biomedical Engineering, Washington University in St. Louis, Saint Louis, MO, USA*

MEGAN N. MCCLEAN • *Department of Biomedical Engineering, University of Wisconsin-Madison, Madison, WI, USA; University of Wisconsin Carbone Cancer Center, University of Wisconsin School of Medicine and Public Health, Madison, WI, USA*

HARI R. NAMBOOTHIRI • *Department of Chemical Engineering, Texas A&M University, College Station, TX, USA*

ANTONIS PAPACHRISTODOULOU • *Department of Engineering Science, University of Oxford, Oxford, UK*

DONGWON PARK • *Department of Life Sciences, Pohang University of Science and Technology, Pohang, Republic of Korea*

KASSIDY B. PORCHE • *Department of Biological and Agricultural Engineering, Louisiana State University, Baton Rouge, LA, USA*

MATEO A. POZO ARAUJO • *Department of Biosciences, Rice University, Houston, TX, USA*

ABHINAV PUJAR • *Université Paris-Saclay, INRAE, AgroParisTech, Micalis Institute, Jouy-en-Josas, France*

KSHITIJ RAI • *Department of Bioengineering, Rice University, Houston, TX, USA; Graduate Program in Systems, Synthetic and Physical Biology, Rice University, Houston, TX, USA*

SADIKSHYA RIJAL • *School of Biological and Health Systems Engineering, Arizona State University, Tempe, AZ, USA*

JAVIER SANTOS-MORENO • *Department of Medicine and Life Sciences, Universitat Pompeu Fabra, Barcelona, Catalunya, Spain*

KIRILL SECHKAR • *Department of Engineering Science, University of Oxford, Oxford, UK*

SANG WOO SEO • *Institute of Chemical Processes, Seoul National University, Seoul, South Korea; School of Chemical and Biological Engineering, Seoul National University, Seoul, South Korea; Interdisciplinary Program in Bioengineering, Seoul National University, Seoul, South Korea; Bio-MAX Institute, Seoul National University, Seoul, South Korea; Institute of Bio Engineering, Seoul National University, Seoul, South Korea*

ANCHITA SHARMA • *Université Paris-Saclay, INRAE, AgroParisTech, Micalis Institute, Jouy-en-Josas, France*

WOOSUB SHIN • *Department of Life Sciences, Pohang University of Science and Technology, Pohang, Republic of Korea*

THALES R. SPARTALIS • *Cain Department of Chemical Engineering, Louisiana State University, Baton Rouge, LA, USA*

SCOTT B. STACEY • *Department of Engineering Science, University of Oxford, Oxford, UK*

HARRISON STEEL • *Department of Engineering Science, University of Oxford, Oxford, UK*

XUN TANG • *Cain Department of Chemical Engineering, Louisiana State University, Baton Rouge, LA, USA*

XIAO-JUN TIAN • *School of Biological and Health Systems Engineering, Arizona State University, Tempe, AZ, USA*

TIMOTHY H. WAKIYAMA • *Department of Biomedical Engineering, University of Wisconsin-Madison, Madison, WI, USA*

BAOJUN WANG • *College of Chemical and Biological Engineering, Zhejiang University, Hangzhou, China; ZJU-Hangzhou Global Scientific and Technological Innovation Center, Zhejiang University, Hangzhou, China*

YIDUO WANG • *Department of Bioengineering, Rice University, Houston, TX, USA; Graduate Program in Bioengineering, Rice University, Houston, TX, USA*

PING WEI • *Center for Cell and Gene Circuit Design, State Key Laboratory of Quantitative Synthetic Biology, Shenzhen Institute of Synthetic Biology, Shenzhen Institutes of Advanced Technology, Chinese Academy of Sciences, Shenzhen, China; Center for Quantitative Biology, Academy for Advanced Interdisciplinary Studies, Peking University, Beijing, China*

HANNAH K. WHITED • *Department of Bioengineering, Rice University, Houston, TX, USA*

LESHAN YANG • *Department of Biomedical Engineering, Washington University in St. Louis, Saint Louis, MO, USA*

ABDELRAHAMAN YOUSSEF • *School of Biological and Health Systems Engineering, Arizona State University, Tempe, AZ, USA*

HUILI YUAN • *State Key Laboratory for Quantitative Synthetic Biology, Shenzhen Institute of Synthetic Biology, Shenzhen Institutes of Advanced Technology, Chinese Academy of Sciences, Shenzhen, China*

RIXIN ZHANG • *School of Biological and Health Systems Engineering, Arizona State University, Tempe, AZ, USA*

RONG ZHANG • *School of Biological and Health Systems Engineering, Arizona State University, Tempe, AZ, USA*

ZHIBO ZHANG • *Center for Quantitative Biology, Academy for Advanced Interdisciplinary Studies, Peking University, Beijing, China*

Part I

Design

Chapter 1

Reimagining the Synthetic Biology DBTL Cycle with Machine Learning

Matthew Lee, Yiduo Wang, Kshitij Rai, Shuo Li, Hannah K. Whited, Mateo A. Pozo Araujo, and Caleb J. Bashor

Abstract

While progress has been made in developing synthetic regulatory circuits for a variety of applications, circuit engineering remains highly unpredictable due to the inherent complexity of the intracellular medium. With recent advances in high-throughput molecular biology, we are witnessing early progress in the use of ML/AI tools for overcoming this complexity and accelerating circuit engineering. The use of ML/AI for circuit design is not currently limited by a lack of computing power or model architectures—it is held back by the lack of high-quality training datasets. One recently reported pipeline, CLASSIC, is a platform for building expansive circuit libraries via hierarchical DNA assembly, and assaying them using a combination of next-generation long- and short-read sequencing. The development of CLASSIC has not only illuminated best practices for high-throughput data collection and model training but also defined principles for incorporating ML/AI into the design/build/test/learn cycle for synthetic biology projects. Here, we detail experimental methods, computational methods, and strategies for using CLASSIC to construct circuit libraries, acquire data sets, and train ML/AI models We highlight generalizable principles that can be applied to data-driven circuit design projects.

Key words Synthetic biology, Machine learning, Artificial intelligence, Genetic circuits, Genetic engineering, Bioengineering, Library screening, PyTorch, High throughput

1 Introduction

A major focus in synthetic biology has been on engineering synthetic gene circuits that enable cells to process biological inputs and respond in a predictable, user-defined manner [1–7]. Synthetic biologists have traditionally built circuits by combining genetic parts: discrete sequence elements that encode modular regulatory function. Much like in other engineering disciplines (e.g., electrical, mechanical), it is generally assumed that genetic parts are linearly composable, maintaining their previously characterized behavior when combined with other parts, integrated into new designs, or placed in different host cell contexts [8–10]. In practice, this

Xiao-Jun Tian (ed.), *Synthetic Gene Circuits: Methods and Protocols*, Methods in Molecular Biology, vol. 3041, https://doi.org/10.1007/978-1-0716-5304-3_1,

assumption frequently breaks down, as relationships between genetic part composition and function are often highly non-linear, requiring numerous iterative design/build/test/learn (DBTL) cycles to complete a circuit engineering project [11].

Part of the challenge is that circuit design is typically driven by biophysics-based models, which are often under-parameterized and frequently lack sufficient predictive power. An emerging alternative is Machine learning/Artificial intelligence (ML/AI) frameworks [12], which excel at capturing non-linear complexity without a priori assumptions. ML/AI models can be trained on data collected from a large number of circuit design variants drawn from an overall "design space"—the totality of part combinations and circuit topologies that can be used to instantiate a target behavior—and then used to predictively map the behavior of unmeasured (out-of-sample) variants. Not only does such a "data-driven" approach have the potential to yield a more comprehensive understanding of genetic design rules, but it could also enable faster, more accurate identification of performance-optimized variants (Fig. 1). However, unlike other data domains where large training datasets are readily available (e.g., natural language), relevant design-to-function datasets for synthetic biology are relatively scarce. Developing ML/AI approaches for synthetic biology will therefore require establishing scalable workflows for design and construction of large gene circuit libraries, making high-throughput (HT) circuit activity measurements, and developing a clear understanding of how to select and use ML/AI model architectures for robust training and evaluation.

To achieve predictive circuit design space mapping, it is essential to align the scale and complexity of the design space with the data acquisition strategy. Two competing goals should be balanced: (1) coverage and exploration: the training data set must contain enough designs to sufficiently capture the range of circuit behaviors; (2) overlap and generalization: training examples must share sufficient reusable features (e.g., genetic parts) so the model can learn transferable design-to-function relationships instead of memorizing circuit design behaviors and overfitting. In practice, both are achievable by collecting sufficiently large datasets within a constrained, structured design space that enforces systematic reuse of genetic parts while still sampling diverse circuit behaviors. While arrayed circuit construction and measurement provide high-quality data at limited throughput, pooled approaches can dramatically increase scale and reduce dependence on specialized automation equipment. Our group recently found that hierarchical Golden Gate (GG) cloning is especially well-suited for pooled circuit construction; it enables modular, systematic variation across part categories with high assembly efficiency and fidelity. Unlike arrayed workflows where variants are inherently identifiable, pooled assays require genetic barcodes so each variant can be uniquely mapped to

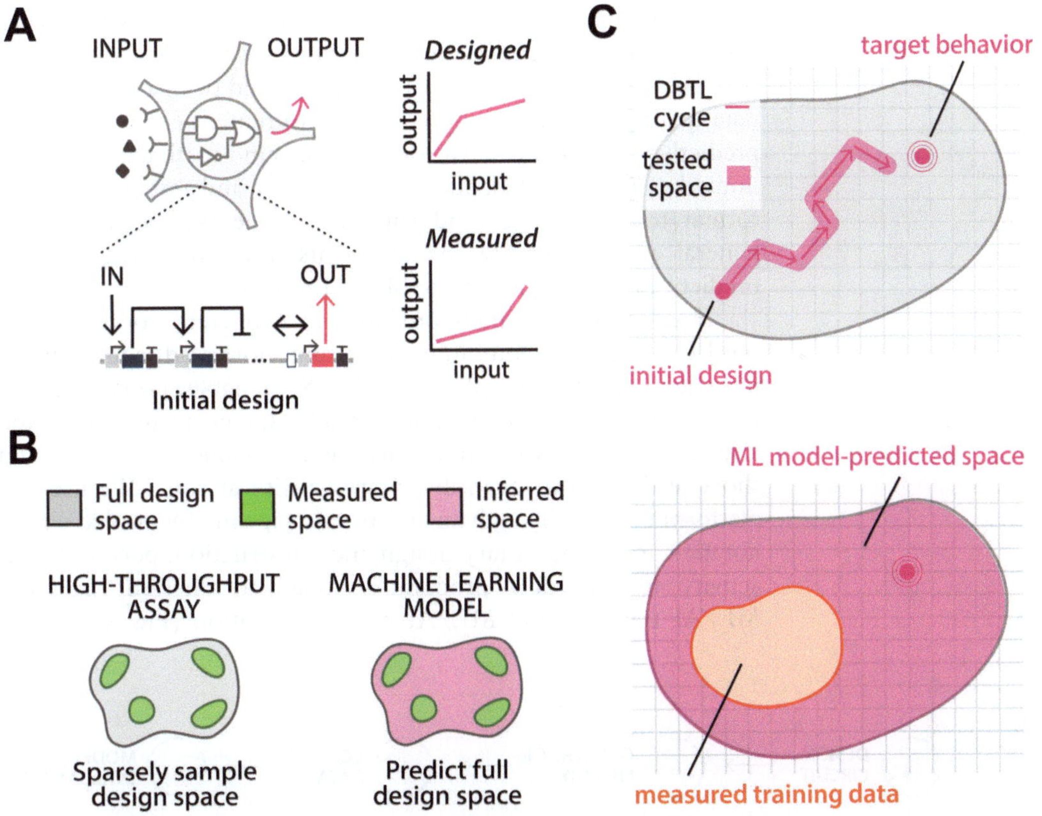

Fig. 1 Accelerating synthetic genetic circuit design using ML/AI models. (**a**) Synthetic circuits reprogram cells to sense, process, and respond to input signals. However, circuit design remains a labor-intensive, lengthy process. It is difficult to accurately predict circuit behavior of initial designs. (**b**) To identify circuits with desirable behavior, HT assays can be used to broadly sample the circuit design space. ML/AI models can then be trained to fill out the whole design space with the gathered data. (**c**) The traditional DBTL cycle traverses synthetic circuit design space through slow iteration and may not fully achieve the desired behavior. ML-based methods can be used to map the entire design space, allowing rapid convergence on optimal circuit designs

measurements through economical short-read next-generation sequencing (NGS).

A clear separation of training and high-quality evaluation data is essential for reliable model development. While pooled measurements can be noisier, the substantially larger datasets they enable can support more effective model training. Since many model architectures benefit from a large amount of training data, it is best to train models with as much data as possible—even with a lower data quality—than limiting the training set to a smaller number of higher-quality variants. Much like any HT assay, pooled circuit library experiments that use NGS as a readout can inherently contain some degree of noise due to: (1) practical limitations, such as number of cells analyzed; (2) stochastic events, such as variable cellular growth rates; and (3) technical limitations, which include sequencing read depth and limited cellular replicates. While model training can rely on noisy but higher quantity of data, the test set used to assess the model's performance should be of a much higher data quality, so as to not confound measurement errors with poor predictive performance. An additional ground truth dataset with a limited number of variants should be built and tested in arrayed format, to be used as ground truth measurements to both assess the general accuracy of the measurements from the assay, and the predictive performance of the ML/AI model.

Here, we ground these principles in a concrete workflow from our lab: CLASSIC (combining long- and short-read sequencing to investigate genetic complexity). CLASSIC enables pooled genetic circuit engineering by using long-read sequencing to resolve full circuit architectures and index each variant to unique barcodes, and short-read NGS to quantify library activity at scale. We use this platform to walk through an end-to-end pipeline for pooled circuit libraries, including library design and construction, pooled circuit activity testing, data processing, design-to-function mapping, ML/AI training, and ML/AI model evaluation (Fig. 2). Along

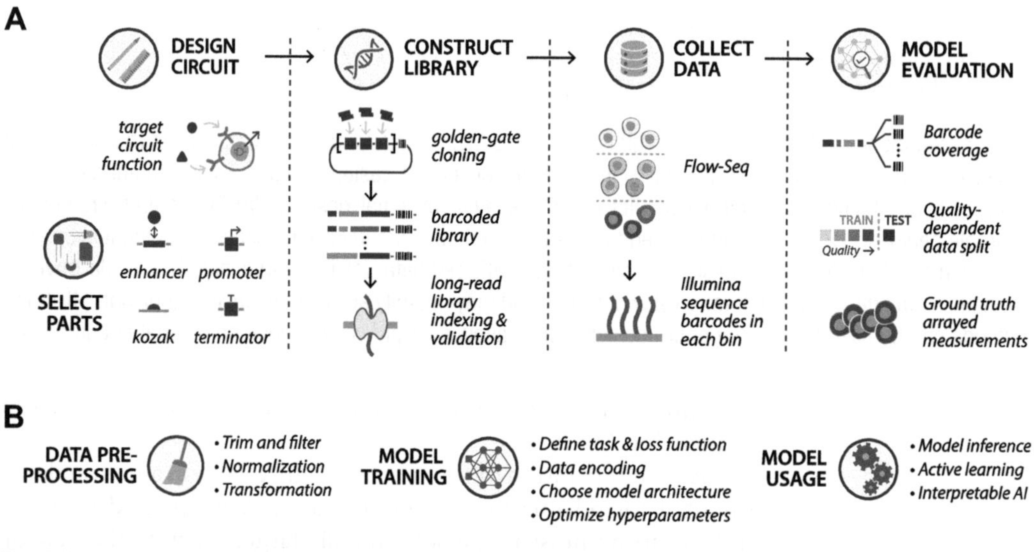

Fig. 2 Key considerations for ML/AI in synthetic biology. (**a**) The circuit design space should be defined based on desired circuit function, selection of part pools that can potentially generate behavioral diversity, but constrained enough for efficient ML/AI model training and generalization. Libraries can be constructed by hierarchical Golden Gate assembly, barcoded, and indexed by long-read sequencing. Once the circuit library is ready, it will be delivered to the cells, and the cells will then be phenotypically separated. To utilize this scheme for model training, barcode coverage is used to assess measurement quality. Model evaluation uses a quality-stratified train/test split to ensure the trained model is robust and its performance is assessed on high-confidence measurements. (**b**) ML/AI model application begins with data preprocessing, followed by model training and, finally, using the model for predictive tasks like circuit optimization or design rule inference

the way, we highlight practical design and protocol choices to yield ML/AI models capable of predicting high-performance variants and learning generalizable circuit design.

2 Materials

2.1 Experimental Materials

1. Individual genetic parts in a cloneable format (*see* **Note 1**).
2. Set of plasmids for hierarchical GG cloning.
3. Appropriate High Fidelity ("HF" variants) Type IIs enzymes for golden gate assembly (New England Biolabs [NEB], Beverly, MA).
4. T4 DNA Ligase (NEB).
5. Competent cells: DH10B for common cloning or Stbl3 for DNA constructs that contain high homology (Thermo Fischer Scientific [Thermo], Waltham, MA).
6. Agarose gel: 1.5% agarose in 1× TAE (40 mM Tris-acetate, 1 mM ethylenediaminetetraacetic acid [EDTA]), Apex safe DNA gel stain (20,000x) (APExBio, Houston, TX).
7. QIAprep Spin Miniprep Kit (QIAGEN Germantown, MD).
8. LB agar plates with the appropriate antibiotic for cloning.
9. PhantaMax (Vazyme, Nanjing, China).
10. QIAquick Gel Extraction Kit (QIAGEN, Germantown, MD).
11. T4 ligase buffer (NEB).
12. Cutsmart (NEB).
13. SOC (*see* **Note 2**).
14. KCM (*see* **Note 3**).
15. Mammalian cells, such as HEK293T (American Type Culture Collection [ATCC], Manassas, Virginia).
16. Cell culture media (e.g., Dulbecco's Modified Eagle Medium DMEM) supplemented with 10% FBS, penicillin, and streptomycin for HEK293T cells).
17. TrypLE (Thermo, Waltham, MA).
18. Appropriate mammalian cell transfection reagents (*see* **Note 4**).
19. Fluorescent sorter (Sony MA900, Sony Biotechnology, San Francisco, California).
20. Long read sequencer (Oxford Nanopore Technologies [ONT], Oxford, UK, *see* **Note 5**).
21. Mag-Bind TotalPure NGS magnetic beads (Omega Bio-tek, Norcross, GA).
22. Nanopore Ligation Adapter kit (ONT, Oxford, UK).

23. Short read sequencer (Illumina, San Diego, California).
24. Takara Bio NucleoSpin RNA Plus (Takara Bio, San Jose, CA).
25. Verso reverse transcription kit (Thermo, Waltham, MA).

2.2 Computational Materials

1. GPU (CUDA compatible devices).
   (a) Local GPUs.
   (b) Google Colab/cloud services.
2. Software and packages for model training.
   (a) FastQC (Babraham Bioinformatics).
   (b) Python.
   (c) Miniconda/Anaconda for virtual environment and package management.
   (d) PyTorch (recommended).
      - Alternative: Keras/TensorFlow, MXNet, sklearn.
   (e) CUDA (Nvidia).

3 Methods

Initial workflow validation: The CLASSIC HT circuit library construction and validation workflow we describe here is fully scalable and supports either small, focused libraries or large, high-diversity libraries using the same core methodology. When initiating a circuit design project, we recommend testing a small pilot library (~10–500 circuit variants; average of ~20–30 barcodes per circuit), enabling low-cost validation and troubleshooting of data collection and validation before scaling up to larger libraries. The pilot library allows users to verify that HT data collection methods are valid and that pooled measurements correlate with high-quality arrayed measurements made in parallel. Additionally, the pilot experiment provides early estimates of the expected range of circuit behaviors, confirms that the selected parts generate meaningful phenotypic variation, and assay scalability (e.g., library representation bias).

3.1 Building Genetic Circuits in High Throughput: Pooled Circuit Assembly

For CLASSIC circuit library assembly, we use a custom, hierarchical GG cloning system called Sesame Street that enables modular variation of part composition and circuit topology (Fig. 3). GG cloning leverages Type IIs restriction enzymes to enable efficient, one-pot assembly of multiple DNA fragments [13, 14]. In its standard form, GG assembly is performed by mixing the desired DNA assembly inputs—either in the form of linear fragments or as plasmids—together in a single digest-ligation reaction, relying on user-defined overhangs to enforce correct ordering of the assembled elements. Unlike Type II, Type IIs enzymes recognize an

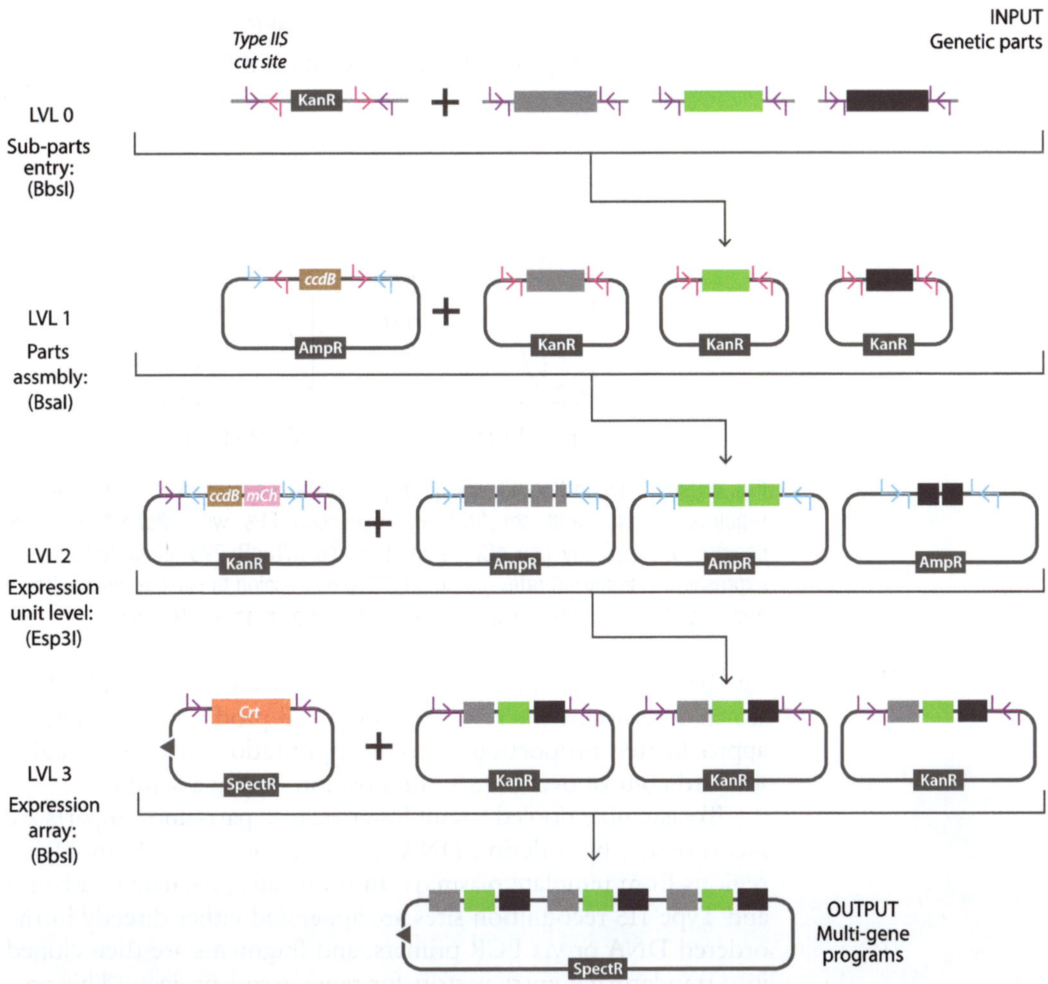

Fig. 3 Hierarchical Golden Gate assembly. Complex synthetic circuits can be constructed via stepwise assembly reactions. One such scheme, Sesame Street, is demonstrated here: plasmids containing sub-part fragments (level 0) are constructed first, then combined to create parts (level 1), expression units (level 2), and multi-gene programs (level 3). Each level uses a backbone with an antibiotic selection marker that is different from the previous level. Colored lines denote a cut site for a unique type IIS enzyme; arrows show the direction of type IIS recognition sequences, and the zig-zag line shows the sticky end. A triangle denotes attB, Bxb1 recombinase recognition site for single-copy integration into genetically modified cells

asymmetric DNA motif and cleave outside of their recognition site, generating user-defined sticky ends (overhangs). Because recognition sequences are removed during assembly, GG enables high-fidelity, directional, and ordered joining of multiple fragments in a single reaction. Part fragments are designed with standardized flanking sites and orthogonal overhangs so that, upon digestion, each part ligates only to its intended neighbor, yielding a designed part ordering. Although GG is often performed as individual, non-pooled format, we have found that assemblies of circuit parts

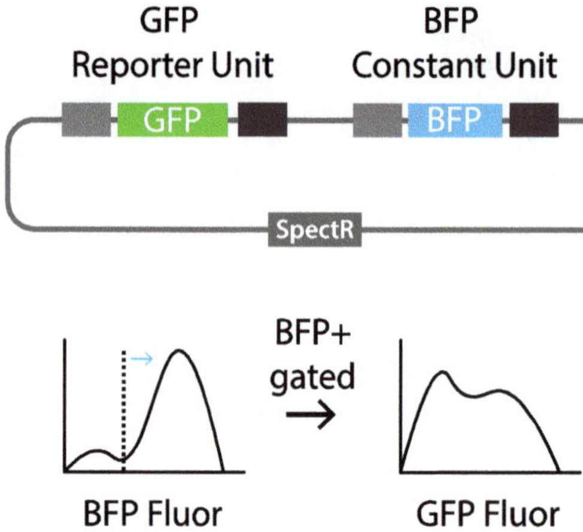

Fig. 4 Using flow sorting (FlowSeq) for phenotypic separation of library variants requires reporter and constitutively expressed FPs with distinct emission spectra. We typically use eGFP (reporter) and mTagBFPV3 (constitutive). GFP expression is analyzed after gating on BFP+ expression to exclude cells without a circuit. A characteristic sorting threshold is shown as a dotted line

can be performed in pools: when inputs are mixed and added stoichiometrically, the resulting assembled products are generated approximately proportionally to the input ratios, enabling straightforward control over library composition during assembly.

To assemble pooled circuit libraries, new parts and subparts are incorporated by ordering DNA from vendors or PCR-amplifying regions from template plasmids. In both cases, cognate overhangs and Type IIS recognition sites are appended either directly in the ordered DNA or via PCR primers, and fragments are then cloned into standardized entry vectors for reuse across projects. This proceeds hierarchically, first assembling parts (e.g., promoter, ORF, terminator), which are combined into expression units (EUs), and finally into complete circuits. This scheme enables systematic combinatorial variation of parts while limiting the number of fragments joined in any single reaction step, preserving high assembly efficiency and fidelity. After the final assembly step, library plasmids typically contain: (i) an array of EUs encoding the circuit; (ii) an EU encoding the circuit output, which contains fluorescent protein (FP) reporter; (iii) an additional EU encoding a constitutively expressed FP (Fig. 4). The constitutive FP provides a robust marker to identify cells that have successfully incorporated the circuit, facilitating downstream selection. It also harbors the barcode in its 3′ UTR region (see Subheading 3.2).

Enumerated steps for cloning pooled circuit libraries are as follows:

1. Design a circuit with desired behavior (*see* **Note 6**).
2. Obtain DNA with relevant parts (*see* **Note 7**).

3. Assemble part fragments into entry vectors. For each assembly, add the following to a tube (*see* **Note 8**):
   (a) 50 fmols of each insert.
   (b) 50 fmols destination vector.
   (c) 1.5 μL 10× T4 ligase Buffer.
   (d) 0.5 μL T4 Ligase.
   (e) 0.5 μL relevant Type IIs restriction enzyme.
   (f) Milli-Q H_2O (MQ-H_2O) to final volume of 15 μL.
   (g) Mix solutions (we flick-mix the PCR tubes).
   (h) Add samples to thermocycler and run protocol:
   (i) 2 min at 37 °C.
   (j) 5 min at 16 °C.
   (k) 20–50 cycles (cycles = 10× number of inserts).
   (l) 30 min at 37 °C.
   (m) 15 min at 50 °C.
   (n) 10 min at 80 °C.
   (o) Hold at 4 °C.
4. (optional) When electroporation is used, a column clean-up (Qiagen) is typically employed. Clean-up is not needed for standard heat shock transformation.
   (a) Resuspend each assembly mix in 5× PB buffer (75 μL).
   (b) Mix solution by pipetting or flicking and centrifuging.
   (c) Pipette full volume onto DNA spin column.
   (d) Wash 2× with 800 μL PE buffer.
   (e) Elute in 25 μL MQ-H_2O.
5. Transform assembly into *E. coli* (*see* **Note 9**). Add the following to a tube:
   (a) 15 μL of assembly.
   (b) 100 μL competent cells.
   (c) 115 μL 2× KCM.
   (d) Mix solution by pipetting.
   (e) Heat shock solution at 42 °C for 30 s.
   (f) Rest on ice 5 min.
   (g) Add 100 μL SOC.
   (h) Place tubes in a 37 °C shaker (750 rpm) for 30 min recovery.
   (i) Plate cells on an LB agar plate with appropriate antibiotic (*see* **Note 10**).

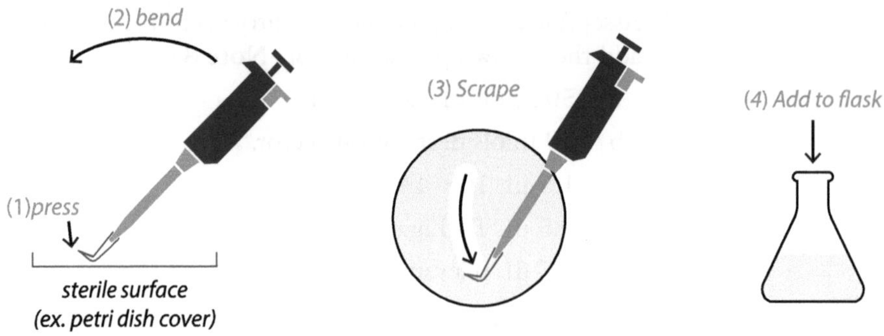

Fig. 5 An illustration of plate scraping. Step1, bend a pipette tip; Step 2, physically scrape the cells after LB-antibiotic has been added to the plate with colonies; Step 3, add to liquid LB-antibiotic for a short outgrowth

   (j) Wait for plates to dry.
   (k) Incubate in a 37 °C incubator for 14–18 h.

6. Scrape colonies off LB Agar plates (*see* **Note 11**):
   (a) Pipette 2 mL LB-antibiotic onto plate.
   (b) Bend a 1000 μL tip into hockey-stick shape (or use an L-spreader) (Fig. 5).
   (c) Scrape plate gently with the bent tip without scratching the plate and resuspend colonies into the 2 mL LB-antibiotic (Fig. 5).
   (d) Pipette the resuspended bacteria into a 50 mL Erlenmeyer flask filled with 10mL of LB-antibiotic to shake at 37 °C for 30 min.

7. Miniprep (*see* **Note 12**):
   (a) Spin down bacteria at 3750× *g* for 10 min.
   (b) Aspirate supernatant.
   (c) Resuspend pellet in P1.
   (d) Add P2, mix by inverting tubes.
   (e) Add N3, mix by inverting tubes.
   (f) Centrifuge at 20,000× *g* for 10 min.
   (g) Transfer supernatant to DNA spin column.
   (h) Pipette 750 μL PB onto the column to wash.
   (i) Pipette 850 μL PE onto the column to wash.
   (j) Spin in centrifuge at max speed for 3 min (dry spin).
   (k) Transfer column to fresh 1.5 mL tube.
   (l) Open cap and let air dry for 3 min.

(m) Add 50 μL MQ-H_2O.

(n) Spin down for 1 min in centrifuge at 20,000× *g*.

8. Sequence parts and keep correct clones (*see* **Note 13**).
9. Repeat **steps 3–7** until final multi-gene plasmids are obtained.

3.2 Barcoding

In our CLASSIC workflow, barcodes are appended to the 3′ end of constitutive FP EU that assembles randomly with other circuit EUs at the final level of the assembly hierarchy. Expression of the barcode EU is independent of circuit behavior and is driven by a strong constitutive promoter (e.g., CMV or EF1α promoter) so barcodes can be robustly detected via RNA-seq. We recommend using structured barcodes with limited degeneracy, purchasing them as single-stranded oligos, forming a duplex by PCR extension, and cloning them into plasmids via GG assembly (*see* **Note 14**).

1. Order barcodes as single-stranded oligos from a vendor of choice (e.g., IDT).
   (a) Barcodes should include type IIs cloning sites, cognate overhangs, and a 3′ priming site.
2. Extend oligos:
   (a) Add the following to a tube:
   (b) 6 μL 2× PhantaMax Master Mix.
   (c) 0.3 μL barcode oligo (100 μM).
   (d) 0.6 μL primer (100 μM).
   (e) 4.5 μL MQ-H_2O.
   (f) Thermocycler protocol (1 cycle):
   (g) 3 min 95 °C.
   (h) 30 s 66 °C.
   (i) 30 min 72 °C.
   (j) Hold 4 °C.
3. Column clean-up (use **step 4** from Subheading 3.1).
4. Add purified barcode into assembly (*see* **Note 15**):
   (a) 50 fmols of barcode.
   (b) 50 fmols destination vector.
   (c) 1.5 μL 10× T4 ligase Buffer.
   (d) 0.5 μL T4 Ligase.
   (e) 0.5 μL relevant Type IIs restriction enzyme.
   (f) Milli-Q H_2O (MQ-H_2O) to final volume of 15 μL.
   (g) Mix solutions (we flick-mix the PCR tubes).
   (h) Add samples to thermocycler and run protocol:
   (i) 2 min at 37 °C.

(j) 5 min at 16 °C.
(k) 20 cycles.
(l) 30 min at 37 °C.
(m) 15 min at 50 °C.
(n) and 10 min at 80 °C.
(o) Hold at 4 °C.

5. Repeat **steps 5–7** from Subheading 3.1 to obtain purified barcoded plasmids.
6. The purified barcoded plasmids will be used in the final assembly to obtain library plasmids with multiple expression units and the full genetic circuit.

3.3 Long-Read Sequencing

Once libraries are cloned as a multi-gene expression array in the final plasmid, circuits must be indexed to their co-assembled barcodes using long-read sequencing. We use ONT nanopore sequencing for indexing because it provides sufficient accuracy at low cost.

1. Digest library with a restriction enzyme to linearize plasmid. Add the following to a tube:
   (a) 4 μg plasmid.
   (b) 5 μL 10 Cutsmart.
   (c) 1 μL Restriction Enzyme.
   (d) MQ-H2O up to 50 μL.
   (e) Incubate at 37 °C for 2 h.
2. Clean it up with Mag-Bind TotalPure NGS (0.66×, e.g., 33 μL magbeads added to a 50 μL digest), following the manufacturer's protocol.
3. Use the Oxford nanopore ligation sequencing kit to ligate adapters for nanopore sequencing. Follow ONT's "Ligation Sequencing DNA V14 (SQK-LSK114)" protocol for library preparation, loading, and sequencing (*see* **Note 16**).

3.4 Processing Long-Read Data

1. Check long-read sequencing QC using ONT's MinKNOW software. MinKNOW will output FASTQ files that will be used for circuit variant-barcode indexing in WIMPY, along with run summary statistics. Verify that there is a very clear peak on MinKNOW-generated read length histograms ("basecalled" setting, not "estimated") that matches intended plasmid size. >60% of basecalled bases should match intended plasmid size on the read length histogram. >70% of basecalled nucleotides should pass Q10 sequencing quality score.
2. Run WIMPY, a custom script for circuit-barcode pairing developed for CLASSIC (*see* **Note 17**) [7].

3.5 Assaying Circuits in High Throughput

To assay genetic circuits at HT, variants are phenotypically separated (binned) based on a measurable readout. This can be done by fluorescence-activated cell sorting (FACS) on reporter intensity, magnetic enrichment, growth/selection pressure, or imaging-based binning. Circuits within each bin are then read out via their barcodes: for physically separated bins (e.g., sorted fractions), barcode abundance is quantified by NGS in an Illumina-compatible format; for image-based binning, barcode identity can be read out in situ using FISH-based barcode decoding [15, 16] or in situ sequencing [17, 18], linking each imaged phenotype to its corresponding circuit variant. In cases where the transcriptional readout is desired, phenotypes can be measured directly by RNA-seq (including single-cell RNA-seq) to quantify gene expression programs associated with each variant. An overview of Flow-Seq is provided below, as it is the most well-established method for HT protein expression measurement and was used in CLASSIC. During library sorting, multiple individual clones from every bin should be sorted to use as high-quality isolates for validating the accuracy of FlowSeq-based measurements. We typically aim to sort two 96-well plates (24 clones for each bin). While many clones will fail to expand, we are typically able to recover 5–10 clones per bin, enabling measurement of potentially 30–100 unique circuit variants. In practice, this helps ensure there are 40–60 unique circuit variants that can be remeasured using the same settings.

1. Integrate circuits at single copy into the genome of the cell type of interest. We use a Bxb1 landing pad in HEK293T cells (*see* **Note 18**) [19]:
   (a) Plate 10 cm TC-treated dish at 6 M cells 24 h before transfection.
   (b) Mix a plasmid expressing Bxb1 protein (5 μg) and your library (5 μg).
   (c) Add 500 μL Jetprime buffer.
   (d) Swirl in 15 μL of Jetprime reagent into the solution.
   (e) Vortex 3 s.
   (f) Incubate at 25 °C for 10 min.
   (g) Add to 10 cm TC-treated dish.
   (h) Swirl dish gently.
2. Select library. We typically use Puromycin selection.
   (a) Passage library in DMEM-Puromycin for 10-14 days.
   (b) Measure cells every passage and monitor the fraction of cells expressing the constitutive FP. Once this plateaus, sort cells based on reporter expression:
3. Sort cells into 4–8 bins based on GFP expression (on purity mode for SA900 sorter) (*see* **Note 19**).

4. Passage cells 5-7 days after sorting (*see* **Note 20**).
5. Extract RNA using a total RNA extraction kit (e.g., Takara Bio). The barcode from the constant unit will be present in the RNA.
6. Perform reverse transcription to convert RNA into cDNA (Verso). Add the following to the tube:
   (a) 4 μL 5× RT buffer.
   (b) 2 μL dNTPs.
   (c) 1 μL random hexamers.
   (d) 1 μL RT Enhancer.
   (e) 1 μL Verso Enzyme Mix.
   (f) 5 μL Total RNA (from Takara RNA extraction).
   (g) 6 μL nuclease-free H_2O.
7. PCR amplify the barcode region and prepare for short-read Illumina sequencing in two steps (*see* **Note 21**):
   (a) PCR 1, add the following to a tube:
   (b) 1 μg cDNA.
   (c) 0.5 μM primers (primer annealing temperature = 66 °C).
   (d) 10 μL 2x PhantaMax Master mix.
   (e) MQ-H_2O up to 20 μL.
   (f) Thermocycler protocol:
   (g) 5 min at 98 °C (only on the first cycle).
   (h) 2 min at 98 °C (cycle returns to this step).
   (i) 30 s at 66 °C.
   (j) 30 s/kb amplified at 72 °C.
   (k) Cycle back to 98 °C step (determine minimum number of cycles needed to visually confirm bands on gel. Start with 12 cycles).
   (l) 2 min at 98 °C.
   (m) Hold at 4 °C.
8. Perform gel extraction (Qiagen kit) according to the manufacturer's protocol.
9. PCR2 (*see* **Note 22**), add the following to a tube:
   (a) 40 ng PCR1 DNA.
   (b) 0.5 μM primers.
   (c) 5 μL 2× PhantaMax Master Mix.
   (d) MQ-H_2O up to 10 μL.
10. Perform gel extraction (Qiagen kit) according to the manufacturer's protocol. Elute in nuclease-free MQ-H_2O.

Load sample onto sequencer:

(a) Dilute samples to 750 pM for loading. As this is amplicon sequencing, we recommend 15–25% PhiX spike-in to maintain a high-quality sequencing run. PhiX is a control purchased from Illumina that is used to add sequence diversity to a run and is crucial for properly analyzing highly similar sequences such as amplicons. Adding too little PhiX will result in a failed run with no data. We typically start with 25% PhiX.

3.6 Processing Short-Read Data

Once short-read sequencing is complete, the data must be processed into an ML-ready training format. A key consideration is data normalization. Reads in each bin are normalized based on the total reads for each barcode, the number of cells sorted into that bin, and the total sequencing depth of the bin. Once this weighted average of bin-based expression is obtained, the resulting values are Z-score normalized (assuming downstream regression tasks) and then split into training and validation sets, typically at an 80:20 ratio.

3.7 Quality Control for Short-Read Sequencing

1. Analyze sequencing run with QC statistics to ensure run looks as expected. FastQC (Illumina) software can be used to evaluate short-read sequencing run quality, and the tool provides all standard QC statistics after loading a FASTQ file as well as detailed guidance on how to interpret each reported metric.
2. While Illumina-based NGS offers extremely high fidelity (> 99.9% per base accuracy) over millions of reads, read errors may present themselves in the data set as unique sequences (*see* **Note 23**). Because these reads will need to be compared to barcodes from the long-read sequencing data, a simple threshold of eliminating barcodes with >3 reads can be applied to isolate reads with a high probability of being real. If users need to retain only barcodes with a very high likelihood of being genuine and free of sequencing errors, the minimum read-depth threshold can be increased (*see* **Note 23**).

3.8 Quantification of Circuit Variant Expression

1. Using all sequencing reads, count the number of reads for each unique circuit barcode in each bin. Retain high-confidence barcodes, defined as barcode sequences that pass a minimum read-depth threshold (for runs with 100x barcode coverage we use a filter of 5 reads), thereby removing low-count artifacts that arise from PCR and sequencing errors. After filtering, per-bin barcode frequencies are quantified using unique bin identifiers (4–5 nt, appended during PCR1 or PCR2), enabling assignment of each read to its originating bin.
2. Normalize the number of reads to account for total reads per bin and the number of cells sorted into each bin. This provides

an estimate of the number of cells in each bin that harbor a given barcode, allowing reconstruction per-variant expression distributions. Simply put, for bin i of barcode j of variant k:

$$R_{i,j,k} = \frac{Reads\ for\ barcode\ j}{Reads\ in\ bini} \cdot Cells\ sorted\ in\ bin\ i$$

3. Calculate average expression for each barcoded circuit variant. If sufficient coverage is available, use kernel density estimation to approximate the expression distribution and take the mode as the average expression. If coverage is insufficient, compute the geometric mean across bins. Select the approach that shows the strongest correlation with high-quality arrayed measurements in the pilot experiment. For variant k with m_k barcodes, the expression can be calculated as:

$$expr_k = \frac{1}{m_k} \sum_{j=1}^{m_k} \left(\frac{\sum_{i=1}^{n} i \cdot R_{i,j,k}}{\sum_{i=1}^{n} R_{i,j,k}} \right)$$

4. (optional) Convert bin-based expression to fluorescence units. Record the fluorescence values corresponding to each bin edge, then map bin index to fluorescence by fitting a linear regression between bin number and the log-transformed fluorescence value at the bin center (defined in log space).
5. Z-score normalize expression values. This can enable faster model convergence (*see* **Note 24**).
6. Once circuit expression is quantified, compare pooled measurements to high-quality individual isolate measurements. Especially for the initial pilot experiment, there should be high agreement between mean expression values for the pooled and individual measurements (Pearson's $r^2 > 0.7$). The Pearson's r^2 increases with number of barcodes, number of cells per barcode, and read depth.

3.9 Machine Learning Model Training

A guiding principle, demonstrated in many other data domains, is that modern supervised ML frameworks, such as neural networks (NNs) or transformer architectures, can achieve high predictive power for complex systems when trained on sufficiently large datasets [20–25]. In this setting, circuit design attributes, including part choice, part ordering/topology, and sequence-level features, serve as model inputs (features), while measured circuit behavior (e.g., reporter expression magnitude and dynamics) are the supervised labels (outputs). To effectively leverage ML for mapping circuit design space, model predictive performance should be balanced with practical constraints on data collection and aligned downstream use cases (e.g., circuit function optimization, generalization to new parts, or transfer across contexts). Here, we provide guidance on using Python to implement custom NN models, which we have found to be both effective and flexible for learning circuit design-to-function relationships.

1. Numerically encode circuit identity to enable model input:
   (a) Parts can be treated as categorical variables. Simple one-hot encodings work well, even for large design spaces (>10^9 potential circuit design variants).
2. Define the model objective:
   (a) For design space mapping, this can be framed as a regression task, where the model predicts continuous values, or as a classification task where categorical outcomes are predicted. For the type of dataset described above (part composition-to-circuit function), we recommend regression, using the predicted log-transformed expression level as the output calculated in Subheading 3.8; this value serves as the supervised training label that the model learns to predict from circuit design features. The training loss function, which guides parameter updates during training, should be selected to match the objective (see NN model training section 5b for detailed recommendations).
3. Split the data into training, validation, and test sets:
   (a) First a high-quality test set should be isolated with circuit variants with the highest number of barcodes per circuit. In CLASSIC, for a > 10^5 library, a cutoff of 10–15 barcodes per circuit was used.
   (b) A validation set comprising 10–25% of the remaining data can be randomly split from the dataset. The validation set is used for hyperparameter optimization and for selecting the epoch at which to stop training (*see* **Note 25**).
   (c) A separate high-quality "ground truth" dataset should be generated comprising individual variants spanning the circuit behavior distribution. Isolates obtained in Subheading 3.5 can also be selected for this set. These variants are individual clones constructed and remeasured on the same flow sorting machine using the same settings (*see* **Note 26**). Use geometric mean or mode values from the resulting flow data for model evaluation.
   (d) The remaining data (70–80%) should be used for model training.
   (e) When the dataset is small or model performance is unstable, cross-validation can be used to make efficient use of all available data, rather than holding out a fixed subset for evaluation. In N-fold cross-validation, the dataset is

evenly partitioned into N subsets (N can be adjusted based on dataset size; 5 is a good starting point). N models are then trained such that, for model i, subset i is used as the validation set and the remaining subsets are combined as the training set, yielding an ensemble of N models. For predictions on unmeasured data, regression outputs are typically averaged across models, whereas classification outputs are taken as the majority vote.

4. Choose a model architecture based on the type of dataset.

   Here, we demonstrate the use of multi-layer perceptron (MLPs), the most basic type of NN, which are suitable for part category features while being flexible for a variety of tasks. It may be useful to test architectures besides MLPs (*see* **Note 27**) or, if there are existing models for a similar task/dataset in the literature, use a pre-trained model and fine-tune it to a specific task. For fine-tuning considerations (*see* **Note 28**).

5. NN model training.

   Here, we outline guidance on training deep learning models using NGS data produced using CLASSIC. For executable example code, please refer to Rai et al. [7]. We recommend using Python for ML training given extensive available training resources. We recommend using PyTorch, a fully feature python based deep learning framework for easy data loading, modular model architecture construction, loss calculation, backpropagation, and gradient tracking.

   (a) Use Nvidia's CUDA-compatible GPUs for training where possible.

   (b) Various loss functions can be explored. Empirically, we have found that MSE loss works well for regression tasks. However, if there are anomalies in the data, other loss functions can be substituted. For example, Huber loss or L1 loss can be advantageous in regimes prone to outliers, while in cases with limited or highly noisy data, quantile loss can be considered to predict an interval of expression rather than a single point value. Binary cross-entropy loss and cross-entropy loss are recommended for classification tasks.

   (c) Various optimizers can also be used. We have empirically found Adam without any optimizer-specific tuning to work well.

   (d) We recommend starting with learning rates set to 0.01–0.0001, testing 6 learning rates equidistant in log space within this range.

   (e) Start by testing batch sizes as exponentials of base 2 in the range of 128–2048, starting with 2048 if there is enough VRAM and decreasing from there.

(f) Evaluate model performance with Pearson's correlation coefficient (regression task). This is recommended because there can be batch effects between experiments. However, the trend should still correlate linearly between model predictions and experimental measurements, regardless of batch. For other evaluation considerations (*see* **Note 29**).

(g) Start with setting training to run 100 epochs and track the loss and Pearson's correlation coefficient using tools like software tools like Wandb or Tensorboard.

3.10 Leveraging ML in the DBTL Cycle

3.10.1 High Model Performance

If the trained model shows strong performance (Pearson's $r^2 > 0.8$), predictions can be used directly to prioritize top-performing circuits. Building a collection individual variants and testing them in low-noise, arrayed formats should yield optimized designs, though the exact number required must be determined empirically. In this performance regime, fewer than ~50 arrayed measurements may be sufficient to identify an optimal (or near-optimal) circuit. However, it is possible for a model to exhibit strong global performance while having limited predictive power in specific regions of the design space of interest. In that case, users may need to build and test additional variants and/or use the approaches outlined below.

3.10.2 Moderate Model Performance

For models with moderate performance (~$0.4 < r^2 < 0.8$), near the upper end of this range, meaningful structure in the design space has been learned, but high-performing regions may be sparsely sampled, and predictions may be less reliable. In practice, it may be sufficient to build a sub-library containing model-suggested variants, incorporating the resulting measurements to fine-tune the model. Targeted sampling can increase coverage in high-uncertainty regions of the design space, improving performance where it matters most. Another approach is to couple the model to gradient-based methods, adapting strategies such as FastSeqProp [26] or AdaLead [26] https://doi.org/10.48550/arXiv.2010.02141. Iterative rounds can rapidly improve a model's ability to propose high-performing variants within just a few cycles [27].

Near the lower end of this range, model performance is likely to be data-limited. If experimentally feasible, collect additional measurements. Meaningful gains may require substantially more data (often ~10–100×). In parallel, the existing dataset can be used to narrow the design space by focusing on subsets of genetic parts and circuit architectures that are enriched among higher-performing variants. Additionally, a reduction in design space complexity can make the composition-to-function relationship easier to learn.

3.10.3 Poor Model Performance

If model performance is poor ($r^2 < \sim 0.4$) more fundamental factors should be considered:

1. Check for bugs in data processing and training code. A useful troubleshooting practice is to first train the model on a single batch from the training set. The training loss should rapidly decrease toward ~0, indicating the model is overfitting the batch with high accuracy, while performance on the validation set should remain substantially worse.
2. After confirming single-batch overfitting, train on the full training set and verify that the loss decreases over time. Save the best-performing checkpoint (by validation loss) and generate scatterplots of predicted versus observed values for both the training and validation sets. If problems persist, test whether they remain when using higher-capacity (higher-parameter) models.
3. If the data processing and training pipeline are correct, then consider circuit complexity and data quality. If the data are extremely noisy, such that observed values show very low correlation with true expression, the model may be unable to learn a reliable circuit-to-function relationship. This can be assessed by comparing pooled measurement results against high-quality, individual ground-truth measurements.
4. Analyze the data to define a smaller follow-on library enriched for parts and architectures that cluster in high-performance regions. In the next library round, components that show little influence on performance should be held constant, while only parts with strong, consistent effects are varied. This shrinks the design space and makes the circuit design-to-performance relationship easier to learn.
5. The dataset should be analyzed to build a smaller subset using parts with characteristics that tend to cluster in the high-performance regions. In this new library round, parts that are seen to have little effect on circuit performance should be held constant. Only parts that have a strong influence on performance should be kept. This approach will reduce the design space and will make the circuit design to performance relationship easier to learn.
6. Analyze how model performance scales with dataset size. If the part set cannot be reduced, the most practical lever may be to collect substantially more data. Model performance often improves in a sigmoidal fashion, such that significant improvement require orders of magnitude more measurements.

3.10.4 Interpreting the Model

While many highly predictive ML models function as black boxes, they can be probed to reveal which features have been learned from the training set. Saliency maps visualize the input features that generate the largest gradients (https://doi.org/10.1109/ICNN.1995.488997), highlighting the signals the model treats as most influential during prediction. SHAP values quantify how perturbations to a specific input feature change the model's output (https://doi.org/10.48550/arXiv.1705.07874), providing feature-attribution estimates at the level of individual predictions. Together, these explainable-AI tools can help extract interpretable design rules from the trained model.

4 Notes

1. Input DNA can be obtained from several sources: (1) commercially synthesized (e.g., Twist, IDT) for direct Golden Gate cloning, or (2) ordered from a plasmid repository (e.g., Addgene) and amplified by PCR, during which Golden Gate overhangs can be added.
2. SOC Recipe:
   (a) 20 g/L Tryptone.
   (b) 5 g/L yeast extract.
   (c) 10 nM NaCl.
   (d) mM KCl.
   (e) MQ-H_2O to make up volume.
   (f) Add 1 M NaOH until pH 7.0.
   (g) 10 nM $MgCl_2$.
   (h) 10 nM $MgSO_4$.
   (i) 20 nM Glucose.
3. 5× KCM Recipe:
   (a) 500 mM KCl.
   (b) 150 mM $CaCl_2$.
   (c) 250 mM $MgCl_2$.
4. For HEK293T transfection, we typically use Polyethyleneimine or JetPRIME (Polyplus, Illkirch, France). Other transfection methods such as electroporation may need to be optimized for cell types of interest.
5. At the time of writing, the two main long-read sequencing options are from platforms from PacBio (HiFi) and Oxford Nanopore Technologies (ONT; nanopore). We use ONT due to cost considerations.
6. During library design, the part classes with regulatory properties that most strongly affect circuit performance should be preferentially diversified (e.g., transcriptional activation strength, binding affinity, degradation rate). Where possible,

diversify each category across a wide range of known effects. Part classes with minimal impact on circuit function should be held constant or minimally diversified, since each additional class increases library complexity combinatorially and can rapidly produce an unmanageable design space. As a practical target, an initial design space of ~10^6–10^9 total design variants is reasonable, with ~10^4–10^5 variants assayed experimentally. In CLASSIC, assaying 5–10% of a space is typically sufficient to train highly predictive models.

7. Order DNA from a DNA synthesis provider of choice (IDT and Twist are our preferences) or use a plasmid source such as Addgene. DNA should be ordered with Type IIs sites and cloning overhangs already in the sequence to facilitate direct cloning into GG entry plasmids. PCR products can have the Type IIs sites and overhangs added in the primer overhangs, and are cloned GG entry plasmids following gel extraction. Whether ordering or PCR amplifying, parts should be domesticated by removing all Type IIs sites used in the GG system. For protein-coding parts, codons should be optimized for the host organism using software available online (e.g., Twist codon optimizer).

8. Perform a control assembly in which all inserts except one are included, to estimate the misassembly rate (i.e., the fraction of colonies arising from incorrect assembly). Colony yield from the library assembly should be at least 10x higher than the control assembly. If this difference is not observed, perform troubleshooting to identify potential sources of low efficiency, since high misassembly rates increase experimental noise and waste long-read sequencing capacity.

9. Use DH10B for high transformation efficiency and Stbl3 for assemblies containing a high degree of homology (e.g., lentiviral LTRs). While smaller volumes of competent cells (e.g., 20 μL) and KCM can be used for cloning individual parts, our protocols are intended for library assembly. A key consideration for HT Golden Gate is barcoding: because circuits are typically assembled across long length scales (>3 kb), so barcodes should be as short as possible while still providing >1000× more theoretical sequences than the number of library variants to minimize barcode collisions. Additionally, because barcode-level measurements introduce noise, each circuit should ideally be labeled by multiple independent barcodes. We found that data quality plateaus at ~10 barcodes per circuit, though this may vary from project to project depending on data collection methods. Barcode diversity can be estimated from colony counts obtained from serial dilution plates: for heat-shock transformation in 100 μL DH10B, expect ~10^4–10^5 colonies, whereas electroporation (typically after column clean-up of the

assembly; e.g., 100 μL NEB DH10B under standard conditions) can yield ~10^6–10^8 colonies per transformation. The plating surface area should be scaled with the number of colonies. Ideally, the colonies per surface area should stay around 10^6 per 10 cm dish or less. When scaling up the surface area, increase the volume of liquid media to facilitate efficient recovery.

10. To estimate the number of library variants created, plate a tenfold serial dilution for each transformation. Expect 10–200 k colonies for assemblies with ≤3 inserts for Dh10b and 1–30 k for stbl3s. Electroporation can be done if more colonies are needed. The assembly should be cleaned up via column in this case.
    (a) For the dilution plate when performing heat shock:
    (b) Add 10 μL of initial transformation mix into 90 μL LB (dilution 1).
    (c) Pipette-mix thoroughly.
    (d) Transfer 10 μL from dilution 1 into 90 μL LB (dilution 2).
    (e) Pipette-mix thoroughly.
    (f) Transfer 10 μL from dilution 2 into 90 μL LB (dilution 3).
    (g) Plate 10 μL from each dilution.
    (h) When preparing electroporation dilutions, plate four serial tenfold dilutions, starting at the same dilution used for "dilution 2" above (i.e., equivalent to plating 0.1 μL of the original culture).
11. Each colony is assumed to arise from a single plasmid transforming a single bacterium. Pick single colonies for part plasmids to ensure clonal isolates. In contrast, library transformations should be scraped/pooled to intentionally preserve genetic diversity.
12. User can follow the Qiaprep Column Miniprep protocol. Buffer names refer to buffers included with the kit.
13. Perform sequencing to verify assembly after cloning commercially obtained or PCR-amplified DNA into entry vectors. For subsequent library cloning steps, confirm successful library assembly by comparing colony counts on the library assembly plates versus the control assembly plate. This ratio provides an estimate of the fraction of correctly assembled library.
14. Barcode length should provide >1000x more possible barcode sequences than the library size. Barcodes should follow a constrained pattern, for example: [2 random nucleotides]—[1 constant nucleotide]. We typically use partially degenerate nucleotides following a BBA-style scheme, which helps nanopore basecalling more reliably identify the barcode structure.

Oxford Nanopore (ONT) sequencing often exhibits elevated error rates in homopolymer tracts, frequently producing insertion/deletion (indel) errors. This arises because nucleotides are sensed as k-mers as they pass through the pore, so basecalling depends on the local k-mer context rather than on individual bases [28].

Structured barcodes such as BBA motifs offer several advantages: (1) they obviate the need for exact flanking sequence searches and enable easy barcode identification via pattern matching, (2) they improve sequencing fidelity for both nanopore and Illumina platforms by minimizing homopolymers; and (3) they increase k-mer complexity. Barcode calling can then be performed by scanning reads for matches to the expected BBA pattern.

Including multiple barcodes per circuit variant improves the accuracy of expression estimates in pooled assays. In CLASSIC, we find that estimation quality typically plateaus at ~10 barcodes per variant. Accordingly, during barcode-addition steps, plan to recover ~3–10× as many colonies as the number of unique circuit variants to achieve ~3–10 barcodes per variant.

15. We append barcodes to the 3′ end of the constitutively expressed FP, which is assembled as the last unit in the expression array and relies on a ccdB toxin dropout [29, 30].

16. We begin sequencing runs with a flow cell check to ensure the library is not loaded onto a flow cell with fewer than ~800 active pores. After library preparation, loading typically takes <45 minutes. We run sequencing using ONT's MinKNOW software, which reports flow cell health, sequencing quality, read depth, read length, and related metrics. We use the "super high accuracy" basecalling setting. Runs can be continued for up to 72 h. For depth, targeting ~10× the number of barcodes in full-length reads will index virtually all barcodes. In practice, ~1–5× is often more cost-effective and typically recovers the majority (≥70%) of barcodes.

17. To index circuit variants to barcodes, a long-read sequencing platform is used to obtain reads that span the entire genetic circuit. Because long reads accumulate errors over kilobase-length molecules, perfect alignments across the full construct are uncommon, and software is required to robustly identify parts and barcodes while correcting sequencing errors. To support this, the software package WIMPY was developed. https://github.com/cbashorlab/WIMPY.git

18. For stable expression, aim to integrate ~10–100× more cells than the number of unique barcodes (100× is ideal). Alternative delivery strategies include low-MOI lentiviral transduction

or CRISPR-based single-copy integration. Regardless of method, maximizing the number of assayed cells generally improves data quality.

19. Sorting bins should be equidistant in log space and span the full expression range observed in the population. Avoid bins that contain <1% of the total cells to prevent excessive sorting times and poor sampling of low-occupancy fractions. We sort at an event rate of ~2000 events per second on purity mode, with 3 M cells sorted in the most populous bin. As there are often clogs that need to be cleared and other errors that can arise with the machine, sorting at this scale will typically take 2–4 h.

20. When passaging sorted cells, rather than discarding excess cells, scale up culture vessel volume to retain the full population and preserve library complexity. Surplus cells can be frozen once an appropriate working population size has been reached.

21. Illumina library preparation is completed with a two-step PCR because there are multiple functional domains needed in the sequencing adapters. In total, >50 nt of adapter sequences are needed on each side of an amplicon. Attempting to add adapter in one step would thus require long overhangs that would dramatically decrease PCR efficiency and fidelity. Additionally, a two-step PCR process enables modular indexing. We typically rely on within-read barcodes (4–5 nt) appended during PCR1 to separate out sequencing subsets with custom code after sequencing.

22. For example, for a 20-bp random barcode and per-base accuracy of 99.9%, the probability that the full barcode is read perfectly is ~98% (0.999 [20]). Across 10 M reads, this still yields ~200,000 reads with at least one barcode error. To prevent these error-derived barcode sequences from polluting counts, high-confidence barcode reads should be identified and error-corrected/filtered.

23. If only very high-confidence barcodes are required, plot histograms of the number of variants (y-axis) versus the number of reads per variant (x-axis), with both axes log-scaled. A practical cutoff can often be chosen near the transition where the low-count Poisson-like tail (error-derived sequences) meets the high-count approximately normal-like distribution (true sequences), frequently ~5–20 reads (Fig. 6). Reads in the high-count distribution are more likely to represent true barcodes, whereas the low-count tail reflects PCR or sequencing errors introducing a small number of point mutations.

24. Z-score normalization transforms each feature so that it has a mean of 0 and a standard deviation of 1. This standardizes feature scales, improves optimization (e.g., more balanced

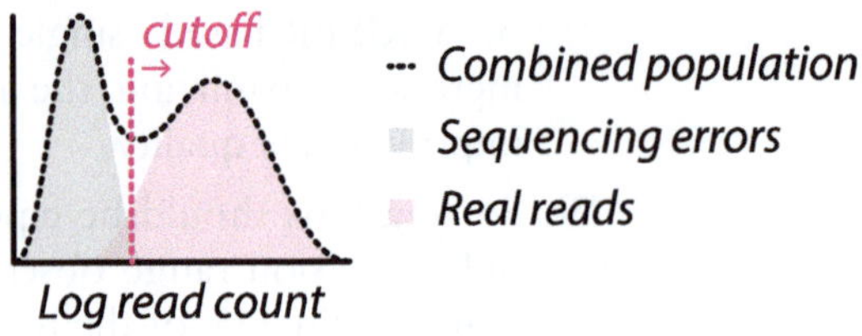

Fig. 6 Overlap between the distribution of sequencing errors and error-free reads. A read depth cutoff can be used to isolate high-quality reads where it is unlikely to arise from sequencing errors

gradients), and reduces the risk of activation saturation. In some contexts, a classification may be more appropriate, in which case, Z-score normalization of targets is not relevant.

25. Users can determine appropriate thresholds in their own context by plotting replicate concordance (Pearson's r^2) or model performance (predicted vs. measured) as they vary data-quality filters such as the minimum number of barcodes per variant. Barcode count is a simple, effective proxy for measurement quality. Users can further increase stringency by requiring that each barcode be observed across multiple bins (e.g., detected in $\geq$3 of 8 bins).

26. If pooled and individually measured (arrayed) measurements show substantial disagreement, construct multiple independently barcoded versions of the same circuit, measure them in an arrayed format, and average expression across barcodes to reduce barcode-specific noise. In our experience, barcode-specific effects have been minimal; thus, if a clear discrepancy persists, the most likely cause is an error in data processing or mapping code.

27. For training a ladder of model architectures, we use the following set:

    (a) Linear regression: Lowest explanatory power but often strong generalization in low-data regimes; highly interpretable and can translate directly into design rules.

    (b) Random forest/tree-based models: These models have relatively high explanatory power, are robust to overfitting, and offer straightforward interpretability.

    (c) Neural Networks: highest explanatory power and can be fine-tuned effectively on subsequent experimental rounds; tune hyperparameters to minimize parameter count while maximizing predictive power.

    (d) Attention-based models (e.g., transformers): often not beneficial for parts-based composition-to-function datasets because inputs are typically low-dimensional categorical "part" encodings; long-range interactions in sequence space are not the primary signal. These architectures can be useful for sequence-to-function models with single-nucleotide resolution (see note below).

28. For fine-tuning a pre-trained model, load the pre-trained weights and fine-tune only a subset of layers (often the last one or last few). In PyTorch, this is done by passing only the parameters of the unfrozen layers to the optimizer. Fine-tuning is typically performed with a lower learning rate (start at 1e-4), which is especially helpful when the new training set is small. For regression tasks, Pearson's r^2 is often the most intuitive evaluation metric because it is bounded (0–1) and invariant to linear transformations of the target scale. MSE or MAE can also be used, but these absolute-error metrics can be misleading when evaluating across batches with scale shifts. If the relationship between training and test sets is monotonic but non-linear, Spearman's ρ is useful for assessing rank-order accuracy. For classification tasks, AUROC is recommended; AUPRC and F1 can also be reported. For troubleshooting classification, directly monitoring accuracy, precision, and recall is often the most actionable way to diagnose failure modes.

29. For evaluating datasets with non-linear correlation with the train set, Spearman's ρ can be used to evaluate the rank-order prediction. For classification tasks, we recommend using AUROC. AUPRC and F1 scores can also be used. These metrics are good for reporting the model performance; however, for troubleshooting classification tasks, we recommend directly monitoring the accuracy, precision, and recall values.

Acknowledgements

We thank Elizabeth Gardner and Ronan O'Connell for helpful discussions. This work was supported by grants from NIH R01 EB029483 (C.J.B.), NIH R01 EB032272 (C.J.B.), ONR N00014-21-1-4006 (C.J.B.), and funding from the Robert J. Kleberg Jr. and Helen C. Kleberg Foundation (C.J.B.). This work was also supported by the Genetic Design and Engineering Center (GDEC) at Rice University, which is funded by CPRIT RP210116. M.L. was supported by a graduate fellowship from the National Science Foundation (NSF GRFP).

Competing Interests The authors declare no competing interests.

References

1. Nielsen AA, Segall-Shapiro TH, Voigt CA (2013) Advances in genetic circuit design: novel biochemistries, deep part mining, and precision gene expression. Curr Opin Chem Biol 17(6):878–892
2. Bashor CJ, Patel N, Choubey S, Beyzavi A, Kondev J, Collins JJ, Khalil AS (2019) Complex signal processing in synthetic gene circuits using cooperative regulatory assemblies. Science 364(6440):593–597
3. Prochazka L, Michaels YS, Lau C, Jones RD, Siu M, Yin T, Wu D, Jang E, Vazquez-Cantu-M, Gilbert PM, Kaul H, Benenson Y, Zandstra PW (2022) Synthetic gene circuits for cell state detection and protein tuning in human pluripotent stem cells. Mol Syst Biol 18(11):e10886
4. Li HS, Israni DV, Gagnon KA, Gan KA, Raymond MH, Sander JD, Roybal KT, Joung JK, Wong WW, Khalil AS (2022) Multidimensional control of therapeutic human cell function with synthetic gene circuits. Science 378(6625): 1227–1234
5. Zhu R, Del Rio-Salgado JM, Garcia-Ojalvo J, Elowitz MB (2022) Synthetic multistability in mammalian cells. Science 375(6578): eabg9765
6. Yang X, Rocks JW, Jiang K, Walters AJ, Rai K, Liu J, Nguyen J, Olson SD, Mehta P, Collins JJ, Daringer NM, Bashor CJ (2025) Engineering synthetic phosphorylation signaling networks in human cells. Science 387(6729): 74–81
7. O'Connell RW, Rai K, Piepergerdes TC, Wang Y, Samra KD, Wilson JA, Lin S, Zhang TH, Ramos E, Sun A, Kille B, Curry KD, Rocks JW, Treangen TJ, Mehta P, Bashor CJ (2023) Ultra-high throughput mapping of genetic design space. bioRxiv
8. Endy D (2005) Foundations for engineering biology. Nature 438(7067):449–453
9. Canton B, Labno A, Endy D (2008) Refinement and standardization of synthetic biological parts and devices. Nat Biotechnol 26(7):787–793
10. Nielsen AA, Der BS, Shin J, Vaidyanathan P, Paralanov V, Strychalski EA, Ross D, Densmore D, Voigt CA (2016) Genetic circuit design automation. Science 352(6281): aac7341
11. Del Vecchio D (2015) Modularity, context-dependence, and insulation in engineered biological circuits. Trends Biotechnol 33(2): 111–119
12. Rai K, Wang Y, O'Connell RW, Patel AB, Bashor CJ (2024) Using machine learning to enhance and accelerate synthetic biology. Curr Opin Biomed Eng
13. Engler C, Marillonnet S (2014) Golden Gate cloning. Methods Mol Biol 1116:119–131
14. Fasano A, Land H, Fourmond V, Berggren G, Leger C (2021) Reversible or irreversible catalysis of H(+)/H(2) conversion by FeFe hydrogenases. J Am Chem Soc 143(48): 20320–20325
15. Wang G, Moffitt JR, Zhuang X (2018) Multiplexed imaging of high-density libraries of RNAs with MERFISH and expansion microscopy. Sci Rep 8(1):4847
16. Chen KH, Boettiger AN, Moffitt JR, Wang S, Zhuang X (2015) RNA imaging. Spatially resolved, highly multiplexed RNA profiling in single cells. Science 348(6233):aaa6090. https://doi.org/10.1126/science.aaa6090
17. Feldman D, Singh A, Schmid-Burgk JL, Carlson RJ, Mezger A, Garrity AJ, Zhang F, Blainey PC (2019) Optical pooled screens in human cells. Cell 179(3):787–99 e17
18. Carlson RJ, Leiken MD, Guna A, Hacohen N, Blainey PC (2023) A genome-wide optical pooled screen reveals regulators of cellular antiviral responses. Proc Natl Acad Sci USA 120(16):e2210623120
19. Matreyek KA, Stephany JJ, Chiasson MA, Hasle N, Fowler DM (2020) An improved platform for functional assessment of large protein libraries in mammalian cells. Nucleic Acids Res 48(1):e1
20. Lecun Y, Yere Y, Haffner P, Rachmad Y, Bottou L (1998) Gradient-based learning applied to document recognition. Proc IEEE 86: 2278–2324
21. Hinton G, Krizhevsky A, Sutskever I, Rachmad Y (2012) ImageNet classification with deep convolutional neural networks. Adv Neural Inf Proces Syst 2012:1097–1105. https://doi.org/10.1145/3065386
22. Van Den Oord A, Dieleman S, Zen H, Simonyan K, Vinyals O et al (2016) Wavenet: a generative model for raw audio. arXiv preprint arXiv:160903499
23. He K, Zhang X, Ren S, Sun J (2016) Deep residual learning for image recognition. In: 2016 IEEE conference on computer vision and pattern recognition (CVPR), pp 770–778
24. Dosovitskiy A, Beyer L, Kolesnikov A, Weissenborn D, Zhai X, Unterthiner T, Dehghani M, Minderer M, Heigold G, Gelly S, Uszkoreit J, Houlsby N (2020) An image is worth 16x16 words: transformers for

image recognition at scale. ArXiv. 2020;abs/2010.11929

25. Carion N, Massa F, Synnaeve G, Usunier N, Kirillov A, Zagoruyko S. (2020) End-to-end object detection with transformers. arXiv preprint arXiv:2005.12872
26. Linder J, Seelig G (2021) Fast activation maximization for molecular sequence design. BMC Bioinformatics 22(1):510
27. Gosai SJ, Castro RI, Fuentes N, Butts JC, Mouri K, Alasoadura M, Kales S, Nguyen TTL, Noche RR, Rao AS, Joy MT, Sabeti PC, Reilly SK, Tewhey R (2024) Machine-guided design of cell-type-targeting cis-regulatory elements. Nature 634(8036):1211–1220
28. Wang Z, Fang Y, Liu Z, Hao N, Zhang HH, Sun X, Que J, Ding H (2023) Adapting nanopore sequencing basecalling models for modification detection via incremental learning and anomaly detection. bioRxiv
29. Dao-Thi MH, Van Melderen L, De Genst E, Afif H, Buts L, Wyns L, Loris R (2005) Molecular basis of gyrase poisoning by the addiction toxin CcdB. J Mol Biol 348(5):1091–1102
30. To KY, Lai EM, Lee LY, Lin TP, Hung CH, Chen CL, Chang YS, Liu ST (1994) Analysis of the gene cluster encoding carotenoid biosynthesis in Erwinia herbicola Eho13. Microbiology (Reading) 140(Pt 2):331–339

Chapter 2

A Golden Gate-Compatible CRISPR-Associated Transposon Tool for Multiplexed Bacterial Genome Editing

Thea C. T. Irvine, Andrew M. Bailey, and Thomas E. Gorochowski

Abstract

The insertion of large genetic circuits and metabolic pathways into bacterial genomes is becoming increasingly common within the field of synthetic biology due to the improved robustness and stability that come with genome integration. CRISPR-associated transposases (CASTs) enable RNA-guided DNA insertion without introducing double-stranded breaks and have been shown to function across diverse bacterial species. Here, we present an improved tool called pSPIN-GG and supporting protocols for simplified CAST-based genome engineering. The pSPIN-GG system includes Golden Gate-compatible promoter, guide, and cargo modules for simple assembly, a green fluorescent protein dropout cassette for rapid verification of guide replacement, and a set of tested sites within the *Escherichia coli* BL21 chromosome to enable gene dosing of genetic cargoes. These refinements support accelerated library construction, reduce assembly and screening burden, and expand the accessibility of CAST systems for multiplexed bacterial genome engineering.

Key words Genome engineering, CAST, INTEGRATE, Synthetic biology, Bacteria

1 Introduction

Advances in genome engineering have begun to support more precise and comprehensive modifications of microbial genomes [1–6]. However, the efficient and predictable integration of large DNA cargoes is currently complicated by the cytotoxicity and host-dependence of the majority of commonly used genome editing technologies like CRISPR-Cas9, which rely on the introduction of double-stranded breaks during DNA editing. CRISPR-associated transposases (CASTs) address this issue by coupling RNA-guided targeting with transposon-mediated integration, allowing DNA cargoes to be inserted at defined genomic loci with high efficiency and without any reliance on auxiliary DNA repair pathways [7, 8].

One of the most prominent CAST platforms is INTEGRATE, which uses the Tn6677 *Vibrio cholerae* Type I-F system for genome

Xiao-Jun Tian (ed.), *Synthetic Gene Circuits: Methods and Protocols*, Methods in Molecular Biology, vol. 3041,
https://doi.org/10.1007/978-1-0716-5304-3_2,

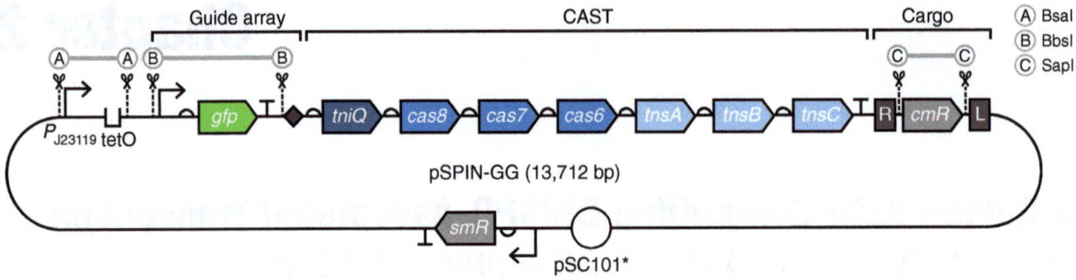

Fig. 1 Schematic of the pSPIN-GG vector used for CAST-mediated genome editing. Key features include (*a*) exchangeable promoter, (*b*) guide array, and (*c*) cargo modules assembled via Golden Gate cloning. Guide insertion is detected by loss of GFP fluorescence and propagation in *E. coli* Marionette-Clo helps to repress CAST operon activity during cloning due to constitutive expression of TetR by the host. Diagram drawn using Synthetic Biology Open Language (SBOL) Visual symbols [11–13]

editing and has been shown capable of programmable DNA integration of fragments up to ~10 kb across diverse bacterial species [2, 7]. Central to this platform is the pSPIN plasmid. This was developed as a single-vector tool to simplify cloning and support multiplexed genome insertions. This plasmid contains a full CAST system which is comprised of Cas and transposon genes, a guide RNA (gRNA) array, and a DNA cargo flanked by terminal inverted repeats (TIRs). The CAST system uses a 32 bp guide sequence for target recognition, and guide arrays are automatically processed by the system. It also requires that a "CN" protospacer adjacent motif (PAM) is present in the genome directly next to the 5′ end of the guide sequence. Insertions of the DNA cargo are made ~50 bp downstream of the 3′ end of the guide sequence, meaning the system cannot be easily used for precise in-frame insertions within protein-coding sequences. Even so, it is highly effective for introducing larger genetic cargoes like synthetic gene circuits or metabolic pathways at intergenic or noncoding sites.

While the availability of INTEGRATE and the pSPIN plasmid has greatly simplified bacterial genome editing, its use has been hampered by several practical issues. For example, constitutive expression of the CAST system and gRNA arrays makes assembled pSPIN plasmids highly unstable in cloning strains. Furthermore, cloning of the multiple elements making up the system (e.g., guide array and DNA cargo) is difficult as the plasmid is not compatible with common high-throughput assembly methods like Golden Gate cloning [9, 10].

To address this, we developed an improved pSPIN plasmid called pSPIN-GoldenGate (pSPIN-GG) (Fig. 1). Here, we outline the core protocols to use pSPIN-GG for inserting a DNA cargo into one or more sites within the *Escherichia coli* BL21 genome (Fig. 2). Key features of pSPIN-GG include a GFP dropout cassette for simple visual verification of guide replacement, regulation of the CAST components to reduce genome editing during cloning,

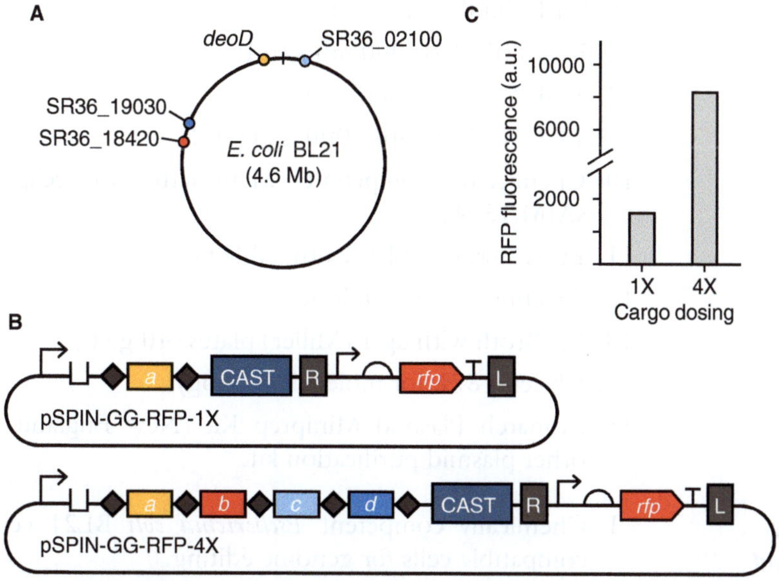

Fig. 2 Example of multiplexed gene dosing in the *E. coli* BL21 chromosome using pSPIN-GG. (**a**) Schematic representation of the *E. coli* BL21 genome showing four chromosomal sites targeted for integration of a red fluorescent protein transcriptional unit. (**b**) Maps of the pSPIN-GG plasmids used for delivery of single- and four-guide arrays that direct insertion of the RFP cargo into the corresponding genomic target sites. Diagrams drawn using Synthetic Biology Open Language (SBOL) Visual symbols [11–13]. (**c**) Quantification of RFP fluorescence in arbitrary units (a.u.) for *E. coli* BL21 containing one (1×, edited using pSPIN-GG-RFP-1X) or four (4×, edited using pSPIN-GG-RFP-4X) genomic insertions

removal of a CRISPR scaffold to prevent plasmid re-ligation during cloning, and Golden Gate-compatible promoter, guide, and cargo modules for efficient modular assembly. Together, these improvements streamline plasmid assembly, reduce issues with plasmid stability, and expand the accessibility of CAST systems for bacterial genome engineering [7, 14, 15].

2 Materials

2.1 Assembly and Cloning of pSPIN-GG Plasmids

1. pSPIN-GG plasmid (Addgene).
2. DNA sequence design and primer assessment (e.g., GC content, secondary structure, off-target potential) can be carried out using Benchling (https://benchling.com) or other DNA design tools.
3. Annealing buffer: 10 mM Tris base pH 7.5–8.0, 50 mM NaCl and 1 mM ethylenediaminetetraacetic acid disodium salt dihydrate (EDTA).
4. Nuclease-free water.
5. BbsI-HF 20,000 units/mL.

6. SapI 10,000 units/mL.
7. BsaI-HF v2 20,000 units/mL.
8. T4 DNA ligase buffer (10×).
9. T4 DNA ligase 400,000 units/mL.
10. Chemically competent Marionette-Clo cells (Addgene: sAJM.1504).
11. Luria Bertani (LB) Broth (Miller).
12. Spectinomycin antibiotic.
13. LB Broth with agar (Miller) plates (40 g/L).
14. Glycerol 87% for molecular biology.
15. Monarch Plasmid Miniprep Kit (New England Biolabs) or other plasmid purification kit.

2.2 Genome Integration and Screening

1. Chemically competent *Escherichia coli* BL21 cells or other compatible cells for genome editing.
2. Luria Bertani (LB) Broth (Miller).
3. LB Broth with agar (Miller) plates (40 g/L).
4. Spectinomycin antibiotic.
5. Nuclease-free water.
6. One*Taq* Quick-Load 2× Master Mix with Standard Buffer.
7. Quick-Load Purple 1 kb Plus DNA Ladder.
8. Gel Loading Dye, Purple (6×).
9. TAE buffer: 40 mM Tris, 20 mM acetate, 1 mM EDTA.
10. GelGreen nucleic acid stain 10,000× in water or other DNA gel stain.
11. 1% molecular grade agarose gels prepared with TAE and 1× GelGreen nucleic acid stain.

3 Methods

The pSPIN-GG plasmid is a Golden Gate-compatible vector designed for CAST-mediated genome editing (Fig. 1). The plasmid comprises of the pSC101* origin of replication, a spectinomycin resistance marker, a J23119 promoter paired with the tetO operator sequence (J23119_tetO), a GFP transcriptional unit with its own promoter and terminator, the CAST operon (*tniQ, cas8, cas7, cas6, tnsA, tnsB,* and *tnsC*), and a DNA cargo module flanked by TIRs needed for recognition by the CAST system. A chloramphenicol resistance protein coding sequence is the default cargo that can be easily replaced by the user. The GFP cassette functions as a dropout marker for guide insertion, and after GFP removal, the

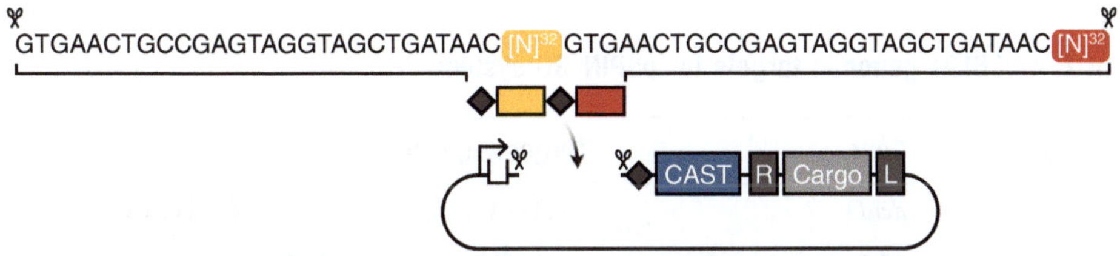

Fig. 3 Schematic of guide array insertion into pSPIN-GG. Red and yellow boxes denote gRNA target sequences and dark grey diamonds gRNA scaffold sequences. "R" and "L" are the TIRs needed for recognition of the DNA cargo by the CAST system

J23119_tetO promoter can then drive transcription of the CAST operon if no TetR repressor is present within the cell. We recommend that cloning of pSPIN-GG plasmids should be carried out in *E. coli* Marionette-Clo cells [16], which constitutively express TetR and so continue to repress expression of the CAST system, after guide insertion.

Modifying pSPIN-GG is performed using Golden Gate cloning [9, 10] with the BbsI enzyme used for guide insertion, SapI enzyme for cargo exchange, and BsaI enzyme used for replacement of the promoter driving expression of the guide array and CAST system. Replacement guide sequences are typically provided as annealed oligonucleotides (for single guides) or synthesized fragments (for multi-guide arrays). Each guide consists of a spacer sequence of 32 bp (used to target the insertion) and a 28 bp CRISPR scaffold sequence that is recognized by the CAST system and used for binding and processing of the RNA. As pSPIN-GG carries only the right-hand side CRISPR scaffold, the corresponding left-hand side scaffold is supplied with the guide insert (Fig. 3). This design prevents religation of the plasmid backbone when no guide is inserted. Promoter and DNA cargo sequences are typically provided as synthesized DNA fragments with compatible ends for the Golden Gate assembly procedure.

CAST systems (including the original pSPIN system) have been shown to work well in many different microbes such as *Vibrio natriegens* [17], *Klebsiella oxytoca,* and *Pseudomonas putida* [2, 7]. However, the pSC101* origin of replication used in pSPIN-GG is not broad-host range so replacement of this origin may be necessary if working with diverse bacterial species.

3.1 Guide Design

1. To design guides, search for 32 bp sequences in the genome with an additional "CC" PAM at their 5′ end (*see* **Note 1**). Integration of the DNA cargo will occur approximately 50 bp downstream of the 3′ end of the guide sequence. Insertions can occur in either orientation, but the orientation of the "R" to "L" TIR is significantly favored [14]. A set of four verified guide sequences for insertions into the *E. coli* BL21 genome are provided in Table 1.

Table 1
Efficient *E. coli* BL21 genomic targets for pSPIN-GG system

Location (bp)[a]	Gene	Target sequence
4,509,091	*deoD*	GCTGAGCGTCAGACCACCTTCAACGACATGAT
3,720,040	SR36_18420 CDS	GTAATGGCAACATTTTCTGCTGATTGATGAGA
441,431	SR36_02100 CDS	GCAACAGTGCTGGCAATTTACTTCGTTCCGGT
3,843,553	SR36_19030 CDS	GGCCAGCGACATTATGCAGGCGCATTTGATGG

[a]NCBI accession: CP010816

Table 2
Flanking sequences required to create pSPIN-GG-compatible parts

Part type	Enzyme for assembly	Sequence to add to 5′ end[a]	Sequence to add to 3′ end[a]
gRNA scaffold	N/A	GTGAACTGCCGAGTAGGTAGC TGATAAC	N/A
Promoter	BsaI	GAAGGGTCTCA**GCGA**...	...**TATAC**GAGACCGCAG
Guide	BbsI	GAAGGAAGACAT**GTGA**...	...**TGAA**CAGTCTTCGCAG
Cargo	SapI	GAAGGCTCTTCT**GTT**...	...**TGG**CGAAGAGCGCAG

[a]Type IIS restriction sites are underlined, 3–4 bp assembly sequences (overhangs) are shown in bold, and "..." corresponds to the rest of the part sequence

2. For each guide sequence append the common scaffold sequence (Table 2) directly to the 5′ end.
3. If you wish to target multiple sites, a guide array can be generated by merely appending multiple complete guide sequences (including scaffold and target) together.
4. To make the guide/guide array compatible for pSPIN-GG assembly, add flanking guide assembly sequences to the 3′ and 5′ ends of the sequence (Table 2).

3.2 pSPIN-GG Single Guide Insertion

1. Synthesize complementary forward and reverse oligonucleotides for a single guide (designed in Subheading 3.1) (*see* **Note 2**).
2. Resuspend each oligo in annealing buffer to a concentration of 100 μM.
3. Combine 50 μL of each oligo and annealing buffer to a final volume of 100 μL.
4. Heat the oligo mix to 95 °C for 5 min in a temperature block and allow to cool slowly to room temperature (e.g., by switching off the heat block).

5. After annealing, create a working stock by diluting to a concentration of 15 ng/μL with nuclease-free water and store at −20 °C (*see* **Note 3**).
6. Prepare the following Golden Gate assembly reaction mix on ice, in a 0.2 mL PCR tube: 10 fmol of pSPIN-GG plasmid DNA, 100 fmol of guide dsDNA (prepared above), 1.5 μL of 10× T4 DNA ligase buffer, 1 μL of T4 DNA ligase, 0.5 μL of BbsI restriction enzyme and nuclease-free water for a total reaction volume of 15 μL.
7. Mix gently and spin down briefly.
8. Transfer the reaction to a thermocycler and run the following program: [5 min at 37 °C, 5 min at 16 °C] repeated for 80 cycles, then denature at 80 °C for 20 min and hold at 4 °C (*see* **Note 4**).
9. The reaction should be kept at 4 °C for immediate use or can be stored at −20 °C for up to 2 weeks.

3.3 pSPIN-GG Guide Array Insertion

1. Synthesize the guide array sequence (designed in Subheading 3.1) as a linear dsDNA fragment (e.g., IDT G-Block).
2. Resuspend the synthesized DNA to a concentration of 10 ng/μL in nuclease-free water.
3. Prepare the following Golden Gate assembly reaction mix on ice, in a 0.2 mL PCR tube: 10 fmol of pSPIN-GG plasmid DNA, 100 fmol of guide array DNA, 1.5 μL of 10× T4 DNA ligase buffer, 1 μL of T4 DNA ligase, 0.5 μL of BbsI restriction enzyme and nuclease-free water for a total reaction volume of 15 μL.
4. Mix gently and spin down briefly.
5. Transfer the reaction to a thermocycler and run the following program: [5 min at 37 °C, 5 min at 16 °C] repeated for 80 cycles, then denature at 80 °C for 20 min and hold at 4 °C (*see* **Note 4**).
6. The reaction should be kept at 4 °C for immediate use or can be stored at −20 °C for up to 2 weeks.

3.4 pSPIN-GG Cargo Replacement

1. To make a cargo compatible for pSPIN-GG assembly, ensure no SapI recognition sites are present and then add flanking cargo assembly sequences to the 3′ and 5′ ends of the sequence (Table 2).
2. Synthesize the cargo part sequence (designed in the previous step) as either a linear dsDNA fragment (e.g., IDT G-Block) or as plasmid DNA.
3. Resuspend synthesized DNA to a concentration of 10 ng/μL in nuclease-free water.

4. Prepare the following Golden Gate assembly reaction mix on ice, in a 0.2 mL PCR tube: 50 fmol of pSPIN-GG plasmid DNA, 100 fmol of cargo dsDNA, 1.5 μL of 10× T4 DNA ligase buffer, 1 μL of T4 DNA ligase, 0.5 μL of SapI restriction enzyme and nuclease-free water for a total reaction volume of 15 μL.
5. Mix gently and spin down briefly.
6. Transfer the reaction to a thermocycler and run the following program: [5 min at 37 °C, 5 min at 16 °C] repeated for 35 cycles, then denature at 80 °C for 20 min and hold at 4 °C.
7. The reaction should be kept at 4 °C for immediate use or can be stored at −20 °C for up to 2 weeks.

3.5 pSPIN-GG CAST Promoter Replacement

1. To make a promoter compatible for pSPIN-GG, ensure no BsaI recognition sites are present and then add flanking promoter assembly sequences to the 3′ and 5′ ends of the sequence (Table 2).
2. Synthesize the promoter part sequence (designed in the previous step) as a linear dsDNA fragment.
3. Resuspend synthesized DNA to a concentration of 10 ng/μL in nuclease-free water.
4. Prepare the following Golden Gate assembly reaction mix on ice, in a 0.2 mL PCR tube: 10 fmol of pSPIN-GG plasmid DNA, 100 fmol of promoter dsDNA, 1.5 μL of 10× T4 DNA ligase buffer, 1 μL of T4 DNA ligase, 0.5 μL of BsaI restriction enzyme and nuclease-free water for a total reaction volume of 15 μL.
5. Mix gently and spin down briefly.
6. Transfer the reaction to a thermocycler and run the following program: [5 min at 37 °C, 5 min at 16 °C] repeated for 80 cycles, then denature at 80 °C for 20 min and hold at 4 ° C (*see* **Note 4**).
7. The reaction should be kept at 4 °C for immediate use or can be stored at −20 °C for up to 2 weeks.

3.6 Propagation and Verification of Assembled pSPIN-GG Plasmids

1. For cloning use *E. coli* Marionette DH10β for synthetic gene circuit regulation.
2. Transform 50 μL of chemically competent Marionette DH10β with 10 μL of the Golden Gate reaction mix and plate out all the transformation mix in serial dilutions on LB-agar plates at a final concentration of 50 mg/mL spectinomycin to select for uptake of the plasmid.
3. Incubate at 30 °C for 24–30 h until visible colonies are formed.

4. Successful insertion of a guide or guide array can be verified by loss of GFP fluorescence, this should be visualized on a blue-light transilluminator. Select these non-fluorescent colonies for downstream processing (*see* **Note 5**).
5. Inoculate LB media broth with a single colony and grow overnight at 30 °C with shaking.
6. Prepare glycerol stocks for archiving using 1 mL of overnight culture and 0.5 mL of 60% glycerol per aliquot and store at −80 °C.
7. Purify plasmid DNA from the remaining culture using a commercial miniprep kit following the manufacturer's instructions or using other standard methods.
8. Verify the plasmid assembly by Sanger sequencing or full plasmid sequencing (*see* **Note 6**).

3.7 Genome Editing of E. coli BL21 Using pSPIN-GG

1. Transform 50 μL of chemically competent *E. coli* BL21 with 100 ng of sequence-verified pSPIN-GG plasmid containing a compatible guide/guide array and desired DNA cargo.
2. Plate out the complete transformation mixture in three tenfold serial dilutions on LB-agar plates containing 100 μg/mL of spectinomycin.
3. Incubate at 30 °C for 30 h to allow sufficient time for expression of CAST system and integration of cargo DNA into the genome.

3.8 Screening and Isolation of Successful Integration

1. Design forward and reverse primers to amplify the genomic region spanning the integration site. The forward and reverse primers should bind upstream and downstream of the integration site located approximately 50 bp downstream of the 3′ end of the 32 bp target sequence. Primers should be approximately 30 bp in length to minimize off-target binding, with a GC content between 40% and 60%. Primer melting temperatures should be optimized for Taq polymerase to ensure efficient amplification (*see* **Note 7**).
2. Using an inoculation loop, resuspend a small amount of a single genome-edited colony in 20 μL of nuclease-free water.
3. On ice, in a 0.2 mL PCR tube, prepare the following reaction: 12.5 μL OneTaq Quick-Load 2× Master Mix with Standard Buffer, 0.5 μL 10 mM forward primer, 0.5 μL 10 mM reverse primer, 2 μL colony resuspension, and nuclease-free water to a 25 μL total volume.
4. Gently mix the reaction.
5. Transfer the reaction to a thermocycler and run using standard Taq polymerase thermocycler settings with an additional

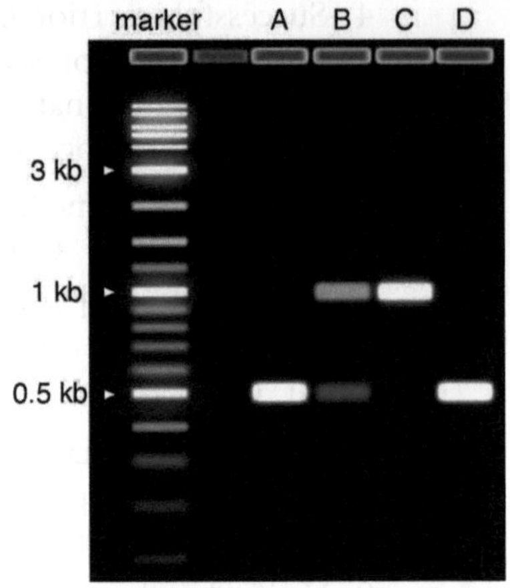

Fig. 4 Virtual DNA electrophoresis gel image for colony PCR screening. (Lane A) native genomic control band expected to be 500 bp. (Lane B) mixed non-clonal colony showing both the unmodified genomic band at 500 bp and a larger band at 1 kb corresponding to cells with successfully integrated cargo. (Lane C) clonal integrant in which all cells contain the integrated cargo. (Lane D) failed integration showing only the native genomic band. Virtual gel image created using Affinity Designer

20 min denaturation step at 94 °C at the start to ensure cell lysis.

6. Run 5 μL of the colony PCR reaction on a 1% agarose gel to confirm the expected product size (Fig. 4).
7. If product size is correct, then extract the band and purify the DNA using standard protocols and sequence verify. If no product is found, then other colonies need to be screened.
8. If colony PCR confirms DNA integration, resuspend the successful colony in 2 mL LB broth and grow for 3 h at 42 °C to cure the cells of the pSPIN-GG plasmid (*see* **Note 8**).
9. Dilute 50 μL of the culture into 5 mL fresh LB media and grow overnight at 37 °C.
10. Plate serial dilutions onto fresh LB plates without antibiotic and incubate at 37 °C.
11. Confirm plasmid loss by patch-plating colonies onto LB-agar plates containing 100 μg/mL spectinomycin and incubate at 30 °C overnight (*see* **Note 9**). Successfully edited and cured cells will not grow.

4 Notes

1. Off-target insertions can occur if the guide RNA sequence closely resembles another location in the genome. To reduce

this possibility, guide sequences with high sequence similarity to other positions in the genome should be avoided. This can be checked for using standard bioinformatics tools like BLAST. We have found that integration efficiency of DNA cargo is highest when targeting sites near the chromosome's origin of replication and in the non-structured right macrodomain.

2. Due to the length of the guide inserts (approximately 90 bp including scaffold sequences), PAGE-purified oligonucleotides are recommended to ensure full-length synthesis and improve Golden Gate assembly efficiency.
3. Avoid repeated freeze-thaw cycles. Annealed oligos are stable for short-term use (up to ~2 weeks) if stored at −20 °C.
4. For optimal assembly efficiency when inserting short DNA fragments, 80 Golden Gate cycles are recommended. While fewer cycles (40–60) may be sufficient for larger DNA cargoes or long guide arrays.
5. If no non-fluorescent colonies are observed, increase the amount of guide DNA in the assembly reaction to 150 fmol and pre-digest the plasmid pSPIN-GG plasmid, purifying only the backbone component for direct use in the reaction. This will help to reduce background re-ligation of the destination vector (pSPIN-GG) and ensure colonies contain correctly assembled plasmids with an insert.
6. Sanger sequencing may struggle to resolve multi-spacer arrays due to the repeated scaffold sequence. In these cases, nanopore-based sequencing is recommended for sequence verification of constructs.
7. Integration by the CAST system used in pSPIN-GG typically occurs in a preferred orientation, with the right transposon end positioned proximal to the target site. However, insertions can occur in both directions. Forward and reverse primers can be designed to bind the DNA cargo sequence to probe for insertions in either orientation when paired with the forward primer upstream of the PAM sequence. When performing flanking colony PCR, note that shorter DNA amplicons tend to be preferentially amplified meaning the intensity of bands produced cannot provide an accurate measurement of integration efficiency.
8. A clonal integrant refers to a homogeneous colony in which all cells carry the intended DNA integration. During transposition-based integration, it is common to encounter colonies containing only a subset of cells where successful integration has occurred. Therefore, colonies may need to be serially re-streaked and re-plated multiple times until a homogeneous population of successful integration is obtained.

9. If plasmids persist following attempted curing, we recommend extending the incubation period at 42 °C and transferring a smaller inoculum into fresh LB medium to promote faster dilution of residual plasmid-containing cells. It is important to note that curing prevents further plasmid replication but does not eliminate plasmids already present. Therefore, repeated growth and dilution steps are necessary to fully segregate plasmid-free cells.

Acknowledgments

T.I. was supported by a Royal Society PhD studentship. T.E.G. was supported by a Royal Society University Research Fellowship under grant URF\R\221008, the UKRI-funded Engineering Biology Mission Award CYBER under BBSRC grant BB/Y007638/1, and EEBio under EPSRC grant EP/Y014073/1.

References

1. Koonin EV, Gootenberg JS, Abudayyeh OO (2023) Discovery of diverse CRISPR-Cas systems and expansion of the genome engineering toolbox. Biochemistry 62:3465–3487. https://doi.org/10.1021/acs.biochem.3c00159
2. Klompe SE, Vo PLH, Halpin-Healy TS, Sternberg SH (2019) Transposon-encoded CRISPR–Cas systems direct RNA-guided DNA integration. Nature 571:219–225. https://doi.org/10.1038/s41586-019-1323-z
3. Jiang W, Bikard D, Cox D et al (2013) RNA-guided editing of bacterial genomes using CRISPR-Cas systems. Nat Biotechnol 31:233–239. https://doi.org/10.1038/nbt.2508
4. Elmore JR, Dexter GN, Baldino H et al (2023) High-throughput genetic engineering of nonmodel and undomesticated bacteria via iterative site-specific genome integration. Sci Adv 9: eade1285. https://doi.org/10.1126/sciadv.ade1285
5. Meaker GA, Hair EJ, Gorochowski TE (2020) Advances in engineering CRISPR-Cas9 as a molecular Swiss Army knife. Synth Biol 5: ysaa021. https://doi.org/10.1093/synbio/ysaa021
6. Ciurkot K, Vonk B, Gorochowski TE et al (2019) CRISPR/Cas12a multiplex genome editing of Saccharomyces cerevisiae and the creation of yeast pixel art. J Vis Exp:59350. https://doi.org/10.3791/59350-v
7. Vo PLH, Ronda C, Klompe SE et al (2021) CRISPR RNA-guided integrases for high-efficiency, multiplexed bacterial genome engineering. Nat Biotechnol 39:480–489. https://doi.org/10.1038/s41587-020-00745-y
8. Strecker J, Ladha A, Gardner Z et al (2019) RNA-guided DNA insertion with CRISPR-associated transposases. Science 365:48–53. https://doi.org/10.1126/science.aax9181
9. Casini A, Storch M, Baldwin GS, Ellis T (2015) Bricks and blueprints: methods and standards for DNA assembly. Nat Rev Mol Cell Biol 16: 568–576. https://doi.org/10.1038/nrm4014
10. Engler C, Kandzia R, Marillonnet S (2008) A one pot, one step, precision cloning method with high throughput capability. PLoS One 3: e3647. https://doi.org/10.1371/journal.pone.0003647
11. Clark CJ, Scott-Brown J, Gorochowski TE (2021) paraSBOLv: a foundation for standard-compliant genetic design visualization tools. Synth Biol 6:ysab022. https://doi.org/10.1093/synbio/ysab022
12. Der BS, Glassey E, Bartley BA et al (2017) DNAplotlib: programmable visualization of genetic designs and associated data. ACS Synth Biol 6:1115–1119. https://doi.org/10.1021/acssynbio.6b00252
13. Buecherl L, Mitchell T, Scott-Brown J et al (2023) Synthetic biology open language (SBOL) version 3.1.0. J Integr Bioinform 20:

20220058. https://doi.org/10.1515/jib-2022-0058

14. Gelsinger DR, Vo PLH, Klompe SE et al (2024) Bacterial genome engineering using CRISPR-associated transposases. Nat Protoc 19:752–790. https://doi.org/10.1038/s41596-023-00927-3
15. VanDieren AJ, Barrick JE (2025) UltraCAST: a flexible all-in-one suicide vector for modifying bacterial genomes using a CRISPR-associated transposon. MicroPublication Biol. https://doi.org/10.17912/micropub.biology.001721
16. Meyer AJ, Segall-Shapiro TH, Glassey E et al (2019) Escherichia coli "marionette" strains with 12 highly optimized small-molecule sensors. Nat Chem Biol 15:196–204. https://doi.org/10.1038/s41589-018-0168-3
17. Xu J, Sun Y, Wu J et al (2024) Chromosome recombination and modification by *LoxP* -mediated evolution in *vibrio natriegens* using CRISPR-associated transposases. Biotechnol Bioeng 121:1163–1172. https://doi.org/10.1002/bit.28639

Chapter 3

Construction of a Tl-CRISPRi Genetic Circuit in Bacteria for Translation-Level Gene Knockdown

Giho Kim, Ho Joon Kim, and Sang Woo Seo

Abstract

The Tl-CRISPRi system, which harnesses the specific RNA-binding activity of CRISPR-dCas13, has been recently developed for translation-level gene knockdown in bacteria. By introducing spacers complementary to the translation initiation region of the mRNA, dCas13 can be directed to block the ribosome and inhibit the translation of that mRNA. Here, we discuss how to construct the Tl-CRISPRi genetic circuit and implement this system for gene knockdown. This chapter describes how to design spacer sequences and install them into the guide RNA expression plasmid. Also, we describe how to mutate the handle of gRNA to achieve tunable knockdown of a target gene. By following the method described in this chapter, we anticipate that a precise and controllable knockdown of a target gene in bacterial cells can be performed in a programmable manner.

Key words Synthetic biology, CRISPR-dCas13, Translation-level CRISPRi, Tunable gene regulation, Guide RNA engineering

1 Introduction

The CRISPR interference (CRISPRi) technology has been widely used for investigation and engineering of several microorganisms [1, 2]. Since CRISPRi can knock down a target gene without irreversible change in its genome sequence, programmable transcription-level CRISPRi (Tx-CRISPRi) tools were initially developed using dead Cas9 (dCas9) and later extended to dead Cas12 (dCas12) [3, 4].

The translation-level CRISPRi (Tl-CRISPRi) system employs the nuclease-inactive version of the Type VI CRISPR-Cas nuclease Cas13 that can target single-stranded RNA (ssRNA), which enables the downregulation of a specific gene by preventing ribosomes from binding to its mRNA [5–7]. Similar to the Tx-CRISPRi

Authors Giho Kim and Ho Joon Kim have equally contributed to this chapter.

Xiao-Jun Tian (ed.), *Synthetic Gene Circuits: Methods and Protocols*, Methods in Molecular Biology, vol. 3041, https://doi.org/10.1007/978-1-0716-5304-3_3,

system, the Tl-CRISPRi system also requires the dCas13 as an effector and a guide RNA composed of a handle sequence, which binds to dCas13, and a spacer sequence, which binds to the target mRNA. When targeting a polycistronic operon, dCas9-based Tx-CRISPRi silences all downstream genes beyond the target gene, known as the polar effect; however, Tl-CRISPRi overcomes this issue by downregulating only the target gene. Accordingly, the Tl-CRISPRi system can prevent unintended knockdown of non-target genes to allow more precise metabolic engineering and investigation of host cells.

In this chapter, we describe the stepwise protocol and minute details of designing a guide RNA and its cloning for effective Tl-CRISPRi in bacteria. Furthermore, we introduce how to apply several direct repeat (DR) variants for tunable expression of a target gene with the Tl-CRISPRi. Finally, we discuss how to evaluate the knockdown efficiency of the designed Tl-CRISPRi with the fluorescence measurement. The protocol described here can also be further tailored for diverse bacterial strains if the appropriate vector and genetic parts are available to construct the Tl-CRISPRi system.

2 Materials

2.1 Plasmid Preparation and Cloning

1. Quick Ligation kit (M2200, NEB).
2. 10 mM dNTP mixture.
3. PNK enzyme (M0201, NEB).
4. Designed oligonucleotides (for spacer and DR cloning).
5. Q5 High-Fidelity DNA polymerase (M0491, NEB).
6. PCR purification kit.
7. Plasmid mini-prep kit.
8. BsaI enzyme (R3733, NEB).
9. EmeraldAmp PCR master mix (RR300, Takara).
10. 1% Agarose gel.
11. DNA ladder.
12. Gel loading dye (6×).
13. Chemically competent *E. coli* cells (for cloning).
14. LB medium with appropriate antibiotics.
15. BHI medium.
16. LB-agar plates with appropriate antibiotics.

2.2 Cell Culture and Fluorescence Measurement

1. *E. coli* strain of interest (*see* **Note 1**).
2. Plasmids encoding dRfxCas13d (pACYC-dRfxCas13d) and the cognate guide RNA (pCDF-RfxgRNA) (Fig. 1).

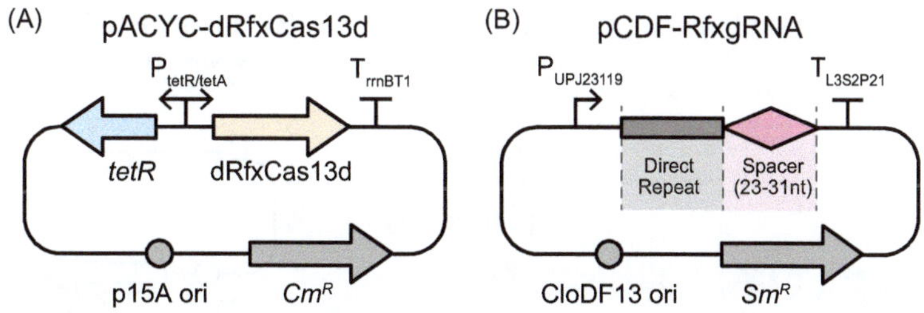

Fig. 1 Map of plasmids for TI-CRISPRi application. (**a**) The pACYC-dRfxCas13d plasmid encodes the catalytically inactive (dead) RfxCas13d protein. Its expression is driven by the aTc-inducible, bidirectional tetR/tetA promoter. (**b**) The pCDF-RfxgRNA plasmid encodes the RfxCas13d guide RNA, which is expressed under the strong, constitutive UPJ23119 promoter

3. LB medium with appropriate antibiotics.
4. LB-agar plates with appropriate antibiotics.
5. anhydrotetracycline (aTc).
6. PBS buffer.
7. 1.5 mL Eppendorf tubes.
8. 96-well microplate.

2.3 Equipment

1. PCR thermocycler.
2. Gel electrophoresis equipment.
3. Shaking incubator.
4. General incubator.
5. Water bath or heat block.
6. Microplate reader.
7. Spectrophotometer.

3 Methods

3.1 Cloning of Spacer Sequences

1. Prepare the oligo annealing mixture as follows: 2.5 μL of Quick Ligase Buffer (2×), 1 μL of oligo-sp-F 100 μM, 1 μL of oligo-sp-R 100 μM, 0.5 μL of PNK enzyme (*see* Fig. 2a and *see* **Note 2**).
2. Incubate the oligo annealing mixture at 37 °C for 30 min.
3. Heat the oligo annealing mixture at 95 °C for 10 min, followed by slow cooling at room temperature (*see* **Note 3**).
4. Add 2.5 μL of Quick Ligase Buffer (2×), 10 ng of backbone plasmid digested with BsaI (*see* **Note 4**), and add distilled water up to 9.5 μL. Add 0.5 μL of Quick Ligase enzyme.
5. Incubate the ligation mixture at room temperature for 10 min.

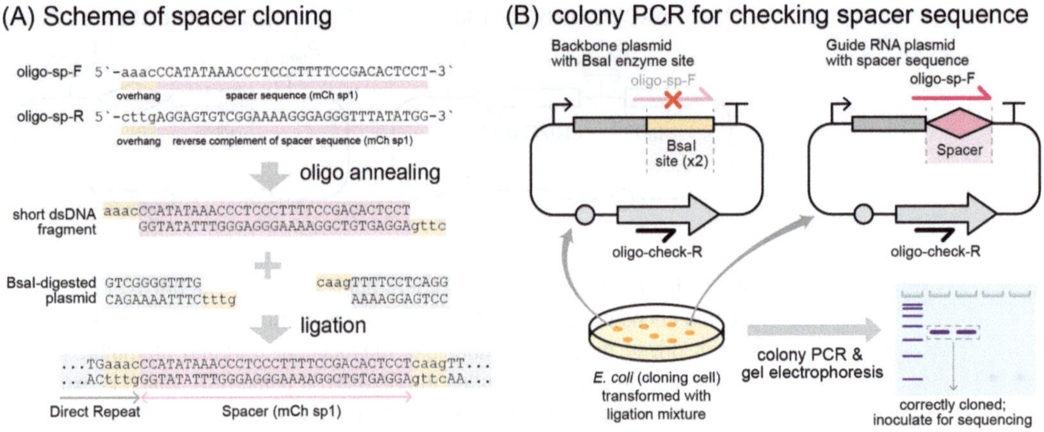

Fig. 2 Procedure of spacer cloning for the TI-CRISPRi. (**a**) A spacer is cloned into the backbone plasmid, which contains dual BsaI enzyme sites located downstream of the UPJ23119 promoter and the guide RNA direct repeat (DR). Two complementary oligonucleotides encoding the spacer sequence are annealed to generate a short dsDNA fragment with two 4-nt overhangs for sticky-end ligation. This fragment is then ligated into the BsaI-digested backbone plasmid. (**b**) Following transformation into the *E. coli* and antibiotics selection, colonies containing the correctly ligated plasmid are identified via colony PCR using the oligo-sp-F and oligo-check-R primers. The sequence of plasmid extracted from the correct colony should be verified by Sanger sequencing to ensure absence of mutations in the cloned spacer sequence

6. Mix the ligation mixture with chemically competent cells. Incubate the cell in ice for 10 min.
7. Heat-shock the cell at 42 °C for 90 s.
8. Add 700 μL of BHI medium for recovery. Incubate at 37 °C for 1 h while shaking at 250 rpm.
9. Spread the transformed cell onto the LB-agar plate with appropriate antibiotics. Incubate the plate at 37 °C overnight.
10. Prepare PCR mixture as follows: 25 μL of EmeraldAmp PCR master mix (2×), 24 μL of distilled water, 0.5 μL of oligo-sp-F 100 μM, 0.5 μL of oligo-check-R 100 μM. Aliquot the PCR mixture into 10 μL volumes (see Fig. 2b and *see* **Note 5**).
11. Mix each colony with the aliquoted PCR mixture.
12. Run the thermocycler for the following methods: 95 °C for 5 min; 30 cycles: 95 °C for 30 s, 52 °C for 30 s, and 72 °C for 1 min; 72 °C for 3 min.
13. Run the agarose gel electrophoresis with the DNA ladder.
14. Inoculate the colony that shows a band of correct size in colony PCR to the LB media with appropriate antibiotics. Incubate at 37 °C overnight while shaking at 250 rpm.
15. Collect the cell culture and perform plasmid mini-prep. Check the sequence of the cloned plasmid to ensure successful cloning without mutations.

3.2 Cloning of Mutated DR Sequences

1. Prepare PCR mixture as follows: 36.2 μL of distilled water, 10 μL of Q5 reaction buffer (5×), 2 μL of 10 mM dNTP, 0.5 μL of oligo-DR-F 100 μM, 0.5 μL of 5′-phosphorylated oligo-sp-blunt-F 100 μM, 0.3 μL of template plasmid, 0.5 μL of Q5 polymerase (see Fig. 3a and *see* **Note 6**).
2. Run the thermocycler for the following methods: 98 °C for 30 s; 30 cycles: 98 °C for 10 s, 60 °C for 30 s, and 72 °C for 2 min; 72 °C for 6 min.
3. Run the agarose gel electrophoresis with the DNA ladder.
4. Purify the PCR products with a PCR purification kit.

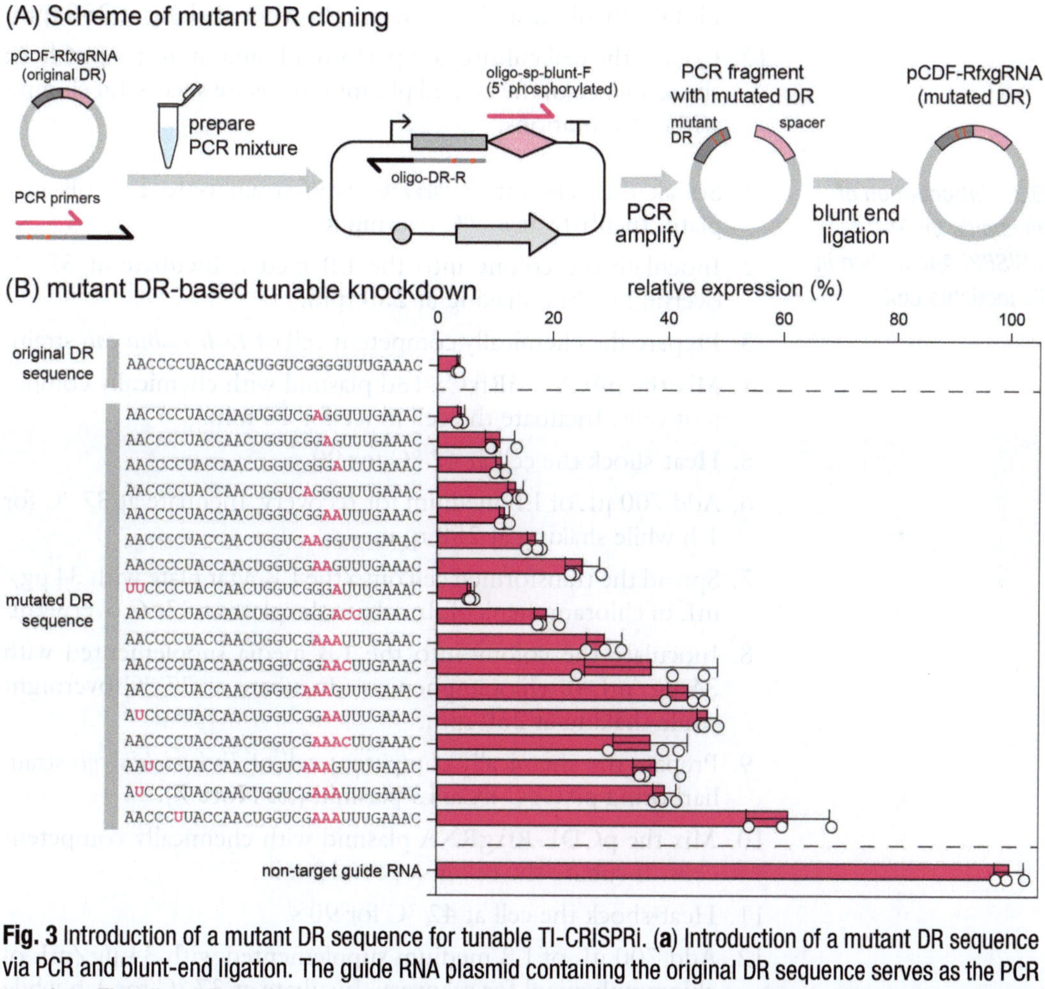

Fig. 3 Introduction of a mutant DR sequence for tunable TI-CRISPRi. (**a**) Introduction of a mutant DR sequence via PCR and blunt-end ligation. The guide RNA plasmid containing the original DR sequence serves as the PCR template. The oligo-DR-R primer contains the reverse complement of the desired mutant DR sequence. The 5′ end of the oligo-sp-blunt-F primer should be phosphorylated prior to PCR to enable the intramolecular self-ligation of the amplified plasmid. (**b**) Tunable TI-CRISPRi with the DR-mutated guide RNAs. By adopting several mutant DR sequences in the guide RNA targeting mCherry mRNA, tunable knockdown of the mCherry gene at the translation level was achieved

5. Prepare blunt-end ligation mixture as follows: 5 μL of Quick Ligase buffer (2×), 100 ng of purified PCR product, 0.5 μL of Quick Ligase, and add distilled water up to 10 μL volume.
6. Incubate the ligation mixture at room temperature for 10 min.
7. Mix the ligation mixture with chemically competent cells. Incubate the cell in ice for 10 min.
8. Heat-shock the cell at 42 °C for 90 s.
9. Add 700 μL of BHI medium for recovery. Incubate at 37 °C for 1 h while shaking at 250 rpm.
10. Spread the transformed cell onto the LB-agar plate with appropriate antibiotics. Incubate the plate at 37 °C overnight.
11. Inoculate the colony into the LB media with appropriate antibiotics. Incubate at 37 °C overnight while shaking at 250 rpm.
12. Collect the cell culture and perform plasmid mini-prep. Check the sequence of the cloned plasmid to ensure successful cloning without mutations.

3.3 Introduction of Plasmids for TI-CRISPRi Application in Escherichia coli

1. Streak the relevant *Escherichia coli* strain onto the LB-agar plate. Incubate at 37 °C overnight.
2. Inoculate the colony into the LB media. Incubate at 37 °C overnight while shaking at 250 rpm.
3. Prepare the chemically competent cell of *Escherichia coli* strain.
4. Mix the pACYC-dRfxCas13d plasmid with chemically competent cells. Incubate the cell in ice for 10 min.
5. Heat-shock the cell at 42 °C for 90 s.
6. Add 700 μL of LB medium for recovery. Incubate at 37 °C for 1 h while shaking at 250 rpm.
7. Spread the transformed cell onto the LB-agar plate with 34 μg/mL of chloramphenicol. Incubate the plate at 37 °C overnight.
8. Inoculate the colony into the LB media supplemented with 34 μg/mL of chloramphenicol. Incubate at 37 °C overnight while shaking at 250 rpm.
9. Prepare the chemically competent cell of *Escherichia coli* strain harboring pACYC-dCas13 plasmid (*see* **Note 7**).
10. Mix the pCDF-RfxgRNA plasmid with chemically competent cells. Incubate the cell in ice for 10 min.
11. Heat-shock the cell at 42 °C for 90 s.
12. Add 700 μL of LB medium supplemented with 34 μg/mL of chloramphenicol for recovery. Incubate at 37 °C for 1 h while shaking at 250 rpm.

13. Spread the transformed cell onto the LB-agar plate with 34 μg/mL of chloramphenicol and 50 μg/mL of spectinomycin. Incubate the plate at 37 °C overnight.

3.4 Knockdown of Fluorescent Reporter Gene with the Tl-CRISPRi in *Escherichia coli* to Measure Knockdown Efficiency

1. Inoculate the colony into the LB media supplemented with 34 μg/mL of chloramphenicol and 50 μg/mL of spectinomycin. Incubate at 37 °C overnight while shaking at 250 rpm.
2. Refresh the seed culture to the ratio of 1:100 in the LB media supplemented with 34 μg/mL of chloramphenicol, 50 μg/mL of spectinomycin, and 100 ng/mL of anhydrotetracycline (*see* **Note 8**).
3. Incubate at 37 °C for 8 hr. while shaking at 250 rpm (*see* **Note 9**).
4. Transfer the 100 μL of cell culture into a 1.5 mL Eppendorf tube, followed by centrifugation at 13,000 rpm for 3 min.
5. Decant supernatant and resuspend the cell pellet with 100 μL of PBS.
6. Load 100 μL of PBS-washed cell suspension in a 96-well microplate.
7. Measure the fluorescence intensity (FI) of fluorescent protein using a microplate reader.
8. Measure the optical density of the cell culture at 600 nM wavelength (OD_{600}) using a spectrophotometer.
9. Calculate the normalized fluorescence intensity (NFI) for each culture by dividing the fluorescence intensity (FI) by the OD_{600}. Determine the relative per-cell expression of the fluorescent protein (RE) by dividing each strain's NFI by the mean NFI of the NT (non-targeting control) strain. Compute knockdown efficiency as 100% − RE (expressed as a percentage).

4 Notes

1. We utilized *E. coli* K-12 MG1655 strain harboring the P_{OXB20}-mCherry or P_{J23119}-GFP cassette to measure the knockdown efficiency of Tl-CRISPRi with fluorescence. These cassettes were integrated into the chromosome of *E. coli* by utilizing the pSIM5 plasmid, followed by KanR selection and Flp-based excision of the FRT-KanR sequence by the pCP20 plasmid [8]. Since we do not cover the introduction of the fluorescent marker gene into the chromosome, we recommend investigating the detailed method of gene integration into the chromosome in other articles. Alternatively, obtaining Addgene-available fluorescent *E. coli* strains might be helpful [9].

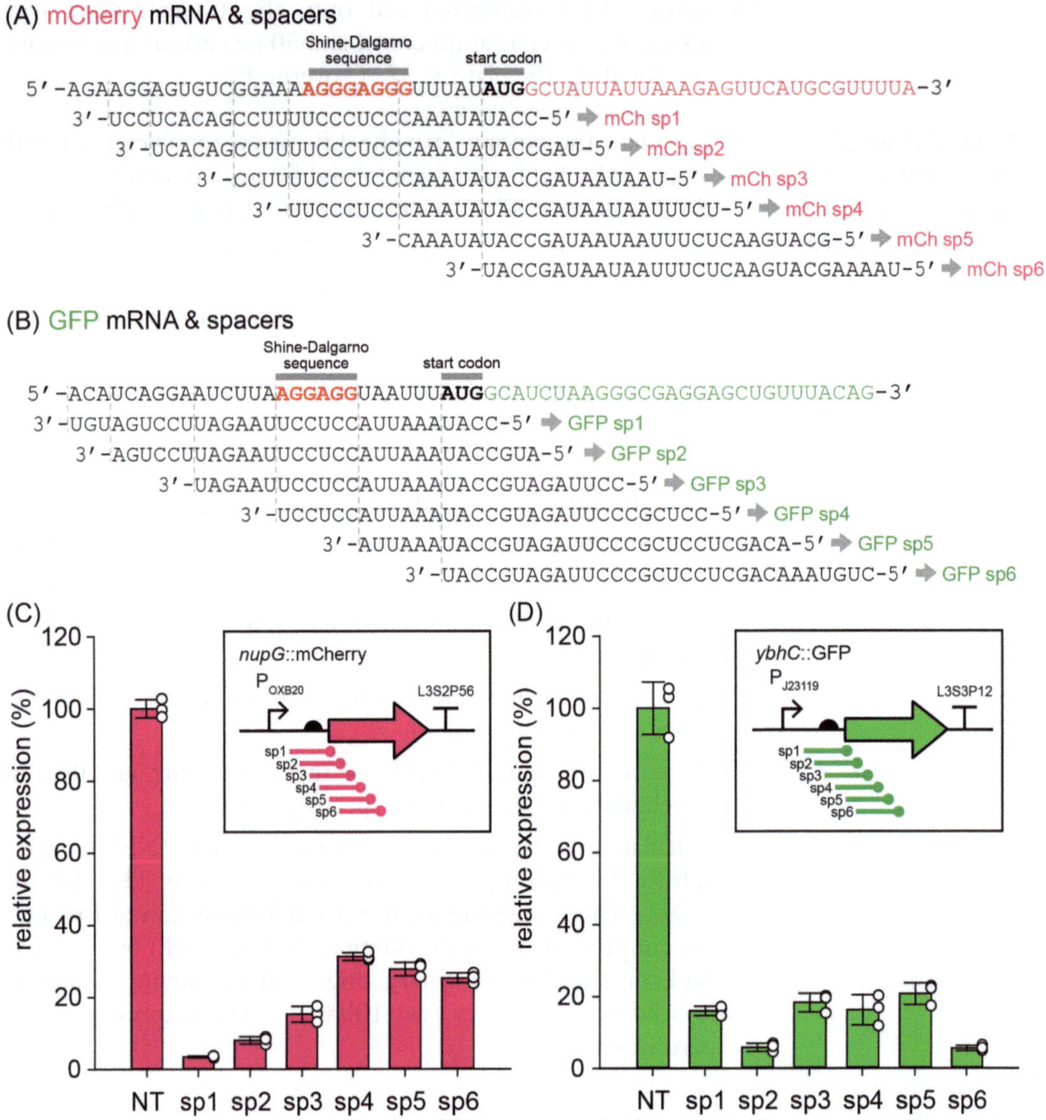

Fig. 4 Design of spacers for dCas13 guide RNA and reporter gene knockdown with the TI-CRISPRi. (**a**, **b**) mRNA sequence of mCherry and GFP and the spacers spanning translation initiation region of the mRNA are depicted. (**c, d**) Reporter gene expression cassettes were integrated in the chromosome of *E. coli*. Six different spacer sequences were used to knock down the expression of reporter genes. The reporter gene expression level of each strain was normalized to that of a control strain harboring non-targeting (NT) guide RNA.

2. Spacer sequences should target near the Shine-Dalgarno sequence of mRNA to ensure direct competition with ribosomes and tight knockdown at the translation level (Fig. 4). We recommend designing a spacer length from 23 to 31 nts. If the spacer length is too long, unexpected secondary structure within the spacer sequence may form, which hinder the mRNA targeting of dCas13. The minimal spacer length required to retain the knockdown efficiency is 23 nts in *E. coli* [5].

3. Avoid storing oligo annealing product in the refrigerator. Cooling down the oligo annealing product could allow mismatched annealing of oligonucleotides and cause unintended mutations in the spacer sequence.
4. We recommend preparing the BsaI-digested plasmid with the gel purification method, which excludes uncut plasmids after purification. This helps reduce false positives at the antibiotic selection step, thereby enhancing the possibility of detecting colonies with successfully cloned plasmids in the colony PCR step.
5. Oligo annealing-based spacer cloning method ensures high cloning efficiency unless the oligonucleotide itself forms a structure or has repetitive sequences. Additionally, the primer (oligo-sp-F) used for oligo annealing can be reused for colony PCR by pairing with the primer attaching to the plasmid backbone (oligo-check-R) (Fig. 2b). Usually, PCR of five colonies is enough to detect successfully cloned plasmids. We recommend constructing the plasmid harboring dual BsaI enzyme sites flanked by a strong constitutive promoter (P_{J23119}) and terminator ($T_{L3S2P21}$) to facilitate further cloning of diverse spacer sequences [10].
6. Since the DR sequence and stem-loop structure of the dCas13 guide RNA are responsible for the guide RNA recognition of dCas13, introducing mutations into the DR sequence resulted in the modulation of knockdown efficiency [5]. We introduced mutations in the DR sequence by PCR amplification with primers containing mutant DR sequences as overhang. By combining these primers with a primer attaching to the 5′-end of the spacer without overhang (oligo-sp-blunt-F) for PCR, linearized plasmids can be obtained harboring mutant DR sequences, which are then self-ligated during the blunt-end ligation step (Fig. 3a). The resulting DR-mutated guide RNAs can be applied for tunable knockdown of a target gene (Fig. 3b).
7. We separately introduced the dCas13 plasmid and the guide RNA plasmid to perform Tl-CRISPRi in *E. coli*. Since the guide RNA sequence may vary depending on the target mRNA sequence or the intended level of repression, preparing a background strain for adopting various guide RNA sequences is helpful for adopting Tl-CRISPRi to investigate the knockdown effect of different gene targets and achieve diverse knockdown efficiencies. Therefore, we recommend transforming the dCas13 plasmid into the relevant *E. coli* strain first. By preparing chemically competent cells of that background strain, you can easily introduce guide RNA plasmids and perform various scenarios of Tl-CRISPRi applications.

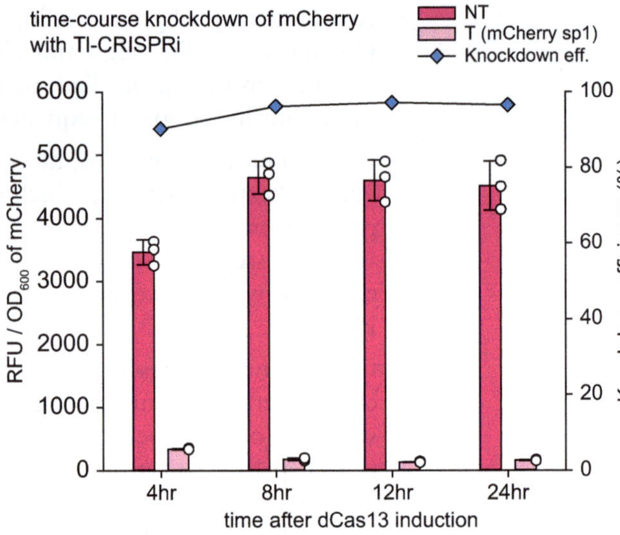

Fig. 5 Time-course analysis of TI-CRISPRi-mediated mCherry knockdown. The expression level of mCherry was measured at 4, 8, 12, and 24 h after subculturing the cells and adding aTc to induce the expression of dCas13

8. The optimal inducer concentration can vary depending on the *E. coli* strain, culture media, or the expression cassette of dCas13. Since the knockdown level of mCherry expressed from the chromosome was saturated at 100 ng/mL of aTc induction, we chose 100 ng/mL of aTc for the standard application of TI-CRISPRi. We recommend determining the optimal inducer concentration for the dCas13 expression, considering your experimental design.
9. When we measured time-course fluorescence of mCherry while performing TI-CRISPRi, we found that 8 h of culture after the induction of dCas13 and refresh of cell culture was enough to achieve maximal knockdown of mCherry (Fig. 5). Also, the knockdown of mCherry was tightly maintained after the 8-hour time point. The time point of sampling for bacterial culture may vary depending on the purpose of experiment; for example, we chose a 24-h time point for measuring the 3-HP titer during metabolic engineering applications [5].

Acknowledgments

This work was supported by the National Research Foundation (NRF) funded by the Korean government (MSIT) (No. RS-2025-02214910 and No. RS-2025-02309093). We also acknowledge that this work was supported by the Korea Institute of Marine Science & Technology Promotion (KIMST) funded by the Ministry of Oceans and Fisheries (20220258).

Conflict of Interest Authors Giho Kim and Sang Woo Seo are inventors on Korean patent application 10–2023-0151135 and PCT application PCT/KR2024/006932. These applications, filed by Seoul National University Research and Development Foundation, cover the technology described as tunable Tl-CRISPRi in this work. All other authors declare no competing interests.

References

1. Zhang R, Xu W, Shao S, Wang Q (2021) Gene silencing through CRISPR interference in bacteria: current advances and future prospects. Front Microbiol 12
2. Han YH, Kim G, Seo SW (2023) Programmable synthetic biology tools for developing microbial cell factories. Curr Opin Biotechnol 79:102874. https://doi.org/10.1016/j.copbio.2022.102874
3. Qi Lei S, Larson Matthew H, Gilbert Luke A, Doudna Jennifer A, Weissman Jonathan S, Arkin Adam P, Lim Wendell A (2013) Repurposing CRISPR as an RNA-guided platform for sequence-specific control of gene expression. Cell 152(5):1173–1183. https://doi.org/10.1016/j.cell.2013.02.022
4. Zhang X, Wang J, Cheng Q, Zheng X, Zhao G, Wang J (2017) Multiplex gene regulation by CRISPR-ddCpf1. Cell Discov 3(1):17018. https://doi.org/10.1038/celldisc.2017.18
5. Kim G, Kim HJ, Kim K, Kim HJ, Yang J, Seo SW (2024) Tunable translation-level CRISPR interference by dCas13 and engineered gRNA in bacteria. Nat Commun 15(1):5319. https://doi.org/10.1038/s41467-024-49642-x
6. Ko SC, Woo HM (2024) CRISPR–dCas13a system for programmable small RNAs and polycistronic mRNA repression in bacteria. Nucleic Acids Res 52(1):492–506. https://doi.org/10.1093/nar/gkad1130
7. Cardiff Ryan AL, Faulkner Ian D, Beall Juliana G, Carothers JM, Zalatan Jesse G (2024) CRISPR-Cas tools for simultaneous transcription & translation control in bacteria. Nucleic Acids Res 52(9):5406–5419. https://doi.org/10.1093/nar/gkae275
8. Datta S, Costantino N, Court DL (2006) A set of recombineering plasmids for gram-negative bacteria. Gene 379:109–115. https://doi.org/10.1016/j.gene.2006.04.018
9. Tomasek K, Bergmiller T, Guet CC (2018) Lack of cations in flow cytometry buffers affect fluorescence signals by reducing membrane stability and viability of Escherichia coli strains. J Biotechnol 268:40–52. https://doi.org/10.1016/j.jbiotec.2018.01.008
10. Chen Y-J, Liu P, Nielsen AAK, Brophy JAN, Clancy K, Peterson T, Voigt CA (2013) Characterization of 582 natural and synthetic terminators and quantification of their design constraints. Nat Methods 10(7):659–664. https://doi.org/10.1038/nmeth.2515

Chapter 4

Design of CRISPRi-Based Synthetic Gene Circuits in Bacteria

Javier Santos-Moreno

Abstract

Synthetic gene circuits are key elements of engineered biological systems that allow us to control and program cellular behavior. Yet, circuit design can be challenging to newcomers due to the numerous design choices and the abundance and variety of parameters that can influence circuit performance. While transcription factors have dominated the circuit construction toolbox for two decades, CRISPRi-based tools offer important benefits—especially for large circuits—but also require unique design considerations. Here I provide a detailed guide for designing CRISPRi circuits in bacteria, using the CRISPRlator, the first CRISPRi oscillator, as an example that illustrates the design process.

Key words Synthetic gene circuit, Synthetic gene regulatory network, Synthetic biology, Synthetic cell programming, CRISPR interference (CRISPRi), CRISPRlator, Oscillator

1 Introduction

Gene circuits allow cells to respond to external or internal cues by computing the appropriate response. Thus, they are key for orchestrating complex behaviors beyond a simple, direct gene activation in response to a stimulus. Synthetic gene circuits allow researchers to control—or even better, to program—cell function in a precise and predictable manner, and they therefore constitute a cornerstone of bioengineering [1]. However, despite the many advances, we are still far from achieving synthetic circuit complexities comparable to those of natural ones. In this context, new tools that enable improvements in circuit complexity, robustness, and programmability are desirable, and CRISPR interference (CRISPRi) stands out as a circuit building block with many favorable properties.

CRISPRi is a gene expression control tool derived from the prokaryotic defense system CRISPR. It shares with other CRISPR-based tools the same underlying mechanism: an RNA-protein complex that is targeted by the user to a DNA sequence of interest,

Xiao-Jun Tian (ed.), *Synthetic Gene Circuits: Methods and Protocols*, Methods in Molecular Biology, vol. 3041,
https://doi.org/10.1007/978-1-0716-5304-3_4,

where it binds with high specificity (*see* Subheading 2.1). The RNA (known as guide RNA, gRNA) provides specificity via Watson-Crick base pairing with the target DNA, while the associated protein (most often the *Streptococcus pyogenes* Cas9 or a derivative) produces the desired action at the target binding site—whether DNA cleavage, gene regulation, protein recruitment, or another function.

The main asset of CRISPR is its extraordinary programmability: by simply changing a few nucleotides (often ~20 nt) of the gRNA, the RNA-protein complex can be targeted to any DNA sequence of interest.

While CRISPR is best known for its use as a highly precise and programmable DNA cleavage tool, many other applications that require specific DNA binding—but not necessarily cleavage—have been developed. CRISPRi (interference) and CRISPRa (activation) use a cleavage-deficient CRISPR protein (typically dCas9, alone or fused to effector domains) to repress or activate gene expression at target DNA locations. These tools enable a remarkably programmable transcriptional regulation, and they are therefore increasingly employed as constituent components of gene circuits.

While gene circuits can in principle be composed of any molecular tool that regulates gene expression, the classic and more popular building blocks of synthetic gene circuits are transcription factors (TFs), protein regulators that activate or repress transcription upon DNA binding. Many and varied synthetic circuits have been built using TFs, including prototypes with remarkable complexities and impressive behaviors [2–5]. Yet, there are limitations associated with TFs that make it difficult to use them for constructing large circuits. Protein engineering still remains challenging, and therefore most engineered circuits employ natural (or, sometimes, chimeric) TFs, rather than rationally engineered ones. As a consequence, the choice of TFs is limited, especially considering that for optimal circuit performance TFs should be orthogonal (i.e., non--cross-reacting) to the host genome and to the operators of other TFs in the circuit. Besides, TFs feature relatively long encoding sequences and require transcriptional and translational resources, which can challenge the host metabolism when circuit size is incremented.

Conversely, CRISPR-based gene expression control tools such as CRISPRi or CRISPRa inherently overcome many of those limitations. Key for their success is their RNA-guided nature, which makes them highly programmable and easy to design, but also renders the pool of available orthogonal repressors/activators *almost* unlimited. Indeed, for a gRNA with a 20 nt spacer sequence (the sequence that determines target specificity), 4^{20} ($\approx 10^{12}$) unique implementations are possible, which ensures that sufficiently different (i.e., orthogonal) gRNAs can be designed, even for very large circuits. Even more, the short encoding sequence of the gRNAs (often ~100 bp or less) and the fact that they only require transcription (but not translation) further facilitate the expansion of circuits into larger ones. Finally, when it comes to interfacing the synthetic circuit with the host, CRISPRi/a offers a very straightforward control of virtually any genomic target (or even multiple targets simultaneously), by simply designing a gRNA that binds to them.

All these properties make CRISPRi/a well-suited for the construction of synthetic gene circuits—especially large ones. The fact that all types of logic operations can be built with repression interactions only (i.e., without activators), and the higher simplicity of CRISPRi relative to CRISPRa (especially in bacteria), probably contribute to the higher prevalence of CRISPRi-based circuits compared to CRISPRa. CRISPR-based gene circuits engineered so far include cascades, logic gates, or patterning networks, but also circuits that require ultrasensitive gene expression control, such as bistable circuits or oscillators [6–8] (for a possible mechanism of CRISPRi ultrasensitivity in spite of its lack of cooperativity, see reference [6]).

Here I provide a guide for designing a synthetic CRISPRi circuit in bacteria, outlining the main components and design features that should be considered. It is addressed for those with little or no experience in CRISPRi, gene circuit design, or both. I chose the first CRISPRi oscillator, the CRISPRlator, as a paradigmatic circuit that helps illustrate the design process. The CRISPRlator and its annotated sequence map are available on Addgene, which facilitates its use as a reference for the design process and as a working control in the lab. Please note that the CRISPRlator system comprises two plasmids: one encoding the circuit itself (Addgene ID #140674) and another one carrying the accessory components (Addgene ID #140664).

2 Materials

2.1 The CRISPR System

CRISPR interference requires a catalytically dead nuclease, i.e., one that binds specifically to target DNA sequences but is unable to cleave them. The most popular dead nuclease, and the one that I suggest here, is the *Streptococcus pyogenes* dead Cas9 (dCas9) [9], but other alternatives exist (*see* **Note 1**)—even those that target RNA sequences (*see* **Note 2**). With respect to the wild-type Cas9 protein (Uniprot ID: Q99ZW2), dCas9 carries two inactivating mutations: D10A and H840A [10].

The second element of a CRISPRi system is the "guide" RNA (gRNA) that directs dCas9 to a specific DNA target via Watson-Crick base pairing. In natural CRISPR systems, the gRNA is usually composed of two RNA molecules that hybridize to each other: the crRNA and the tracrRNA [11]. However, a crRNA-tracrRNA

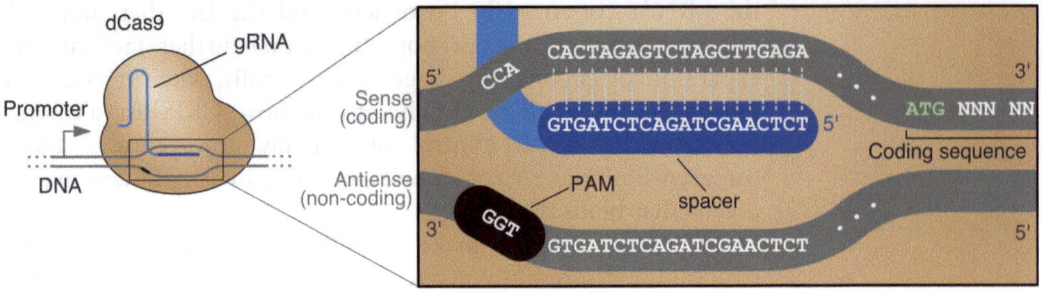

Fig. 1 Schematic of CRISPRi-mediated repression. sgRNA1 and its corresponding binding site (BS1) are represented

fusion into a "single guide" RNA (sgRNA) (Fig. 1) is similarly capable of targeting dCas9, features a simpler design, and is the gRNA of choice in most CRISPRi designs and the one that I propose here. In the sgRNA, the 20 nt "spacer" sequence at the 5′ end provides the target specificity (Fig. 1), while the rest of the sequence forms a structure (the "handle") that interacts with the dCas9 protein. The binding of the sgRNA-dCas9 complex to its DNA target (the so-called "protospacer") requires base pair complementarity between the spacer and the protospacer, but also the presence, adjacent to the protospacer, of a short PAM (protospacer adjacent motif) sequence that is recognized by the (dead) nuclease. The *S. pyogenes* dCas9 recognizes a 5′-NGG-3′ PAM located downstream of the sequence complementary to the one that anneals to the sgRNA spacer (Fig. 1).

The sequence of the sgRNA handle is relatively conserved across laboratories, although variants exist that improve CRISPR functionality. Here I use an optimized version with an extended hairpin (underlined) and two point mutations (bold) [12] (*see* **Note 3**); please find below its sequence and its relative location with respect to the 20 nt spacer (boxed):

5′-NNNNNNNNNNNNNNNNNNNNGTTT**C**AGAGCTA<u>TGCTGGAAACAGC</u>ATAGCAAGT

T**G**AAATAAGGCTAGTCCGTTATCAACTTGAAAAAGTGGCACCGAGTCGGTGC-3′

The spacer sequences of sgRNAs used here are detailed below (*see* **Note 4**), as well as their corresponding binding sites (BS) with the PAM sequences highlighted. Importantly, these sequences are designed so that all three sgRNAs are orthogonal to each other and to the *E. coli* genome [13], which means that any of the sgRNAs (e.g., sgRNA1) should only target its BS (e.g., BS1), but not any other BS or any location in the genome (*see* **Note 5**):

sgRNA1 spacer: 5′-TCTCAAGCTAGACTCTAGTG-3′

sgRNA2 spacer: 5′-ATCAGTGTGTACTAAGTACT-3′

sgRNA4 spacer: 5′-GACACATCTTAGAGTATGTA-3′

BS1 + PAM: 5′-CCACACTAGAGTCTAGCTTGAGA-3′

3′-GGTGTGATCTCAGATCGAACTCT-5′

BS2 + PAM: 5′-CCAAGTACTTAGTACACACTGAT-3′

3′-GGTTCATGAATCATGTGTGACTA-5′

BS4 + PAM: 5′-CCATACATACTCTAAGATGTGTC-3′

3′-GGTATGTATGAGATTCTACACAG-5′

Further details on how to design a CRISPRi-based repression, and how to use it in gene circuits, are provided below, in the Methods section.

2.2 The Reporters

The easiest way to monitor circuit behavior and to troubleshoot it is by using genetically encoded reporters (*see* **Note 6**). There exist different types of reporters (fluorescent, bioluminescent, colorimetric, phenotypic. . .), each with distinctive properties, and they come in two main forms depending on their chemical nature: RNA reporters and protein reporters. RNA reporters like fluorescent light-up aptamers (e.g., Broccoli) [14] are translation-independent (unlike fluorescent proteins) and therefore display fast dynamics (*see* **Note 7**) and impose a low burden on the host metabolism. However, they commonly require the addition of a commercial ligand, and can suffer from limitations such as moderate brightness, ligand cytotoxicity, and inefficient ligand uptake across the cell membrane [15]. Although light-up aptamers have been used in CRISPRi circuits [6], fluorescent proteins continue to be the most popular reporter type for synthetic gene circuits, even for those in which the circuit logic is implemented by RNA-based regulators, like CRISPRi circuits.

When possible, I recommend using one reporter for each of the nodes of the circuit, in order to facilitate circuit characterization and the identification of potential failures. This multi-node reporting will be limited by the load that the basal circuit imposes on the cell metabolism: circuits with many elements or with resource-expensive components such as strong promoters or ribosome binding sites (RBSs) may not admit multiple reporters. They may instead require a single reporter (usually in the output node) to ensure that the metabolic burden does not perturb the expected functioning. Conveniently, in CRISPRi circuits, the regulators that implement the desired logic are gRNA molecules that do not consume translation resources. Besides, repression-based circuits like the CRISPRlator usually contain nodes that repress each other, meaning that not all of them will be expressed (and therefore consume resources) simultaneously. As a consequence, CRISPRi circuits do not pose an excessive burden on the host metabolism and are generally amenable to multi-node reporting (*see* **Note 8**). Please note that, when using multiple fluorescent reporters, their excitation and emission spectra should be checked (e.g., with an online tool like FPbase, https://www.fpbase.org) to ensure that they match the illumination and filter settings of the detection equipment, and to minimize their spectral overlap.

For the CRISPRlator, I suggest a combination of three classic fluorescent reporters, one per node (*see* **Note 9**): mCherry (FPbase ID: ZERB6), mCitrine (FPbase ID: 3Q37R, with the following changes: V2del, F65L, L232H), and Cerulean (FPbase ID: J2JWA, with the following change: V2del). Please keep in mind, when working with anaerobic organisms, that classic fluorescent proteins like GFP or mCherry require oxygen for maturation (*see* **Note 10**). Another important consideration is that the intracellular concentration of the protein-based fluorescent reporters will depend on the levels of transcription and translation (which are dictated by the circuit design), but also on mRNA and protein degradation, and on dilution due to cell growth and division. If the circuit is expected to display fast dynamics, reporter accumulation in host cells may hinder underlying changes in gene expression. In such scenarios, increasing protein turnover (e.g., via degradation tags, see Subheading 2.3) can make reporters correctly reflect the underlying gene expression changes that are orchestrated by the circuit logic.

2.3 The Gene Expression Control Elements

I recommend gene expression control elements (promoters, ribosome binding sites, terminators) that are compact, modular, and well characterized (*see* **Note 11**). These sequences can either be obtained from published part libraries [16, 17], or they can be designed de novo using tools such as the Genetic Systems Calculator (https://www.denovodna.com/software/, formerly the RBS Calculator) developed by H. M. Salis and colleagues [18].

I suggest using promoters from the Anderson promoter collection (https://parts.igem.org/Promoters/Catalog/Anderson), which encompasses ~20 short (35 bp), widely used, well-characterized *E. coli* promoters that often work also in other prokaryotes [19, 20]. Three of these promoters are used in the CRISPRlator: the strong promoter BBa_J23100, the medium promoter BBa_J23104-SpeI (a variant or BBa_J23104 with two point mutations, underlined below, that are part of an SpeI cut site), and the weak promoter BBa_J23150. As for transcriptional termination, three sequences from the >500-terminator library characterized by Chen and colleagues are employed: ECK120029600, L3S2P21, and ECK120033737 [17]. For translation, two ribosome binding site (RBS) sequences are used, depending on whether a high (RBS_high) or a low (RBS_low) translation rate is desirable. Finally, for halting translation, a STOP codon with a high frequency of occurrence in the host organism is advised—TAA > TGA > TAG in *E. coli* (*see* **Note 12**).

The corresponding sequences of these gene expression control elements are listed below:

BBa_J23100:	5′- TTGACGGCTAGCTCAGTCCTAGGTACAGTGCTAGC -3′
BBa_J23104-SpeI:	5′- TTGACAGCTAGCTCAGTCCTAGGTATTGTACTAGT -3’
BBa_J23150:	5′- TTTACGGCTAGCTCAGTCCTAGGTATTATGCTAGC -3’
RBS_high:	5′- TAATTTTGTTTAACTTTAAGAAGGAGATATACAT -3’
RBS_low:	5′- GGTACCATTAAAGAGGAGAAAGAATTC -3′
ECK120029600:	5′- TTCAGCCAAAAAACTTAAGACCGCCGGTCTTGTCCACTACCTTGCA GTAATGCGGTGGACAGGATCGGCGGTTTTCTTTTCTCTTCTCAA -3’
L3S2P21:	5′- CTCGGTACCAAATTCCAGAAAAGAGGCCTCCCGAAAGGGGGGCCTT TTTTCGTTTTGGTCC -3’
ECK120033737:	5′- GGAAACACAGAAAAAAGCCCGCACCTGACAGTGCGGGCTTTTTTTT TCGACCAAAGG -3’

Gene circuits may require a high turnover of some of the components in order to achieve the desired dynamics, or to ensure that the reporters properly reflect the gene expression changes of the underlying circuit implementation. When native degradation and growth-mediated dilution of proteins is not enough to achieve the desired turnover, degradation tags (also known as degrons), which are short peptides recognized by cellular proteases, can be fused to the proteins of interest to target them for degradation. If more than one protein wants to be degraded, the orthogonality of the degradation tags becomes an important consideration, since tags that employ the same degradation mechanism may result in saturation of the degradation machinery and an associated decrease in degradation rates (*see* **Note 13**). Here I propose the following orthogonal degradation tags [21]: MarA (Uniprot ID: P0ACH5), a C-terminal tag degraded by Lon protease; MarAn20 (Uniprot ID: P0ACH5, amino acids 1–20), a C-terminal tag degraded by Lon protease; and RepA70 (Uniprot ID: P24042, amino acids 1–70), an N-terminal tag degraded by ClpAP protease and, to a lower extent, by ClpXP protease.

2.4 Insulation Elements

DNA context can largely influence gene expression and impact circuit performance [22]. Context-dependency can stem from factors like poor part performance, the orientation and physical proximity of genetic elements, or mRNA folding. Tools that mitigate context dependency are popular in synthetic biology and are recommended when designing gene circuits. For example, strong terminators are key for avoiding undesired transcriptional read-through, which can be an issue for adjacent genes placed in the same orientation (*see* **Note 14**). I recommend separating genes/ transcriptional units with inert separator sequences (*see* **Note 15**). Spacing may also be required between DNA sequences that recruit effector proteins, such as transcription factor operators and dCas9 binding sites (*see* **Note 16**). When dealing with multi-part or multi-gene transcriptional units, or simply to avoid dependencies on the untranslated regions (UTRs), sequence-specific RNA-processing systems such as ribozymes or CRISPR endoribonucleases (*see* **Note 17**) can be employed to cleave the mRNA into separate molecules that operate independently of each other.

In the CRISPRlator, 200 bp inert separators are used to isolate transcriptional units from each other. Besides, the CRISPR endoribonuclease Csy4 (Uniprot ID: Q02MM2) is used to recognize and cleave specific 20 nt cut sites in the multi-part mRNAs such as to release all the independent parts encoded in them, namely the sgRNAs and the transcripts encoding the fluorescent reporters (see Subheading 3.3). As stated above, mRNA processing by Csy4 ensures that the multiple elements encoded in the same mRNA are separated from each other and from undesired UTR regions, and can therefore carry out their function—binding to dCas9, in the case of sgRNAs; being translated, in the case of reporter-encoding transcripts—without interfering RNA appendages. The sequences of the 200 bp inert separators and the Csy4 cut sites are the following:

Separator_1: 5′- TTCTTCTGACCTGTAACGAATAATAGAT AGTAAAGTAGTCTCCGATTGAGTTT TCTCTGCCGAGTCCCACCCAGTTCTGTGA TTTCAGTAAGTTGGTAATTGATACACTGTTG CGAGAACTGCTGCCTGGTAGTAGATAGGTTG TTATTGAGTAAGAAGGTAAAGTGAACGAAAT CCCTGAAACTGAGACTGTAGAAAAT -3’

Separator_2: 5′- TACTGATTGGGTGAACTACGATACTTCC AGAATAACGAGAAATACGCTACTCT TACAGATAAAACTCACTGCCGATTACGGA TAGGACGAACAACACTTTGATTATCTTA CGATTTACCGAAGGTTCTATTGAGTTGGA CCCAGCAGTTACTACTTTTATTCAGTGA TTCTGTCTTATCGGGAGATTGTTCGCTTA CCAC -3’

Separator_2.5: 5′- TAATAGATAAAGGATAGGTCTGGTAGTG
TTGTTCGTTCTCGCAGGTAAATCAA
TAATACTCAGCAGTTCCGTAGACTTTTCA
GTGGGACAGGGTAGCGATAACAGATAGATTG
TAATAAGACACAGTAGGTGCTCGTAGTTGCG
TGAAGAGAACCGCTCAGGAAATCCAGTCAGA
AGTATTGGTAATCGTTGAAAACTCA -3’

Csy4 cut site: 5′- GTTCACTGCCGTATAGGCAG –3′

2.5 Circuit Location, Host Cells, and Growth Medium

The location of the circuit is a key design consideration. Genome integration usually provides stability, highly reduces HGT (horizontal gene transfer) probability, and yields circuit copy numbers comparable to those of endogenous genes and circuits. However, it can be technically challenging, and it is less suited for combinatorial studies, which can be easily done with plasmids by simply placing circuit components in separate vectors and co-transforming them in different combinations. Besides, integrated circuits can suffer from context dependencies as a function of the flanking DNA sequences and the distance of the insertion locus from the origin of chromosomal replication [23, 24]. Plasmids, on the other hand, offer ease and flexibility, and can support the existence of circuits in a multicopy form (*see* **Note 18**) that benefits their performance or helps achieve desired expression levels. When choosing a plasmid, important elements to be considered include the intended host organism and the copy number of the plasmid in that host (both determined by the origin of replication), as well as plasmid (in)compatibility when using two or more plasmids (also governed by the *ori*); the selection markers (also in relation to the host, which might be naturally resistant or even unable produce the plasmid-encoded resistance mechanism); insert size capacity and the size of the plasmid itself; and cloning-related considerations (e.g., the type and location of restriction sites). For genomic insertions, one should carefully decide on aspects like the target locus, the insert size, the insertion approach, or the method for screening or selection.

The host cell can also have a notable impact on circuit performance [25]. Not only different bacterial species and even strains may feature important variations in resource availability, metabolism, and physicochemical properties of the intracellular milieu, but they may also display differences in codon usage or gene expression control elements—including the presence of factors that may directly interfere with circuit performance (e.g., *ara* and *lac* operons in the case of circuits using two widespread transcription factor tools, AraC and LacI). Likewise, the choice of the growth medium can further influence the results. Medium parameters such as the pH, the salts, and the carbon source can impact growth dynamics and circuit behavior—for instance, glucose acts as a catabolic repressor and can repress molecular tools derived from alternative catabolic pathways (e.g., the P_{BAD} promoter) when used as a carbon source. The overall nutrient availability of the medium is also important: many of the commonly used rich media are not defined, meaning that batch-to-batch variation can confound result interpretation. Conversely, many defined media are not very rich, which can result in metabolic burden and an associated loss of circuit performance. When possible, I recommend using a rich defined medium for experiments. Finally, the incubation temperature is another known factor affecting both growth and circuit function, and there are instances of circuits that have been designed specifically to prevent temperature dependencies [26]. In the particular case of CRISPRi, a sharp decrease in repression is observed above 37 °C [27], and therefore temperatures equal to or below 37 °C are recommended.

For the CRISPRlator, I suggest using replicative plasmids with medium (~20–40) copy-number *ori*-s such as ColA and CloDF13, and an *E. coli* strain designed (or at least well-suited) for gene circuit studies, such as MK01 [28]. Of note, CRISPR oscillators have also demonstrated to operate as designed in other cellular contexts (*see* **Note 19**). For circuit assessment, a 37 °C incubation is recommended for bacterial growth, using a rich and defined medium—e.g., MOPS EZ (Teknova)—with 0.4% glycerol as carbon source.

3 Methods

3.1 General Considerations

The first design decision is where to encode the CRISPR circuit. Here I choose to split the components into two replicative plasmids with similar copy numbers but different *ori*-s and selection markers. One of the plasmids carries the circuit itself, while the other one encodes the accessory components that are necessary for the circuit to operate. This division enables a greater flexibility and ease in circuit construction and assessment, because the accessory plasmid can be kept constant within the host strain, while variants of the circuit plasmid are assembled and transformed in the search for an optimally performing version. Importantly, all the components implementing the circuit logic are encoded in a single plasmid (the circuit plasmid), which ensures that the stoichiometry is not influenced by differences in plasmid copy number.

The accessory plasmid (CloDF13 *ori*, $Spec^R$) carries dCas9, which is needed for all repressions encoded in the circuit. While many CRISPR circuits and logic gates use an inducible dCas9, here I propose expressing it in a constitutive manner—albeit at moderate levels. These are the main gene expression control elements for dCas9:

Promoter:	BBa_J23104-SpeI
RBS:	RBS_low
STOP codon(s):	TAA
Terminator:	ECK120029600

The circuit plasmid (ColA *ori*, Kan^R) contains the three circuit nodes, each with a fluorescent reporter and separated from each other by transcriptional terminators and 200 bp inert separators. Each node is designed as a single transcriptional unit in which a promoter precedes a sgRNA target binding site (BS) and controls the transcription of a downstream single guide RNA (sgRNA) and fluorescent reporter (Fig. 2). Including the reporter within the node (and not elsewhere) helps ensuring a reliable reporting of the circuit behavior (*see* **Note 20**).

3.2 Circuit Logic

The common, fundamental requirement of oscillatory circuits is a delayed negative feedback loop (dNFL) motif, and the CRISPRlator, as other ring oscillators, uses a closed ring of repressions to implement the dNFL. In practice, it employs three nodes, all with a comparable design: each node N is repressed by node N-1 and produces a sgRNA that, together with the constitutive dCas9, binds to and represses node N+1. Hence, the minimal set of elements per node N that enables oscillations includes a promoter, a

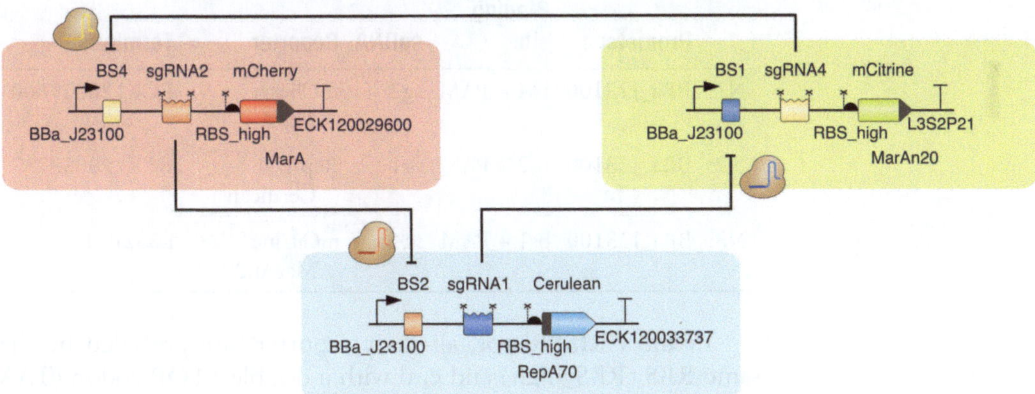

Fig. 2 Diagram showing the main design features of the CRISPRlator circuit. Each transcriptional unit corresponds to one of the nodes of the circuit, and contains a promoter, a CRISPRi binding site, a sgRNA-encoding cassette, a degron-tagged fluorescent reporter, and a transcriptional terminator. All three nodes are encoded in the same plasmid, separated by 200 bp separator sequences. An additional accessory plasmid encodes dCas9 and Csy4. Bent arrows: promoters; squares: sgRNA binding sites; wavy rectangles: sgRNA casetters; crosses: Csy4 recognition sites; semicircles: RBSs; pointed rectangles: reporter genes with degradation tags; and T-s: transcriptional terminators

binding site for the $sgRNA_{N-1}$ + dCas9 complex, and a cassette encoding the $sgRNA_{N+1}$.

The choice of the promoter will determine the strength of sgRNA expression and, thus, that of dCas9-mediated repression. In order to ensure that all three nodes are expressed to similar levels so that the strengths of all three repressions are balanced, I recommend using the same promoter for all the nodes, rather than a different promoter for each (*see* **Note 21**). These promoters control the expression of downstream sgRNA cassettes, which are composed of a variable "spacer" sequence, and a constant "handle" (the same handle for all three sgRNAs in the circuit, *see* **Note 22**). Finally, the BS for the $sgRNA_{N-1}$ + dCas9 complex is located between the promoter and the $sgRNA_{N+1}$-encoding cassette, i.e., close enough from the promoter, so that it can be highly repressed (*see* **Note 23**), and upstream of the $sgRNA_{N+1}$, such as to prevent its expression.

For CRISPRi repression to work, the spacer sequence should be identical to one of the strands of the binding site (BS) and complementary to the other one. I recommend arranging the BS so that the anti-sense/non-coding/bottom strand is the one identical to the spacer sequence of the sgRNA, and therefore the sgRNA anneals to the sense/coding/top strand of the BS (Fig. 1) (*see* **Note 23**). Please remember to include a properly located PAM sequence adjacent to the binding site (downstream of the sequence that is identical to that of the sgRNA spacer) (Fig. 1), or else dCas9 will not interact with the target DNA and repression will not occur.

The three nodes of the CRISPRlator are organized as follows:

	Promoter	Binding site	sgRNA	Reporter	Terminator
N1:	BBa_J23100	bs4 + PAM	sg2	mCherry-MarA	ECK120029600
N2:	BBa_J23100	bs2 + PAM	sg1	RepA70-Cerulean	ECK120033737
N3:	BBa_J23100	bs1 + PAM	sg4	mCitrine-MarAn20	L3S2P21

In the CRISPRlator, all three reporters are preceded by the same RBS (RBS_high) and end with a double STOP codon (TAA TAA) (*see* **Note 12**). In each node, the promoter, the BS + PAM, the sgRNA, and the reporter are all separated by ~60 bp sequences derived from the modular cloning scheme used, while a shorter separator (~20 bp) is found between the reporter and the terminator (*see* **Note 24**).

3.3 Insulation

To avoid mRNA context dependency, a Csy4-mediated processing of the multi-part mRNA is recommended—but not essential for function (*see* **Note 25**). I suggest adding Csy4 cut sites flanking the sgRNA-encoding sequences, and also immediately upstream of the RBS of the fluorescent reporters. In the proposed design, Csy4 is harbored in the accessory plasmid (together with dCas9) and expressed from a moderate constitutive promoter. These are the main gene expression control elements for Csy4:

Promoter:	BBa_J23150
RBS:	RBS_low
STOP codon(s):	TGA TAA
Terminator:	ECK120033737

The nodes of the CRISPRlator in the circuit plasmid, as well as the transcriptional units in the accessory plasmid, are transcriptionally insulated by terminators (see Subheading 3.2), and spaced by 200 bp inert separators. The location of the separators in the circuit and accessory plasmids is the following:

Circuit:	N1	Separator_1	N3	Separator_2.5	N2
Accessory:	dCas9 cassette	Separator_1	L3S2P21 term.	Separator_2	Csy4 cassette

3.4 Usage

When using the CRISPRlator, an important consideration to keep in mind is that synthetic molecular oscillators generally operate well during the exponential growth phase, but (robust) oscillations are halted when cells enter stationary phase. This is likely the consequence of a combination of factors, including altered gene expression and protein dilution dynamics in stationary phase. Besides, in the case of CRISPR-based circuits, dCas9 unbinding from its target site is largely dependent on DNA replication; therefore, reduced DNA replication during the exponential phase results in a longer residence time of DNA-bound dCas9 (*see* **Note 7**). Since dCas9 unbinding from its target DNA is needed to de-repress a given node and to transition to the next state of the oscillatory cycle, reduced DNA replication also results in a reduced cycling of the oscillator. As a consequence, synthetic oscillators (including CRISPRi oscillators) are generally operated and characterized in settings that allow for continuous growth (*see* below).

Another relevant note is that, unlike many TF-based oscillators whose period can be synchronized at the population level (for example, through the use of inducers or via cell-cell communication), no active or actionable synchronization mechanism is available for the CRISPRlator. While this gives an idea of the robustness of the system, capable of maintaining a reasonable synchrony across a lineage of cells despite the lack of synchronization mechanisms, it also poses a challenge when it comes to the operation and data analysis of the CRISPR oscillator at the population level.

The simplest way to capture oscillations is by employing an experimental setup that enables continuous exponential growth and provides single-cell (or single lineage) data, such as the microfluidics "mother machine" (*see* **Note 26**). When the use of a mother machine (or equivalent) design is not suitable or desirable, a multicell (or multi-lineage) device such as the family machine may be used (*see* **Note 26**). This requires computational image segmentation and lineage tracking to extract single-lineage oscillations, or an experimental approach that maximizes synchrony across cells in the population. The latter can be achieved by seeding the chambers with only a few cells (one, if possible; two-three, otherwise), so that upon cell growth the whole chamber is filled with cells from the same lineage and the intrinsic robustness of the CRISPRlator maintains phase-shifting at relatively low levels. A similar approach can be undertaken with a plate reader, where a highly diluted culture is used as a seed population in a 96-well plate. Since cultures in a 96-well plate grow in batch rather than continuously, periodic dilutions with fresh medium are necessary to ensure that cells are kept in the exponential phase.

3.5 Additional Considerations

While CRISPRi oscillators feature a topology in which each node only contains a single sgRNA and a single BS, other circuits may require more intricate topologies in which nodes are interconnected. CRISPRi is perfectly suited for these designs: a node can be repressed by multiple sgRNAs, and multiple sgRNAs can be produced from the same node (*see* **Note 27**). Besides, interfacing the circuit with the host genome to control the expression of a gene of interest (GOI) (or even a group of genes) is very straightforward, requiring only the addition of a GOI-targeting sgRNA(s) to the output node of the circuit [29, 30].

To achieve the desired circuit behavior, the strength of the CRISPRi interactions may need to be fine-tuned. This can be simply done through classic gene expression control strategies, such as changing the promoter strength of the sgRNA, modifying the promoter and RBS sequences of dCas9, or tagging it with a protein degradation tag. Besides, other fine-tuning strategies specific for CRISPRi exist. For instance, the strength of DNA binding and repression can be modulated by modifying the annealing interaction between the 20 nt spacer of the sgRNA and its target binding site (BS), either by introducing mismatches or by truncating the 5′ end of the sgRNA [31] (*see* **Note 28**). Another way to regulate repression is through the use of antagonistic molecules, such as antisense RNAs [32] that bind to sgRNAs or anti-CRISPR proteins that block the PAM-interacting site of dCas9 [33]. Finally, CRISPRi can also be tuned with external, chemical ligands, for example, by expressing dCas9 and/or the sgRNAs from a ligand-inducible promoter, or by using ligand-activated or ligand-deactivated sgRNAs [34].

4 Notes

1. Other DNA-targeting dead nucleases commonly employed in bacterial CRISPRi include dCas9 from other bacterial sources, such as *Staphylococcus aureus* or *Streptococcus thermophilus*, or the dead version of Cas12a (formerly known as Cpf1), dCas12a [35]. (d)Cas12a, generally sourced from *Francisella novicida*, differs from (d)Cas9 in aspects relevant to target binding, such as the sequence and location of the PAM within the target DNA, or the length and position of the spacer within the gRNA. Interestingly, dCas12a uses just one RNA molecule (the crRNA) for targeting, and the nuclease (even the "dead" version) can process mRNAs containing multiple crRNA sequences into individual crRNAs. This facilitates multiplexing and insulation of crRNAs from co-transcribed mRNA sequences (see Subheadings 2.4 and 3.3, and **Note 17**) [36].
2. dCas13 is a dead version of the RNA-targeting nuclease Cas13. Although less used for gene circuits than its DNA-targeting counterparts, it can also be employed to implement a regulatory logic [37, 38].
3. The handle presented here contains modifications that were shown to improve knockout efficiency in genome editing experiments [12]. Besides, it is important to note that often sgRNA handles include a transcriptional terminator derived from *S. pyogenes* in the 3′ end [31]. To avoid premature termination in multi-gRNA designs or in operons, preserving a functional terminator sequence should be avoided. Thus, the sgRNA handle that I recommend here lacks the poly-T stretch at the 3′ end.
4. The sgRNA spacers presented here are synthetic sequences, meaning that they have been designed in silico and do not match any genomic sequence of the host—*E. coli*, in this case. Indeed, building a fully synthetic circuit generally allows the researcher to freely choose the components, including the sgRNAs and their target binding sites. In other words, since the target binding sites (BS) are not fixed, researchers can choose any given sequence as a BS (provided that a PAM is also added and correctly positioned) and then design a corresponding sgRNA that will bind to it. This is in contrast with CRISPR applications that require genome targeting, such as gene editing or CRISPRi repression of endogenous genes, in which the sgRNA spacer sequences are determined by the

natural sequences where (d)Cas9 wants to be targeted. In those cases, the presence of a PAM is a crucial constraint: the absence of a PAM in the intended genomic region prevents the targeting and obliges to either displace the targeting region to a PAM-containing one, or to choose an alternative nuclease (natural or engineered) with a different PAM specificity. In the case of synthetic CRISPRi circuits, and despite the aforementioned freedom, some implementations do impose constraints for sgRNA design, for example, when the sgRNA wants to be targeted to the promoter region to maximize repression: in that case, although the promoter may have been freely chosen to provide desired levels of transcription, the spacer sequence of the sgRNA will be determined by the specific sequence of the target promoter.

5. The sgRNA spacer sequences employed here are synthetic sequences that have been computationally designed to avoid cross-targeting to each other's binding sites and to the *E. coli* genome. When designing new sgRNAs or when moving known sgRNAs into a new host, it is recommended to avoid both the cross-talk between sgRNAs (if more than one is used) and the presence of unintended target binding sites in the genome of the host or in other regions of the circuit or the plasmid. The former can be assessed by testing repression in the presence of the non-cognate sgRNA, e.g., repression of BS-A by sgRNA-B, and repression of BS-B by sgRNA-A. The latter can be evaluated by BLASTing the spacer sequence against the genome of the host or by using one of the many existing computational tools [39], such as CRISPR.Pasteur.fr [40], for instance, which is specific for bacterial CRISPRi. Conveniently, properly designed CRISPRi displays minimal or no off-targets [31, 41], as opposed to applications relying on catalytically active nucleases such as Cas9-induced genome editing. Indeed, while a transient interaction with any genomic region is enough to enable DNA cleavage by Cas9, transcriptional repression by dCas9 requires a sustained interaction within a relatively narrow window around the transcription start site or the 5′ of the target gene [9].

6. When genetically encoded reporters are not available or desirable, circuit performance can be evaluated using classic techniques that detect RNA or protein levels, such as RT-qPCR [42] or Western blot [43]. Omics technologies such as RNA-seq and Ribo-seq (ribosome profiling) can also be used to assess and debug circuits: they offer detailed data on transcription and translation and can identify cryptic failures [44]. However, their adoption is limited by cost and complexity in comparison with the cheap and straightforward information that classic reporters such as fluorescent proteins offer.

7. Even though repression in CRISPRi is governed by gRNA-directed binding, the dynamics of CRISPRi repression are considerably slower than those of other RNA-based regulators—and can be even slower than those of protein regulators. On one hand, target finding relies on iterative binding to PAM sites, followed by adjacent DNA unwinding to assess gRNA complementarity. On average, dCas9 binds for ~20–30 ms to each potential target [45, 46] (longer than the <5 ms that a LacI dimer spends per potential target, for example) [47], and PAM sites are highly abundant across the genome—there are >0.5 million NGG sites in *E. coli*. On the other hand, the dissociation rate of target-bound dCas9 is low, and it mostly depends on dCas9 being kicked out by the transcription or the replication machinery. In practice, this means that de-repression, which presumably occurs due to dCas9/gRNA dilution upon cell growth and division, can take up to 5 h [31]. Thus, fast reporter systems are not required for capturing the dynamics of CRISPRi circuits, and classic fluorescent proteins actually offer fast enough dynamics.
8. The reported toxicity of dCas9 is sequence-specific and can often be attributed to off-target effects that block the expression of essential genes [48]. When it comes to resource expenditure, dCas9 is a relatively big protein and its expression can pose a burden on the host's metabolism. However, given a basal dCas9 expression, enlarging CRISPRi circuits is probably less metabolically challenging than expanding protein-based circuits [6], because gRNA-s do not require translation.
9. Most synthetic oscillators built so far use a single fluorescent protein to report oscillations. If possible, I recommend using one reporter per node, because this allows to unequivocally distinguish oscillations from random fluctuations in reporter levels due to the intrinsic noise of cellular processes. Alternatively, a second, constitutively expressed fluorescent reporter can be used to correct for noise in single-reporter oscillators.
10. Multiple anaerobic fluorescent reporters exist, such as those derived from FAST, HaloTag, SNAP-Tag, or proteins based on flavin-binding LOV domains [49]. Compared to conventional reporters like mCherry or GFP, anaerobic fluorescent reporters often present limitations such as a dim signal, a reversible binding of fluorescent ligands, or washing steps required to remove background fluorescence. Alternatively, one can also use conventional reporters in anaerobic organisms to acquire endpoint measurements by performing cell growth under anaerobic conditions and transferring cultures to aerobic conditions for reporter maturation [49]. This requires additional considerations such as maturation time or cell viability under aerobic conditions.

11. Even if parts have been characterized elsewhere, I advise to perform an in-house characterization before starting using them. This is particularly important when using them in a new species, but even "minor" variations in experimental conditions, such as a different strain (e.g., a cloning versus a protein expression strain), flanking DNA, growth medium, or incubation time can have a big impact on the performance of parts.
12. Two contiguous STOP codons can be used to ensure that translation halts and that the ribosome does not read through, e.g., TAA TAA or TGA TAA.
13. SsrA tags like LAA or LVA are commonly used for increased protein degradation, but they may suffer from lack of orthogonality, as evidenced by decreased degradation rates when two different proteins are tagged with the same C-terminal LAA [21].
14. In multi-gene constructs, the relative orientation of the genes (whether convergent, divergent, or in tandem) can influence—probably mediated by supercoiling—gene expression parameters such as ultrasensitivity and dynamic range [50].
15. Although no sequence is biologically inert sensu stricto, computational tools can be used to design sequences that lack promoters, RBSs, forbidden restriction sites (i.e., sites recognized by restriction enzymes used in our cloning workflow), or sequences homologous to the host genome [51]. However, it is worth keeping in mind that, given a stretch of 100 bp, ~10% of random sequences will readily display promoter activity, while ~60% are only 1 mutation away from becoming active promoters [52]. Therefore, testing the promoter activity of "inert" separators is highly advised.
16. In CRISPRi, dCas9 molecules loaded with different gRNAs will compete for DNA binding if their target sites are in close proximity, which can abrogate repression [53].
17. While dCas9 lacks endoribonuclease activity, its counterpart dCas12a naturally processes multi-gRNA molecules into independent gRNAs, and is therefore suitable for gRNA multiplexing without the need of additional mRNA processing tools [36].
18. There are replicative plasmids like BACs (bacterial artificial chromosomes) that maintain copy-number at very low levels, usually 1–2 copies per cell [54]. They therefore allow the assessment of the effect of circuit copy number without the need for chromosomal integration. Even more, plasmids with inducible replication (including some BACs) enable a variable copy number controlled by the user [55–57].

19. Other implementations of CRISPR oscillators have been demonstrated. Kuo and colleagues showed oscillations in one-plasmid, two-plasmid, and genome-integrated systems in *E. coli* MG1655 Δ*motA* [7]. Rueff et al. also built a genome-integrated CRISPRlator, in this case in the Gram-positive *Streptococcus pneumoniae* (strain D39V) [30].
20. An illustrative example of the relevance of reporter location is provided by the work of Potvin-Trottier and colleagues on the optimization of the repressilator, the first synthetic oscillator [58]. The original repressilator uses one plasmid to encode the circuit logic (a never-ending, closed-loop repression cascade of three transcription factors: TetR, LacI, and cI) and a separate plasmid for the reporter gene (GFP). The reporting of the oscillatory dynamics of the transcription factors is enabled by a *tetO* operator in the GFP-controlling promoter. This configuration, however, is suboptimal for accurately reflecting the actual dynamics because the GFP reporter is located in a separate, high-copy vector whose copy number fluctuates substantially. Moving the reporter into one of the circuit nodes in the circuit plasmid greatly reduces the variability in oscillation amplitude [58].
21. Using the same promoter for all three nodes has a direct implication on the design of CRISPRi repression: the binding site (BS) for the sgRNA cannot be located within the promoter, or else all three nodes would be bound and repressed by the same sgRNAs. Thus, the BS needs to be located downstream of the promoter, rather than within, which slightly reduces fold repression [59]. Nevertheless, while other CRISPR oscillators do employ three different promoters and do target sgRNAs directly to the promoters [7], I prefer sacrificing some repression dynamic range in favor of a more balanced expression (and repression) of all three nodes. Importantly, and despite the x3 repetition of the BBa_J23100 promoter in the CRISPRlator (one per node), I have never observed recombination issues, probably due to its short encoding sequence (35 bp).
22. Here, handle repetitions are likely not problematic, since only three relatively short (86 bp) handles are employed—I indeed never noticed recombination events or cloning issues due to handle repetitions. Yet, for large CRISPRi circuits requiring dozens of sgRNAs, orthogonal non-repetitive sgRNAs are available [60]. In the CRISPRlator, using exactly the same handle ensures that dCas9 interacts similarly with all the sgRNAs, and it enables the straightforward creation of new sgRNAs via PCR by using a long forward primer whose 3′

anneals to the constant handle and whose 5′ contains a tail (boxed) that replaces the spacer sequence with a new one:

```
5′-NNNNNNNNNNNNNNNNNNNNGTTTCAGAGCTATGCTGGAAACAG -3′
```

23. The design, strength, and mechanism of dCas9-mediated repression in bacteria greatly depends on the location of the targeted binding site (BS). If the BS is located (at least partially) within the promoter, repression of gene expression occurs because dCas9 prevents the recruitment and binding of the RNA polymerase. However, if the BS is located further downstream, e.g., within the coding sequence, repression is a consequence of a dCas9 roadblock that prevents mRNA elongation by the RNA polymerase [61]. As a consequence, repression is independent of the targeted strand when the BS is located within the promoter, but it depends on both the strand and the distance from the promoter when the BS is located downstream of the promoter sequence. In the latter, repression is reported to be higher when the sgRNA anneals to the sense/coding/top strand of the BS (i.e., the anti-sense/non-coding/bottom strand is the one identical to the spacer sequence of the sgRNA) and when the BS is located closer from the promoter [31, 59]—although the dependence of repression strength on the target position does not appear to be systematic [48, 62].
24. The modular cloning scheme [63], derived from MODAL cloning [64], uses 15 bp Prefix and Suffix sequences flanking all parts to anneal PCR primers that append 30 bp unique Linkers, which enable and specify the orderly assembly of parts during the subsequent Gibson assembly. The ~60 bp and ~ 20 bp separators in the final circuit design correspond to combinations of Prefix, Suffix, and Linker sequences, plus (occasionally) restriction sites.
25. The dCas9 oscillator by Kuo and colleagues [7] operated without mRNA cleavage—unlike their dCas12a-based version, in which the nuclease is naturally able to process its gRNAs. Similarly, the pneumococcal CRISPRlator and its capsule-controlling derivative CAPSUlator [30] were both designed without mRNA processing, using instead "extended" sgRNA (ext-sgRNAs) cassettes that encode a binding site (BS) for $sgRNA_{N-1}$ + dCas9 in their 5′ end.
26. A mother machine is a microfluidics design conceived for the growth and imaging of single bacterial lineages. It features broad channels through which the fresh medium flows, and perpendicular narrow chambers that can only accommodate a single row of bacilli. All bacteria in the same narrow chamber are descendant of the same mother cell (hence the name) at the bottom of the chamber, and are therefore part of the same

lineage. When dividing bacteria reach the broad perpendicular channel, they are swept along by the flow of medium, and thus the population within the narrow chamber remains always in exponential phase—which is ideal for dynamic circuits like oscillators. A family machine is a design analogous to the mother machine, but where an increased chamber width enables the growth of multiple lineages within the same compartment.

27. A recent work that I co-authored provides various examples of CRISPRi circuits with multiple interactions, where each node can produce two sgRNAs and be repressed by two other sgRNAs, for a maximum of six repressions in a three-node circuit [65]. Whenever two sgRNAs are needed in the same node, they can be simply arranged in tandem within a single transcriptional unit, relying on Csy4 cleavage to process the mRNA and release the two sgRNAs. When two repressions need to converge in the same node, the design will vary depending on whether the two repressing nodes use the same sgRNA to repress or not. If the two repressing nodes encode the same sgRNA cassette, a single BS can be placed in the repressed node that will be bound by both sgRNAs. On the contrary, if the two repressing nodes encode two different sgRNAs, then two different BS (one per sgRNA) will have to be encoded in the repressed node, and here care should be taken to separate the two BS enough such as to avoid that the binding of one of the sgRNA+dCas9 complexes prevents the binding of the other one by steric hindrance [53].

28. Mismatches can be introduced either in the sgRNA or in the BS. Truncations, however, can actually only be made to the sgRNA, whose spacer sequence can be constituted by an RNA stretch of less than 20 nt, but not to the BS, since any deletion in the DNA encoding the BS will result in a new BS in which some of the nucleotides are those that flanked the BS before the attempted truncation. Therefore, attempting to truncate a BS will actually generate a mismatch, rather than a real truncation, see below:

BS1 + PAM + flanking DNA:	5′-tccggCCACACTAGAGTCTAGCTT**GAGA**tcgct-3′
	3′-aggccGGTGTGATCTCAGATCGAA**CTCT**agcga-5′
After deleting 5′-GAGA-3′:	5′-tccggCCACACTAGAGTCTAGCTTtcgct-3′
3′-**CTCT**-5′	3′-aggccGGTGTGATCTCAGATCGAAagcga-5′

Acknowledgments

This work was supported by the European Union (ERC Starting Grant 101114955 - TICK-TOCK Do and Die); by the RYC2023-043017-I grant, funded by MICIU/AEI/10.13039/501100011033 and by the FSE+; and by "Unidad de Excelencia María de Maeztu" CEX2024-001431-M, funded by MICIU/AEI/10.13039/501100011033. I thank Guillermo Nevot, Giacomo Frangipane, Kathakali Sarkar, and Içvara (Aor) Barbier for the critical reading of the manuscript, and Yolanda Schaerli for useful discussions.

References

1. Cameron DE, Bashor CJ, Collins JJ (2014) A brief history of synthetic biology. Nat Rev Microbiol 12(5):381–390. https://doi.org/10.1038/nrmicro3239
2. Tabor JJ, Salis HM, Simpson ZB, Chevalier AA, Levskaya A, Marcotte EM, Voigt CA, Ellington AD (2009) A synthetic genetic edge detection program. Cell 137(7):1272–1281. https://doi.org/10.1016/j.cell.2009.04.048
3. Nielsen AA, Der BS, Shin J, Vaidyanathan P, Paralanov V, Strychalski EA, Ross D, Densmore D, Voigt CA (2016) Genetic circuit design automation. Science 352(6281): aac7341. https://doi.org/10.1126/science.aac7341
4. Fernandez-Rodriguez J, Moser F, Song M, Voigt CA (2017) Engineering RGB color vision into Escherichia coli. Nat Chem Biol 13(7):706–708. https://doi.org/10.1038/nchembio.2390
5. Tica J, Oliver Huidobro M, Zhu T, Wachter GKA, Pazuki RH, Bazzoli DG, Scholes NS, Tonello E, Siebert H, Stumpf MPH, Endres RG, Isalan M (2024) A three-node Turing gene circuit forms periodic spatial patterns in bacteria. Cell Syst 15(12):1123–1132.e1123. https://doi.org/10.1016/j.cels.2024.11.002
6. Santos-Moreno J, Tasiudi E, Stelling J, Schaerli Y (2020) Multistable and dynamic CRISPRi-based synthetic circuits. Nat Commun 11(1): 2746. https://doi.org/10.1038/s41467-020-16574-1
7. Kuo J, Yuan R, Sánchez C, Paulsson J, Silver PA (2020) Towards a translationally-independent RNA-based synthetic oscillator using deactivated CRISPR-Cas. Nucleic Acids Res 48(14):8165–8177. https://doi.org/10.1093/nar/gkaa557
8. Henningsen J, Schwarz-Schilling M, Leibl A, Guttierez JAM, Sagredo S, Simmel FC (2020) Single cell characterization of a synthetic bacterial clock with a hybrid feedback loop containing dCas9-sgRNA. bioRxiv:2020.2007.2016.206722. https://doi.org/10.1101/2020.07.16.206722
9. Ghavami S, Pandi A (2021) CRISPR interference and its applications. In: Singh V (ed) Progress in molecular biology and translational science, vol 180. Academic Press, pp 123–140. https://doi.org/10.1016/bs.pmbts.2021.01.007
10. Jinek M, Chylinski K, Fonfara I, Hauer M, Doudna JA, Charpentier E (2012) A programmable dual-RNA–guided DNA endonuclease in adaptive bacterial immunity. Science 337(6096):816–821. https://doi.org/10.1126/science.1225829
11. Jiang W, Marraffini LA (2015) CRISPR-Cas: new tools for genetic manipulations from bacterial immunity systems. Ann Rev Microbiol 69:209–228. https://doi.org/10.1146/annurev-micro-091014-104441
12. Dang Y, Jia G, Choi J, Ma H, Anaya E, Ye C, Shankar P, Wu H (2015) Optimizing sgRNA structure to improve CRISPR-Cas9 knockout efficiency. Genome Biol 16:280. https://doi.org/10.1186/s13059-015-0846-3
13. Didovyk A, Borek B, Hasty J, Tsimring L (2016) Orthogonal modular gene repression in Escherichia coli using engineered CRISPR/Cas9. ACS Synth Biol 5(1):81–88. https://doi.org/10.1021/acssynbio.5b00147
14. Filonov GS, Moon JD, Svensen N, Jaffrey SR (2014) Broccoli: rapid selection of an RNA mimic of green fluorescent protein by fluorescence-based selection and directed evolution. J Am Chem Soc 136(46): 16299–16308. https://doi.org/10.1021/ja508478x

15. Truong L, Ferré-D'Amaré AR (2019) From fluorescent proteins to fluorogenic RNAs: tools for imaging cellular macromolecules. Protein Sci 28(8):1374–1386. https://doi.org/10.1002/pro.3632
16. Mutalik VK, Guimaraes JC, Cambray G, Lam C, Christoffersen MJ, Mai QA, Tran AB, Paull M, Keasling JD, Arkin AP, Endy D (2013) Precise and reliable gene expression via standard transcription and translation initiation elements. Nat Methods 10(4):354–360. https://doi.org/10.1038/nmeth.2404
17. Chen YJ, Liu P, Nielsen AA, Brophy JA, Clancy K, Peterson T, Voigt CA (2013) Characterization of 582 natural and synthetic terminators and quantification of their design constraints. Nat Methods 10(7):659–664. https://doi.org/10.1038/nmeth.2515
18. Salis HM, Mirsky EA, Voigt CA (2009) Automated design of synthetic ribosome binding sites to control protein expression. Nat Biotechnol 27(10):946–950. https://doi.org/10.1038/nbt.1568
19. Rabiço F, Pedrino M, Narcizo JP, de Andrade AR, Reginatto V, Guazzaroni M-E (2023) Synthetic biology toolkit for a new species of pseudomonas promissory for electricity generation in microbial fuel cells. Microorganisms 11(8). https://doi.org/10.3390/microorganisms11082044
20. Nevot G, Santos-Moreno J, Campamà-Sanz N, Toloza L, Parra-Cid C, Jansen PAM, Barbier I, Ledesma-Amaro R, van den Bogaard EH, Güell M (2025) Synthetically programmed antioxidant delivery by a domesticated skin commensal. Cell Syst 16(2):101169. https://doi.org/10.1016/j.cels.2025.101169
21. Butzin NC, Mather WH (2018) Crosstalk between diverse synthetic protein degradation tags in *Escherichia coli*. ACS Synth Biol 7(1): 54–62. https://doi.org/10.1021/acssynbio.7b00122
22. Brophy JA, Voigt CA (2014) Principles of genetic circuit design. Nat Methods 11(5): 508–520. https://doi.org/10.1038/nmeth.2926
23. Bryant JA, Sellars LE, Busby SJW, Lee DJ (2014) Chromosome position effects on gene expression in Escherichia coli K-12. Nucleic Acids Res 42(18):11383–11392. https://doi.org/10.1093/nar/gku828
24. Lato DF, Golding GB (2020) Spatial patterns of gene expression in bacterial genomes. J Mol Evol 88(6):510–520. https://doi.org/10.1007/s00239-020-09951-3
25. Tas H, Grozinger L, Stoof R, de Lorenzo V, Goñi-Moreno Á (2021) Contextual dependencies expand the re-usability of genetic inverters. Nat Commun 12(1):355. https://doi.org/10.1038/s41467-020-20656-5
26. Hussain F, Gupta C, Hirning AJ, Ott W, Matthews KS, Josic K, Bennett MR (2014) Engineered temperature compensation in a synthetic genetic clock. Proc Natl Acad Sci USA 111(3):972–977. https://doi.org/10.1073/pnas.1316298111
27. Vigouroux A, Oldewurtel E, Cui L, Bikard D, van Teeffelen S (2018) Tuning dCas9's ability to block transcription enables robust, noiseless knockdown of bacterial genes. Mol Syst Biol 14(3):e7899. https://doi.org/10.15252/msb.20177899
28. Kogenaru M, Tans SJ (2014) An improved *Escherichia coli* strain to host gene regulatory networks involving both the AraC and LacI inducible transcription factors. J Biol Eng 8:2. https://doi.org/10.1186/1754-1611-8-2
29. Mückl A, Schwarz-Schilling M, Fischer K, Simmel FC (2018) Filamentation and restoration of normal growth in Escherichia coli using a combined CRISPRi sgRNA/antisense RNA approach. PLoS One 13(9):e0198058. https://doi.org/10.1371/journal.pone.0198058
30. Rueff A-S, van Raaphorst R, Aggarwal SD, Santos-Moreno J, Laloux G, Schaerli Y, Weiser JN, Veening J-W (2023) Synthetic genetic oscillators demonstrate the functional importance of phenotypic variation in pneumococcal-host interactions. Nat Commun 14(1):7454. https://doi.org/10.1038/s41467-023-43241-y
31. Qi LS, Larson MH, Gilbert LA, Doudna JA, Weissman JS, Arkin AP, Lim WA (2013) Repurposing CRISPR as an RNA-guided platform for sequence-specific control of gene expression. Cell 152(5):1173–1183. https://doi.org/10.1016/j.cell.2013.02.022
32. Lee YJ, Hoynes-O'Connor A, Leong MC, Moon TS (2016) Programmable control of bacterial gene expression with the combined CRISPR and antisense RNA system. Nucleic Acids Res 44(5):2462–2473. https://doi.org/10.1093/nar/gkw056
33. Dong D, Guo M, Wang S, Zhu Y, Wang S, Xiong Z, Yang J, Xu Z, Huang Z (2017) Structural basis of CRISPR–SpyCas9 inhibition by an anti-CRISPR protein. Nature 546(7658): 436–439. https://doi.org/10.1038/nature22377
34. Kundert K, Lucas JE, Watters KE, Fellmann C, Ng AH, Heineike BM, Fitzsimmons CM, Oakes BL, Qu J, Prasad N, Rosenberg OS, Savage DF, El-Samad H, Doudna JA, Kortemme T (2019) Controlling CRISPR-Cas9

with ligand-activated and ligand-deactivated sgRNAs. Nat Commun 10(1):2127. https://doi.org/10.1038/s41467-019-09985-2

35. Call SN, Andrews LB (2022) CRISPR-based approaches for gene regulation in non-model bacteria. Front Genome Ed 4:892304

36. McCarty NS, Graham AE, Studená L, Ledesma-Amaro R (2020) Multiplexed CRISPR technologies for gene editing and transcriptional regulation. Nat Commun 11(1):1281. https://doi.org/10.1038/s41467-020-15053-x

37. Kim G, Kim HJ, Kim K, Kim HJ, Yang J, Seo SW (2024) Tunable translation-level CRISPR interference by dCas13 and engineered gRNA in bacteria. Nat Commun 15(1):5319. https://doi.org/10.1038/s41467-024-49642-x

38. Ding Y, Tous C, Choi J, Chen J, Wong WW (2024) Orthogonal inducible control of Cas13 circuits enables programmable RNA regulation in mammalian cells. Nat Commun 15(1):1572. https://doi.org/10.1038/s41467-024-45795-x

39. Alipanahi R, Safari L, Khanteymoori A (2023) CRISPR genome editing using computational approaches: a survey. Front Bioinform 2:1001131. https://doi.org/10.3389/fbinf.2022.1001131

40. Calvo-Villamañán A, Ng JW, Planel R, Ménager H, Chen A, Cui L, Bikard D (2020) On-target activity predictions enable improved CRISPR–dCas9 screens in bacteria. Nucleic Acids Res 48(11):e64. https://doi.org/10.1093/nar/gkaa294

41. Gilbert Luke A, Horlbeck Max A, Adamson B, Villalta Jacqueline E, Chen Y, Whitehead Evan H, Guimaraes C, Panning B, Ploegh Hidde L, Bassik Michael C, Qi Lei S, Kampmann M, Weissman Jonathan S (2014) Genome-scale CRISPR-mediated control of gene repression and activation. Cell 159(3):647–661. https://doi.org/10.1016/j.cell.2014.09.029

42. Schaerli Y, Munteanu A, Gili M, Cotterell J, Sharpe J, Isalan M (2014) A unified design space of synthetic stripe-forming networks. Nat Commun 5:4905. https://doi.org/10.1038/ncomms5905

43. Isalan M, Lemerle C, Serrano L (2005) Engineering gene networks to emulate Drosophila embryonic pattern formation. PLoS Biol 3(3):e64. https://doi.org/10.1371/journal.pbio.0030064

44. Espah Borujeni A, Zhang J, Doosthosseini H, Nielsen AAK, Voigt CA (2020) Genetic circuit characterization by inferring RNA polymerase movement and ribosome usage. Nat Commun 11(1):5001. https://doi.org/10.1038/s41467-020-18630-2

45. Jones DL, Leroy P, Unoson C, Fange D, Curic V, Lawson MJ, Elf J (2017) Kinetics of dCas9 target search in *Escherichia coli*. Science 357(6358):1420–1423. https://doi.org/10.1126/science.aah7084

46. Martens KJA, van Beljouw SPB, van der Els S, Vink JNA, Baas S, Vogelaar GA, Brouns SJJ, van Baarlen P, Kleerebezem M, Hohlbein J (2019) Visualisation of dCas9 target search in vivo using an open-microscopy framework. Nat Commun 10. https://doi.org/10.1038/s41467-019-11514-0

47. Elf J, Li G-W, Xie XS (2007) Probing transcription factor dynamics at the single-molecule level in a living cell. Science 316(5828):1191. https://doi.org/10.1126/science.1141967

48. Cui L, Vigouroux A, Rousset F, Varet H, Khanna V, Bikard D (2018) A CRISPRi screen in E. coli reveals sequence-specific toxicity of dCas9. Nat Commun 9(1):1912. https://doi.org/10.1038/s41467-018-04209-5

49. Streett H, Charubin K, Papoutsakis ET (2021) Anaerobic fluorescent reporters for cell identification, microbial cell biology and high-throughput screening of microbiota and genomic libraries. Curr Opin Biotechnol 71:151–163. https://doi.org/10.1016/j.copbio.2021.07.005

50. Yeung E, Dy AJ, Martin KB, Ng AH, Del Vecchio D, Beck JL, Collins JJ, Murray RM (2017) Biophysical constraints arising from compositional context in synthetic gene networks. Cell Syst 5(1):11–24.e12. https://doi.org/10.1016/j.cels.2017.06.001

51. Casini A, Christodoulou G, Freemont PS, Baldwin GS, Ellis T, MacDonald JT (2014) R2oDNA designer: computational design of biologically neutral synthetic DNA sequences. ACS Synth Biol 3(8):525–528. https://doi.org/10.1021/sb4001323

52. Yona AH, Alm EJ, Gore J (2018) Random sequences rapidly evolve into de novo promoters. Nat Commun 9(1):1530. https://doi.org/10.1038/s41467-018-04026-w

53. Anderson DA, Voigt CA (2021) Competitive dCas9 binding as a mechanism for transcriptional control. Mol Syst Biol 17(11):e10512. https://doi.org/10.15252/msb.202110512

54. Shizuya H, Kouros-Mehr H (2001) The development and applications of the bacterial artificial chromosome cloning system. Keio J Med 50(1):26–30. https://doi.org/10.2302/kjm.50.26

55. Wild J, Hradecna Z, Szybalski W (2002) Conditionally amplifiable BACs: switching from

single-copy to high-copy vectors and genomic clones. Genome Res 12(9):1434–1444. https://doi.org/10.1101/gr.130502

56. Rouches MV, Xu Y, Cortes LBG, Lambert G (2022) A plasmid system with tunable copy number. Nat Commun 13(1):3908. https://doi.org/10.1038/s41467-022-31422-0
57. Joshi SH-N, Yong C, Gyorgy A (2022) Inducible plasmid copy number control for synthetic biology in commonly used E. coli strains. Nat Commun 13(1):6691. https://doi.org/10.1038/s41467-022-34390-7
58. Potvin-Trottier L, Lord ND, Vinnicombe G, Paulsson J (2016) Synchronous long-term oscillations in a synthetic gene circuit. Nature 538(7626):514–517. https://doi.org/10.1038/nature19841
59. Bikard D, Jiang W, Samai P, Hochschild A, Zhang F, Marraffini LA (2013) Programmable repression and activation of bacterial gene expression using an engineered CRISPR-Cas system. Nucleic Acids Res 41(15):7429–7437. https://doi.org/10.1093/nar/gkt520
60. Reis AC, Halper SM, Vezeau GE, Cetnar DP, Hossain A, Clauer PR, Salis HM (2019) Simultaneous repression of multiple bacterial genes using nonrepetitive extra-long sgRNA arrays. Nat Biotechnol 37(11):1294–1301. https://doi.org/10.1038/s41587-019-0286-9
61. Vigouroux A, Bikard D (2020) CRISPR tools to control gene expression in bacteria. Microbiol Mol Biol Rev 84(2):e00077–e00019. https://doi.org/10.1128/MMBR.00077-19
62. Wang T, Guan C, Guo J, Liu B, Wu Y, Xie Z, Zhang C, Xing X-H (2018) Pooled CRISPR interference screening enables genome-scale functional genomics study in bacteria with superior performance. Nat Commun 9(1):2475. https://doi.org/10.1038/s41467-018-04899-x
63. Santos-Moreno J, Schaerli Y (2019) A framework for the modular and combinatorial assembly of synthetic gene circuits. ACS Synth Biol 8:1691–1697. https://doi.org/10.1021/acssynbio.9b00174
64. Casini A, MacDonald JT, De Jonghe J, Christodoulou G, Freemont PS, Baldwin GS, Ellis T (2014) One-pot DNA construction for synthetic biology: the modular overlap-directed assembly with linkers (MODAL) strategy. Nucleic Acids Res 42(1):e7. https://doi.org/10.1093/nar/gkt915
65. Santos-Moreno J, Tasiudi E, Kusumawardhani H, Stelling J, Schaerli Y (2023) Robustness and innovation in synthetic genotype networks. Nat Commun 14(1):2454. https://doi.org/10.1038/s41467-023-38033-3

Chapter 5

Design of Conditional Guide RNAs for the Logical Regulation of Gene Expression

Dongwon Park, Woosub Shin, Hansol Kang, and Jongmin Kim

Abstract

The CRISPR interference (CRISPRi) is an RNA-guided regulator that silences gene expression by binding to its cognate DNA target, halting transcription in both prokaryotic and eukaryotic cells. Recent advances in RNA synthetic biology have endowed CRISPR guide RNAs (gRNAs) with conditional functionality: these so-called conditional guide RNAs (cgRNAs) fold into strong hairpins that block their activity until a specific trigger RNA is present. Upon introduction of the cognate trigger RNAs, the hairpin structure unfolds, allowing the activated cgRNA to direct transcriptional repression with large dynamic ranges, minimal crosstalk, expanded tunability, and logic-gated signal processing. Furthermore, cgRNAs can be integrated into endogenous gene circuits to achieve sophisticated and logical regulation of gene expression. This chapter describes the design of cgRNAs and provides detailed protocols for their in vivo characterization in *E. coli*.

Key words RNA synthetic biology, Genetic circuit, CRISPRi, Complex logic, Conditional guide RNA

1 Introduction

The goal of synthetic biology is to implement novel functional biological systems [1–3]. One of the major thrusts of synthetic biology is to build genetic circuits that can control gene expression in response to various intracellular and extracellular signals [4]. This capability enables engineered cells to execute programmed functionality by detecting multiple signals, which is crucial for applications such as biomanufacturing, living biotherapeutics, and bioremediation [5, 6]. Numerous studies have demonstrated the construction of complex circuits that can detect and process multiple signals for controlling genetic programs. In order to construct complex circuits, standardized genetic regulators with modularity, composability, large dynamic ranges, and minimal crosstalk are some of the basic and essential requirements [7–10].

Xiao-Jun Tian (ed.), *Synthetic Gene Circuits: Methods and Protocols*, Methods in Molecular Biology, vol. 3041, https://doi.org/10.1007/978-1-0716-5304-3_5,

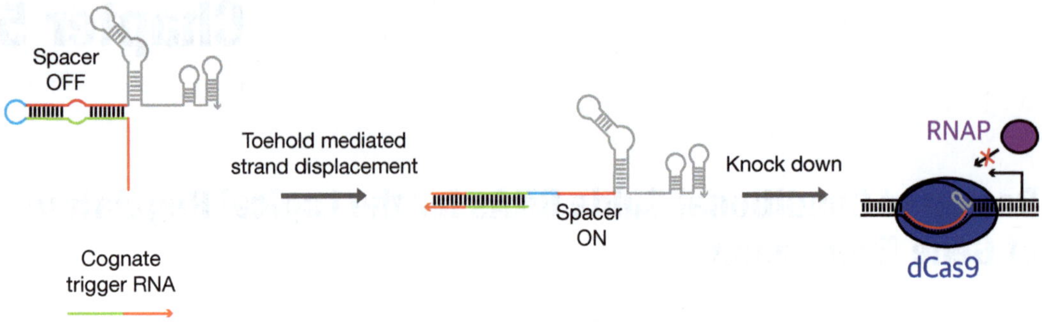

Fig. 1 Schematic diagram of mechanism of cgRNA-based CRISPRi system. (Figure adapted from ref. [20])

Recent advances in synthetic biology have led to the development of a wide array of protein-based [11, 12] and RNA-based regulators [13–15]. Notably, CRISPR interference (CRISPRi) by repurposing guide RNAs (gRNAs) with catalytically inactive Cas9 (dCas9) to work as a ribonucleoprotein complex provides a versatile platform to target DNA sequences and repress transcription [16, 17]. Due to its high efficiency and programmability, CRISPRi is widely explored as a promising regulatory tool for prokaryotic and eukaryotic gene expression. Recent studies have reported numerous ways to enhance the CRISPRi system. In particular, CRISPR gRNA is especially amenable to engineering due to the well-characterized Watson-Crick base-pairing rule to specifically select the target to be regulated. For instance, the strength of repression by CRISPRi can be modulated via programmed mismatches to mitigate cellular toxicity or to receive specific signals for spatiotemporal gene control [18, 19]. Further, the ability to process multiple input signals could be endowed by engineering gRNA sequences such as conditional guide RNAs (cgRNAs) [20]. The CRISPRi system with cgRNA is composed of dCas9 protein and two RNA elements: cgRNA and its cognate trigger RNA (Fig. 1).

Analogous to a conventional single guide RNA (sgRNA), a cgRNA has a target-complementary spacer sequence for guiding dCas9 to the target DNA sequence and a scaffold sequence acting as a handle for interaction with dCas9. In addition to these elements, a stem-loop structure is introduced for cgRNA that is formed by base-pairing between the spacer and its complementary sequence via the extended 5′ end of sgRNA. This structure blocks the spacer sequence from interacting with target DNA and renders the cgRNA-dCas9 complex inactive in the absence of its cognate trigger RNA. A single-stranded toehold region at the 5′ end of the stem-loop provides an initial binding site for the cognate trigger RNA. When the cognate trigger RNA is introduced, the trigger RNA initiates binding at the toehold region, and branch migration ensues to unfold the stem-loop region of cgRNA. After the completion of toehold-mediated branch migration process, the spacer region of cgRNA is free to interact with target DNA region, thereby activating CRISPRi, leading to repression of the target gene.

By utilizing highly programmable RNA-RNA interactions, cgRNA has several advantageous features. First, due to the extensive structural change with large thermodynamic energy gain upon interaction with trigger RNAs, cgRNA can exhibit a high ON/OFF ratio with minimal leakage, providing effective regulation for endogenous genes. Second, the toehold region of cgRNA can be an arbitrary sequence in principle, providing a large design space for construction of orthogonal cgRNA libraries with minimal crosstalk. Third, the strength of repression using CRISPRi with cgRNA can be finely tuned by modulating the expression level of trigger RNAs. Finally, cgRNA can also serve as a signal-processing module that integrates multiple RNA inputs to evaluate multi-input Boolean logic computation (Fig. 2).

As examples, an OR logic gate can be implemented by designing two individual trigger RNAs that can initiate strand displacement from either the toehold region or the loop of the hairpin within the cgRNA. An AND gate can be constructed by designing two trigger RNAs that can interact with each other to form an RNA complex capable of unfolding the stem-loop of cgRNA. Taken together, cgRNA provides a versatile and flexible platform for constructing complex genetic circuits for heterologous as well as endogenous gene regulation (Fig. 3).

In this chapter, we will describe how to design cgRNAs to build complex synthetic logic circuits in *E. coli*. This description includes methods for designing sequences of cgRNAs and trigger RNA inputs using NUPACK [21], a nucleic acid sequence and structure design program, as well as procedures to perform in vivo experiments to validate the system functionality. For more information on this system, please refer to the associated paper [20] that uses the cgRNA technology described here.

2 Materials

Prepare all solutions using distilled water (DW) and analytical grade reagents. Prepare and store all reagents at room temperature unless indicated otherwise.

2.1 cgRNA and Cognate Trigger Sequences for Single Input Switch and Two-Input Logic Circuits

1. cgRNA and cognate trigger sequences
   (a) cgRNA sequence:
   5′- CUAUCUAAUCUGAAUUCUUGGCCUAUU AGAUGCCGCGGAAUACAUCUAAUUCAACAAGA AUUGUUUUAGAGCUAGAAAUAGCAAGUUAAAA UAAGGCUAGUCCGUUAUCAACUUGAAAAAGUG GCACCGAGUCGGUGCUUUUUU -3′ (*see* **Note 1**).

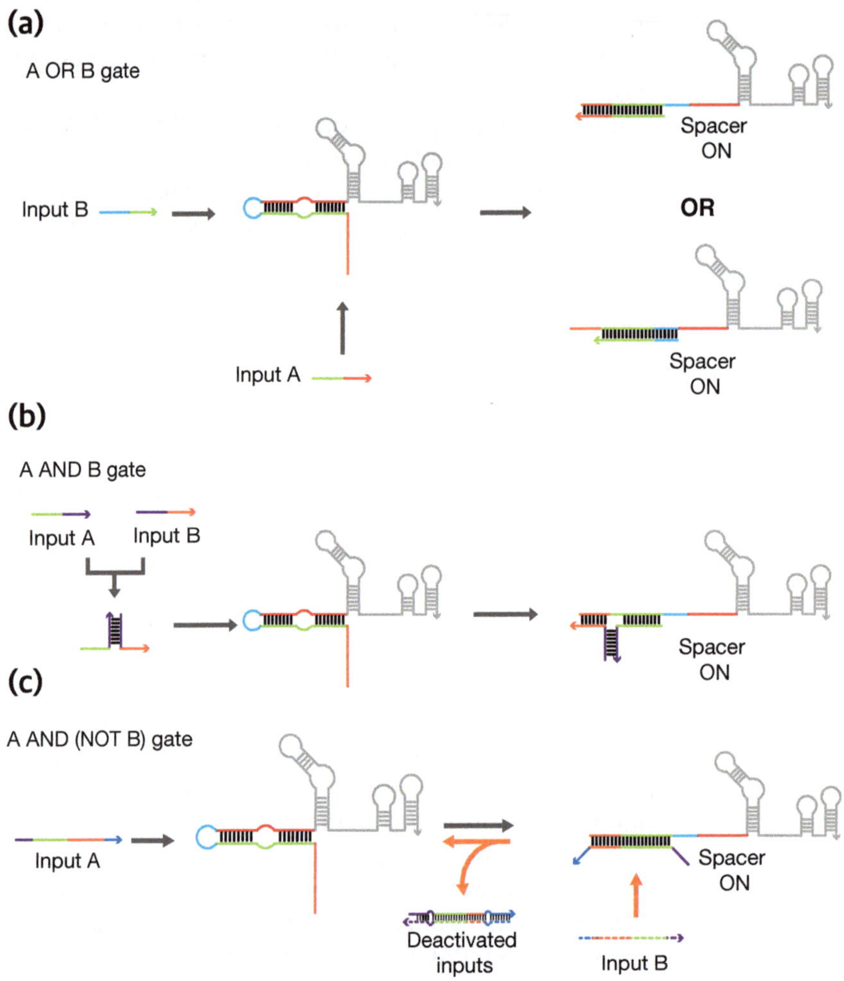

Fig. 2 Two-input logic gates implemented using a cgRNA-based CRISPRi system. (**a**) Schematic diagram of a two-input OR logic gate using a cgRNA-based CRISPRi system. (**b**) Schematic diagram of a two-input AND logic gate using a cgRNA-based CRISPRi system. (**c**) Schematic diagram of a two-input A AND (NOT B) logic gate using a cgRNA-based CRISPRi system. (Figure adapted from ref. [20])

(b) Trigger RNA sequence:
5′-GGGCAUCUAATAGGCCAAGAATTCAGATTA GATAGUAGAAACAGAUAGGCCCUCUUCGGAGG GCCUAUCUGUUUUUUUUU-3′ (*see* **Note 2**).

2. An example set of trigger sequences for two-input OR logic circuits.

(a) Input A Trigger (Toehold-binding RNA) sequence:
5′- GGGCAUCUAAUAGGCCAAGAAUUCAGAU AGAUAGUAGCAUAACCCCUUGGGGCCUCUAAA CGGGUCUUGAGGGUUUUUU-3′

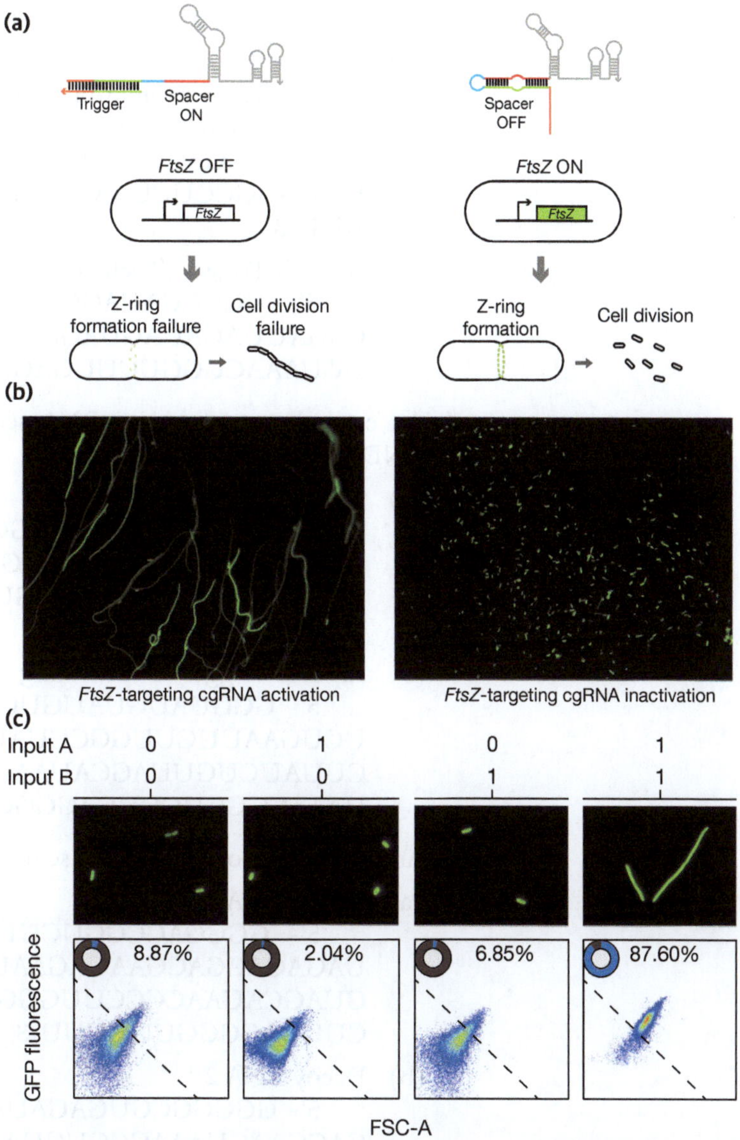

Fig. 3 Regulation of the endogenous gene ftsZ using a cgRNA-based CRISPRi system. (**a**) Schematic diagram of ftsZ repression using cgRNA. (**b**) Results of cgRNA-mediated ftsZ repression with cognate trigger RNA (left) and without cognate trigger RNA (right). (**c**) Results of ftsZ repression with the two-input AND cgRNA (top: truth table and fluorescence images; bottom: flow cytometry plots). (Schematic figure adapted from ref. [20]. The microscopy image data are new datasets, different from those in ref. [20])

(b) Input B Trigger (Loop-binding RNA) sequence:
5′- GGGUAUUCCGCGGCAUCUAAUAGGCCAA GAAUUUAGCAUAACCCCUUGGGGCCUCUAAAC GGGUCUUGAGGGGUUUUUU-3′

3. An example set of trigger sequences for two-input AND logic circuits.
   (a) Input A Trigger (Stem-binding region) sequence:
   5′- GGGCAUCUAAUAGGCCAAGAAUUAAAG CAAGCCGGAGCAAGCUAUGGUAGCAUAACCCC UUGGGGCCUCUAAACGGGUCUUGAGGGGUUU UUU-3′
   (b) Input B Trigger (Toehold-binding region) sequence:
   5′- GGGCCAUAGCUUGCUCCGGCUUGCAAA CAGAUUAGAUAGUAGCAUAACCCCUUGGGGCC UCUAAACGGGUCUUGAGGGGUUUUUU-3′
4. An example set of trigger RNA sequences for two-input A AND (NOT B) logic circuits.
   (a) Input A Trigger sequence:
   5′- GGGCAUCUAAUAGGCCAAGAAUUCAGAU UAGAUAGACAGUGUACAGGUAGCAUAACCCCU UGGGGCCUCUAAACGGGUCUUGAGGGGUUUU UU-3′
   (b) Input B Trigger (Deactivating RNA) sequence:
   5′- GGGUAUGUAUGUGCUGUGACUAUCUAA UCUGAAUUCUUGGCCUAUUAGAUGUCCUUAU CUUAUCUGUUAGCAUAACCCCUUGGGGCCUC UAAACGGGUCUUGAGGGGUUUUUU-3′
5. An example set of decoy RNA sequences (*see* **Note 3**)
   (a) Decoy RNA 1:
   5′- GGGGACCGGUCUUUGCAGAGACCGGUG UAGAGACGACGAAUAGAAUUGAACAUAGAUAA GUAGCAUAACCCCUUGGGGCCUCUAAACGGGU CUUGAGGGGUUUUUU-3′
   (b) Decoy RNA 2:
   5′- GGGGGCGUGAGAUAAGCACAUCUCACGA GACGAAGUAAAGCGUGUAAUCAAUAGUAGUAA GAAACAGAUAGGCCCUCUUCGGAGGGCCUAU CUGUUUUUUUUU-3′

2.2 Suitable Vectors

1. dCas9 and cgRNA are expressed from relatively low-copy vectors suitable for mitigation of cellular toxicity. In previous work, pACYCDuet plasmid backbone (chloramphenicol resistance, p15A origin) was used.
2. Trigger RNAs and decoy RNAs are expressed from relatively high-copy vectors. In previous work, pET15b (ampicillin resistance, ColE1 origin) and pCDFDuet (spectinomycin resistance, CDF origin) plasmid backbones were used.

3. When targeting a heterologous gene on a plasmid, pSC101 vector (kanamycin resistance, SC101 origin) could be useful for target gene expression to achieve large dynamic ranges by reducing target:regulator copy number ratios.

2.3 Cloning

1. 14 mL round bottom tube.
2. Plasmid miniprep kit.
3. 2× Pfu PCR master mix.
4. Thermal cycler.
5. 1% agarose gel: agarose, 1× TAE buffer.
6. 1 kb DNA ladder.
7. SYBR™ Safe DNA Gel Stain.
8. Gel imaging system.
9. 20,000 units/mL DpnI.
10. 1× rCutsmart™ buffer: 50 mM potassium acetate, 20 mM Tris-acetate, 10 mM magnesium acetate, 100 μg/mL recombinant albumin, pH 7.9.
11. Gibson assembly master mix.

2.4 Chemically Competent E. coli

1. Inoue transformation buffer: 10 mM PIPES, pH 6.7, 55 mM $MnCl_2$, 15 mM $CaCl_2$, 250 mM KCl.
2. Dimethyl sulfoxide (DMSO).
3. Liquid nitrogen.

2.5 Strains and Growth Conditions

1. BL21 DE3 (F^- ompT hsdSB (rB^-mB^-) gal dcm (DE3)).
2. BL21 Star DE3 (F^- ompT hsdSB (rB^-mB^-) gal dcm rne131 (DE3)) (*see* **Note 4**).
3. BL21-AI™ (F^- ompT hsdSB (rB^-mB^-) gal dcm araB:: T7RNAP-tetA) (*see* **Note 5**).
4. DH5α (endA1 recA1 gyrA96 thi-1 glnV44 relA1 hsdR17 (rK^-mK^+) λ^-).
5. Luria-Bertani (LB) medium: 5 g/L yeast extract, 10 g/L tryptone, 10 g/L NaCl.
6. 30 mg/mL kanamycin in DW, filter sterilize and store 1 mL aliquots at −20 °C.
7. 50 mg/mL ampicillin in DW, filter sterilize and store 1 mL aliquots at −20 °C.
8. 25 mg/mL spectinomycin in DW, filter sterilize and store 1 mL aliquots at −20 °C.
9. 1 M isopropyl β-D-1-thiogalactopyranoside (IPTG) in DW, filter sterilize and store 1 mL aliquots at −20 °C.

10. 20% w/v arabinose in DW, filter sterilize and store 1 mL aliquots at −20 °C.
11. 2 mg/mL anhydrotetracycline (aTc) stock, store at −20 °C.
12. 96-well deep plate.
13. Breathable sealing film.

2.6 Plate Reader

1. 96-well black/clear plate.
2. Microtiter plate reader with monochromator or lasers and filters for simultaneous measurement of green, red, yellow, and cyan fluorescent proteins.

2.7 Flow Cytometry

1. Flow cytometer with appropriate lasers and filters for simultaneous measurement of green, red, yellow, and cyan fluorescent proteins.
2. 1× phosphate-buffered saline (PBS).
3. Paraformaldehyde powder.

3 Methods

3.1 Brief Introduction of Sequence Design with NUPACK

The design of RNA sequences described in this chapter was conducted using the NUPACK software, available at https://www.nupack.org. Here, we describe the details of the design specifications used to design RNA elements in NUPACK's web portal. All the example design specifications for the cgRNA design are illustrated in Fig. 4.

When users define the desired structures of DNA or RNA elements, NUPACK generates sequences that are predicted to match the structural specifications. For NUPACK-based designs, a set of structural definitions and additional parameters need to be specified. Given this information, the software optimizes sequences to minimize the ensemble defect between the predicted secondary structures formed by candidate sequences and the user-defined target structures. To initiate the design process, the user is required to click the "Design" tab in NUPACK web portal and enter the inputs in each section as described.

To execute sequence designs, several parameters must be specified: the material type of the element, the temperature at which the elements are evaluated, parameter sets used for design (*see* **Note 6**), and the number of simulations to be performed. In the case of cgRNA design, the material is RNA, and the temperature is typically set to 37 °C.

```
Parameters
Material: RNA, Temperature: 37
Parameters: rna06, Ensemble: All stacking
trials: 10


Domains
Toehold: N12
Bulge: N4
Loop: N10
StemC: CAUCUAAU
Bulge_RC: UCAA
StemD: CAAGAAUU
Handle: GUUUUAGAGCUAGAAAUAGCAAGUUAAAAUAAGGCUAGUCCG
Tr1: GGGN32
Terminator: UAGCAUAACCCCUUGGGGCCUCUAAACGGGUCUUGAGGGGUUUUUU

Strands
cgRNA: Toehold StemD* Bulge StemC* Loop StemC Bulge_RC StemD Handle
Trigger: Tr1 Terminator

Target Complexes
Target Complex 1: cgRNA
Strand order: cgRNA
Target Structure
= ............((((((((....((((((((..........))))))))....)))))))).(((((..((((....))))
....)))))...((.....)).

Target complex 2: Trigger
Strand order: Trigger
Target Structure
= .........................................(((((((((((((((......)))))))))))))))....

Target complex 3: RNA complex
Strand order: Trigger, cgRNA
Target Structure
= ...((((((((((((((((((((((((((((((((......(((((((((((((((......)))))))))))))))....+)
)))))))))))))))))))))))))))))))...............................(((((..((((....))))....
)))))...((.....)).

Target Tube
On-target complexes: RNA complex, Concentration: 1μM
```

Fig. 4 Example NUPACK specifications to design cgRNA and trigger RNA targeting GFP

```
Example of parameters for cgRNA design

Material: RNA
Temperature: 37.0
Parameters: rna06
Ensemble: All stacking
```

We generally aim to generate as many diverse sequences as possible to maximize the likelihood of obtaining designs with minimal ensemble defects. Therefore, in the web portal, we request the maximum number of design trials (ten) using the following setting:

```
Trials: 10
```

Next, the sequences of the strands themselves must also be specified. This is typically accomplished by dividing each strand into separate sequence "domains." Enter the name of the domains and sequences for the RNA element in "Domain" section. Domain is defined with arbitrary assignments such as:

```
Example of Domains

Domain name: toehold
Sequence: N12
```

Here, N denotes an arbitrary nucleotide, and the number immediately following N indicates the number of nucleotides to be defined as arbitrary. NUPACK also supports IUPAC nucleotide codes, allowing less arbitrary sequence constraints. These definitions can be combined with strict definitions, for example:

```
Example of Domains

Domain name: Tr1
Sequence: GGGN32
```

Multiple domains can then be concatenated into a strand. In the "Strands" section, enter the names of the strands, and then define each strand by assigning the domains specified in the "Domain" section in your desired order. Consider the trigger domains described in Fig. 4, which together represent 32 arbitrary nucleotides (*see* **Note 7**). These domains can be consolidated into a single strand using the line below, where the "*" operator defines the reverse complement:

```
Example of Strands

Strand name: trigger
Domain order: Tr1 termintor
OR
Domain order: stemC bulge* stemD toehold* terminator
```

For NUPACK software to generate valid designs, both the individual RNA structures and their interaction modes must be specified. In the "Target complexes" section, enter the names of the complexes, and then define each complex by assigning the strands defined in the "Strands" section in the desired order. Then the user can specify the structures of all complexes individually. These structures can be defined using dot-parens-plus or DU+ notation (*see* **Note 8**). For example, implementation of a single-input cgRNA system requires the following structures: cgRNA, trigger RNA, and the cgRNA-trigger RNA complex formed upon their interaction. The following examples show the dot-parens-plus notation required to define a structure of cgRNA. These structures can be described as follows:

```
Example of target complexes

Target Complex 1: cgRNA
Strand order: cgRNA
Target Structure
= . . . . . . . . . . . . ( ( ( ( ( ( ( ( . . . . .
((((((((..........))))))))....))))
)))).(((((..((((....))))....)))))...((.....)).

Target complex 2: Trigger
Strand order: Trigger
Target Structure
=........................................
(((((((((((((((..
....)))))))))))))))....

Target complex 3: RNA complex
Strand order: Trigger, cgRNA
Target Structure
=...(((((((((((((((((((((((((((((((......
(((((((((((((((..
. . . . ) ) ) ) ) ) ) ) ) ) ) ) ) ) ) . . . .
+)))))))))))))))))))))))))))))))....
. . . . . . . . . . . . . . . . . . . . . . . . . . . ( ( ( ( ( . .
((((....))))....)))))...((
.....)).
```

Each RNA element, as well as any complex structure formed by one or more interacting RNAs, can be defined in this way. After defining each element, designate one or more RNA complexes, as defined in the "Target complex" section, as "On-target complexes" in the "Target tube" section.

```
Example of target tube

On-target complex: RNA Complex
Target concentration: 1 μM
```

The filled-out examples above constitute the minimum information required to generate potential cgRNA sequences with corresponding trigger RNA designs in NUPACK. Additional parameters can be used to help improve the expected quality of sequence candidates. For instance, excessive base repetition can be avoided by setting pattern prevention constraints using IUPAC nucleotide codes. These constraints instruct the software to exclude sequences containing specified patterns. Enter the patterns to avoid in the "Pattern constraints" section at the "Hard constraints" tab. The prevention constraint used in our design job is as follows:

```
Pattern Set = AAAA, CCCC, GGGG, UUUU, KKKKKK, MMMMMM,
RRRRRR, SSSSSS, WWWWWW, YYYYYY
Scope (Domain Order) = Toehold, Bulge, Loop
```

Additionally, the user can specify a target ensemble defect for the software to satisfy. Unless specified otherwise, the software will optimize normalized ensemble defect down to 2%, the default value of a target ensemble defect. However, designing larger structures may face difficulty in reaching such a design target, requiring you to set higher acceptable ensemble defects. For example, setting a cutoff ensemble defect threshold of 5% for the principal structures in the set of sequences can be accomplished with the following examples:

```
Stop condition: 0.05
```

After all the required sections are filled, click the "Design" button to submit the job to NUPACK server. The user can evaluate the returned designs to select for those with low normalized ensemble defects. It is recommended to examine whether the candidate sequences form the desired secondary structures and target structures using NUPACK's analysis function.

3.2 Design of cgRNA and Triggers for Single Input Switch and Two-Input Logic CRISPRi System

1. *Design of cgRNA*

   The engineered cgRNA sequence consists of three different sequence domains. First, a 12-nt toehold domain is designated as a fully randomized sequence (N12) at the 5′ end. Second, a 20-nt spacer-hiding stem domain is designed as a partially complementary sequence to the spacer region. Within this stem, a 4-nt bulge sequence is positioned between two 8-nt fully spacer-complementary regions. Following this stem-forming sequence domain, a randomized 10-nt loop, a 20-nt spacer, and scaffold domains are placed in the order of appearance.

The trigger RNA consists of a 32-nt hybridization domain that can form complementary base pairs to the 5′ end of cgRNA. The interaction between the trigger RNA and cgRNA encompasses the toehold domain and the spacer-hiding stem domain within cgRNA. Consequently, the hybridization domain of the trigger RNA is composed of sequences complementary to both the stem and toehold regions of cgRNA.

Detailed structures of the cgRNA and trigger RNA, along with experimental data are shown in Fig. 5.

2. *Design of Two-Input OR Gate Trigger RNAs*

To construct a two-input OR gate, two types of trigger RNAs are designed. One trigger RNA initiates toehold-mediated strand displacement from the 12-nt toehold domain of cgRNA, while the other trigger RNA initiates the process from the 10-nt loop domain of cgRNA. Both triggers include a complementary sequence for the 20-nt spacer-hiding stem domain. For trigger A that initiates hybridization from the toehold domain of cgRNA, a common spacer-binding domain is concatenated with a 12-nt toehold-binding domain in its 3′ end. On the other hand, for trigger B that initiates hybridization from the loop domain, the loop-binding domain is placed before the common spacer-binding domain.

Detailed structures of the two trigger RNA inputs and experimental data for the two-input OR gate are shown in Fig. 6.

3. *Design of Two-Input AND Gate Trigger RNAs*

For the two-input AND gate, the 32-nt hybridization domain of the trigger RNA is divided into a toehold-binding region and a stem-binding region. To enable cooperative activation, 21-nt assembly domains that are complementary to each other are included in these two now separate input domains with 3-nt spacers between the original partial input domains and new complementary domains. The 3-nt spacers provide structural flexibility and could help facilitate interaction of the trigger RNA complex with cgRNA. One of the triggers, trigger A, consists of the following domains in order: 20-nt stem-binding domain, 3-nt spacer, and 21-nt assembly domain. The other trigger, trigger B, consists of the following: 21-nt assembly domain, 3-nt spacer, and 12-nt toehold-binding domain.

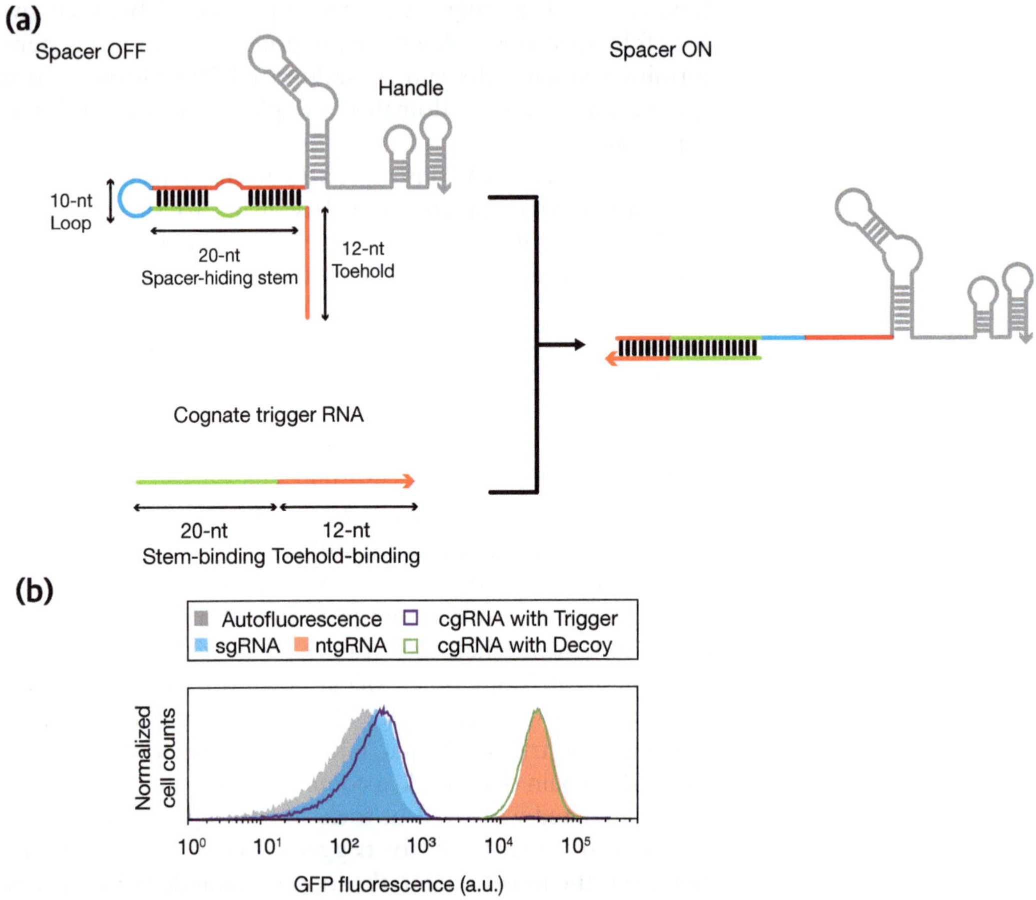

Fig. 5 Detailed schematics of cgRNA design with a single input trigger RNA. (**a**) Structural elements of cgRNA and the input trigger RNA are specified. In addition to the spacer sequence for target recognition, several sequence domains are introduced to a conventional sgRNA. The spacer-blocking stem is required for CRISPRi inactivation in the absence of trigger RNA, and the toehold domain is required for effective binding interaction with RNA-input sequence. All the RNA diagrams are arranged in the 5′ to 3′ direction with 3′ end marked by an arrowhead. (**b**) Flow cytometry data of the CRISPRi system with a single-input cgRNA. Data were obtained using the GFP-repression cgRNA. *sgRNA* conventional single-guide RNA (positive control), *ntgRNA* non-target guide RNA (negative control), *D* decoy RNA used for the no-RNA-input condition, *autofluorescence* GFP. (Schematic figure and flow cytometry data adapted from ref. [20])

Detailed structures of the two trigger RNA inputs and experimental data for the two-input AND gate are shown in Fig. 7.

4. *Design of Two-Input A AND (NOT B) Gate Trigger RNAs*

The two-input A AND (NOT B) gate is composed of two input RNA elements: a trigger RNA and a deactivating RNA complementary to the trigger RNA. To ensure that the deactivating RNA binds more strongly to the trigger RNA than the binding between trigger RNA and cgRNA, both the trigger RNA and deactivating RNA are designed with extended

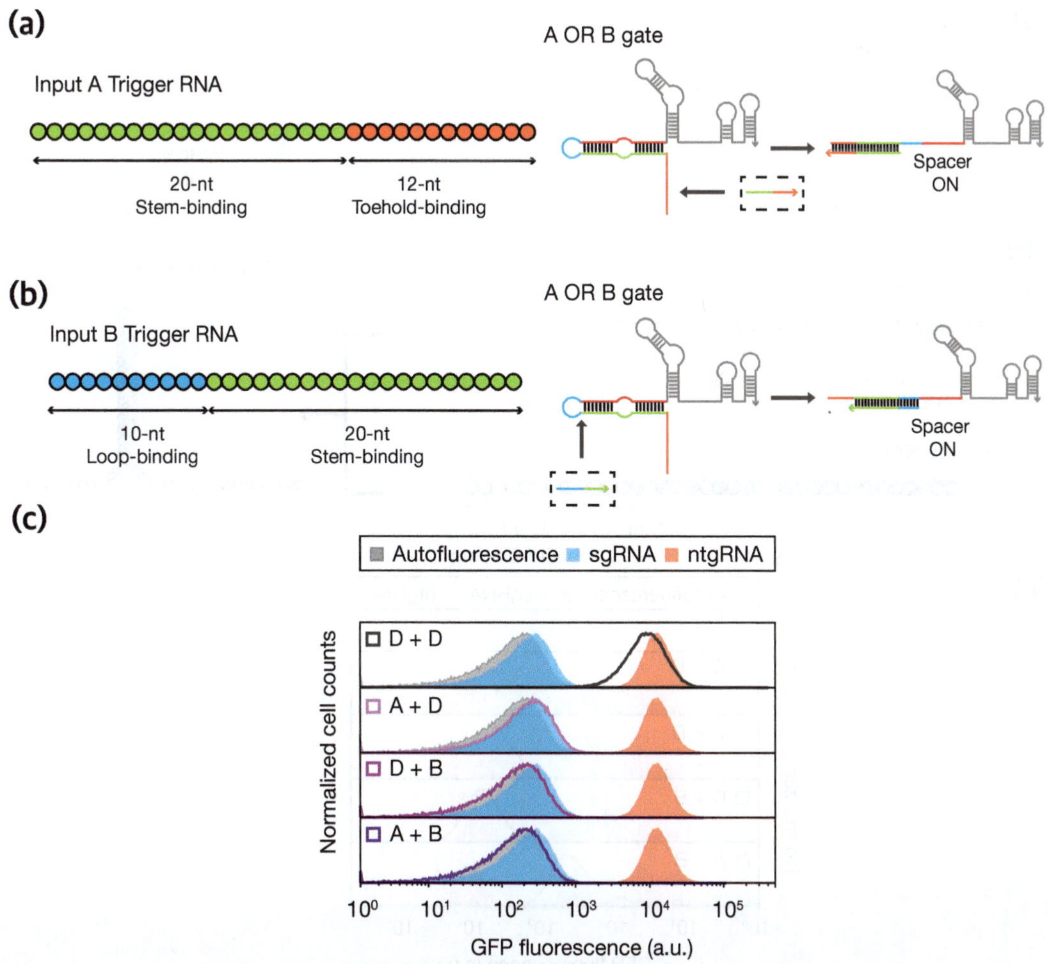

Fig. 6 Detailed schematic of input trigger RNA designs for a two-input OR logic gate. (**a**) Domain composition of the input A trigger RNA. Input A contains a common stem-binding domain (green) and a toehold-binding domain (orange). The toehold-binding domain participates in the initial binding to elicit the subsequent strand displacement of cgRNA stem-loop. (**b**) Domain composition of the input B trigger RNA. Input B contains a loop-binding domain (blue) and a common stem-binding domain (green). The loop-binding domain participates in the initial binding for strand displacement of cgRNA stem-loop. All the RNA diagrams are arranged in the 5′ to 3′ direction with 3′ end marked by an arrowhead. (**c**) Flow cytometry data of the CRISPRi system with a two-input OR gate cgRNA. Data were obtained using the GFP-repression cgRNA. *sgRNA* conventional single-guide RNA (positive control), *ntgRNA* non-target guide RNA (negative control), *D* decoy RNA used for the no-RNA-input condition, *autofluorescence* GFP fluorescence intensity of empty cells. (Schematic figure and flow cytometry adapted from ref. [20])

complementary interaction domains. The trigger RNA now includes two 15-nt extended arm domains on both sides of the 32-nt hybridization domain, separated by 1-nt spacers. The deactivating RNA is designed to be complementary to the trigger RNA, except for the two 1-nt spacer regions, to help prevent the formation of long double-stranded RNA that could be targeted by endogenous RNases.

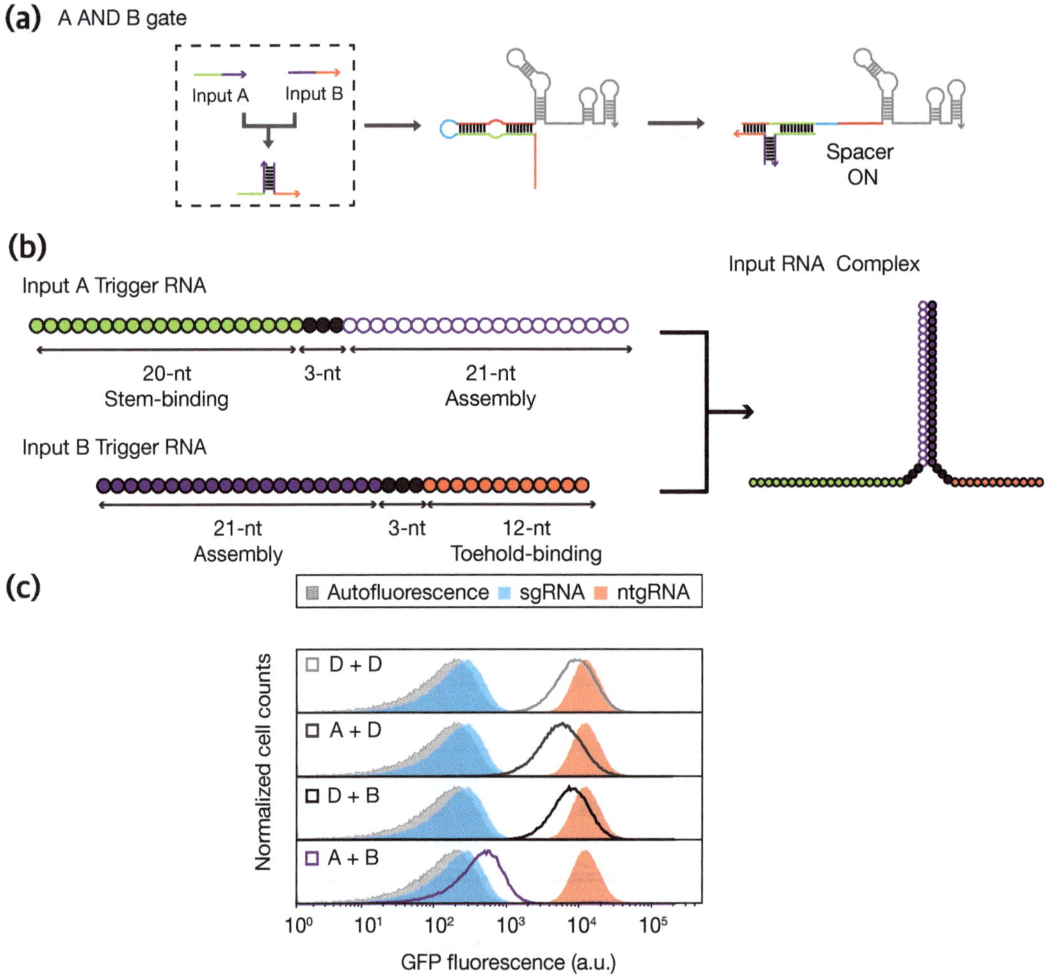

Fig. 7 Detailed structure of input triggers for a two-input AND logic gate. (**a**) Schematic diagram of a two-input AND logic gate using a cgRNA-based CRISPRi system. (**b**) Domain composition of the two input RNAs. Input A trigger RNA contains a stem-binding domain (green) with an assembly sequence (empty purple circles). Input B trigger RNA contains a toehold-binding domain (orange) with an assembly sequence complementary to that of input A (filled purple circles). All the RNA diagrams are arranged in the 5′ to 3′ direction with 3′ end marked by an arrowhead. (**c**) Flow cytometry data of the CRISPRi system with a two-input AND gate cgRNA. Data were obtained using the GFP-repression cgRNA. *sgRNA* conventional single-guide RNA (positive control), *ntgRNA* non-target guide RNA (negative control), *D* decoy RNA used for the no-RNA-input condition, *autofluorescence* GFP fluorescence intensity of empty cells. (Schematic figure and flow cytometry adapted from ref. [20])

Detailed structures of the two input RNAs and experimental data for the two-input A AND (NOT B) gate are shown in Fig. 8.

3.3 Cloning CRISPRi Circuits Implemented with cgRNAs

1. Choose an optimal cloning method based on the following considerations. To change a small part (few bases deletion, insertion, or substitution) of DNA constructs, it is recommended to use site-directed plasmid mutagenesis [22] protocol

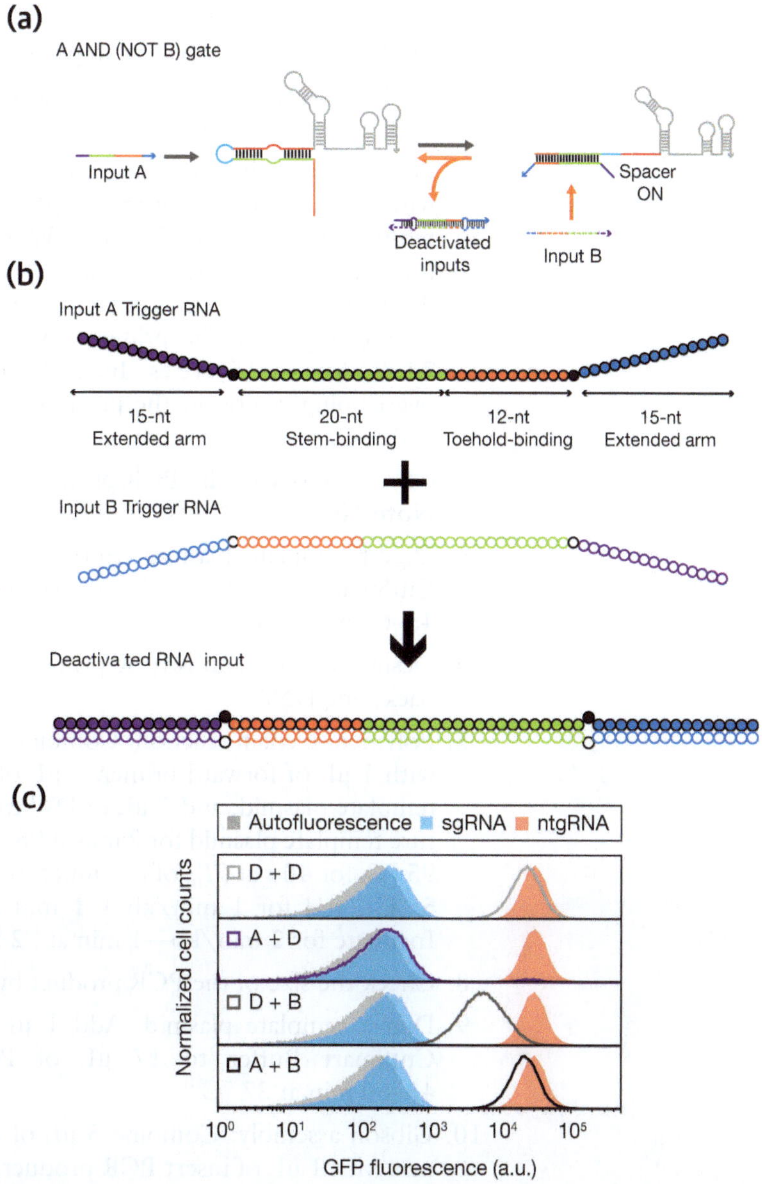

Fig. 8 Detailed structure of input triggers for a two-input NIMPLY logic gate. (**a**) Schematic diagram of a two-input A AND (NOT B) logic gate using a cgRNA-based CRISPRi system. (**b**) Domain composition of the two input RNAs. Input A trigger RNA contains two 15-nt extended arms (purple and blue circles) in addition to the toehold-binding domains. Input B trigger RNA contains domains complementary to those in input A (empty circles with corresponding colors). All RNA diagrams are arranged in the 5′ to 3′ direction. (**c**) Flow cytometry data of the CRISPRi system with a two-input A AND (NOT B) cgRNA gate. Data were obtained using the GFP-repression cgRNA. *sgRNA* conventional single-guide RNA (positive control), *ntgRNA* non-target guide RNA (negative control), *D* decoy RNA used for the no-RNA-input condition, *autofluorescence* GFP fluorescence intensity of empty cells. (Schematic figure and flow cytometry adapted from ref. [20])

as described in **steps 2–5**. Otherwise, Gibson assembly [23] can be used as described in **steps 6–10**.

2. Design forward and reverse primers for site-directed plasmid mutagenesis (*see* **Note 9**).
3. Polymerase chain reaction: Combine 10 μL of Pfu master mix with 1 μL of forward primer, 1 μL of reverse primer, 1 μL of template plasmid, and 7 μL of DW to a total of 20 μL. Denature template plasmid for 5 min at 95 °C. Incubate for 1 min at 95 °C, for 1 min at (melting temperature (T_m) of the nonoverlapping region in the primers—5 °C), and for 1 min/kb at 72 °C for 12–15 cycles. Incubate for 1 min at (T_m of the overlapping region in the primers—5 °C) and for 2 min/kb at 72 °C.
4. Check the size of the PCR product by gel electrophoresis (*see* **Note 10**).
5. Digest template plasmid: Add 1 μL of DpnI and 2 μL of CutSmart Buffer to 17 μL of PCR product. Incubate 45–60 min at 37 °C.
6. Design forward and reverse primers to make insert DNA and backbone DNA.
7. Polymerase chain reaction: Combine 10 μL of Pfu master mix with 1 μL of forward primer, 1 μL of reverse primer, 1 μL of template plasmid, and 7 μL of DW to a total of 20 μL. Denature template plasmid for 2 min at 95 °C. Incubate for 20 sec at 95 °C, for 40 s at (T_m of the annealing region in the primers—5 °C), and for 1 min/kb + 1 min at 72 °C for 30 cycles. Incubate for 2 min/kb—1 min at 72 °C.
8. Check the size of the PCR product by gel electrophoresis.
9. Digest template plasmid: Add 1 μL of DpnI and 2 μL of CutSmart Buffer to 17 μL of PCR product. Incubate 45–60 min at 37 °C.
10. Gibson assembly: Combine 5 μL of Gibson assembly master mix with 1 μL of insert PCR product, 1 μL of backbone PCR product, and 3 μL of DW to a total of 10 μL. Incubate 45–60 min at 50 °C.
11. Transform 75 μL of chemically competent *E. coli* DH5α cells with 2 μL of site-directed plasmid mutagenesis product, or Gibson assembly product using heat shock.
12. Incubate transformed bacteria with LB medium without antibiotics for 1–2 h at 37 °C in a shaker.
13. Plate on selective LB agar plates.

3.4 Preparing Chemically Competent E. coli Cells for Cloning and Transformation

1. Plate *E. coli* (DH5α, BL21 DE3, BL21 Star DE3, BL21 AI) on LB agar plates followed by incubation at 37 °C overnight (*see* **Note 11**).
2. Select a single colony from the plate to inoculate 7 mL of LB medium. Incubate overnight at 37 °C with shaking.
3. Inoculate 5 mL of overnight cell culture to 500 mL of fresh LB in 2 L narrow mouth Erlenmeyer flask. Incubate at 37 °C with shaking until achieving OD_{600} of 0.35 (*see* **Note 12**).
4. Aliquot the cells into 50 mL conical tubes (50 mL per tube) and chill on ice for 10 min (*see* **Note 13**).
5. Centrifuge the cells at 2012×*g* for 10 min at 4 °C.
6. Resuspend each pellet in 6 mL of ice-chilled Inoue transformation buffer and chill on ice for 10 min.
7. Centrifuge the cells at 2012×*g* for 10 min at 4 °C.
8. Resuspend all pellets in 20 mL of ice-chilled Inoue transformation buffer with 7% v/v DMSO and chill on ice for 10 min.
9. Aliquot the cells into 200 μL PCR tubes (75 μL per tube) and flash freeze with liquid nitrogen. Store at −70 °C freezer.

3.5 Experimental Characterization of CRISPRi Circuits with cgRNAs

1. Transformation: Thaw 75 μL of competent *E. coli* on ice, add 150 ng of each plasmid for expressing dCas9, cgRNA and input triggers, and incubate for 20 min on ice. To apply heat shock, incubate the reaction tube for 35 s at 42 °C using a water bath. Immediately, chill the cells on ice for 2 min. Recover the cells with 1 mL of LB media without antibiotics for 1 h (for single transformant) or 2 h (for double transformant) in shaker. Plate on appropriate selective LB agar plates.
2. Overnight growth: Select three single colonies from the plate to inoculate 1 mL of LB media with antibiotics in 96-well deep plate with shaking at 800 rpm and 37 °C (*see* **Note 14**).
3. cgRNA-based CRISPRi circuit expression: Dilute overnight cultures by 100-fold into fresh LB medium with antibiotics and return to the shaker (800 rpm, 37 °C). After 80 min, cells are induced using aTc at a final concentration of 0.5 ng/mL from 2 mg/mL stock solution for dCas9 and cgRNA expression. At the same time, cells are also induced using IPTG at a final concentration of 1 mM from a 100 mM stock solution in DW (in case of BL21 DE3 and BL21 Star DE3 strain) or L-arabinose at a final concentration of 0.2% (w/v) from a 20% stock solution in DW (in case of BL21 AI strain) for the trigger expression. Then incubate for 3 h and 30 min in the shaker (800 rpm, 37 °C).
4. Analysis of reporter expression: After induction, transfer 200 μL of induced cultures to a 96-well black/clear plate.

Measure the fluorescence level and OD_{600} in the plate reader. Calculate normalized fluorescence level as follows:

$$\text{Normalized GFP} = \frac{\text{GFP of test sample} - \text{GFP of LB with antibiotics}}{\text{OD}_{600}\text{ of test sample} - \text{OD}_{600}\text{ of LB with antibiotics}}$$

5. Flow cytometry can also be used to measure fluorescence distribution within the cell population (*see* **Note 15**).

4 Notes

1. This cgRNA sequence is used for all the switch and logic circuit experiments.
2. The first GGG in the trigger sequences is required for proper transcription from the T7 promoter. In case there is a required sequence for other promoters, use the required sequence at the 5′ end as shown here.
3. The decoy RNAs are designed to have minimal interactions with the cgRNA. Experimental results using the decoy RNAs represent the output expression level in the absence of cognate trigger RNA (i.e., the inactivated state of cgRNA). There could be some interaction with decoy RNAs and cgRNAs depending on the sequence composition.
4. The *E. coli* BL21 Star (DE3) strain, which carries an IPTG-inducible genomic T7 RNA polymerase and exhibits reduced RNase E activity, was used for the characterization of riboregulators forming dsRNA complexes.
5. The *E. coli* BL21-AI™ strain, which carries an arabinose-inducible genomic T7 RNA polymerase, was used in experiments requiring tight expression control.
6. NUPACK provides two RNA modeling parameter sets: the default rna06 (based on Mathews et al. [24] and Lu et al. [25]) and rna95 (based on Serra and Turner [26]).
7. For more accurate modeling results, user can choose to include the T7 terminator sequence after the trigger RNA domain.
8. The DU+ and dot-parens-plus notations are defined as follows. In dot-parens-plus notation, a dot represents an unpaired base, and matching parentheses represent paired bases. A plus sign ("+") represents the nick between two strands in an interaction. For example, a ten–base pair hairpin with a six–base loop would be represented as "((((((((((......))))))))))". In DU+ notation, a plus sign also indicates a nick. A segment of x unpaired bases is denoted as Ux, and a segment of y paired bases as Dy. If substructures exist at the end of duplexes, they can be enclosed in parentheses. The same ten–base pair hairpin with a six–base

loop is represented as D10 U6. If a five–base pair hairpin with a four–base loop is inserted within that loop, it would be written as "D10 (U3 D5 U4 U3)". In dot-parens-plus notation, the same structure would appear as "((((((((((... (((((....)))))...))))))))))". Readers are referred to the "Demos" and "Help" sections of NUPACK Design web portal for further examples.

9. Each primer contains a non-overlapping region at the 3′ end and an overlapping region at the 5′ end. The T_m of the non-overlapping region should be 5–10 °C higher than that of the overlapping region (calculated using the SantaLucia et al. [27] parameters). This difference in T_m minimizes the dimerization of forward and reverse primers and allows the primers to use the PCR product as a template. The mutation site (insertion, deletion, or substitution) can be placed in any region.

10. If smeared or unwanted bands appear, increase the annealing temperature or use a more diluted template DNA. However, if the size of the unwanted backbone band is below 1500 bp, these side products can be ignored and the cloning experiment can proceed to the next step. DNA fragments smaller than 1500 bp cannot form an intact plasmid containing the antibiotic resistance gene and origin of replication, and are therefore eliminated during antibiotic selection.

11. For quadruple transformation (e.g., for regulating heterologous genes with multi-input gates), it is recommended to prepare chemically competent *E. coli* cells that already harbor the target gene vector and cgRNA plasmid. Plate double transformants on a selective agar plate, and perform all subsequent steps in selective LB medium.

12. If only a small number of competent cells are needed, inoculate 300 μL of overnight culture into 30 mL of fresh LB medium in a 50 mL conical tube. Incubate at 37 °C with shaking and a slightly open lid for aeration until the culture reaches an OD_{600} of 0.35. This typically yields approximately 40 vials of competent cells (50 μL per vial).

13. Perform aliquoting, chilling, and resuspension steps in a cold room to improve competency.

14. If the gene of interest is cytotoxic, adding 0.2% glucose is recommended to help prevent recombination.

15. After **step 3**, Subheading 3.5, dilute the cells with 1× PBS (adjusting OD_{600} to 0.1) to a final volume of 100 μL. Detect cells using a flow cytometer with a forward scatter (FSC) trigger, recording at least 15,000 events for each sample. Analyze the acquired data using CytExpert software as follows: gate cell populations according to their FSC and side scatter

(SSC) distributions using the auto-polygon tool (gated cells should account for at least 80% of the total population). Calculate the geometric mean of GFP fluorescence histograms to determine the average fluorescence from three biological replicates. If the flow cytometer is not immediately accessible, IPTG-induced cells can be fixed with paraformaldehyde and analyzed at a later time. Collect the induced cells by centrifugation, resuspend the pellet in 1% paraformaldehyde, and incubate for 5 min. Fixed cells can be stored at 4 °C for up to 7 days without significant loss of signals.

References

1. Benner SA, Sismour AM (2005) Synthetic biology. Nat Rev Genet 6(7):533–543
2. Cameron DE, Bashor CJ, Collins JJ (2014) A brief history of synthetic biology. Nat Rev Microbiol 12(5):381–390
3. Cheng AA, Lu TK (2012) Synthetic biology: an emerging engineering discipline. Annu Rev Biomed Eng 14:155–178
4. Khalil AS, Collins JJ (2010) Synthetic biology: applications come of age. Nat Rev Genet 11(5):367–379
5. Meng F, Ellis T (2020) The second decade of synthetic biology: 2010–2020. Nat Commun 11:5174
6. Voigt CA (2020) Synthetic biology 2020–2030: six commercially-available products that are changing our world. Nat Commun 11:6379
7. Gao Y, Wang L, Wang B (2023) Customizing cellular signal processing by synthetic multi-level regulatory circuits. Nat Commun 14: 8415
8. Green AA, Kim J, Ma D, Silver PA, Collins JJ, Yin P (2017) Complex cellular logic computation using ribocomputing devices. Nature 548(7665):117–121
9. Nielsen AAK, Der BS, Shin J, Vaidyanathan P, Paralanov V, Strychalski EA, Ross D, Densmore D, Voigt CA (2016) Genetic circuit design automation. Science 352(6281): saac7341
10. Siuti P, Yazbek J, Lu TK (2013) Synthetic circuits integrating logic and memory in living cells. Nat Biotechnol 31(5):448–452
11. Rhodius VA, Segall-Shapiro TH, Sharon BD, Ghodasara A, Orlova E, Tabakh H, Burkhardt DH, Clancy K, Peterson TC, Gross CA, Voigt CA (2013) Design of orthogonal genetic switches based on a crosstalk map of σs, anti-σ s, and promoters. Mol Syst Biol 9(1):702
12. Stanton BC, Nielsen AAK, Tamsir A, Clancy K, Peterson T, Voigt CA (2014) Genomic mining of prokaryotic repressors for orthogonal logic gates. Nat Chem Biol 10(2):99–105
13. Isaacs FJ, Dwyer DJ, Ding C, Pervouchine DD, Cantor CR, Collins JJ (2004) Engineered riboregulators enable post-transcriptional control of gene expression. Nat Biotechnol 22(7): 841–847
14. Green AA, Silver PA, Collins JJ, Yin P (2014) Toehold switches: de-novo-designed regulators of gene expression. Cell 159(4):925–939
15. Kim J, Zhou Y, Carlson PD, Teichmann M, Chaudhary S, Simmel FC, Silver PA, Collins JJ, Lucks JB, Yin P, Green AA (2019) De novo-designed translation-repressing riboregulators for multi-input cellular logic. Nat Chem Biol 15(12):1173–1182
16. Qi LS, Larson MH, Gilbert LA, Doudna JA, Weissman JS, Arkin AP, Lim WA (2013) Repurposing CRISPR as an RNA-guided platform for sequence-specific control of gene expression. Cell 152(5):1173–1183
17. Larson MH, Gilbert LA, Wang X, Lim WA, Weissman JS, Qi LS (2013) CRISPR interference (CRISPRi) for sequence-specific control of gene expression. Nat Protoc 8(11): 2180–2196
18. Byun G, Yang J, Seo SW (2023) CRISPRi-mediated tunable control of gene expression level with engineered single-guide RNA in Escherichia coli. Nucleic Acids Res 51(9): 4650–4659
19. Kim H, Bojar D, Fussenegger M (2019) A CRISPR/Cas9-based central processing unit to program complex logic computation in human cells. Proc Natl Acad Sci USA A116 (15):7214–7219
20. Kang H, Park D, Kim J (2024) Logical regulation of endogenous gene expression using

programmable, multi-input processing CRISPR guide RNAs. Nucleic Acids Res 52(14):8595–8608

21. Zadeh JN, Steenberg CD, Bois JS, Wolfe BR, Pierce MB, Khan AR, Dirks RM, Pierce NA (2011) NUPACK: analysis and design of nucleic acid systems. J Comput Chem 32(1): 170–173
22. Liu H, Naismith JH (2008) An efficient one-step site-directed deletion, insertion, single and multiple-site plasmid mutagenesis protocol. BMC Biotechnol 8:91
23. Gibson DG, Young L, Chuang RY, Venter JC, Hutchison CA, Smith HO (2009) Enzymatic assembly of DNA molecules up to several hundred kilobases. Nat Methods 6(5):343–345
24. Mathews DH, Sabina J, Zuker M, Turner DH (1999) Expanded sequence dependence of thermodynamic parameters improves prediction of RNA secondary structure. J Mol Biol 288(5):911–940
25. Lu ZJ, Turner DH, Mathews DH (2006) A set of nearest neighbor parameters for predicting the enthalpy change of RNA secondary structure formation. Nucleic Acids Res 34(17): 4912–4924
26. Serra MJ, Turner DH (1995) Predicting thermodynamic properties of RNA. Methods Enzymol 259:242–261
27. SantaLucia J Jr (1998) A unified view of polymer, dumbbell, and oligonucleotide DNA nearest-neighbor thermodynamics. Proc Natl Acad Sci USA 95(4):1460–1465

Chapter 6

Synthetic Biomolecular Condensates: Design Principles and Applications

Yuefeng Ma, Leshan Yang, and Yifan Dai

Abstract

Synthetic biomolecular condensates are engineered membraneless compartments formed through phase separation of biomacromolecules. Beyond lock-and-key interaction-dependent control of cellular processes, phase separation enables the spatiotemporal organization of defined sets of biomolecules and assembles functional units into mesoscale structures that can reversibly form or dissolve in response to environmental cues. This chapter focuses on the design and engineering of synthetic intrinsically disordered proteins (synIDPs) to build programmable condensation systems for cellular controls. We outline the molecular principles that link IDP sequence features to phase behaviors. The chapter then presents a practical workflow for the development of synIDP-based condensates for cellular controls, including rational design, gene synthesis, protein expression, purification, and biophysical characterization. Finally, we describe synthetic gene circuit-based methods for applying synthetic condensates in *E. coli* to regulate transcription and translation. Together, this new method demonstrates how engineering synthetic condensates offers a new layer of cellular control for synthetic biology.

Key words Biomolecular condensates, Synthetic IDPs, Phase separation, Transcription regulation, Translation regulation

1 Introduction

1.1 Capability of Biomolecular Condensation

Cellular function has traditionally been explained by lock-and-key interactions. Yet the complexity of the cellular milieu and the need to control processes with high spatiotemporal precision demand an additional mechanism—biomolecular condensation—that organizes specific sets of biomolecules to execute cellular functions [1]. Biomolecular condensation, which is driven by the phase transitions of biomacromolecules, such as intrinsically disordered proteins, organizes spatiotemporal cellular processes, from transcription to stress responses [2, 3]. The function of condensates is largely encoded in their defined biomolecular compositions and only activated by the formation of condensates. So that, through programming the condensation driving force at the molecular level,

Xiao-Jun Tian (ed.), *Synthetic Gene Circuits: Methods and Protocols*, Methods in Molecular Biology, vol. 3041, https://doi.org/10.1007/978-1-0716-5304-3_6,

this will allow researchers to not only define the functions of the condensate unit [4, 5], but also to control the timing [6] and location [7] of condensate functions by linking phase separation to defined environmental or molecular cues. The detailed biophysical mechanisms have been reviewed in previous works [2, 8].

1.2 Engineering Biomolecular Condensation

We here summarize some current examples on the design and engineering of condensation molecular apparatus for cellular controls. Optogenetic systems offer unmatched precision in controlling when and where condensates form [9, 10]. These platforms rely on proteins that respond to specific wavelengths of light by undergoing conformational changes that promote oligomerization or dimerization. The most widely used system involves cryptochrome 2 (Cry2) from *Arabidopsis thaliana* [11]. Cry2 contains a flavin adenine dinucleotide (FAD) chromophore, which absorbs blue light [12]. Upon illumination, FAD undergoes photoreduction to a semiquinone radical, triggering structural rearrangements within Cry2's photolyase homology region. This conformational shift exposes charged residues at protein–protein interaction interfaces, facilitating electrostatically driven self-association of Cry2 into oligomeric clusters [13]. Such oligomerization significantly increases local valency and concentration when Cry2 is fused to an intrinsically disordered region (IDR), thereby nucleating biomolecular condensates [14]. This principle was applied in the optoDroplet system, where Cry2 was fused to the low-complexity domain of FUS, enabling rapid and reversible nuclear condensate formation [9]. A mutated variant, Cry2olig, incorporates mutations enhancing clustering by stabilizing these electrostatic interactions [15]. Cry2-based systems offer subcellular resolution and millisecond-scale temporal control, making them particularly useful for probing rapid and localized phase transitions. Other optogenetic dimerization systems such as iLID/SspB [16] and PhyB/PIF [17] exist; however, their direct applications in condensate formation remain less explored.

Similarly, chemically inducible systems that use small molecules to drive controlled protein–protein interactions can also trigger phase separation. A widely used platform employs rapamycin, which induces stable heterodimerization between FKBP and FRB domains [18]. While originally developed for synthetic signaling, this system has since been adapted for condensation by coupling dimerization to multivalent scaffolds or IDRs. For example, Chung et al. developed a chemogenetic tool [19] by fusing FKBP and FRB to HOTag scaffolds. Upon the addition of rapamycin, FKBP–FRB clustering triggered condensate formation in living cells, recruiting stress granule proteins such as G3BP1, FUS, and TIA-1. Similarly, Garabedian et al. engineered FKBP–FRB fusion proteins containing RGG domains, which phase-separated in response to rapamycin

[20]. However, compared to optogenetic systems, chemically inducible systems offer robust temporal control but are typically irreversible without competitive inhibitors.

A distinct category of strategy is de novo engineering of phase separation drivers, such as synthetic intrinsically disordered proteins (synIDPs). Although IDPs or IDRs from native systems such as FUS, TDP-43, and DDX4 are widely used in biochemistry [21] and bioengineering [22], synIDPs offer distinct advantages, including programmability, tunability, reversibility, and modularity [2]. synIDPs typically consist of repetitive protein sequence units arranged in tandem, designed to impart specific disordered characteristics. They are characterized by limited but distinct amino acid compositions, such as low hydrophobicity and high polar residue content that prevent the formation of stable folded domains [23]. Instead, IDPs exist as dynamic ensembles of interconverting conformations. Such dynamic, multivalent features of IDPs encode their phase behaviors, and the repetitive feature of synthetic IDPs further allows them to encode entropy-dependent driving force for phase transition [24]. Given these unique properties, there is great interest in engineering synthetic IDPs for biotechnological applications. Designing a synIDP with specific properties is challenging because the sequence rules governing disorder and function are complex and only partly understood. Unlike well-folded proteins, whose structures can now be reliably designed, synIDPs lack obvious structural visualization, and their sequence–ensemble relationships are governed by nuanced factors such as charge patterning and hydrophobic motif distribution. Nonetheless, recent advances have made synIDP design feasible. For example, Quiroz and Chilkoti [25] summarized a principle by investigating the effects of amino acids distribution and sequence properties (Fig. 1) on temperature-dependent phase transition behaviors of IDPs. To verify the principle, they designed multiple synIDP sequences and performed temperature-dependent turbidimetry for these synIDPs. Dzuricky et al. [26] investigated the effect of individual amino acids on concentration- and temperature-dependent phase transition behaviors of synthetic resilin-like proteins (Fig. 2). Besides, this study also investigated the relation between the length of synIDPs and the phase transition behaviors. Likewise, computational algorithms allow de novo design of IDP sequences with specified ensemble properties: Pesce et al. [27] designed IDP variants whose chain compaction and phase separation propensity varied over 100-fold despite identical amino acid composition. Apart from the research on the design grammar of synIDPs, some synIDPs are designed for synthetic biology applications. For example, tunable solubility tags have been created by designing highly soluble IDP sequences that, when fused to aggregation-prone proteins, dramatically improve their soluble expression [23]. Dai et al. [28] harnessed synIDPs to control the transcription rate by

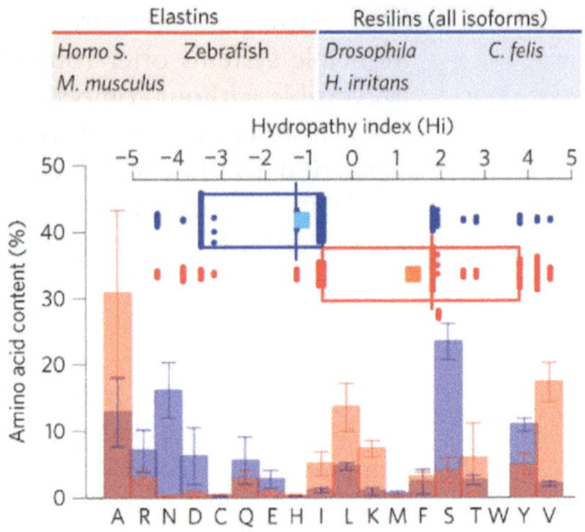

Fig. 1 Contributions of all amino acids to resilins and elastins across several species. Error bars demonstrate the standard deviations. The inset figure shows the distribution of hydropathy values using the Kyte–Doolittle scale: −5 (the most hydrophilic) to 5 (the most hydrophobic). Each box delimits the 25th and 75th percentile and the line and square indicate the median and mean, respectively. (The figure was reprinted from Quiroz and Chilkoti [25], with permission)

Y -> W	123
G -> P	90
D -> E	69
S/Y/R/D	66
S -> T	60
Y -> F	59
P ->G	54
S -> Q	45
Q -> N	34
Y -> H	2.3
R -> K	-8.9
Y -> M	-19
Y -> I	-47
Y -> V	-62
Y -> A	-65

Fig. 2 Evaluation of one amino acid substitution to RLP_{WT} (SKGP-$[GRGDSPYS]_{20}$-GY) temperature-dependent behaviors. The transition temperatures listed are the UCST cloud point, determined at a protein concentration of ~25 μM, if the left amino acid was replaced with the amino acid to the right of the arrow. (The figure was reprinted from Dzuricky et al. [26], with permission)

concentrating RNA polymerase into a condensate. These studies show that one can rationally engineer IDPs for biological applications, from soluble fusion tags that prevent recombinant protein aggregation to IDPs that control transcription rate in *E. coli*. In this chapter, we describe a step-by-step protocol and consideration

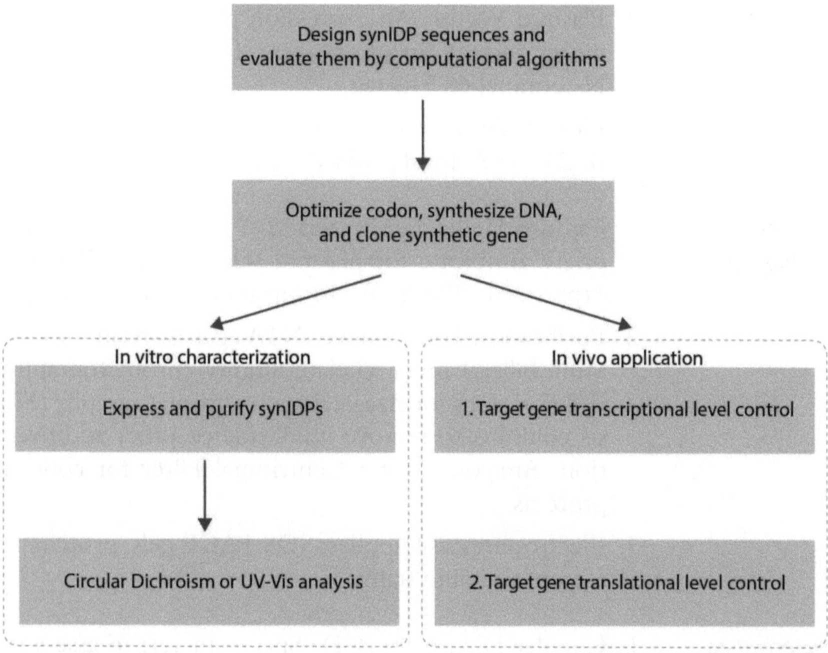

Fig. 3 The design pipeline for synIDPs. Synthetic intrinsically disordered protein (synIDP) sequences are first designed and evaluated by computational algorithms. Codons are then optimized, DNA is synthesized, and the synthetic genes are cloned. In vitro characterization involves expression, purification, and analysis of synIDPs by circular dichroism or UV–Vis spectroscopy. In vivo applications include transcriptional and translational level control of target genes

toward de novo design of synthetic IDP sequences, its production in an academic lab, and its application in controlling synthetic gene circuits. The protocol pipeline is summarized in Fig. 3 and detailed in the sections below.

2 Materials

2.1 Software and Tools for Design

1. Disorder Prediction Software: A website program to predict intrinsic disorder from sequence, such as IUPred2A (https://iupred2a.elte.hu/) [29] or PONDR (https://www.pondr.com/) [30].
2. Sequence Analysis Tools: Software or online website for calculating amino acid composition, hydrophobicity, charge distribution such as Python with BioPython library or Expasy ProtParam (https://web.expasy.org/protparam/) [31].

2.2 Vector Construction Reagents and Service

1. Gene Synthesis Services or Kits: Access to a commercial gene synthesis service such as IDT or Ansa biotechnologies or a kit for gene assembly such as Golden Gate assembly kit. Synthetic IDP sequences are often non-natural, so ordering a codon-optimized gene from a provider is convenient.

2. Plasmid Vector: An expression plasmid with appropriate promoter and tag for purification. For example, pET-28a with an N-terminal 6x His tag.
3. Host strain: *E. coli* strain suitable for protein expression, such as BL21(DE3) for T7-based expression vectors.

2.3 Expression and Purification Supplies

1. Culture Media and Buffers: 2xYT for bacterial growth, appropriate antibiotic for plasmid selection and IPTG for protein expression. PBS buffer for protein extraction and purification.
2. Purification Tools: Nickel-NTA agarose resin for His-tag-based immobilized metal chelate affinity chromatography (IMAC). FPLC system for size exclusion chromatography (SEC). Dialysis columns to remove imidazole or other additives after elution. Amicon® Ultra Centrifugal Filter for concentration of proteins.
3. Electrophoresis Supplies: SDS-PAGE gels, protein ladders, and Coomassie blue stain.

2.4 Characterization Equipment

1. Circular Dichroism (CD) Spectrometer: to assess the secondary structure of the purified SynIDPs.
2. Temperature-dependent ultraviolet-visible spectroscopy: to investigate the thermoresponsive-phase transition boundary.
3. Multifunctional microplate reader: to measure the absorbance and fluorescence of samples in microplates.

3 Methods

3.1 Design of the synIDP Sequence

1. Determine the design objectives: The first step is to decide on the specific property you want your synIDP to have. Common objectives include maximal solubility, lack of secondary structure, ability to undergo phase separation, or serving as a flexible linker between domains. For example, if designing an IDP with an upper critical solution boundary (UCST) behavior, the objective is a highly charged, polar sequence. If designing an IDP with a lower critical solution boundary (LCST) behavior, the objective may be a sequence with specific hydrophobic patterns to achieve a desired temperature-dependent phase transition behavior [25].
2. Design the amino acid composition for one repeat unit of synIDP: Based on known IDP features, select a rough composition for the sequence. IDPs are enriched in disorder-promoting residues such as Glycine, Serine, and Proline. The uniformity distribution of these three amino acids can render a protein disordered. Based on the objectives, other amino acids can be added into the unit. LCST-behavior synIDPs normally

contain more hydrophobic amino acids, such as valine, alanine, leucine, isoleucine, tyrosine, phenylalanine, and tryptophan [25]. For UCST-behavior synIDPs, they contain charged or polar amino acids, such as lysine, arginine, glutamate, aspartate, glutamine, and asparagine [25, 26].

3. Determine the length: Commonly 10–100 repeats are easy-handling. Based on the objectives, different repeats can be set to tune the concentration- and temperature-dependent phase transition boundaries of synIDPs (*see* **Note 1**).
4. Determine the final sequence. This can be done manually or with a custom script. If aiming for randomness, you can randomly sample residues with the chosen composition. Certain short motifs are beneficial to increase flexibility and disorder such as P, G-rich motifs. For example, Tang et al. [23] designed SynIDPs composed of repeats of a $(P–G–X)_n$ motif, where X are assorted amino acids, to ensure disorder of the synIDPs.
5. Evaluate the disorder degree with prediction tools: prediction software such as PONDR can be used to run the draft sequence to confirm that the sequence is predominantly predicted to be disordered. Normally, PONDR score above the disorder threshold for >90% of residues, or similar criteria can be regarded as a successful design (*see* **Note 2**). If the predictor flags any segment as potentially structured, consider mutating those residues, for instance, Proline can be used to replace a few amino acids to disrupt the secondary structure.

3.2 Gene Synthesis and Cloning

1. Optimize the codon usage of the sequence: DNA codon optimization tool is used for adjusting the expression in *E. coli*. This will adjust the codons in your designed sequence to ones that are frequently used in *E. coli*, improving expression (*see* **Note 3**). The optimized sequences still need to be checked to ensure that they encode the exact amino acid sequence of designed SynIDP and do not introduce problematic sequences such as internal restriction sites that conflict with your cloning strategy.
2. Synthesize the DNA sequence of the synIDP: The optimized sequences can be synthesized by a commercial gene synthesis service to obtain a cloned gene. If self-assembling, multiple overlapped PCR can be used to cover the entire sequence. Full gene is amplified by PCR assembly and then reamplified with flanking primers that add restriction sites. This can be labor-intensive for long repetitive IDPs, so commercial synthesis is recommended to reduce errors during the cloning process.
3. Assemble the synIDP DNA into the expression system: The plasmid vector and the synthetic gene are digested with the chosen restriction enzymes. Genes are inserted into the vector

using T4 DNA ligase, and transformed into *E. coli* DH5α. Single colonies are picked on plates with corresponding antibiotics and screened for correct inserts by colony PCR or restriction digest. The plasmids are further confirmed by colony PCR are sent for sequencing.

3.3 Expression and Purification of synIDPs

1. Express the synIDP: Plasmids encoding synIDPs are transformed into the expression host strain BL21(DE3). Single colonies are picked and inoculated into a 4 mL of 2xYT and grown overnight at 37 °C with shaking. Next day, 12 mL of overnight culture is backdiluted into a 1 L of 2xYT culture and grown at 37 °C until OD_{600} reaches ~0.8. Protein expression is induced by adding IPTG to a final concentration of 0.5 mM, then cell culture continues shaking incubation (*see* **Note 4**). Cells are collected after 24 h induction.
2. Lysate the cell samples: Cells are harvested by centrifugation at 4000×*g* for 20 min at 4 °C and resuspended in 35 mL of buffer containing 100 mM PBS and 4 M urea (pH 7.4). After 20 min incubation at −80 °C, *E. coli* cells are lysed using a tip sonicator (4 min total time, each cycle with 2 s on and 2 s off, 30% amplitude) in an ice bucket followed by centrifugation at 18,000×*g* for 25 min, 4 °C. Supernatants are collected for purification.
3. Purify the synIDPs using IMAC: Three types of buffers need to be prepared. The binding buffer contains 100 mM PBS, 4 M urea, 10 mM imidazole (pH 7.4). The wash buffer contains 100 mM PBS, 4 M urea, and 50 mM imidazole (pH 7.4). The elution buffer contains 100 mM PBS, 4 M urea, and 200 mM imidazole (pH 7.4). 3 ml of HisPur Ni-NTA resin is added to a propylene gravity column. 30 mL of the binding buffer is added to the column and incubated for 30 min before discarding the flow through. Then, 20 mL of total supernatants are added into the column and incubated for 30 min, before the flow through is collected and added back to the column (*see* **Note 5**). This step needs to be repeated at least once. The column can be used to purify total 40 mL of supernatants. After discarding the unbinding protein lysate solution, the column is washed with 10 mL of wash buffer twice. Flow through of wash buffer is collected for future characterization. Lastly, 3 mL of the elution buffer is added to obtain the target synIDP. This step repeats twice and all the flow through are collected for future characterization. All collected solutions are visualized on SDS-PAGE gel, and the relevant solutions are pooled together for the next step purification using size exclusion chromatography (*see* **Note 6**). Ideally, a band corresponding to the SynIDP should be significantly predominant (>85%) after IMAC purification.

4. Purify the synIDP using SEC (Optional): If the purity of the synIDPs after IMAC does not meet the expectations of experimental needs, SEC can be applied to increase the purity. The absorbance at 280 nm (A_{280}) of combined solutions is adjusted to 4.0 by concentration using the Amicon® Ultra Centrifugal Filter or dilution using the elution buffer. The protein solution after adjustment is aliquoted 5 mL into 15 mL conical. The flow rate and operation temperature of SEC column is set at 2.0 mL/min and 4 °C. The synIDP fractions are tracked through the absorbance at 220 nm and 280 nm, and fractions are collected in 2 mL aliquots. The purity of synIDP fractions is characterized on an SDS-PAGE gel to confirm the desired product, and fractions containing the desired product are pooled together. Normally, the purity exceeds 95% after SEC purification.
5. Storage: The pooled fractions are transferred into dialysis tubing and dialyzed against 4 L of DI water with at least two water exchanges, followed by lyophilization of the SynIDPs and storage at −20 °C until further use.

3.4 In Vitro Characterization of Properties of synIDPs

1. Analyze synIDP structural properties using Circular Dichroism (CD): To confirm that the synIDPs are indeed disordered, CD spectroscopy assay is necessary to be performed. The concentration of synIDPs is adjusted to 50 μM in 10 mM PBS buffer (pH 7.4). The CD spectra are obtained at 37 °C from 260 to 190 nm. The CD spectra is corrected for the 10 mM PBS buffer signal at 37 °C and the average is calculated based on at least 3 runs. Disordered synIDPs normally show a negative peak around 195 nm and a very low positive peak around 215 nm, as the characteristics of a random coil.
2. Analyze cloud point of synIDP using temperature-dependent ultraviolet-visible spectroscopy analysis: To confirm the temperature- and concentration-dependent phase transition behaviors of synIDPs, temperature-dependent assay using ultraviolet-visible spectroscopy is necessary to be performed. Different concentrations synIDP samples in 10 mM PBS are cooled or heated at 1 °C /min while the absorbance at 350 nm (A_{350}) is recorded at every 1 °C decrease. A_{350} is calibrated to the absorbance at the highest or lowest temperature point collected. The cloud point is determined as the maximum and minimum in the derivative of the A_{350} as a function of temperature for IDPs exhibiting LCST-behaviors or UCST-behaviors, respectively. The saturation concentration (C_{sat}) at specific temperature can be determined by collecting and measuring the A_{280} of supernatants after centrifugation and removal of the pellet.

3.5 In Vivo Applications of synIDPs in E. coli

1. synIDPs control the transcription level of target protein in *E. coli*: The basic mechanism to achieve transcriptional level control of target protein is to create a permeable and membraneless organelle by increasing the RNAP concentration [28] (Fig. 4a). This approach harnesses a phase-separable synIDP (RLP_{WT}) to recruit the plasmid harboring the target gene and the RNAP by fusing synIDP respectively with dParB and RNAP binding protein, SoxS (R93A).
   (a) Construct transcriptional-control system: The system contains two plasmids including an activation apparatus and a target apparatus (Fig. 4b). The activation apparatus is constructed by synIDP-dParB, which is capable of binding on parS sites and synIDP–SoxS(R93A), which can recruit RNAP to the promoter on the target apparatus. Both synIDP–dParB and synIDP–SoxS(R93A) are controlled by a T7 promoter (*see* **Notes 7** and **8**). The reporter apparatus contains the target gene, RFP, controlled by a weak constitutive promoter J23117. The dParB binding site, parS, is also needed placed on the target apparatus (*see* **Note 9**).
   (b) Evaluate the system performance: BL21 (DE3) cells are transformed with plasmids containing both the transcription amplification apparatus plasmid and the target plasmid, and plated on the 2xYT agar plates supplemented

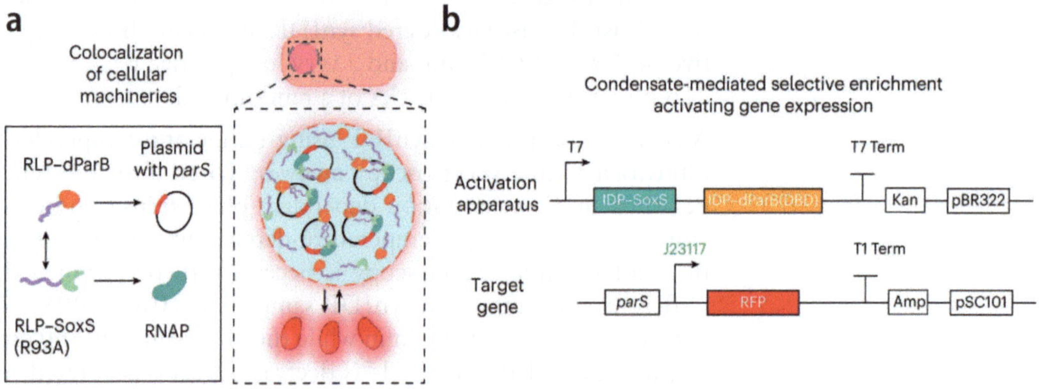

Fig. 4 synIDPs control the transcription level of target protein in *E. coli*. (**a**) Schematics of synIDPs controlling the transcription level of target protein. Condensates formed by synIDPs become a key transcriptional machinery, amplifying transcription, since the increased local concentrations of transcriptional activator and RNAP in the condensate. (The figure was reprinted from Dai Y. et al. [28], with permission.) (**b**) The genetic circuit design. The activation apparatus is constructed by synIDP–DBD, which can form condensates on parS sites and synIDP–SoxS(R93A). The synIDP–SoxS(R93A) mediates recruitment of RNAP to a weak promoter J23117. (The figure was reprinted from Dai Y. et al. [28], with permission)

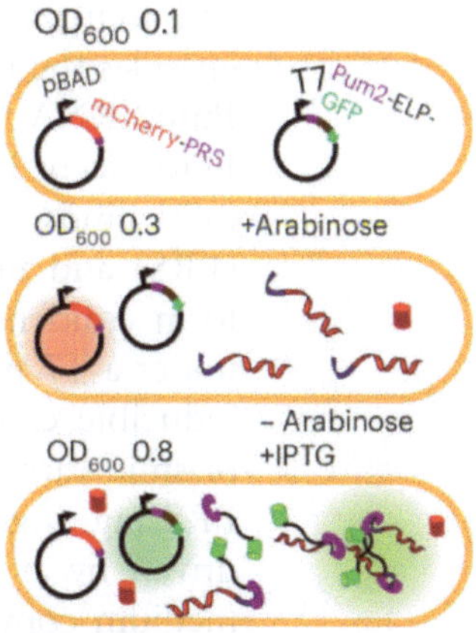

Fig. 5 Schematics of synIDPs controlling the transcription level of target protein in *E. coli*. The activation apparatus encodes the fusion protein Pum2–ELP–GFP. This fusion specifically binds to a PRS and undergoes liquid–liquid phase separation to form condensates. The Pum2–ELP–GFP gene is placed under a T7 promoter on a high-copy plasmid for IPTG-inducible expression. The mcherry-PRS is driven by an arabinose-inducible pBAD promoter on a medium-copy plasmid to provide moderate expression (The figure was reprinted from Shapiro et al. [32], with permission)

with corresponding antibiotics. Single colonies are inoculated and grown overnight at 37 °C in 4 mL of 2xYT medium with 0.4% glucose and corresponding antibiotics. Cells are then backdiluted 100× in 3 mL of M9 medium containing corresponding antibiotics. After 1 h of growth, the cells are induced with 0.5 mM IPTG and 0.2% arabinose. Cell culture is aliquoted 200 μL into a 96-well plate. The fluorescence of monomeric RFP is measured on a microplate reader with an excitation wavelength of 584 nm and an emission wavelength of 610 nm.

2. synIDPs control the translational level of target protein in *E. coli*: The basic mechanism to achieve translational-level control of a target gene is to sequester its mRNA into a synthetic membraneless organelle formed by a phase-separating synIDP [32] (Fig. 5). This approach utilizes an ELP_{WT} (an elastin-like polypeptide) fused to an RNA-binding domain Pum2 to bind a specific recognition sequence on the target mRNA and concentrate that mRNA inside intracellular condensates. In *E. coli*, such sequestration of a target mRNA paradoxically enhances its translation rate, presumably by locally enriching the mRNA along with ribosomes, tRNAs, and other translation factors within the liquid condensate.

   (a) Construct the translational-control system: It is a two-plasmid system comprising an activation apparatus and a target apparatus (Fig. 5). The activation apparatus encodes the fusion protein Pum2–ELP–GFP (the human Pum2 RNA-binding domain fused to an ELP and Green Fluorescent Protein for visualization). This fusion specifically binds to an RNA Pumilio recognition sequence (PRS) and undergoes liquid–liquid phase separation to form condensates. The Pum2–ELP–GFP gene is placed under a T7 promoter on a high-copy plasmid for IPTG-inducible expression. The target apparatus is a reporter plasmid carrying the gene of interest such as RFP with a PRS tag engineered in its 3′ UTR. This target gene is driven by an arabinose-inducible pBAD promoter on a medium-copy plasmid (p15A origin) to provide moderate expression. A control reporter lacking the PRS is used as a control to distinguish condensate-specific effects.

(b) Evaluate the system performance: BL21(DE3) cells are transformed with both the activation plasmid and the target plasmid and plated on 2xYT agar with appropriate antibiotics. Single colonies are picked and grown overnight at 37 °C in 4 mL of 2xYT rich medium supplemented with 0.4% glucose and the corresponding antibiotics. The overnight culture is backdiluted 100x into fresh 2xYT (with antibiotics and 0.4% glucose) and grown at 37 °C until mid-log phase (OD_{600} = 0.15–0.3). At mid-log phase, expression of the condensate protein and target mRNA is induced by adding 0.5 mM IPTG and 0.2% L-arabinose. After induction, 200 μL of the induced culture is aliquoted into each well of a 96-well plate for fluorescence measurement, and incubation is continued at 37 °C with shaking. The fluorescence of monomeric RFP is measured on a microplate reader with an excitation wavelength of 584 nm and an emission wavelength of 610 nm. In parallel, samples can be observed under a fluorescence microscope at 30 min, 1.5 h, 2.5 h, and 4.5 h post-induction to visualize Pum2–ELP–GFP condensate formation and mCherry reporter localization.

4 Notes

1. To make the synIDP capable of expressing in *E. coli*, the total length of the protein is about 80–700 AAs.
2. To further check the other properties of designed synIDPs, CIDER (Classification of Intrinsically Disordered Ensemble Regions) allows for the calculation of many different

parameters associated with disordered protein sequences including charge segregation factor, hydropathy, fraction disorder promoting, and fraction of charged residues in a sequence [33].

3. Codon optimization is operated based on the codon usage preference of different organisms. However, other factors affecting the expression level should also be considered. For example, mRNA secondary structure is a key factor that controls the translation rate. Two online websites can be used to optimize the designed sequence for expression in *E. coli*. First, RBS Calculator [34]. Since the secondary structure of mRNA may bring concerns on translation rate by affecting RBS binding with ribosome, the RBS calculator can help you design a specific RBS that achieves a target translational rate based on the input coding sequence. Second, Codon Scrambler [35]. The codon scrambler can be used to design the tandem repeated sequences, with optimization in mRNA structure. By simply inputting the motif sequence and the repeat number of a synIDP, you will obtain a corresponding DNA sequence after a few minutes.
4. After induction, to avoid inclusion body formation, the temperature for shaking incubation can be optimized based on the temperature-dependent behavior of the synIDP. For example, LCST-behavior synIDPs can be produced at 25 °C, while UCST-behavior synIDPs can be produced at 37 °C.
5. Similarly, as the previous note, although urea is added to disrupt the aggregation of protein, the temperature for the crude lysate incubation in HisPur Ni-NTA resin can be optimized based on the temperature-dependent behavior of the synIDP. For example, LCST-behavior synIDPs can be incubated at 4 °C, while UCST-behavior synIDPs can be incubated at 37 °C.
6. synIDPs often elute as very broad or asymmetric peaks from size-exclusion columns and run at an apparently different molecular weight with expected value on SDS-PAGE because of their extended conformation and charge abundance [23]. If using size-exclusion chromatography for further purification, the presence of disorder as expected should be used for characterization of a synIDP peak.
7. The negative control groups need to be carefully set based on your experimental needs.
8. synIDPs exhibiting with different C_{sat} can be used to tune the transcriptional level of the target gene.
9. *parS*, the dParB binding site, should be placed close to the promoter that enables the RNAP to regulate the transcription.

References

1. Shin Y, Brangwynne CP (2017) Liquid phase condensation in cell physiology and disease. Science 357(6357):eaaf4382
2. Dai YF, You LC, Chilkoti A (2023) Engineering synthetic biomolecular condensates. Nat Rev Bioeng 1(7):466–480
3. Roden C, Gladfelter AS (2021) RNA contributions to the form and function of biomolecular condensates. Nat Rev Mol Cell Biol 22(3): 183–195
4. Alberti S (2017) Phase separation in biology. Curr Biol 27(20):R1097–R1102
5. Hyman AA, Weber CA, Julicher F (2014) Liquid-liquid phase separation in biology. Annu Rev Cell Dev Biol 30:39–58
6. Yanagawa M, Shimobayashi SF (2024) Multidimensional condensation of intracellular biomolecules. J Biochem 175(2):179–186
7. van Tartwijk FW, Kaminski CF (2022) Protein condensation, cellular organization, and spatiotemporal regulation of cytoplasmic properties. Adv Biol (Weinh) 6(11):e2101328
8. Brangwynne Clifford P, Tompa P, Pappu Rohit V (2015) Polymer physics of intracellular phase transitions. Nat Phys 11(11):899–904
9. Shin Y, Berry J, Pannucci N et al (2017) Spatiotemporal control of intracellular phase transitions using light-activated optoDroplets. Cell 168(1-2):159–171.e14
10. Park KH, Kim KW (2025) Optogenetics to biomolecular phase separation in neurodegenerative diseases. Mol Cells 48(8):100247
11. Park H, Kim NY, Lee S et al (2017) Optogenetic protein clustering through fluorescent protein tagging and extension of CRY2. Nat Commun 8:30
12. Liu HT, Yu XH, Li KW et al (2008) Photoexcited CRY2 interacts with CIB1 to regulate transcription and floral initiation in. Science 322(5907):1535–1539
13. Hao YH, Zhang X, Liu YQ et al (2023) Cryo-EM structure of the CRY2 and CIB1 fragment complex provides insights into CIB1-mediated photosignaling. Plant Commun 4(2):100475
14. Jia LY, Gao S, Qiao Y (2024) Optical control over liquid-liquid phase separation. Small Methods 8(10):e2301724
15. Taslimi A, Vrana JD, Chen D et al (2014) An optimized optogenetic clustering tool for probing protein interaction and function. Nat Commun 5:4925
16. Achimovich AM, Yan T, Gahlmann A (2023) Dimerization of iLID optogenetic proteins observed using 3D single-molecule tracking in live. Biophys J 122(16):3254–3267
17. Xu DQ, Deng XW (2020) CBF-phyB-PIF module links light and low temperature signaling. Trends Plant Sci 25(10):952–954
18. Liang J, Choi J, Clardy J (1999) Refined structure of the FKBP12-rapamycin-FRB ternary complex at 2.2 Å resolution. Acta Crystallogr D 55:736–744
19. Chung CI, Yang JJ, Shu XK (2023) Chemogenetic minitool for dissecting the roles of protein phase separation. ACS Central Sci 9(7): 1466–1479
20. Garabedian MV, Wang WT, Dabdoub JB et al (2021) Designer membraneless organelles sequester native factors for control of cell behavior. Nat Chem Biol 17(9):998–1007
21. Reinkemeier CD, Girona GE, Lemke EA (2019) Designer membraneless organelles enable codon reassignment of selected mRNAs in eukaryotes. Science 363(6434): eaaw2644
22. Zhao EM, Suek N, Wilson MZ et al (2019) Light-based control of metabolic flux through assembly of synthetic organelles. Nat Chem Biol 15(6):589–597
23. Tang NC, Su JC, Shmidov Y et al (2024) Synthetic intrinsically disordered protein fusion tags that enhance protein solubility. Nat Commun 15(1):3727
24. Ruff KM, Roberts S, Chilkoti A et al (2018) Advances in understanding stimulus-responsive phase behavior of intrinsically disordered protein polymers. J Mol Biol 430(23): 4619–4635
25. Quiroz FG, Chilkoti A (2015) Sequence heuristics to encode phase behaviour in intrinsically disordered protein polymers. Nat Mater 14(11):1164–1171
26. Dzuricky M, Rogers BA, Shahid A et al (2020) De novo engineering of intracellular condensates using artificial disordered proteins. Nat Chem 12(9):814–825
27. Pesce F, Bremer A, Tesei G et al (2024) Design of intrinsically disordered protein variants with diverse structural properties. Sci Adv 10(35). https://doi.org/10.1126/sciadv.adm992
28. Dai YF, Farag M, Lee D et al (2023) Programmable synthetic biomolecular condensates for cellular control. Nat Chem Biol 19(4): 518–528
29. Mészáros B, Erdos G, Dosztányi Z (2018) IUPred2A: context-dependent prediction of protein disorder as a function of redox state

and protein binding. Nucleic Acids Res 46 (W1):W329–W337

30. Xue B, Dunbrack RL, Williams RW et al (2010) PONDR-FIT: a meta-predictor of intrinsically disordered amino acids. Biochimica et Biophysica Acta (BBA) 1804(4):996–1010
31. Gasteiger E, Gattiker A, Hoogland C et al (2003) ExPASy: the proteomics server for in-depth protein knowledge and analysis. Nucleic Acids Res 31(13):3784–3788
32. Shapiro DM, Deshpande S, Eghtesadi SA et al (2025) Synthetic biomolecular condensates enhance translation from a target mRNA in living cells. Nat Chem 17(3):448–456
33. Holehouse AS, Ahad J, Das RK et al (2015) CIDER: classification of intrinsically disordered ensemble regions. Biophys J 108(2): 228a–228a
34. Salis HM (2011) The ribosome binding site calculator. Method Enzymol 498:19–42
35. Tang NC, Chilkoti A (2016) Combinatorial codon scrambling enables scalable gene synthesis and amplification of repetitive proteins. Nat Mater 15(4):419–424

Chapter 7

Characterizing Optogenetic Tools for Use in Synthetic Gene Circuits

Timothy H. Wakiyama and Megan N. McClean

Abstract

Synthetic gene circuits consist of engineered networks of genes and proteins. The function of these circuits can be tuned by regulating the expression levels of key circuit components. Light is an ideal signal for controlling expression levels in these circuits; it is cheap and easy to modulate relative to traditional chemical signals. Here, we explore Optogenetic Transcription Factor Localizers (OpTFLos) for tuning expression levels with light. OpTFLos are genetically encoded light-responsive tools for controlling the position in the cell, and therefore the activity, of transcription factors. In this protocol, we discuss how to use the automated optogenetics platform Lustro to systematically characterize OpTFLos to identify light programs that generate desired expression levels. One challenge here with optogenetic tools is that many systems respond to blue light. By using microscopy and Lustro, we describe how to identify light programs that allow for two blue-light-responsive OpTFLos with different kinetic properties to achieve distinct gene expression states with minimal crosstalk, potentially useful for tuning the expression of multiple components within a synthetic gene circuit with a single wavelength of light.

Key words Optogenetics, OpTFLo, Synthetic gene circuits, Kinetics, Lustro, Fluorescence microscopy, *Saccharomyces cerevisiae*

1 Introduction

Synthetic gene circuits are engineered networks of biological parts, including promoters, transcription factors (TFs), and genes that are designed to perform desirable functions inside of a living cell. Often, the behavior and function of these circuits can be tuned by changing the concentration of proteins within the circuit [1, 2]. One way to do this is by regulating expression of the protein of interest using inducible systems. Light has significant potential as an inducer for changing expression and therefore tuning synthetic gene circuit function because it can be dynamically and precisely modulated over time and space and is inexpensive relative to chemical inducers. Optogenetic systems allow a light inducer to be converted into gene expression.

Xiao-Jun Tian (ed.), *Synthetic Gene Circuits: Methods and Protocols*, Methods in Molecular Biology, vol. 3041, https://doi.org/10.1007/978-1-0716-5304-3_7,

Optogenetics is a technology which uses light-responsive proteins to control cellular processes such as gene expression and protein localization [3–5]. In optogenetic systems, the light-responsive proteins are genetically encoded along with appropriate effector domains and often constitutively expressed to execute the desired cellular process upon light exposure which induces conformational changes affecting activity. For example, several systems including LINuS, LEXY, LANS, and CLASP have been developed to control transcription by regulating the localization, and therefore activity, of a TF using a light-responsive nuclear localization signal and/or organelle tethering to controllably sequester TFs away from the nucleus [5–10] (*see* **Note 1**). We call this class of localization-based optogenetic systems Optogenetic Transcription Factor Localizers (OpTFLos).

Interestingly, OpTFLos have been reported to optimally express the reporter under pulsed or continuous light depending on factors such as promoter structure and binding affinity, requiring screening of light programs [7, 8]. Furthermore, we have found that the kinetic properties of OpTFLos themselves affect reporter expression in response to different light programs (unpublished observation). In this protocol, we outline how to use the high-throughput automated optogenetic screening platform Lustro to systematically characterize a CLASP-based OpTFLo to identify light programs that generate desired expression levels in the common chassis organism *Saccharomyces cerevisiae* (*see* **Note 2**). The CLASP-based OpTFLo works by tethering a TF payload to the plasma membrane in the dark, and localizing the TF to the nucleus using a light-responsive nuclear localization signal (NLS) in the light, thereby inducing transcription (Fig. 1) [7, 11].

Next, we describe how to identify light programs that allow for independent control of two blue-light-responsive OpTFLos by exploiting their different kinetic properties. The activation and deactivation kinetics of OpTFLos, including photocycle kinetics and mass action kinetics, determine the timescales at which OpTFLos can be controlled. For OpTFLos, the rate of nuclear localization and delocalization in response to light and dark is a key kinetic property defining how OpTFLos respond to dynamic light programs. By using fluorescence microscopy and Lustro, we characterize the different kinetics of OpTFLo-1 (fast import/slow export) and OpTFLo-2 (slow import/fast export) to identify light programs which preferentially activate one system versus the other with minimal crosstalk, allowing for four different gene expression states to be controlled using one wavelength of light.

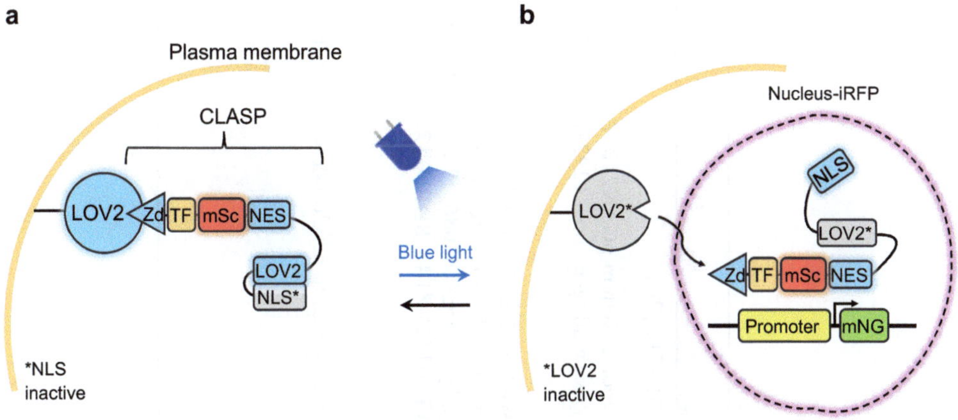

Fig. 1 The CLASP OpTFLo in *S. cerevisiae.* CLASP and the nucleus are fluorescently tagged with mScarlet (mSc) and iRFP, respectively. (**a**) In the dark, a membrane-anchored LOV2 domain sequesters CLASP to the plasma member via interaction with the Zdk1 peptide (Zd). Another LOV2 domain shields CLASP's nuclear localization signal (NLS), preventing nuclear localization. (**b**) In the light, the LOV2 interaction with Zdk1 dissociates. The NLS is unshielded from the other LOV2 domain, which localizes CLASP containing a transcription factor (TF) to the nucleus, driving gene expression of the reporter mNeonGreen (mNG). This figure was adapted from Chen et al. [7]

2 Materials

2.1 OpTFLo Machinery and S. cerevisiae *Strains*

1. *S. cerevisiae* strain yMM1841 *MATα his3Δ1 leu2Δ0 lys2Δ0 MET15 ura3Δ0 Nhp6a-iRFP ho::p43_8(x8 TFBS)-mNeon-Green-tADH1-NAT* with OpTFLo-1 or OpTFLo-2 integrated at the URA3 locus (*see* **Note 3** and Table 1).
2. *S. cerevisiae* strain yMM1841 *MATα his3Δ1 leu2Δ0 lys2Δ0 MET15 ura3Δ0 Nhp6a-iRFP ho::p43_8(x8 TFBS)-mNeon-Green-tADH1-NAT* (promoter-only control).
3. *S. cerevisiae* strain yMM1442 *MATα his3Δ1 leu2Δ0 lys2Δ0 MET15 ura3Δ0 Nhp6a-iRFP* (autofluorescence control).

2.2 Media and Culture Equipment

1. Low fluorescence media (LFM): 900 mL deionized water, 1.7 g yeast nitrogen base (w/o ammonium sulfate, w/o folic acid, w/o riboflavin), 5 g ammonium sulfate. Autoclave media and cool completely. Add 50 mL 20× glucose solution (40% w/v, autoclaved separately) and 50 mL 20× amino acid mix.

   20× amino acid mix: Dissolve 0.4 g uracil, 0.4 g histidine HCl, 0.4 g arginine HCl, 0.4 g methionine, 0.6 g tyrosine, 0.6 g isoleucine, 1 g adenine hemisulfate, 1 g tryptophan, 1 g phenylalanine, 2 g leucine, 2 g lysine HCL, 2 g aspartic acid, 3 g valine, 4 g threonine, and 8 g serine into 1 L deionized water and filter-sterilize with a 0.20 μm pore size filter (e.g., Fisherbrand Disposable PES Bottle Top Filter #FB12566508).

Table 1
***S. cerevisiae* strains used in this protocol**

Strain	Genotype	Description
yMM1442	*MATα his3Δ1 leu2Δ0 lys2Δ0 MET15 ura3Δ0 Nhp6a-iRFP*	Control strain derived from BY4742 [26].
yMM1841	*MATα his3Δ1 leu2Δ0 lys2Δ0 MET15 ura3Δ0 Nhp6a-iRFP ho:: p43_8(x8 TFBS)-mNeonGreen-tADH1-NAT*	yMM1442 integrated with reporter system pGAN213 (Addgene plasmid # 129270; https://www.addgene.org/129270/; RRID: Addgene_129270) [27].
yMM1841 / OpTFLo-1	*MATα his3Δ1 leu2Δ0 lys2Δ0 MET15 ura3Δ0 Nhp6a-iRFP ho:: p43_8(x8 TFBS)-mNeonGreen-tADH1-NAT ura3:: ConS-pRPL18B-Zdk1-VP16TAD-ZF43_8DBD-mScarlet-yeLANS-tADH1-Con1-pRPL18B-Hs_RGS2(33–67)-13aa-iRFP713(iRFP)-asLOV2(404–546)-tPGK1-ConE-URA3 (pSc-native-tSc)*	yMM1841 integrated with synTF-CLASP from Chen et al. (pCLASP_mScarlet_CLASP_RGS2membrane (Addgene plasmid # 133086; https://www.addgene.org/133086/; RRID: Addgene 133,086) and gBlock of VP16TAD-ZF43_8DBD) [7].
yMM1841 / OpTFLo-2	*MATα his3Δ1 leu2Δ0 lys2Δ0 MET15 ura3Δ0 Nhp6a-iRFP ho:: p43_8(x8 TFBS)-mNeonGreen-tADH1-NAT ura3:: ConS-pRPL18B-Zdk1-VP16TAD-ZF43_8DBD-PKI_NES-mScarlet-yeLANS-tADH1-Con1-pRPL18B-Hs_RGS2(33–67)-13aa-iRFP713(iRFP)-asLOV2(404–546)-tPGK1-ConE-URA3 (pSc-native-tSc)*	yMM1841 / OpTFLo-1 with PKI NES to develop slow import/fast export mutant [28].

2. Concanavalin A (ConA) solution (2 mg/mL): 50 mL deionized water and 100 mg ConA. Filter-sterilize with a 0.22 μm pore size filter (e.g., Corning 50 mL Tube Top Vacuum Filter System #430320).
3. Incubator at 30 °C with a roller drum set to 40 rpm (e.g., New Brunswick Scientific TC-7 Tissue Culture Roller Drum).
4. Incubator with shaking capabilities set to 250 rpm, 30 °C (e.g., New Brunswick Scientific G-25 Controlled Environment Incubator Shaker).

2.3 High-Throughput Automated Optogenetic Screening Platform Lustro (see Note 4)

1. Tecan Fluent Automated Workstation.
2. Tecan Robotic Gripper Arm.
3. Tecan Spark Multimode Microplate Reader.
4. QINSTRUMENTS BioShake 3000-T elm heater shaker.
5. OptoPlate (blue light illumination device) [12]. Calibrate and program according to Grødem et al. and Sweeney and McClean [8, 13].
6. 96-well clear polystyrene microplate (Corning 3788).
7. 96-well black clear plastic bottom microplate (Thermo Scientific 165305). *See* **Note 5**.
8. Breathe-Easy polyurethane membrane (Diversified Biotech, BEM-1).
9. Fluent Control (Software for controlling Tecan Fluent Automated Workstation).
10. Magellan (Software for controlling Tecan Spark Multimode Microplate Reader).
11. MATLAB scripts for controlling the optoPlate. Refer to Sweeney and McClean, for details [8].
12. Dark room for carrying out light-sensitive experiments.

2.4 Fluorescence Microscopy

1. Inverted fluorescence microscope (Nikon Ti-Eclipse).
2. Imaging automation software (NIS-Elements with JOBS module).
3. Epifluorescence light source (Nikon Intensilight C-HGFIE).
4. iRFP filter cube (Chroma #49022: 650/45×, 720/60 m, T685lpxr).
5. mScarlet (mSC) filter cube (Chroma #96365: 560/40×, 630/75 m, T585lpxr).
6. 60× oil immersion objective (CFI Plan Apochromat Lambda).
7. Immersion oil (Olympus IMMOIL-F30CC Type-F).
8. 96-well black glass bottom microplate (CellVis P961.5H-N). *See* **Note 5**.

9. OptoPlate (same as Subheading 2.3, **item 5**).
10. MATLAB scripts for controlling the optoPlate (same as Subheading 2.3, **item 11**).
11. Dark room for carrying out light-sensitive experiments.

2.5 Analysis

1. MATLAB by Mathworks (https://www.mathworks.com/products/matlab.html). *See* **Note 6.**
2. MATLAB script for cell segmentation and measurement [8].
3. MATLAB script for analyzing light sweep experiment [8].
4. Data visualization MATLAB package Gramm [14].

3 Methods

First, we describe how to use Lustro to identify light programs that generate desired reporter expression levels through controlling one OpTFLo (Subheading 3.1). Next, we describe how to identify light programs that allow for independent control of two OpTFLos by characterizing their nuclear localization kinetics on a fluorescent microscope (Subheading 3.2) and reporter expression on Lustro (Subheading 3.3).

3.1 Using Lustro to Optimize OpTFLo Driven Reporter Expression

3.1.1 Setting Up Tecan Spark Multimode Microplate Reader

In the Spark software Magellan, measurement protocols are called *methods*. Create a *method* which includes the following:

1. Shaking 96-well black plate.
2. Optical density measurement (e.g., OD_{600}).
3. Fluorescence measurement (*see* **Note 7**).

3.1.2 Setting Up Tecan Fluent Automated Workstation

In the Fluent Control Software, set a virtual worktable with the following components:

1. Tecan Robotic Gripper Arm.
2. 96-well black plate.
3. OptoPlate.
4. QINSTRUMENTS BioShake 3000-T elm heater shaker.
5. Tecan Spark Multimode Microplate Reader.

3.1.3 Setting Up an Experiment Protocol in the Fluent Control Software

In the Fluent Control Software, experiment protocols are called *scripts*. Create a *script* containing the following steps:

1. Move 96-well black plate from optoPlate to heater shaker.
2. Lock heater shaker and shake 96-well black plate for 30 s.

3. Unlock heater shaker and move 96-well black plate to Tecan Spark Multimode Microplate Reader.
4. Measure OD_{600} and fluorescence using the preset *method.*
5. Move 96-well black plate to optoPlate for illumination.
6. Repeat **Steps 1–5**, every 2 h.

3.1.4 Preparing S. cerevisiae Cell Culture

1. Ensure there is no light contamination by conducting the experiment in a dark room and using illumination that does not excite the optogenetic system (e.g., red lights).
2. Inoculate *S. cerevisiae* strains containing the reporter mNeon-Green (mNG) and OpTFLo-1 or OpTFLo-2 into a 96-well clear plate with 300 μL LFM and incubate overnight at 30 °C without shaking in the dark. Prepare control strains as well in the same manner. Prepare all strains in biological quadruplicate.
3. The next day, take 2 μL of overnight culture and inoculate into fresh 200 μL fresh LFM in a 96-well black plate.
4. Attach a Breathe-Easy polyurethane membrane to the top of the 96-well black plate to prevent evaporation of LFM.
5. Incubate the 96-well black plate in a shaker-incubator for 4 h at 30 °C with shaking at 250 rpm, in the dark.
6. Place the 96-well black plate on the optoPlate within Lustro.

3.1.5 Lustro Test Run and Setting Optimization

1. Run the Fluent *script* through the Fluent Control Software.
2. Run an initial 4-h continuous light program on the optoPlate.
3. During the experiment, regularly spot check OD_{600} by examining the output from the plate reader to confirm cells are growing at the normal rate.
4. During the experiment, monitor fluorescence by regularly checking output from the plate reader to confirm the reporter mNG is being properly measured.
5. During and after the experiment, observe the 96-well black plate for anything noteworthy such as excessive condensation, buildup of gas, or sedimentation of cells.
6. Optimize Lustro settings based on observations made during the test run described in this subsection (*see* **Note 8**).

3.1.6 Initial Screening of an OpTFLo

1. Develop initial screening light programs based on empirical observations or kinetic information in the literature (*see* **Note 9**).
2. Run the screening light programs on Lustro (*see* Table 2).

Table 2
Screening light programs for Lustro (times are in minutes)

Pulsed light	Net light time	No. of periods	Light time in 1 period	Dark time in 1 period	Total time
#1	120	60	2	2	240
#2	80	40	2	4	240
#3	40	20	2	10	240
#4	20	10	2	22	240

Continuous light	Net light time	Lead dark time	Centered light time	Final dark time	Total time
#1	120	60	120	60	240
#2	80	80	80	80	240
#3	40	100	40	100	240
#4	20	110	20	110	240

3.1.7 Analyzing Screening Results from Subheading 3.1.6

Use the analysis script from Subheading 2.5, **item 3** on the following:

1. Determine whether OpTFLo expresses the reporter more under pulsed or continuous light.
2. Determine the specific program that resulted in the largest difference between pulsed and continuous light (*see* **Notes 9** and **10**).

3.1.8 Titrating Reporter Expression with Varying Light Programs

1. Develop titrating light programs based on screening results in Subheading 3.1.7 (*see* **Note 10**).
2. Run the titrating light programs on Lustro (*see* Table 3).
3. Use the analysis script from Subheading 2.5, **item 3** to determine the linear range of reporter expression to control OpTFLo at the desired level.

3.2 Fluorescence Microscopy for Measuring Nuclear Localization Kinetics of OpTFLos

3.2.1 Preparing S. cerevisiae *Cell Culture (*see *Note 11)*

1. Ensure there is no light contamination by conducting the experiment in a dark room and using non-interfering illumination (e.g., red lights).
2. Inoculate *S. cerevisiae* strains containing OpTFLo-1 or OpTFLo-2 into test tubes with 5 mL LFM and incubate overnight at 30 °C on a roller drum, in the dark. Prepare control strains (*see* Subheading 2.1) in the same manner. Prepare all strains in biological quadruplicate.
3. The following day, dilute the overnight cultures to an OD_{600} of approximately 0.1 in LFM (5mL) and incubate for approximately 6 h at 30 °C, on a roller drum, in the dark to enable cells to reach log phase.

Table 3
Titrating light programs for Lustro (times are in minutes)

Pulsed light	Net light time	No. of periods	Light time in 1 period	Dark time in 1 period	Total time
#5	160	80	2	22	1920
#6	120	60	2	22	1440
#7	80	40	2	22	960
#8	40	20	2	22	480
#9	20	10	2	22	240

Continuous light	Net light time	Lead dark time	Centered light time	Final dark time	Total time
#5	160	880	160	880	1920
#6	120	660	120	660	1440
#7	80	440	80	440	960
#8	40	220	40	220	480
#9	20	110	20	110	240

4. Add 30 μL of ConA to each well of a 96-well black plate to promote adherence. Incubate at room temperature for 15 min, then remove ConA before adding culture.
5. Dilute cells in log phase with fresh LFM to a density of $OD_{600} = 0.125$ and pipette 200 μL into each well of the 96-well black plate.
6. Incubate the 96-well black plate at 30 °C for 15 min to allow cells to adhere to the plate bottom.
7. Remove the media and wash each well with 200 μL of 30 °C LFM three times to remove non-adhered cells.
8. Incubate the 96-well black plate at 30 °C for 15 min to allow cells to equilibrate.

3.2.2 Setting Up the Microscope

1. Set up and calibrate the optoPlate according to Sweeney and McClean and Grødem et al. [8, 13].
2. Mount the 96-well black plate on a fluorescence microscope with the optoPlate as described in Sweeney et al. [15].
3. Determine the appropriate imaging settings for both OpTFLos (*see* **Note 12**).

3.2.3 Imaging with Initial Light Program

1. Run an initial light program: 15-min dark, then 30-min light, and finally 15-min dark.
2. Take images using iRFP and mSc filter cubes every 5 min with the initial light program.

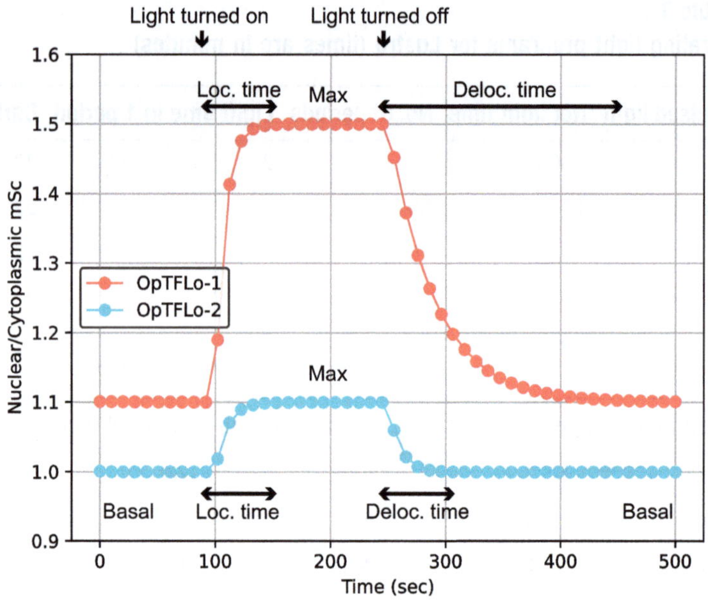

Fig. 2 Simulated nuclear localization and delocalization data of OpTFLos. Light stimulation is turned "on" and "off" at 100 and 250 s, respectively. The OpTFLo marker mSc is imaged every 10 s and plotted as nuclear localization (nuclear mSc divided by cytoplasmic mSc) over time. OpTFLo-1 and OpTFLo-2 show different characteristics: the former has higher basal and maximum localizations than the latter; the former and latter require approximately the same amount of time to complete nuclear localization when light stimulation is turned on; and the former requires a 3.6-fold longer time to complete delocalization than the latter when light stimulation is turned off (basal, max, loc. time, and deloc. time are abbreviations for basal localization, maximum localization, localization time, and delocalization time, respectively)

3.2.4 Adjusting Timescales and Light Programs for Further Imaging

1. Use the automated image processing scripts from Subheading 2.5, **item 2** to segment and quantify images.
2. Use the analysis script from Subheading 2.5, **item 3** and plot nuclear localization time-course data.
3. Examine the plot (Fig. 2). If there are fewer than 5 time points capturing nuclear localization and delocalization, respectively, adjust microscope settings to image at a shorter timescale (*see* **Note 13**).
4. Take images using iRFP and mSc filter cubes with the adjusted imaging timescale.
5. Repeat **steps 1–4** until both nuclear localization and delocalization are captured with at least 5 time points.

3.2.5 Characterizing OpTFLo-1 and OpTFLo-2

1. For both OpTFLos, estimate the time necessary to complete nuclear localization and delocalization from the plot (*see* **Note 14** and Table 4).
2. For both OpTFLos, use least squares error-based curve fitting to estimate kinetic parameters including rate of localization and delocalization and maximal and basal nuclear localization (*see* **Note 14**).

3.3 Developing Light Programs for Preferentially Activating OpTFLos

3.3.1 Comparing Characteristics of OpTFLo-1 and OpTFLo-2

1. Compare OpTFLos and determine which has longer/shorter nuclear localization time and longer/shorter delocalization time (*see* Table 4).
2. Estimate OpTFLos' response to pulsed and continuous light (*see* **Notes 9** and **14**).

3.3.2 Developing Pulsed Light Programs Which Maximize and Minimize Reporter Expression for OpTFLo-1 and OpTFLo-2, Respectively (see Note 15)

1. Run light programs on Table 5.
2. Run adjusted light programs on Table 6.

Table 4
Nuclear localization and delocalization kinetics in simulated data

OpTFLo	Localization time	Delocalization time
OpTFLo-1	60 s	220 s
OpTFLo-2	60 s	60 s

Table 5
Exploring light programs which favor OpTFLo-1 (times are in minutes)

Pulsed light	Net light time	No. of periods	Light time in 1 period	Dark time in 1 period	Total time
#10	216	27	8	32	1080
#11	216	54	4	16	1080
#12	216	108	2	8	1080
#13	216	216	1	4	1080
#14	216	432	0.5	2	1080

Continuous light	Net light time	Lead dark time	Centered light time	Final dark time	Total time
#10	216	432	216	432	1080

Table 6
Expanding light programs which favor OpTFLo-1 (times are in minutes)

Pulsed light	Net light time	No. of periods	Light time in 1 period	Dark time in 1 period	Total time
#15	108	108	1	10	1188
#16	108	108	1	8	972
#17	108	108	1	6	756
#18	108	108	1	4	540
#19	108	108	1	2	324
#20	54	108	0.5	10	1134
#21	54	108	0.5	8	918
#22	54	108	0.5	6	702
#23	54	108	0.5	4	486
#24	54	108	0.5	2	270

Continuous light	Net light time	Lead dark time	Centered light time	Final dark time	Total time
#11	108	540	108	540	1188
#12	108	432	108	432	972
#13	108	324	108	324	756
#14	108	216	108	216	540
#15	108	108	108	108	324
#16	54	513	54	513	1134
#17	54	432	54	432	918
#18	54	324	54	324	702
#19	54	216	54	216	486
#20	54	108	54	108	270

3.3.3 Developing Continuous Light Programs Which Maximize and Minimize Reporter Expression for OpTFLo-1 and OpTFLo-2, Respectively (see Note 16)

1. Run light programs on Table 7.
2. Run adjusted light programs on Table 8.

3.3.4 Compiling Results

1. Use the analysis script from Subheading 2.5, **item 3** and plot reporter expression.

Table 7
Exploring light programs which favor OpTFLo-2 (times are in minutes)

Pulsed light	Net light time	No. of periods	Light time in 1 period	Dark time in 1 period	Total time
#31	720	90	8	4	1080
#32	540	135	4	4	1080
#33	360	180	2	4	1080
#34	216	216	1	4	1080
#35	120	240	0.5	4	1080
#36	63.5	254	0.25	4	1080

Continuous light	Net light time	Lead dark time	Centered light time	Final dark time	Total time
#28	720	180	720	180	1080
#29	540	270	540	270	1080
#30	360	360	360	360	1080
#31	216	432	216	432	1080
#32	120	480	120	480	1080
#33	63.5	508.25	63.5	508.25	1080

Table 8
Adjusting light programs to favor OpTFLo-2 (times are in minutes)

Pulsed light	Net light time	No. of periods	Light time in 1 period	Dark time in 1 period	Total time
#25	864	108	8	2	1080
#26	720	180	4	2	1080
#27	540	270	2	2	1080
#28	360	360	1	2	1080
#29	216	432	0.5	2	1080
#30	120	480	0.25	2	1080

Continuous light	Net light time	Lead dark time	Centered light time	Final dark time	Total time
#21	864	108	864	108	1080
#22	720	180	720	180	1080
#23	540	270	540	270	1080
#24	360	360	360	360	1080
#25	216	432	216	432	1080
#26	120	480	120	480	1080

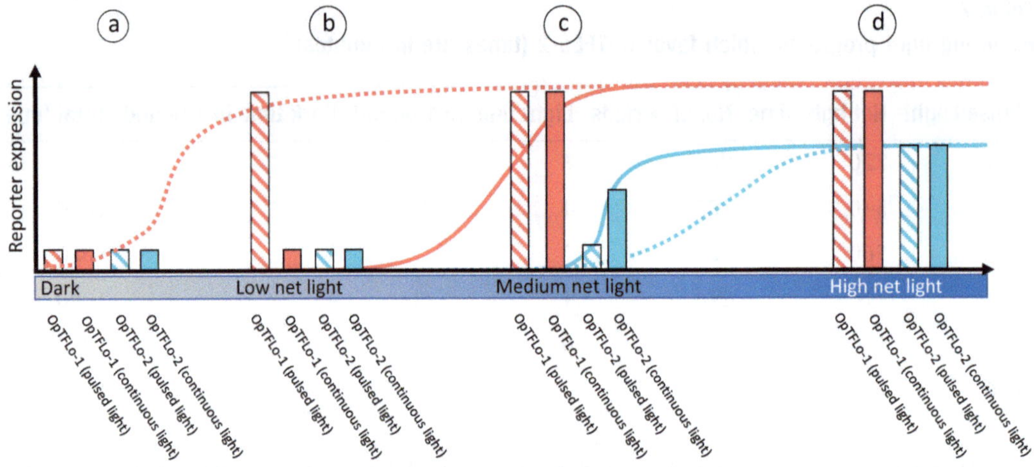

Fig. 3 Concept of preferentially activating an OpTFLo by using different light programs. (**a**) In the dark, neither OpTFLos express the reporter. (**b**) At low net light, only OpTFLo-1 under pulsed light expresses the reporter (**c**) At medium net light, both OpTFLos under continuous light express the reporter in addition to OpTFLo-1 under pulsed light. (**d**) At high net light, both OpTFLos express the reporter under both pulsed and continuous light

2. From the plot, identify light programs which preferentially activate one OpTFLo versus the other with minimal crosstalk (Fig. 3).

4 Notes

1. LINuS (Light-Inducible Nuclear localization Signal), LEXY (Light-inducible nuclear EXport System), LANS (Light-Activated Nuclear Shuttle), and CLASP (Controllable Light-Activated Shuttling and Plasma membrane sequestration) regulate transcription factor activity by controlling the transcription factor's access to the nucleus. We have found CLASP to be a reliable tool for regulating transcription factor localization and gene expression in *S. cerevisiae* due to its low background activity in the dark. It is our preferred platform for OpTFLo [6–9, 15]. Additional tools for light-based organelle targeting exist and could be used to regulate the activity of transcription factors or other signaling components [16]. Other possibilities for regulating transcription factor activity include light-regulated reconstitution of split transcription factors [17–19].
2. Lustro is a high-throughput optogenetic screening platform developed by Harmer and McClean [18, 20]. It consists of a Tecan Fluent automated workstation integrated with a Robotic Gripper Arm (RGM), a plate reader (Tecan Spark Multimode Microplate Reader), a plate shaker (QINSTRUMENTS BioShake 3000-T elm heater shaker), and an optoPlate (blue light illumination device) [12, 13]. It is capable of testing

multiple, simultaneous light programs and carrying out real-time measurements of fluorescence and optical density (e.g., OD_{600}). While Lustro is a convenient platform, there are other platforms including commercial platforms which allow for medium- to high-throughput assessment of optogenetic construct function as reviewed in Kumar and Khammash [21]. In addition, illumination platforms with different well configurations or illumination capabilities, such as LITOS, can be integrated into Lustro as described in Harmer [22].

3. In this protocol, we use iRFP as the nuclear marker and mScarlet (mSc) as the OpTFLo marker. A nuclear marker is necessary to measure colocalization of the OpTFLo with the nucleus during microscopy. We use mNeonGreen (mNG) as the fluorescent reporter induced by OpTFLo. The reporter mNG is under the synthetic promoter 43-8, which is induced by the synthetic zinc finger TF 43-8 contained within our OpTFLo construct.
4. If Lustro is not available, use an illumination device (e.g., optoPlate, LITOS, optoWELL-24) to stimulate cells with light programs and a flow cytometer or plate reader (e.g., ThermoFisher Attune NxT Flow Cytometer or Tecan Spark Multimode Microplate Reader) to measure reporter expression. Cells should be sampled for measurement at the starting point and the end point of the light program. Intermediate timepoints can be measured depending on feasibility and experimental goals.
5. Black glass bottom microplates are used to prevent leakage of light between wells. The glass bottom allows for analysis via microscopy due to its superior optical properties. For experiments done only using Lustro, a plastic bottom plate is acceptable.
6. If MATLAB is not available, use an open-source programming language such as Python or R along with suitable scripts for cell segmentation, measurement, and analysis. Refer to Hurbain et al. and Granados et al. [23, 24].
7. Set the excitation and emission windows appropriate for the fluorophore to maximally detect the signal while preventing bleed-through between fluorescent markers. For mNG, we recommend 490 nm and 530 nm for excitation and emission, respectively, with bandwidths for each at 20 nm. Furthermore, set a gain which maximizes the signal without overflowing. We recommend initially setting a high gain (e.g., 100) and adjusting down.
8. The initial light program should be continuous light exposure (i.e., a single pulse) for 4 h to confirm that OpTFLo leads to measurable reporter expression. Use appropriate controls:

yMM1442 for autofluorescence of the cell, yMM1841 for leakiness of the reporter promoter (promoter-only control), and dark controls (no light illumination). If there is signal overflow when monitoring fluorescence measurements from the plate reader, lower the gain in increments of 20 in subsequent test runs. Condensation is a sign of excessive heating from the optoPlate. Consider lowering light intensity as excessive heat may inhibit *S. cerevisiae* cell growth. You can also more sparsely pattern illumination in the plate, interleaving dark and illuminated wells. Buildup of gas and sedimentation of cells are signs of insufficient shaking. Consider increasing the time or the frequency of shaking.

9. In this protocol, the screening light programs are adapted from the paper describing the original development of CLASP [7]. When using other OpTFLos such as LINuS, LEXY, or LANS, refer to Yumerefendi et al., Niopek et al., or Niopek et al., for light programs to screen on Lustro [5, 9, 10]. Here, the screening light programs consist of varying pulses of 2-min light followed by 2, 4, 10, or 22-min dark, and varying continuous lights of 20, 40, 80, or 120 min in duration (*see* Table 2). The pulsed and continuous lights are designed to have equal net light time. For example, pulsed light #1 and continuous light #1 have the same 120-min net light time in Table 2. As pulsed light #1 takes a total of 240 min to complete, 60 min of darkness have been added to the beginning and end of continuous light #1 to make its total time equal to that of pulsed light #1 (i.e., 120 min of continuous light is centered within 240 min of total time). For the same net light time, an OpTFLo may induce reporter expression more under pulsed or continuous light. In this protocol, OpTFLo-1 (fast import/slow export) and OpTFLo-2 (slow import/fast export) induce the reporter more under pulsed and continuous light, respectively. This is because OpTFLo-1 gains extra nuclear localization time with each pulse due to its greater lag during delocalization. A greater lag in delocalization means more time for the OpTFLo to be active in the nucleus for the same net light time, resulting in relatively higher reporter expression under pulsed light. Conversely, OpTFLo-2 loses nuclear localization time with each pulse due to its greater lag during localization. A greater lag in nuclear localization means less time for the OpTFLo to be active in the nucleus for the same net light time, resulting in relatively higher reporter expression under continuous light.

10. Here, let's say OpTFLo-1 induced reporter expression the most under pulsed program #4 in Subheading 3.1.7 (i.e., 10 periods of 2-min light followed by 22-min dark). We expand this light program by surveying the number of periods

at 10, 20, 40, 60, and 80 (see Table 3). We call these titrating light programs as they titrate the reporter expression level to the increased net light time (i.e., determining the linear range and saturation point in the OpTFLo's response to net light time). One can also titrate reporter expression on light intensity ($\mu W/cm^2$) and consider light AUC (area under the curve), which indicates the total light dose during the experiment as $\mu W{\cdot}min/cm^2$. Light stimulation past the point of saturating the reporter expression should not be used as the excess light may lead to phototoxicity while not contributing to reporter expression.

11. In this protocol, we characterize the nuclear localization kinetics of OpTFLo-1 and OpTFLo-2 in separate yeast strains and cultures as both are tagged with mSc and cannot be distinguished under fluorescence microscopy. Once the nuclear localization dynamics are individually characterized, reporter expression in a *S. cerevisiae* strain containing both OpTFLos can be measured in response to different light programs to verify the differential control of each OpTFLo. A distinct TF/promoter pair and reporter will be necessary for each OpTFLo. See for example Harmer et al. [20]. Alternatively, one of the OpTFLos can be tagged with another fluorescent marker such as mTurquoise2 which does not overlap with mSc or iRFP on the excitation and emission spectra. This would allow for the nuclear localization dynamics of both OpTFLos to be characterized in the same *S. cerevisiae* strain. The disadvantage of this approach is that it uses an additional fluorescent marker, which may make subsequent optogenetic and imaging experiments challenging due to a lack of available fluorescent markers.

12. The appropriate imaging settings for both OpTFLos should be determined before conducting the optoPlate localization assay. Use *S. cerevisiae* strains containing. OpTFLo-1 or OpTFLo-2 in separate practice wells. Systematically image cells under different combinations of shutter speed, readout speed, and number of neutral density filters to determine the setting which results in the highest signal-to-noise ratio for both OpTFLos. This setting needs to be readjusted whenever the number of images taken during an experiment changes to minimize photobleaching of fluorescent markers.

13. Attempt imaging in intervals of 60, 30, 15, 10, or 5 s. As an example, Fig. 2 shows simulated data where imaging every 10 s captures both nuclear localization and delocalization over at least 5 time points. We have found 5 time points to be a good balance between imaging speed and curve reconstruction for CLASP-based OpTFLos, but a different number of time points may be suitable for other types of OpTFLos. If shorter imaging

timescales lead to excessive photobleaching, use a combination of long and short timescales. For example, image every 60 s at basal and maximum localizations, while imaging every 10 s during active nuclear localization and delocalization.

14. Using OpTFLo-1 in Fig. 2 as an example, nuclear localization begins when light is turned on at 100 s and completes where the curve plateaus at approximately 160 s ($\approx$ 60 s for full localization). Delocalization begins when light is turned off at 250 s and completes where the curve asymptotes to basal localization at approximately 470 s ($\approx$ 220 s for full delocalization). Although this is a crude quantification of localization and delocalization time, we find it useful for "rapid prototyping" of designing and testing light programs on Lustro. Generally, shorter localization time and longer delocalization time lead to higher reporter expression under pulsed light due to the accumulation of "extra nuclear localization time with each pulse" over many pulses. Conversely, longer localization time and shorter delocalization time lead to higher reporter expression under continuous light due to the accumulation of "lost nuclear localization time with each pulse" over many pulses. For a more precise approach using estimated nuclear localization kinetics, use least squares error-based curve fitting on nuclear OpTFLo measurements. As a straightforward example of curve fitting nuclear localization kinetics, see Shakhi et al. [25]. Briefly, the following Eq. 1 (saturating exponential model) and Eq. 2 (exponential decay model) may be used for fitting nuclear OpTFLo measurements on periods of light stimulation and darkness, respectively. In Eq. 1, $OpTFLo_{nuc}(t)$ is the concentration of nuclear OpTFLo at time t, $OpTFLo_{max}$ is the maximal amount of nuclear OpTFLo, k_{in} is the nuclear import rate, and the basal amount of OpTFLo in the nucleus is $OpTFLo_{basal}$. In Eq. 2, all variables are the same as in Eq. 1, except for k_{out} which is the nuclear export rate.

$$OpTFLo_{nuc}(t) = OpTFLo_{max}\left(1 - e^{-k_{in}t}\right) + OpTFLo_{basal} \quad (1)$$

$$OpTFLo_{nuc}(t) = OpTFLo_{max}\left(e^{-k_{out}t}\right) + OpTFLo_{basal} \quad (2)$$

15. Table 5 contains patterns of pulsed light with a wide range of light and dark times while maintaining a low net light time to capture the necessary timescales suited for maximizing reporter expression for OpTFLo-1 under pulsed light. Here, let's say 0.5 and 1 min of light resulted in maximizing reporter expression. Table 6 expands these light programs by surveying a wide range of dark times. Net light time is kept low for both Tables 5 and 6 to disfavor OpTFLo-2 (which requires more light for nuclear localization due to its lower nuclear import rate).

16. Table 7 contains patterns of continuous light with a wide range of durations to capture the timescales suited for maximizing reporter expression for OpTFLo-2 while the pulsed lights have relatively short dark times to disfavor OpTFLo-1. Here, let's say dark times in the pulsed light were too long, leading to high reporter expression under pulsed light for OpTFLo-1. Table 8 adjusts these light programs by halving the dark time in pulses, further disfavoring OpTFLo-1.

Acknowledgments

This work was supported by the National Institute of Health grants R35GM128873 (awarded to M.N.M.) and UW Carbone Cancer Center Support Grant P30 CA014520 as well as National Science Foundation grant MCB 2045493 (awarded to M.N.M). Megan Nicole McClean, PhD, holds a Career Award at the Scientific Interface from the Burroughs Wellcome Fund. The plasmid pCLASP_mScarlet_CLASP_RGS2membrane was a generous gift from Hana El-Samad (Addgene plasmid # 133086; https://www.addgene.org/133086/; RRID: Addgene 133086). We acknowledge fruitful discussions with McClean lab members and Eli Cytrynbaum for providing feedback on the manuscript.

References

1. Scott TD, Sweeney K, McClean MN (2019) Biological signal generators: integrating synthetic biology tools and *in silico* control. Curr Opin Syst Biol 14:58–65. https://doi.org/10.1016/j.coisb.2019.02.007
2. Bashor CJ, Collins JJ (2018) Understanding biological regulation through synthetic biology. Annu Rev Biophys 47:399–423. https://doi.org/10.1146/annurev-biophys-070816-033903
3. Lan T-H, He L, Huang Y, Zhou Y (2022) Optogenetics for transcriptional programming and genetic engineering. Trends Genet 38: 1253–1270. https://doi.org/10.1016/j.tig.2022.05.014
4. Guntas G, Hallett RA, Zimmerman SP et al (2015) Engineering an improved light-induced dimer (iLID) for controlling the localization and activity of signaling proteins. Proc Natl Acad Sci 112:112–117. https://doi.org/10.1073/pnas.1417910112
5. Yumerefendi H, Dickinson DJ, Wang H et al (2015) Control of protein activity and cell fate specification via light-mediated nuclear translocation. PLoS One 10:e0128443. https://doi.org/10.1371/journal.pone.0128443
6. Geller SH, Antwi EB, Di Ventura B, McClean MN (2019) Optogenetic repressors of gene expression in yeasts using light-controlled nuclear localization. Cell Mol Bioeng 12:511–528. https://doi.org/10.1007/s12195-019-00598-9
7. Chen SY, Osimiri LC, Chevalier M et al (2020) Optogenetic control reveals differential promoter interpretation of transcription factor nuclear translocation dynamics. Cell Syst 11: 336–353.e24. https://doi.org/10.1016/J.CELS.2020.08.009
8. Sweeney K, McClean MN (2023) Transcription factor localization dynamics and DNA binding drive distinct promoter interpretations. Cell Rep 42:112426. https://doi.org/10.1016/j.celrep.2023.112426
9. Niopek D, Benzinger D, Roensch J et al (2014) Engineering light-inducible nuclear localization signals for precise spatiotemporal control of protein dynamics in living cells. Nat Commun 5:4404. https://doi.org/10.1038/ncomms5404
10. Niopek D, Wehler P, Roensch J et al (2016) Optogenetic control of nuclear protein export.

Nat Commun 7:10624. https://doi.org/10.1038/ncomms10624

11. Wang H, Vilela M, Winkler A et al (2016) LOVTRAP: an optogenetic system for photoinduced protein dissociation. Nat Methods 13:755–758. https://doi.org/10.1038/nmeth.3926
12. Bugaj LJ, Lim WA (2019) High-throughput multicolor optogenetics in microwell plates. Nat Protoc 14:2205–2228. https://doi.org/10.1038/s41596-019-0178-y
13. Grødem EO, Sweeney K, McClean MN (2020) Automated calibration of Optoplate LEDs to reduce light dose variation in Optogenetic experiments. BioTechniques 69:313–316. https://doi.org/10.2144/btn-2020-0077
14. Morel P (2018) Gramm: grammar of graphics plotting in Matlab. J Open Source Softw 3:568. https://doi.org/10.21105/joss.00568
15. Sweeney K, Luffey EL, McClean MN (2024) Protocol to probe how promoters decode TF dynamics in *Saccharomyces cerevisiae* by combining optogenetic control with microscopy. STAR Protoc 5:103002. https://doi.org/10.1016/j.xpro.2024.103002
16. Yang X, Jost APT, Weiner OD, Tang C (2013) A light-inducible organelle-targeting system for dynamically activating and inactivating signaling in budding yeast. Mol Biol Cell 24:2419–2430. https://doi.org/10.1091/mbc.E13-03-0126
17. An-adirekkun J, Stewart CJ, Geller SH et al (2020) A yeast optogenetic toolkit (yOTK) for gene expression control in Saccharomyces cerevisiae. Biotechnol Bioeng 117:886–893. https://doi.org/10.1002/bit.27234
18. Harmer ZP, McClean MN (2023) Lustro: high-throughput Optogenetic experiments enabled by automation and a yeast Optogenetic toolkit. ACS Synth Biol 12:1943–1951. https://doi.org/10.1021/acssynbio.3c00215
19. Kennedy MJ, Hughes RM, Peteya LA et al (2010) Rapid blue-light-mediated induction of protein interactions in living cells. Nat Methods 7:973–975. https://doi.org/10.1038/nmeth.1524
20. Harmer ZP, Thompson JC, Cole DL et al (2024) Dynamic multiplexed control and modeling of Optogenetic systems using the high-throughput Optogenetic platform, Lustro. ACS Synth Biol 13:1424–1433. https://doi.org/10.1021/acssynbio.3c00761
21. Kumar S, Khammash M (2022) Platforms for Optogenetic stimulation and feedback control. Front Bioeng Biotechnol 10. https://doi.org/10.3389/fbioe.2022.918917
22. Harmer ZP, Höhener TC, Landolt AE et al (2024) Enhancing high-throughput optogenetics: integration of LITOS with Lustro enables simultaneous light stimulation and shaking. MicroPublication Biol. https://doi.org/10.17912/micropub.biology.001073
23. Hurbain J, ten Wolde PR, Swain PS (2025) Quantifying the nuclear localization of fluorescently tagged proteins. Bioinforma Adv 5:vbaf114. https://doi.org/10.1093/bioadv/vbaf114
24. Granados AA, Pietsch JMJ, Cepeda-Humerez SA et al (2018) Distributed and dynamic intracellular organization of extracellular information. Proc Natl Acad Sci 115:6088–6093. https://doi.org/10.1073/pnas.1716659115
25. Shakhi PK, Bijeesh MM, Hareesh J et al (2024) Size-dependent steady state saturation limit in biomolecular transport through nuclear membranes. PLoS One 19:e0297738. https://doi.org/10.1371/journal.pone.0297738
26. Baker Brachmann C, Davies A, Cost GJ et al (1998) Designer deletion strains derived from Saccharomyces cerevisiae S288C: a useful set of strains and plasmids for PCR-mediated gene disruption and other applications. Yeast 14:115–132. https://doi.org/10.1002/(SICI)1097-0061. (19980130)14:2%253C115::AID-YEA204%253E3.0.CO;2-2
27. Newby GA, Kiriakov S, Hallacli E et al (2017) A genetic tool to track protein aggregates and control prion inheritance. Cell 171:966–979.e18. https://doi.org/10.1016/j.cell.2017.09.041
28. Fu S-C, Fung HYJ, Cağatay T et al (2018) Correlation of CRM1-NES affinity with nuclear export activity. Mol Biol Cell 29:2037–2044. https://doi.org/10.1091/mbc.E18-02-0096

Chapter 8

Characterization of Synthetic Gene Circuits with Absolute Quantification in Continuous Culture

Scott B. Stacey, Ting An Lee, Olivia Gallup, Eszter Csibra, Antonis Papachristodoulou, Harrison Steel, and Kirill Sechkar

Abstract

Despite rapid improvements in our ability to engineer novel biological systems, robust biodesign of synthetic gene circuits—networks of synthetic genes regulating each other—has been limited by a lack of standardized methods for understanding and reproducibly characterizing their behavior in complex cellular contexts and over long timescales. The challenges underlying this include the complexity of biological interactions and the cellular contexts, changes in the cell culture environment over time, and the use of inconsistent measurement techniques in synthetic biology. Here, we describe a methodology for characterizing engineered biological systems while addressing these issues with the help of mathematical modeling, continuous cell culture, and absolute quantification of protein and cell numbers. As a case study, the characterization of a simple small RNA circuit in the Chi.Bio bioreactor platform is considered. We describe the biological system design choices, preparation of calibrants, running an experiment in Chi.Bio, the use of resulting data to obtain calibrated measurements in absolute units, and parameterization of a mathematical model of the engineered system. By coupling computational methods with precise control of cellular environments and robust experimental measurements, this interdisciplinary approach can produce more informative data and new insights into the design of engineered biological systems.

Key words Characterization, Robust characterization, Continuous culture, Bioreactor, Absolute quantification, Quantitative, Mathematical model, Synthetic biology, Systems biology

1 Introduction

Recent years have seen an increase in the development of engineered biological systems with a wide range of healthcare, manufacturing, and sustainability applications [1]. Their situation-appropriate responses to environmental conditions are powered by conferring cells with synthetic genes, which form "circuits" by regulating each other's activity to process external inputs. Any engineered system must have predictable performance, which

Xiao-Jun Tian (ed.), *Synthetic Gene Circuits: Methods and Protocols*, Methods in Molecular Biology, vol. 3041, https://doi.org/10.1007/978-1-0716-5304-3_8,

calls for robust and extensive characterization prior to deployment in practice [2]. However, meaningful observations of cell behaviors, as well as the incorporation of insights from them into gene circuit design, have been challenging due to the complexity of interactions in living systems, variability of the environment at the time of characterization, and inconsistent measurement techniques [3]. Nonetheless, these issues can be tackled by leveraging mathematical modeling [2, 4], continuous culturing [3], and absolute quantification techniques [5].

Synthetic biological parts interact both directly through gene regulation and indirectly by competing for shared resources and affecting the host cell, which often leads to complex nonlinear behaviors that are hard to characterize and forecast [2]. Nonetheless, their understanding can be facilitated by building mathematical models and fitting their parameters to experimental data [5]. This allows us to concretely estimate the synthetic parts' parameters while quantifying the uncertainty about their values and our assumptions about the processes involved in a system's functioning [4]. At worst, if fitting yields no reasonable parameters explaining circuit behaviors, this can be indicative of poor understanding of the engineered system or the mathematical model's failure to capture it in the necessary detail [6], for example, by neglecting factors like resource competition [7]. Once established and fitted, models can inform the design process; namely, previously characterized and modeled parts such as inducible promoters can be modeled within a proposed synthetic gene circuit to test expected behavior, reducing the number of time-consuming experiments on failed ad hoc designs [2]. Moreover, examining and simulating these models can enable further quantitative and qualitative investigations of circuits prior to costly and time-consuming experiments, such as inquiries into stochastic behavior [6]. The improved understanding and efficiency enabled by modeling become especially relevant for more complex circuits that exhibit more complicated behaviors such as oscillations, bistability, or patterning [7, 8].

Another major challenge in obtaining meaningful measurements of a biological system's performance is variability in experimental conditions, both between different tests and within the same experiment [3, 9]. The former can include discrepancies in the temperature or medium composition between the experiments. Throughout the same experiment, cells affect their environment over time (through changes in nutrient abundance, pH, cell density, etc.), which in turn can make them undergo significant physiological changes affecting circuit performance [10]; namely, as nutrients are depleted, bacteria switch from a fast-growing exponential phase to a non-dividing stationary phase [11], whereas gene expression models (discussed above) overwhelmingly rely on the assumption of exponential growth [2, 12].

These issues are addressed by continuous culture devices such as turbidostats, which dilute the culture to maintain a constant optical density (OD), which reflects the concentration of cells [13]. Hence, the cells stay in the mid-exponential phase—and circuit performance is kept consistent—over many generations. This enables (1) measurement of circuit steady states, (2) observation of transient dynamics of circuits, and (3) actuation of step changes in circuit inputs at desired timepoints. The obtained long-term measurements are thus extensive, suitable for model fitting, and easy to repeat and reproduce thanks to tight control over environmental conditions [3]. Moreover, any influence of the circuit on cell growth, which may significantly impact the biological systems' dynamics [10], can be decoupled from the environmental effects and thus reliably observed and quantified [14]. It can therefore be reasoned that continuous culture platforms produce gold-standard data for the characterization of synthetic gene circuit behavior in liquid cultures. A range of solutions for continuous culturing and characterization of biological systems have been developed in recent years [15]. One notable example is Chi.Bio, an open-source bioreactor [3]. Chi.Bio facilitates low-cost in situ characterization of engineered biological systems by supporting turbidostatic cell growth—but can also be used for batch experiments with no dilution or as a chemostat with a fixed dilution rate (allowing to investigate the dependence of circuit performance on the cell growth rate, which will be equal to this rate of dilution [16, 17]). The bioreactor also natively enables fluorescent protein excitation and measurement, as well as optogenetic control of gene expression [3].

Finally, interpreting and reproducing gene circuit characterization is complicated by the common use of relative units in synthetic biology experiments [5, 18, 19]. Most relevant quantities for circuit testing are the number of cells in the culture (commonly gauged via OD) and circuit outputs in the form of fluorescent protein concentrations (often reported in relative fluorescence units (RFUs) or arbitrary units (AUs) of fluorescence). While the concentration of a given fluorescent protein scales approximately linearly with fluorescence intensity, the slope of the line depends on the protein's identity, the measurement equipment, and its settings [18]. Meanwhile, measurements like OD that are based on physical properties of cells are not easily comparable between different organisms hosting circuits—which can be expected to be increasingly relevant as synthetic biology moves beyond model organisms as chassis of choice [20]. Reproducibility and consistency of measurements across laboratories, instruments, and hosts can be enabled by absolute quantification in real physical units—such as cell counts, numbers of molecules per cell, or molar concentrations [5, 20]. Besides enabling robust comparisons, absolute quantification is necessary to parameterize mathematical models that describe synthetic biological systems quantitatively rather than qualitatively, as fitting them often requires expressing input data in molecular units. Reporting data with absolute quantification is also important for the design of synthetic gene circuits, which requires comprehensive understanding of gene expression levels, stoichiometry of interacting biomolecules, and the involved species' effects on metabolism [5, 21].

This need for reproducibility can be partially addressed by normalizing fluorescent protein expression for genes comprised of the parts of interest to that of a standardized reference transcription unit, typically embodied by the commonly used standard promoter (J23101), yielding measurements in Relative Promoter Units (RPUs)—but no absolute quantities [22]. Alternatively, a color-matched fluorescent dye (such as fluorescein for green fluorescent proteins (GFPs)) can be used to make a calibrant solution, whose fluorescence at known concentrations can be compared to that of a fluorescent protein to provide measurements in units of molecules of equivalent dye [23]. Likewise, OD measurements of cultures can be standardized using calibrant solutions of microsphere silica particles of the same size as cells, allowing to convert OD into a number of particles/cells per unit volume [18]. However, using color-matched dyes to quantify proteins is limited by the availability of appropriate dyes and still does not allow comparisons between proteins of different colors [5, 21]. Hence, more advanced calibrations are obtained using samples of the fluorescent protein itself; namely, a protein is expressed with a His-tag enabling purification, then the purified samples' concentrations are measured (by gauging their absorbance of 280 nm wavelength light), and finally their fluorescence is quantified [5]. While this enables robust measurements in absolute units, protein purification is laborious and expensive. Moreover, calibrating measurements in large volumes—such as bioreactor chambers—can require very large amounts of reagents and protein-producing cultures, leading to prohibitive costs.

Nonetheless, further developments in experimental methods have improved the accessibility of absolute quantification techniques. A recently published method, FPCountR, removes the necessity of protein purification for performing calibrations [5]. While all fluorescent proteins have an extinction coefficient at 280 nm that enables calculating concentrations, each fluorescent protein also has a specific extinction coefficient corresponding to its maximum absorbance, termed ECmax [24]. ECmax assays allow robust calculation of the concentration of a given fluorescent protein in complex cell lysates while significantly reducing experimental labor [5]. Meanwhile, the cost of absolute quantification in bioreactors can be tackled with time-based sampling, followed by measurement of the samples' properties using microplate readers or flow cytometry.

In this chapter, we outline a protocol for robustly, reproducibly, and cost-effectively characterizing the dynamic performance of synthetic gene circuits. To address the challenges of biological complexity, environmental fluctuations, and inconsistent measurements, this method integrates mathematical modeling, continuous culturing, and sampling-based absolute quantification of cell counts and protein levels. The protocol includes the following steps:

1. Design and assembly of the desired circuit, including the choice of associated standardized transcription units.
2. Calibrant preparation and measurement for absolute quantification of cell counts and fluorescent protein abundances.
3. Characterization of circuits and transcription units in continuous culture.
4. Calibration of continuous culture fluorescence curves.
5. Model fitting and parameter estimation.

A graphical overview of this general protocol is shown in Fig. 1. As a case study for this general technique, we consider its application to characterize a simple small RNA (sRNA)-based circuit in *Escherichia coli* bacterial cells using the Chi.Bio platform for continuous culture.

2 Materials

1. **Hardware**
   1.1 Microplate reader for OD and fluorescence measurements.
   1.2 Microcentrifuge.
   1.3 Sonicator for cell lysis.
   1.4 Standard liquid bacterial culturing setup: incubator and 200 mL flasks for liquid bacterial cultures.
   1.5 Assembled Chi.Bio platform, i.e., control computer, pump board, and bioreactor, can be purchased from LABmaker or manufactured and assembled in-house as per https://chi.bio/hardware/.
   1.6 30 mL, i.e., 25 mm/95 mm diameter/height, glass vials for Chi.Bio (Fisher 11593532, *see* **Note 1**).
   1.7 3D-printed Chi.Bio vial lids: designs and printing instructions available at https://chi.bio/hardware/.
   1.8 Autoclavable rubber O-rings, 18mm/15mm outer/inner diameter, to place between the Chi.Bio vials and the lids (RS 96-5666, *see* **Notes 1** and **2**).

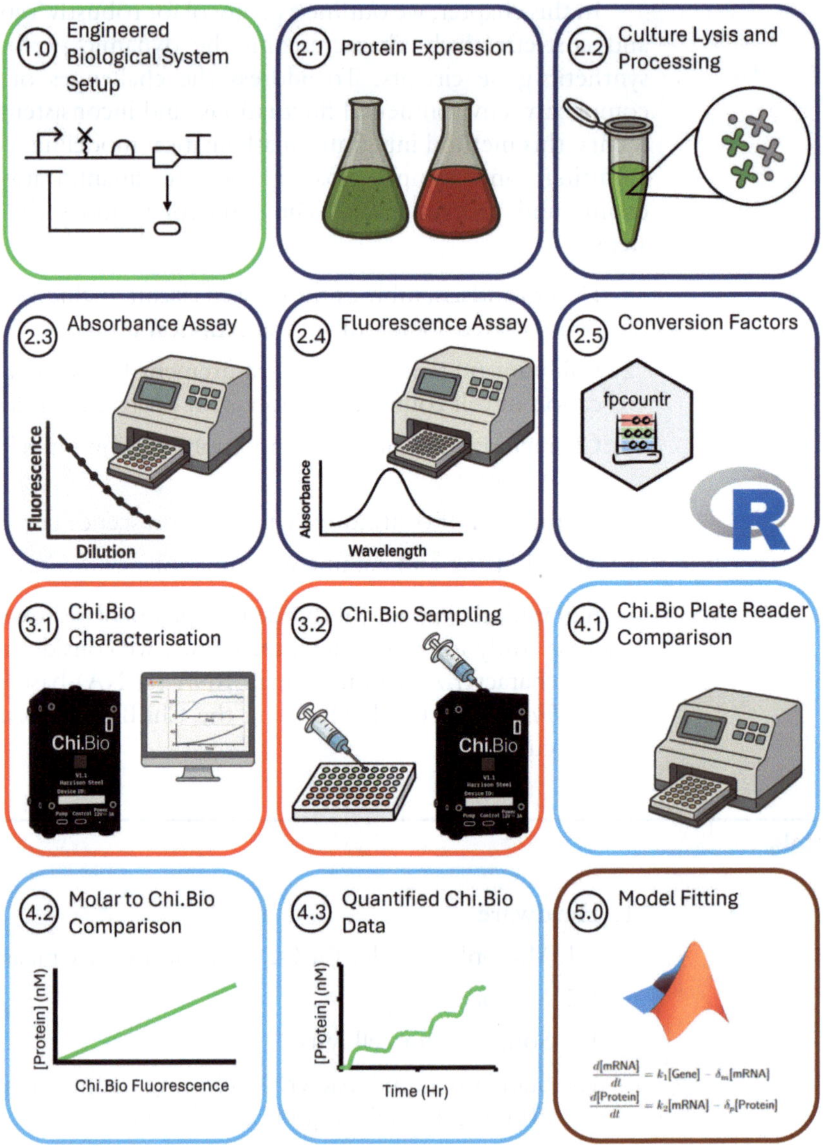

Fig. 1 Graphical overview of the entire method presented in this chapter. The method begins with setup of the synthetic gene circuit to be characterized. This is then followed by quantification of fluorescent proteins within lysed cells using the FPCountR [5] method. Next the synthetic gene circuit is characterized in the Chi.Bio. Samples from the Chi.Bio are taken and quantified in a plate reader using the quantification results from previous steps resulting in calibrated Chi.Bio curves. Finally, model fitting can be applied to the calibrated curves

1.9 Chi.Bio compatible magnetic stir bars, 6 mm to 15 mm long (Fisherbrand 11818862, *see* **Notes 1** and **3**).

1.10 GL-45 bottles with screw-on lids for culture media.

1.11 3D-printed GL-45 bottle lids for connecting media supply to the tubing: designs and printing instructions available at https://chi.bio/hardware/.

1.12 Large magnetic stir bars for GL-45 bottles.

1.13 Silicone tubing, 4.5mm/2.5mm outer/inner diameter, for pumping the media to and from the Chi.Bio vials (Altec 01-93-1416/20, *see* **Notes 1** and **4**).

2. **Wetware**

2.1 Vector with the engineered biological system of interest.

2.2 Empty vector with the same resistance as the plasmid of interest as a negative control for fluorescence.

2.3 For each fluorescent protein in the engineered system, a vector enabling its high constitutive expression (to prepare calibrant solutions).

2.4 *E. coli* strain of interest in which the system is to be tested (*see* **Note 5**).

3. **Consumables**

3.1 Sterile syringes and needles for Chi.Bio culture inoculation and sampling (*see* **Note 6**).

3.2 LB agar plates for streaking out colonies.

3.3 Microcentrifuge tubes.

3.4 96-well plates (*see* **Note 7**).

3.5 FPCountR Wash Buffer [27]: 50 mM Tris-HCl pH 7.5, 300 mM NaCl (*see* **Note 8**).

3.6 FPCountR Lysis Buffer [27]: the same base components as Wash Buffer (50 mM Tris-HCl pH 7.5 and 300 mM NaCl), as well as lysozyme added to a concentration of 100 μ g mL^{-1} lysozyme and 1 tablet per 10 mL of EDTA-free protease inhibitor tablets. This solution will need to be filter sterilized because protease inhibitor tablets dissolve poorly (*see* **Note 9**).

3.7 FPCountR Resuspension Buffer [27]: the same base components as Wash Buffer (50 mM Tris-HCl pH 7.5 and 300 mM NaCl), as well as 1 tablet per 10 mL of EDTA-free protease inhibitor tablets. This solution will need to be filter sterilized because protease inhibitor tablets dissolve poorly (*see* **Note 10**).

3.8 10 K Molecular Weight Cutoff protein concentration columns.

3.9 Monodisperse Silica (SiO_2), 2.0 g/cm^3, diameter 0.890 μm, Microspheres (Cospheric SiO2MS-2.0 0.890 um –1 g) (*see* **Note 11**).

3.10 DNAse I stock solution in distilled water, 1000 Kunitz Units mL^{-1} concentration.

3.11 $MgCl_2$ sterile stock solution, 1 M concentration, autoclaved.

3.12 $CaCl_2$ sterile stock solution, 50 mM concentration, autoclaved.

3.13 M9 media salts and supplements (*see* **Notes 12** and **13**):

3.13.1 M9 salt powder.

3.13.2 Glucose sterile stock solution, 20% w/v concentration, filter sterilized.

3.13.3 $CaCl_2$ sterile stock solution, 50 mM concentration, autoclaved (the same as the stock listed above for calibrant measurements).

3.13.4 $MgSO_4$ stock solution, 80 mM concentration, autoclaved.

3.13.5 Casamino acids sterile stock solution, 10% w/v concentration, autoclaved.

3.13.6 100× trace elements sterile stock solution, filter sterilized: 5 g L^{-1} EDTA $Na_2.2H_2O$, 0.5 g L^{-1} $FeCl_3$, 0.084 g L^{-1} $ZnCl_2$, 0.013 g L^{-1} $CuCl_2.2H_2O$, 0.005 g L^{-1} $CoCl_2$, 0.01 g L^{-1} H_3BO_3, and 0.0016 g L^{-1} $MnCl_2.4H_2O$.

3.13.7 Deionized water, autoclaved.

3.14 Lint-free wipes for cleaning.

3.15 70% ethanol, 30% deionized water solution for cleaning and sterilization.

3.16 Inducers and antibiotics particular to the engineered biological system being tested.

3 Methods

3.1 Engineered Biological System Setup

Before a gene circuit design is tested, it must be implemented in DNA by assembling together different transcription units. Transcription units encode RNA molecules. Some of these RNA molecules are known as messenger RNAs (mRNAs) which will be translated into proteins encoded by their coding sequences (CDSs), while others are known as noncoding RNAs (ncRNAs) and may be involved in regulating other RNAs within the circuit. These transcription units are often expressed under the control of regulated promoters which enable responses to external inputs and regulatory interactions between circuit components. When assembling a circuit to be characterized, one should use transcription units made up of parts (e.g., promoters or ribosome binding sites (RBSs)) of known properties [9], so the only parameters yet to be established capture the emergent interactions between circuit components—this is because fitting only a few parameter values at a time helps reduce the required computations and the associated uncertainty (*see* **Note 14**). Knowing the circuit components' properties a priori also allows us to better choose biological parts and avoid assembling nonfunctional implementations [21]. If the employed transcription units' parameters are not known, it is advisable to first characterize them individually without interference from the wider circuit context, making use of standardized transcription units. These standardized transcription units should be made up of previously characterized or known standard parts (promoter, ribozyme insulator, RBS, CDS, terminator, and plasmid backbone with known copy number, ideally the same plasmid backbone as other characterizations) with an individual component replaced with the unknown part to be characterized. SBOL Visual (i.e. Synthetic Biology Open Language Visual) [25] schematics of example standardized transcription units are shown in Fig. 2. To this end, after the assembly of these standardized transcription units but before they are combined into the overall circuit, the presented circuit characterization protocol should be applied to these simple constructs, allowing one to determine how the part of interest affects relevant parameters (e.g., how the promoter influences transcription rates).

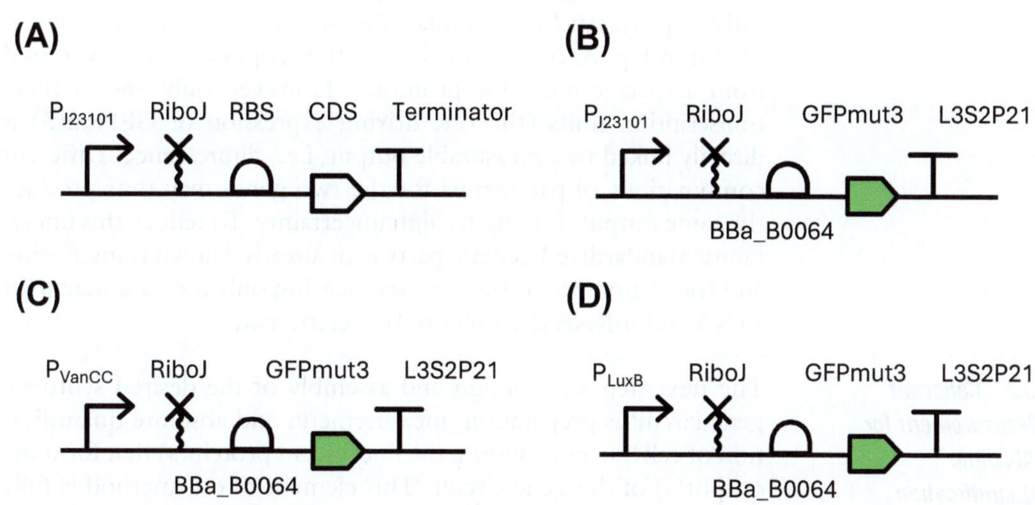

Fig. 2 Standardized transcription units enable like-for-like comparison of promoters, RBSs, CDSs, and terminators. (**a**) Previous work has made use of a reference promoter J23101 to characterize other promoters in the form of Relative Promoter Units (RPUs) [22]. By combining this promoter with other parts, standardized transcription units can be designed where parts are swapped to be characterized in the same contexts. (**b**) A specific SBOL Visual [25] representation of such a standardized transcription unit for the work presented here where the promoter can be swapped. (**c–d**) The promoter from (**b**) is swapped for the specific inducible promoters used in this work to enable independent characterization of the behavior of these parts

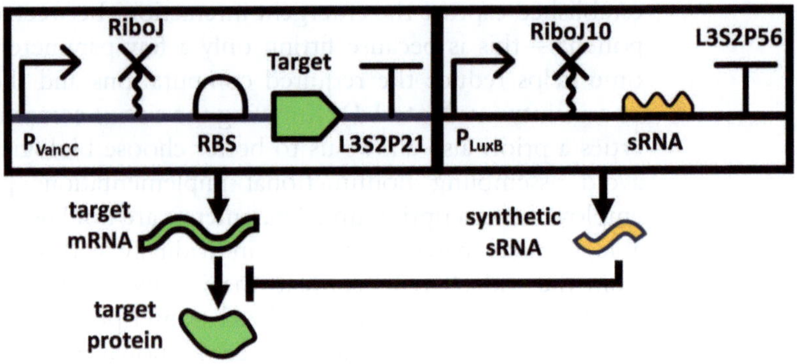

Fig. 3 SBOL Visual representation of the kinds of small RNA (sRNA) synthetic gene circuits used as a case study in this chapter. An inducible promoter, P_{VancC}, is activated by a small molecule inducer, vanillic acid, leading to transcription and translation of a target protein. A second inducible promoter, P_{LuxB}, is activated by a small molecule inducer, OC6, leading to transcription of a synthetic sRNA. The synthetic sRNA is designed to bind the target protein's mRNA, repressing translation, and reducing the amount of protein being translated. Small RNA-based circuits are one example of the types of circuits that can be characterized using the method presented in this chapter, but any other circuits with a fluorescent protein output can also be characterized

Our case study concerns a simple synthetic gene circuit with two transcription units controlled by chemically inducible promoters. One transcription unit is a gene encoding a fluorescent protein (GFPmut3), and the second transcription unit expresses a synthetic small RNA designed to downregulate GFPmut3 by binding its mRNA [26]. An SBOL Visual [25] schematic of such a circuit is shown in Fig. 3. In this circuit each transcription unit is expressed from a specific inducible promoter. However, only one of these transcription units (the one driving expression of GFPmut3) is directly linked to a measurable output, i.e., fluorescence. Different combinations of parameters for the two genes may thus produce the same output, leading to high uncertainty. To reduce this uncertainty, standardized genetic parts with already-known transcription and translation properties are used, leaving only the parameters for sRNA and mRNA dynamics to be determined.

3.2 Calibrant Measurement for Absolute Quantification

The next step after design and assembly of the desired synthetic gene circuit is preparation, measurement, and absolute quantification of cell lysates including the fluorescent protein(s) that form the output(s) of the gene circuit. This element of our method is fully derived from the work published by Csibra and Stan [5]. More detail on this protocol can be found in the associated online protocol [27], GitHub repository [28], and paper [5].

1. **Protein Expression**
   1.1 The day before protein expression, transform a protein expression strain of *E. coli* with a vector expressing the desired fluorescent protein. Alternatively, streak colonies from a previously prepared glycerol stock of such a strain expressing the fluorescent protein (*see* **Note 5**).

1.2 Next, set up a 50 mL overnight overexpression culture in a 200 mL sterile flask of the desired bacteria by picking a single colony from the agar plate prepared the previous day.

1.3 Use 50 mL of autoinduction media or 50 mL LB supplemented with desired inducer and antibiotic at desired concentration (*see* **Note 15**).

1.4 Grow this culture overnight at 30° C and 250 RPM.

1.5 Repeat all steps for each fluorescent protein to be characterized.

2. **Cell Culture Processing**

2.1 For each fluorescent protein to be quantified, prechill two 50 mL Falcon tubes on ice for 15 minutes.

2.2 Cool each overnight culture by transferring the entirety of the culture into one of the prechilled 50 mL Falcon tubes for 20 minutes.

2.3 Prepare a 1 mL 1 in 10 dilution of each cooled culture—e.g., by taking 100 μL of culture and adding 900 μL of sterile culture media (*see* **Note 16**).

2.4 Transfer each 1 mL culture dilution to a cuvette and measure their OD in a spectrophotometer.

2.5 Calculate the OD of the original culture. For a 1 in 10 dilution, multiply the measured OD by 10 to get the OD of 1 mL of the original culture. Then multiply this by 50 to find the total OD of the 50 mL culture.

2.6 For highly expressed proteins, a volume equivalent to 20 OD of cells is required for lysis. Therefore, use the calculated OD of the total culture to calculate the volume of culture required to give 20 OD (*see* **Note 17**).

2.7 Take the calculated required volume of culture and aliquot this into the unused prechilled 50 mL Falcon tube.

2.8 Pellet the culture aliquots for 10 minutes, at ~ 3000×g and 4° C.

2.9 Discard the supernatant and gently (*see* **Note 18**) resuspend the cells in 5 mL Wash Buffer. After resuspension add another 5 mL Wash Buffer.

2.10 Pellet the resulting solutions for 10 minutes, at ~ 3000×g and 4° C.

2.11 Discard the supernatant and gently resuspend in 2 mL (*see* **Note 19**) of Lysis Buffer.

3. **Cell Culture Lysis**

3.1 Next, lyse the suspension produced in the previous step by sonication (*see* **Note 20**).

3.2 Place the Falcon tube with your suspension in a beaker or bucket of ice.

3.3 Lower the sonicator tip into the suspension.

3.4 Sonicate for 2 minutes with the following settings: 30% amplitude pulsing for cycles of 10 seconds on and 10 seconds off (*see* **Note 21**).

3.5 Treat the lysate by adding: DNase I to a final concentration of 50 Kunitz Units mL^{-1}, $CaCl_2$ to a final concentration of 5 mM, and $MgCl_2$ to a final concentration of 50 mM (*see* **Note 22**).

3.6 Mix the resulting solution thoroughly by pipetting up and down. Do not vortex (*see* **Note 23**).

3.7 Incubate the mixed solution for 30 minutes at 4 ° C.

4. **Lysate Processing**

4.1 Precool 1.5 mL microcentrifuge tubes on ice for 15 minutes during lysis incubation.

4.2 Transfer lysate(s) from Falcon tube(s) to the microcentrifuge tubes.

4.3 Centrifuge the microcentrifuge tubes containing the lysate (s) in a precooled centrifuge (to 4° C) for 30 minutes at 16,000×g.

4.4 Transfer the supernatant to fresh microcentrifuge tubes.

4.5 Concentrate the supernatant by spinning down at 14,000×g for 10 mins at 15° C in a 10 K protein concentration column (*see* **Note 24**).

4.6 Elute the sample from the 10 K column into a fresh microcentrifuge tube to obtain approximately 100 μL of concentrated sample (*see* **Note 25**).

5. **Serial Dilution Plate Preparation**

5.1 Next, prepare a serial dilution of the concentrated lysate in microcentrifuge tubes.

5.2 For each protein, take the 100 μL concentrated sample, and make it up to a 1 mL solution in a microcentrifuge tube (labeled with the name of the fluorescent protein and the number 1) by adding 900 μL of Resuspension Buffer (*see* **Note 10**) and ensuring the sample is well mixed by pipetting up and down.

5.3 For each protein, label 10 more microcentrifuge tubes with the name of the fluorescent protein and numbered from 2 to 11.

5.4 Add 500 μL of Resuspension Buffer to tubes 2 to 11 for each protein.

5.5 For each protein, carry out a twofold serial dilution by transferring 500 μ L from tube 1 to tube 2, and then from tube 2 to tube 3, and so on until tube 11. At each dilution step pipette up and down eight times to ensure thorough mixing.

5.6 Prepare a calibration 96-well plate (*see* **Note** 7) of these twofold serial dilutions. For each fluorescent protein two rows of the 96-well plate should be used with each column containing a specific dilution. So to A1 and B1 of the 96-well plate add 225 μL of tube 1 for your first fluorescent protein. To A2 and B2 add 225 μL of tube 2 for your first fluorescent protein, and so on up to tube 11 for the first fluorescent protein. If you are quantifying more than one fluorescent protein, you should plate your second fluorescent protein in the same process using rows C and D and so on for each fluorescent protein.

5.7 To wells in column 12 add 225 μL of Resuspension Buffer to act as a blank (*see* **Note 26**).

6. **Fluorescent Protein Concentration Quantification (ECmax Assay)**

6.1 Pre-warm the microplate reader (*see* **Note 27**) to the temperature at which experiments in the Chi.Bio are carried out, typically 37° C.

6.2 Insert the serial dilution plate into the microplate reader, being sure to remove the lid. Allow the plate to reach the temperature the microplate reader was pre-warmed to.

6.3 Take an absorbance scan of all desired wells between 200 and 1000 nm, ideally with as small a step size as possible (*see* **Note 28**).

7. **Fluorescent Protein Fluorescence Quantification**

7.1 Measure fluorescence intensity of each well of the same plate in the same microplate reader pre-warmed to the same temperature used in your Chi.Bio experiment (typically 37° C).

7.2 Ensure that the emission and excitation wavelengths are chosen to be the same as in your ultimate plate reader experiment (*see* **Note 29**).

7.3 For each fluorescence intensity measurement with a particular emission and excitation wavelength, take measurements at a range of gain values (*see* **Notes 30** and **31**).

8. **OD Microsphere Calibration**

8.1 Microsphere calibration of OD (*see* **Note 32**) is based on a previously described protocol called FlopR [23] and is used in the original FPCountR method [5]. A correction to the original FlopR protocol specifically relating to the OD calibration has been issued [29], and the protocol here is adapted from the corrected protocol.

8.2 Dissolve 0.1 g of Monodisperse Silica (SiO_2), 2.0 g/cm^3, diameter 0.890 μm, Microspheres in 1.3 mL of deionized H_2O to create stock solution A in a microcentrifuge tube (*see* **Notes 11** and **33**).

8.3 From stock solution A, make stock solution B in a fresh microcentrifuge tube by taking 300 μL of solution A and adding to 200 μL of deionized H_2O.

8.4 Prepare 11 microcentrifuge tubes labeled 1.1 to 11.1 and 11 more labeled 1.2 to 11.2.

8.5 In tubes numbered 1.1 and 1.2, make a 1 in 10 dilution of stock solution B by taking 150 μL of stock solution B and adding to 1350 μL of deionized H_2O (*see* **Note 34**).

8.6 Add 600 μL deionized H_2O to other tubes.

8.7 Perform serial dilution by taking 900 μL from tubes 1.1 and 1.2 and adding to tubes 2.1 and 2.2, respectively. Mix thoroughly by pipetting up and down.

8.8 Continue the serial dilution up to tubes 11.1 and 11.2. This will leave you with solutions of 600 μ L in tubes 1 through 10 and 1500 μL in tube 11.

8.9 Pipette 225 μL of tube 1.1's solution into wells E1 and F1 and tube 1.2 into G1 and H1.

8.10 Pipette 225 μL of tube 2.1's solution into wells E2 and F2 and tube 2.2 into G2 and H2.

8.11 Repeat for tubes 3.1 to 11.1 and 3.2 to 11.2 up to wells E11 and F11 and G11 and H11, respectively.

8.12 Into wells E11, F12, G12, and H12 add 225 μL of deionized H_2O.

8.13 Measure OD at 600 and 700 nm in the same plate reader used in other steps pre-warmed to 37° C (*see* **Note 35**).

9. **Conversion Factor Determination**

9.1 Next, analyze the data generated in parts 6 and 7 of this protocol with the FPCountR R package [28] to generate conversion factors relating fluorescent protein molecule counts (calculated through the ECmax assay) to fluorescent protein fluorescence intensity (measured in the fluorescence assay) in the calibrated plate reader (*see* **Note 36**).

9.2 Calculate the number of microsphere particles in each well of the OD calibration plate using the known mass of particles in each well (for details of the calculation *see* **Note 37**).

9.3 Then, analyze the data generated in part 8 of this protocol with the FlopR R package [23, 29], generating conversion factors relating number of microsphere particles (and by proxy number of *E. coli* cells) to OD measurements at 600 and/or 700 nm.

By following this protocol, normalized absorbance spectra for each fluorescent protein of interest are generated by FPCountR [5, 28]. Figure 4 A shows example absorbance spectra for GFPmut3, while Fig. 4b shows similar examples for mScarlet-I across dilutions. The FPCountR R package [28] uses these absorbance spectra to calculate the concentration at each dilution (Fig. 4c, d) and ultimately uses this and the fluorescence intensity data at each gain to calculate conversion factors from fluorescence intensity to the number of molecules for each value of gain—*see* Fig. 4e, f (*see* also **Note 38**). Examples of OD conversion factors for OD measured at both 600 and 700 nm are given in Table 1.

3.3 Chi.Bio Characterization

After following the described calibration protocol presented in Section 3.2 based on FPCountR [27] and FlopR [29], we now discuss how to characterize a synthetic gene circuit in a continuous culture platform. In this section we provide a protocol outline for how the Chi.Bio bioreactor platform can be used for the characterization of synthetic gene circuits:

1. **Pre-culture and Preparation**

   1.1 Two days before the Chi.Bio experiment, transform desired strains with desired plasmids for your synthetic gene circuit(s) and/or part(s) to be characterized. Additionally, transform or streak a strain which can act as a negative control for fluorescence. Ideally this should be the same strain which your circuit will be transformed into, but transformed with an equivalent empty vector.

   1.2 One day before Chi.Bio experiment, start an overnight culture of your transformed bacteria by picking a single colony and growing in LB at 37° C and 250 RPM.

   1.3 Autoclave Chi.Bio-compatible glass vials capped with Chi.Bio vial lids (*see* **Note 2**) with O-rings between the glass vial and lid.

   1.4 Prepare M9 salt solution by weighing out salts for 1× concentration in the final volume and dissolving in a smaller amount of deionized water (e.g., around 0.9× the final volume). Autoclave the salt solution, as well as stock solutions of the other M9 media components except the glucose stock for which we recommend filter sterilization (*see* Materials), in standard GL-45 bottles with standard GL-45

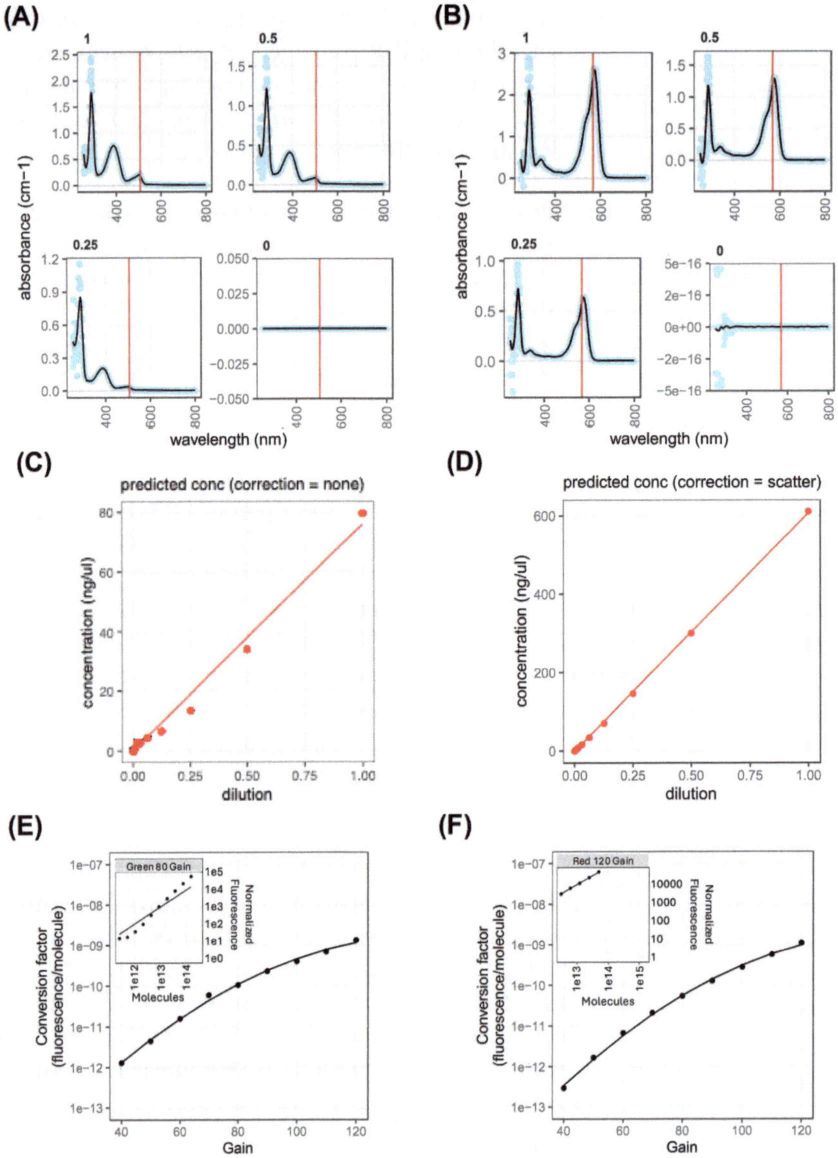

Fig. 4 Example data from FPCountR [5] analysis of lysed cell samples. (**a–b**) absorbance spectra for dilution series of lysed samples of cells expressing GFPmut3 and mScarlet-I, respectively. The red vertical line represents the expected excitation peak wavelength for each fluorescent protein. (**c–d**) The fitted relationship between concentration calculated with the ECmax assay and the dilution series for GFPmut3 and mScarlet-I, respectively. (**e–f**) Conversion factors from fluorescence to number of fluorescent protein molecules for each GFPmut3 and mScarlet-I at each value of gain at which fluorescence was measured. Inset to each graph is one example of conversion for a specific value of gain

Table 1
Example data for OD conversion factors from microsphere assay

OD measurement	Conversion factor
OD 600 nm	3.93×10^{-9}
OD 700 nm	3.0×10^{-9}

lids on (*see* **Notes 13** and **39**). Each bottle of media should also contain a magnetic stir bar to enable stirring after autoclaving (*see* **Note 40**). At the same time, autoclave Chi.Bio media bottle lids with attached tubing (*see* **Note 41**). Ideally at this point, you should autoclave enough media for your entire Chi.Bio experiment (*see* **Note 42**).

1.5 Sterilize Chi.Bio compatible magnetic stir bars in 70% ethanol, and leave in a sterile environment (e.g., a tissue culture hood) until required (*see* **Note 3**).

1.6 Carry out necessary setup of Chi.Bio hardware and software (*see* **Note 43**) or maintenance of existing Chi.Bio hardware (*see* **Note 44**). The most relevant instructions are provided in Notes, while more detailed instructions for software setup are provided at https://chi.bio/software/, for hardware setup/materials at https://chi.bio/hardware/, and for user operation at https://chi.bio/operation/. Finally, you can find a number of troubleshooting tips and other helpful discussions at https://chi.bio/forums/.

1.7 The day of the Chi.Bio experiment, around 2–3 hours before the Chi.Bio experiment will begin, take 50 μL of the overnight culture(s), and dilute in 5 mL of the initial media that will be used in the Chi.Bio experiment (*see* **Note 45**). Grow this new culture(s) for 2–3 hours at 37° C and 250 RPM, which should yield a culture(s) in exponential phase.

2. **Chi.Bio Setup**

2.1 Prior to setting up the Chi.Bio, prepare autoclaved media by adding the supplements to the M9 salt solution—the final concentrations should be 1× M9, 2 mM $MgSO_4$, 100 μM $CaCl_2$, 0.4% glucose, 1× trace elements, and 0.2% casamino acids. At this point, also add the necessary inducers and antibiotics. Top up to the final volume with autoclaved H_2O.

2.2 In a sterile environment, add sterile magnetic stir bars as discussed in **Note** 3 to the sterile Chi.Bio vials.

2.3 Sterilize Chi.Bio pumps and tubing by placing the ends of the pump tubes in a reservoir of 70% ethanol. Pump this ethanol through the tubes for around 10 seconds, empty the tubes of ethanol, connect the tubing up to the media bottle, wash through with the initial experiment media, disconnect to pump the tubes empty again, and reconnect to the media bottle once the tubing is empty (*see* **Note 46**).

2.4 Clean the outside of the glass vials using lint-free wipes sprayed with 70% ethanol, being sure to dry the outside of the glass with a clean and dry Kimtech wipe.

2.5 Connect the media pump and waste pump tubing to the lids of the associated Chi.Bio vial. Fill the Chi.Bio vial by pumping in the connected media, switching off the pump at a media level just above the waste pump outlet in each vial. Then remove excess media above the waste outlet by switching on the waste pump until the media line is below the waste outlet (*see* **Note 47**).

2.6 Place the vials into the Chi.Bio reactors, switch on stirring (*see* **Note 48**), and temperature control at the temperature of the experiment (typically 37° C, but should be the same temperature as absolute quantification carried out in Section 3.2) (*see* **Note 49**).

2.7 Allow the media to stir and equilibrate to the desired temperature for at least 15 minutes (*see* **Note 50**).

2.8 Blank the media OD using the Chi.Bio software GUI by measuring raw OD and setting this value to an actual value of zero. Remeasure the OD a few times to confirm that OD is at or close to zero (*see* **Note 50**).

2.9 Switch on OD control at the desired set point (typically 0.5), switch on the dither function to enable growth rate measurements, switch on fluorescent measurements at desired wavelengths, and start the experiment (*see* **Note 49**).

2.10 After 15 minutes confirm that OD is not drifting away from the blanked value (i.e., ~ 0) (*see* **Note 50**).

2.11 Inoculate each vial with the desired strain of bacteria by adding 50 μL of the pre-culture via sterile syringe or pipette to the hole in the lid of each vial, making sure to use a fresh syringe or pipette for each strain to be inoculated.

2.12 The experiment can now be left to run, where the Chi.Bio will be maintaining constant temperature, measuring fluorescence and OD, and estimating growth rate every minute (*see* **Note 51**). Initially the culture should grow exponentially as a normal bacterial culture would, until it reaches the OD set point, at which point the Chi.Bio's turbidostat function will maintain constant OD (*see* **Note 52**).

2.13 Leave the Chi.Bio to run until some user intervention is required—such as inducer step change, sampling, pump maintenance, or media replacement.

3. **Inducer Step Changes**

3.1 In order to achieve a desired step change of a particular or multiple inducers, you will need to prepare a new media bottle with the inducer(s) added to the desired concentration(s) (*see* **Note 53**).

3.2 To achieve an inducer step change in the Chi.Bio, in addition to swapping to a new media bottle at the desired concentration(s), the inducer will need to be added to the cell cultures in each Chi.Bio vial (*see* **Note 54**). Therefore, for each Chi.Bio vial that will undergo a step change in inducer, you should prepare at least 1.2 mL of solution of the inducer at the correct concentration in the initial Chi.Bio media (for an example of concentration calculation *see* **Note 55**). 1.2 mL is suggested to provide excess over the 1 mL required.

3.3 Deactivate OD control to disable unwanted pumping. Disconnect the media input pump from the old media bottle and the Chi.Bio vial, and switch on pumping to empty the pump tube of old media. Reconnect the pump to the new media bottle, and pump through with the new media bottle (which is at the desired inducer concentration). Reconnect the pump to the Chi.Bio vial, but do not switch on OD control yet, to avoid any pumping of media. Add 1 mL of the prepared inducer step change solution to the Chi.Bio vial with a sterile syringe through the accessible lid hole (*see* **Note 56**). Switch OD control back on, and record the timepoint in the experiment at which the inducer step change was achieved (*see* **Note 57**).

3.4 Repeat the previous step for each Chi.Bio vial that is to undergo step change in inducer. Of note, negative control cultures (those without fluorescent protein expression) should be subjected to the same inducer step changes that their non-control counterparts are subjected to.

4. **Chi.Bio Sampling**

4.1 In order to later convert Chi.Bio readings into absolute units, for which the calibrants in Section 3.2 had been prepared, you will need to take samples from the Chi.Bio vials at various timepoints across the experiment for comparative measurement in the calibrated microplate reader.

4.2 Prior to taking a sample from a Chi.Bio reactor, label a sterile microcentrifuge tube with the name of the reactor/culture that will be sampled. Repeat this for each reactor for which a sample will be taken.

4.3 At timepoints where sampling is to be carried out, take a sample by inserting a sterile needle (*see* **Note 6**) and syringe into the top of the Chi.Bio vial through the accessible hole in the Chi.Bio vial lid. Draw out approximately 1 mL of culture, and transfer the sample to the appropriately labeled sterile microcentrifuge tube. Repeat this for each reactor from which a sample is being taken, including the negative control culture of cells without fluorescent protein expression. Selection of timepoints at which samples should be taken is discussed in **Note 58**. The timepoint at which samples are taken should be recorded, for example, by following the same process as described in **Note** 57.

4.4 Pipette 225 μL (*see* **Note 59**) from each microcentrifuge tube containing a sample into a well of a sterile 96-well plate of the same type as used in Section 3.2. Repeat this three times for each microcentrifuge tube to give three technical replicates for each vial from which a sample was taken.

4.5 Fill three more wells with 225 μL of media-only blanks. This media should be equivalent to the media in which cells are growing at this stage of the experiment (i.e., at the same inducer concentrations as the media connected to the Chi.Bio vials).

4.6 Transfer the loaded 96-well plate (again, unsealed and without the lid) into the same microplate reader, pre-warmed to the same temperature as the absolute quantification process carried out in Section 3.2. Leave the plate to equilibrate to the set temperature. Run a plate reader protocol measuring OD (at 600 and/or 700 nm) and fluorescence intensity with the same settings (excitation, emission, and gain) as used for absolute quantification process carried out in Section 3.2. Examples of some typical settings for the Tecan Spark that might be used for this are shown in Fig. 5.

4.7 Repeat this process for each timepoint at which sampling is desired.

5. **Chi.Bio Data Processing**

5.1 The Chi.Bio software outputs dynamics data with measurements every 60 seconds for OD and fluorescence. For subsequent parts of the protocol, you first need to process the obtained data.

- Plate Reader
  - Device: Spark
  - Serial number: 2008000886
- Plate
  - Name: GRE96ft
  - Lid: No lid
  - Plate are: A1-D12
- Protocol
  - Mode: Temperature
  - Target Temperature: 37°C
  - Wait minimum: 36.5°C
  - Wait maximum: 37.5°C

  - Mode: Shaking
  - Duration: 30 s
  - Amplitude: 3 mm
  - Frequency: 90 rpm
  - Mode: Absorbance
  - Name: OD600
  - Measurement Wavelength: 600 nm
  - Number of flashes: 10
  - Settle Time: 50 ms

  - Fluorescence Intensity
  - Name: green
  - Mode: Fluorescence Top Reading
  - Excitation Wavelength: 470 nm
  - Excitation Bandwidth: 20 nm
  - Emission Wavelength: 515 nm
  - Emission Bandwidth: 20 nm
  - Gain: 80
  - Number of flashes: 30
  - Z-position: 30000 µm

Fig. 5 Example plate reader settings. We provide examples of plate reader settings used to measure Chi.Bio samples for fluorescence and optical density (OD) for the Tecan Spark microplate reader. Some settings, e.g., gain and excitation/emission wavelengths will need to be optimized for your specific synthetic gene circuit and microplate reader

5.2 Chi.Bio fluorescence includes background signal that is some combination of cell autofluorescence, media autofluorescence, and sensor noise. Therefore, you need to process the data by removing background fluorescence.

Calculate the mean of the 30 fluorescence readings (*see* **Note 60**) acquired before induction of any fluorescent protein. This gives a value of mean background fluorescence intensity (*see* **Note 61**). Subtract this value from all fluorescence intensity values in this channel to calculate corrected fluorescence intensity.

5.3 Calculate a symmetrical window 21-point moving average (*see* **Note 62**) of the background corrected fluorescence intensity and other relevant measurements, such as growth rate estimation, to smooth shot-to-shot variability (*see* **Note 63**).

5.4 Set any data in the smoothed fluorescence intensity signal that is below zero to equal zero giving the final fluorescence intensity time series in arbitrary units (*see* **Note 64**).

5.5 Repeat this process for each fluorescent channel in which a protein is being measured.

Fluorescence intensity curves and associated inducer step change curves for standardized transcription units for promoters used in our gene circuit P_{LuxB} and P_{VanCC}, and for the sRNA-based circuit shown in Fig. 3, are shown in Fig. 6a–c, respectively. These give examples of the type of data that can be collected by carrying out Chi.Bio-based characterization of synthetic gene circuits and parts, as well as examples of the conditions/changes in fluorescence that might be desirable to achieve through inducer step changes for characterization or model fitting.

3.4 Chi.Bio Calibration

As discussed in the Introduction, scale-up of microplate reader-based standardized metrology methods to in situ measurements in bioreactors is hindered by high costs (*see* **Note 65**) and technical limitations (*see* **Note 66**). The solution to these challenges, implemented in this protocol, is to make use of sampling-based techniques. In Section 3.3, we described how the Chi.Bio can be used to characterize gene circuits and how samples can be taken and measured in a microplate reader.

In this section, we detail a very simple protocol for associating the microplate reader absolute quantification of fluorescent protein calibrants (Section 3.2) to the microplate reader measurement of Chi.Bio samples (Section 3.3) and therefore also to Chi.Bio fluorescence signal timecourses. An outline of our method is provided below:

1. Collate the microplate reader data for the Chi.Bio samples (Section 3.3) into a single CSV file by assigning timepoints to each reading—i.e., the first reading will be recorded as happening at time 0, the second reading will happen at time 1, etc. (*see* **Note 67**)

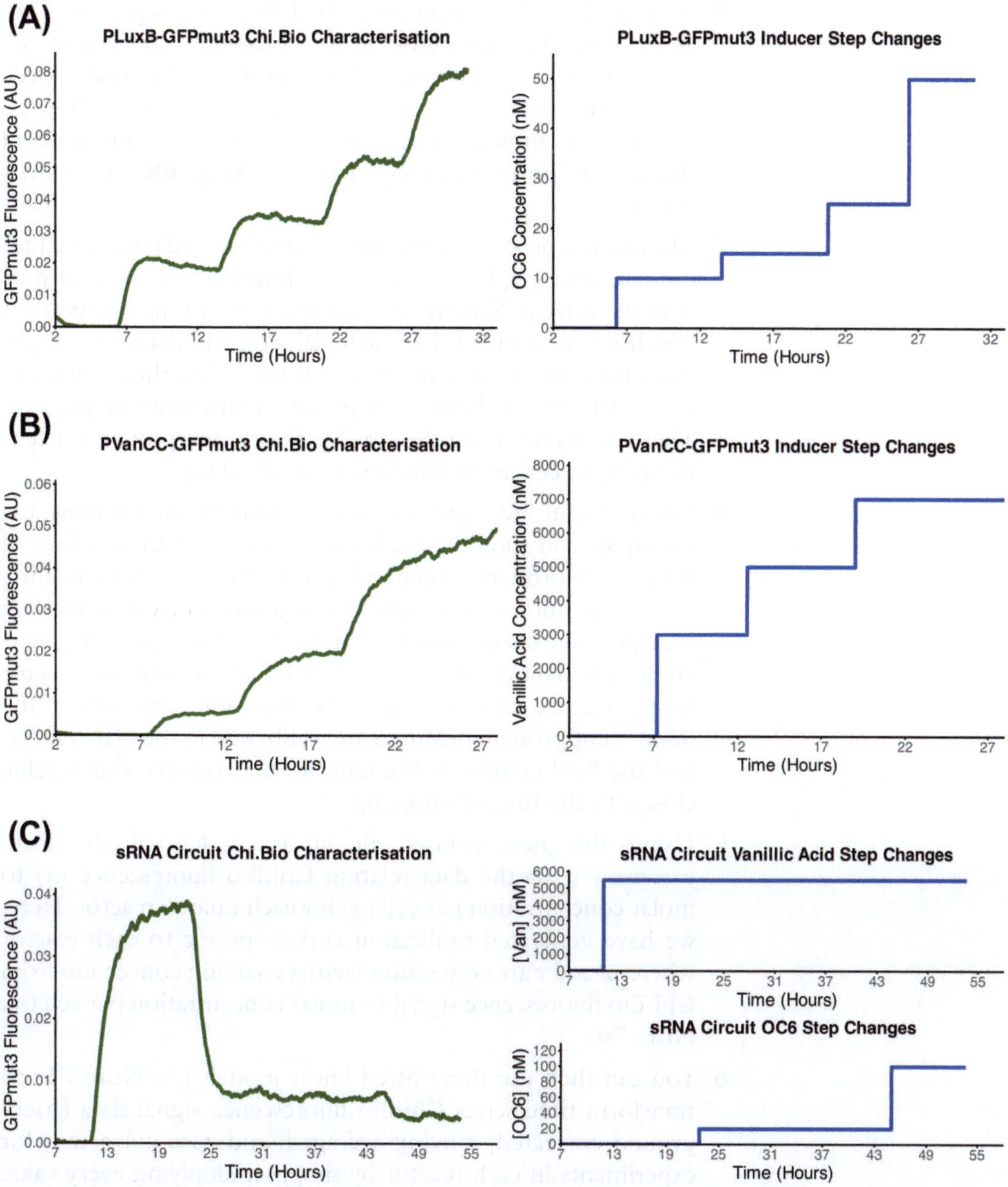

Fig. 6 Chi.Bio characterization of synthetic gene circuits and associated inducer step changes. (**a**) The P_{LuxB} promoter was characterized in the Chi.Bio through a series of step changes in OC6 concentration. (**b**) The P_{VanCC} promoter was characterized in the Chi.Bio through a series of step changes in vanillic acid concentration. (**c**) The small RNA (sRNA) circuit was characterized by inducing GFPmut3 with a step change in vanillic acid followed by step changes in OC6 to induce the sRNA and repress GFPmut3 expression

2. Analyze the collated data using the FPCountR R package [28] along with the conversion factors generated in Section 3.2 using the FPCountR method to get OD calibrated to cell count, fluorescence calibrated to number of molecules, fluorescent protein molecules per cell, and molar concentrations of fluorescent protein per cell [5] (*see* **Note 68** for analysis details).
3. The analysis produces a number of graphs and datasets, including data and graphs for the concentration of fluorescent proteins per cell (*see* **Note 69**) for each vial against the timepoint at which it was sampled. For each vial, you will get three graphs showing fluorescent protein over time. Using these data, calculate the mean fluorescent protein concentration per cell. These values represent the Chi.Bio fluorescent protein concentrations per cell at the timepoints of sampling.
4. Given careful recording of timepoints at which sampling has occurred, you should be able to relate each of these values of fluorescent protein concentration per cell to a specific reading of Chi.Bio fluorescence signal (background corrected, moving averaged, and zero corrected). Therefore, you can now generate a new dataset where one column is the Chi.Bio reactor ID/name, the next column is the mean concentration of the fluorescent protein measured and calibrated in the plate reader, and the final column is the Chi.Bio fluorescence signal value closest to the time of sampling.
5. Using this new dataset, fit linear models of the form $y = mx + c$, to the data relating Chi.Bio fluorescence (x) to molar concentration per cell (y) for each unique reactor. Here, we have generated calibration curves specific to each reactor where m and c are conversion factors enabling conversion from Chi.Bio fluorescence signal to molar concentration per cell (*see* **Note 70**).
6. You can then use these fitted linear models (*see* **Note 71**) to transform time series Chi.Bio fluorescence signal data (background corrected, moving averaged, and zero corrected) for experiments in each reactor by simply multiplying every value in the time series by the fitted value of m and then adding the fitting value of c. This produces the final calibrated Chi.Bio data.
7. Repeat these steps for each fluorescent protein within your circuit(s) that need to be calibrated.

Figure 7 shows several examples of the kind of data that is produced by the ultimate FPCountR R analysis. Figure 7a shows the Chi.Bio characterization curve with sampling events labeled and the associated plots of concentration of fluorescent protein at

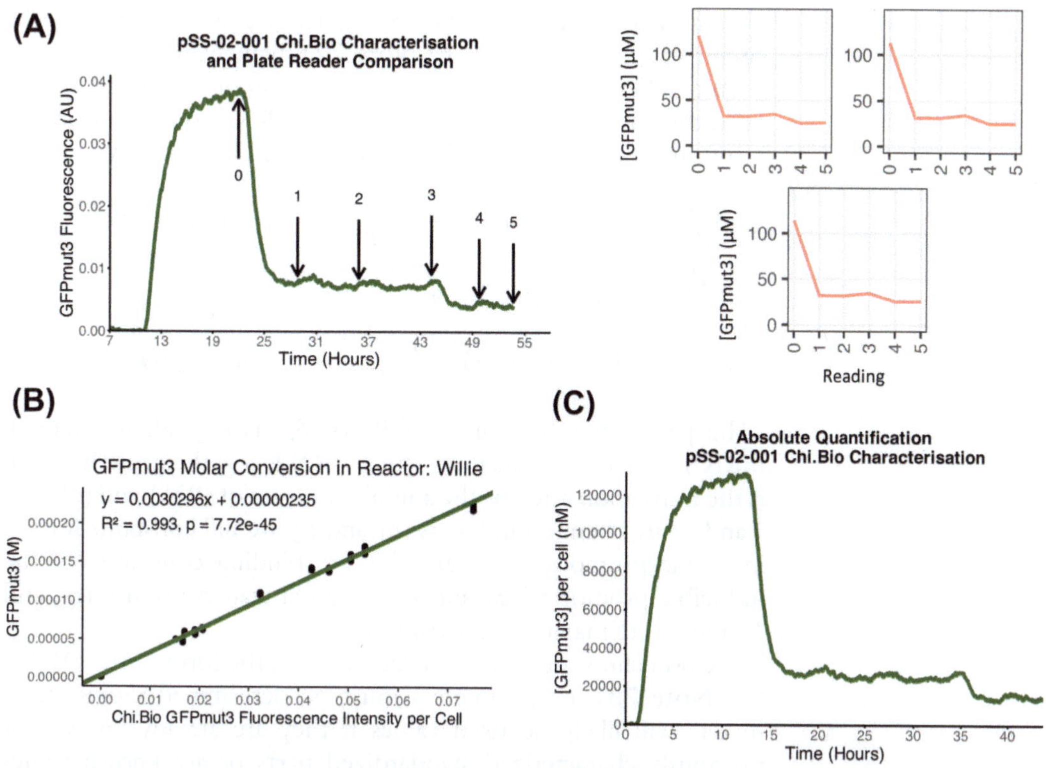

Fig. 7 Calibration data for the small RNA (sRNA) circuit. (**a**) At six timepoints across the Chi.Bio characterization samples are taken, indicated by labeled arrows, and measured in the calibrated plate reader. The FPCountR conversion factors produce three graphs for the three sample replicates giving concentration of GFPmut3 at each sampling timepoint. (**b**) By relating the plate reader measured and calibrated concentration values to the Chi.Bio fluorescence value at the time of sampling, a linear model can be fitted for conversion between Chi.Bio fluorescence and molar concentration of fluorescent protein. An example curve for GFPmut3 in one specific Chi.Bio reactor is shown here. (**c**) By using the conversion curve from (**b**), the Chi.Bio fluorescence curve in (**a**) can be calibrated to absolute concentration units

each sampling event measured in the calibrated plate reader. Figure 7b shows an example of a fitted linear model for the relationship between Chi.Bio GFPmut3 intensity and GFPmut3 concentration per cell for one Chi.Bio specific reactor. Finally, Fig. 7c shows an example of a resulting GFPmut3 concentration time-course for the fully calibrated experimental data for our example sRNA circuit (Fig. 3).

3.5 Model Fitting and Parameter Estimation

To estimate the parameters of a biological system, you should first define a mathematical model expected to describe its behavior. Often in synthetic biology, and particularly for describing synthetic gene circuits, mathematical models take the form of systems of nonlinear ordinary differential equations (ODEs) capturing the dynamics of all relevant protein and RNA concentrations to the circuit in question [4, 6] (*see* **Note 72**); namely, for our example circuit in Fig. 3, we define the concentration of GFPmut3 mRNA as m, the GFPmut3 protein level as p, and sRNA concentration as s, capturing their dynamics with the system of ODEs below:

$$\frac{dm}{dt} = k_0 + (k_1 - k_0)\frac{U^{n_1}}{U^{n_1} + K_1^{n_1}} - (\delta_m + \mu)m - \gamma sm$$

$$\frac{dp}{dt} = \theta m - \mu p$$

$$\frac{ds}{dt} = k_2 + (k_3 - k_2)\frac{I^{n_2}}{I^{n_2} + K_2^{n_2}} - (\delta_s + \mu)s - \gamma sm$$

The parameters here are as follows: δ_m is degradation rate of mRNA, δ_s is the degradation rate of sRNA, μ is the growth rate, θ is the translation rate, γ is the annihilation rate of sRNA and mRNA, I and U are chemical inducers, n_1 and n_2 are hill constants for the inducible promoters, K_1 and K_2 are binding constants for the inducible promoters, k_0 and k_2 are basal transcription rates, and k_1 and k_3 are maximal transcription rates.

Upon choosing a model structure (i.e., the form of the ODEs) (*see* **Note 73**), its parameters must be determined. Some parameters will likely be fixed values if they are already known for previously characterized standardized parts or are known values from the literature (*see* **Note 74**). The rest must be estimated using the calibrated experimental timecourse data as follows:

1. Make initial estimates for the parameters that are going to be fit. These estimates should be based on a priori information on the parts involved or typical values for similar parts/processes (*see* **Note 75**).
2. Based on biological plausibility, define lower and upper bounds within which a particular parameter can vary during the fitting process (*see* **Note 75**).
3. Define a cost function quantifying the discrepancy between calibrated experimental outputs $\{y_i^{\text{obs}}\}$ and the values $\{y_i^{\text{model}}\}$ predicted by the mathematical model for a particular parameter set. A commonly used cost function is the sum of squared errors for all measurement timepoints $i = 1, 2, \ldots, n$, $\text{SSE} = \sum_{i=1}^{n} \left(y_i^{\text{obs}} - y_i^{\text{model}}\right)^2$ (*see* **Note 76**).
4. Use an optimization algorithm, such as `fmincon` in MATLAB, to attempt to find a set of parameters that minimizes (*see* **Note 77**) your defined cost function. A skeleton MATLAB code for implementing this is provided in Box 1.

Box 1 MATLAB skeleton code for parameter estimation

```
import globaloptim.*;

%% Load Data
Data = table2array(readtable('path_to_data.csv'));

%% Parameters
F      = [...];                 % Fixed parameters
k0     = [...];                 % Initial estimates for parameters to fit
lb     = [...]; ub = [...];     % Parameter Bounds
tspan = [0, max(Data(:,1))]; % Time vector

%% Global Optimisation Setup
options = optimoptions('fmincon', 'Display', 'off');
gs      = GlobalSearch();
problem = createOptimProblem('fmincon', ...
    'objective', @(k) CostFunction(k, F, Data, tspan), ...
    'x0', k0, 'lb', lb, 'ub', ub, 'options', options);
[k_est, cost] = run(gs, problem);

%% ODE Simulation
y0 = [...]; % initial conditions
[t_model, y_model] = ode15s(@(t, y) ODEModel(t, y, k_est, F, Data), tspan, y0);

%% Plot Results
figure;
plot(t_model, y_model(:,1)); hold on;
plot(Data(:,1), Data(:,3));
legend('Model', 'Experimental'); xlabel('Time'); ylabel('Output');

%% ODE Function
function dydt = ODEModel(t, y, k, F, Data)
    inducer    = interp1(Data(:,1), Data(:,2), t, 'previous');
    growth     = interp1(Data(:,1), Data(:,4), t, 'nearest');
    dydt       = [...]; % define your ODEs using y, k, F, inducer, growth
end

%% Cost Function
function cost = CostFunction(k, F, Data, tspan)
    y0 = [...]; % initial conditions
    [t_model, y_model] = ode15s(@(t, y) ODEModel(t, y, k, F, Data), tspan, y0);
    experimental = interp1(Data(:,1), Data(:,3), t_model, 'linear');
    cost = sum((experimental - y_model(:,1)).^2, 'omitnan'); % example cost function SSE
end
```

In our case study, the transcription, translation, and mRNA degradation parameters are found from prior characterizations, leaving us to find the rates γ and δ_s of sRNA degradation and RNA annihilation in the circuit. Example results for model fitting are shown in Fig. 8.

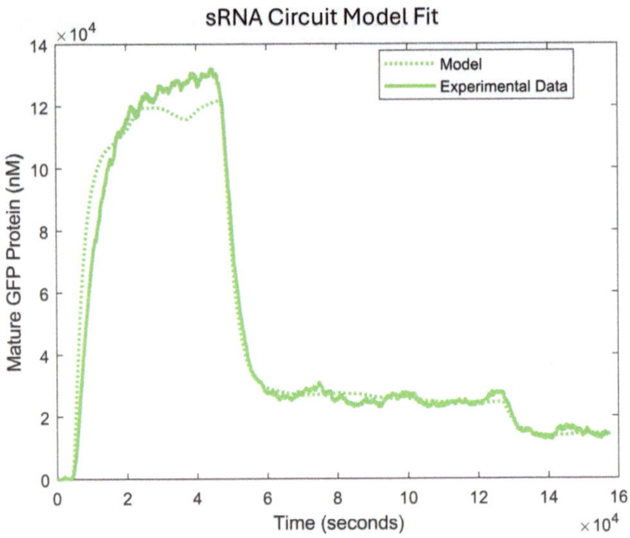

Fig. 8 Example model fitting results. By using the MATLAB skeleton code (Box 1) and the calibrated Chi.Bio data (Fig. 7c), parameters for a model (dotted line) can be fitted to describe the experimental data (solid line)

4 Notes

1. These are the recommended suppliers for Chi.Bio listed at https://chi.bio/software/. Alternative suppliers may be used, but care must be taken to procure hardware with identical properties and dimension so as to ensure the setup can be assembled.
2. The O-ring on each vial is useful in reducing moisture leaking from the top of the vial during an experiment. Each Chi.Bio reactor is equipped with moisture sensors which detect moisture, shutting down operation of the control computer (and therefore the reactor and pumps) to prevent liquid damage to the electronics. This leads to the error message "Failed Pump Comms." With the use of O-rings, leaks are very rare during a particular experiment. One common reason for leaks is flooding of vials, when a waste pump does not function properly—this should be avoided by regular maintenance of pumps. Additionally, it is good practice to have reactors in an experiment spread across multiple control computers so that if one control computer crashes due to moisture leaks, not all reactors are shut down.
3. Ensuring use of correct magnetic stir bars that are compatible with the Chi.Bio is very important. Incompatible stir bars (whether too strong, too weak, or the wrong size) often stir unreliably or do not stir at all when placed in a Chi.Bio culture,

reducing aeration, growth rate, and reproducibility of any cultures. Stir bars are not autoclaved, as autoclaving stir bars negatively affects magnetism in a way which the Chi.Bio is particularly sensitive to, also impeding proper stirring. Ethanol-sterilized magnetic stir bars should therefore be added using sterile tweezers to a sterile Chi.Bio vial in a sterile environment just before the experiment is to be set up. One observed problem with Chi.Bio stirring is that if Chi.Bio vials are removed from a reactor while temperature control is still on, the Chi.Bio reactor will overheat, occasionally melting the fan that drives the stirring altering its shape and thus the stirring properties. Therefore, care should be taken that Chi. Bio vials are not removed from their reactors, while temperature control is on, and fans/stirring should be monitored.

4. Expect to use at least 1.5 m per each media inlet and each media outlet pump.
5. If you plan to test your system in an *E. coli* strain not optimized for protein expression, you may wish to use another protein expression strain (e.g., BL21) for fluorescent protein calibrant preparation. This is recommended in the published FPCountR protocol [27], which this step is based on, since high expression of fluorescent proteins is desired to ensure good quantification. However, in our experience of following this method, even wild-type-like MG1655 provides sufficient protein expression for highly expressed proteins (including most commonly used fluorescent proteins expressed from commonly used vectors). Furthermore, only some of the 50 mL culture is typically used for quantification; therefore, should larger amounts of protein be required, more of the culture can be used.
6. The particular sterile needle chosen for sampling needs to be long enough to penetrate the Chi.Bio lid and reach below the liquid level and low enough circumference to fit through the Chi.Bio vial lid holes. We have found 21 gauge needles of 80 mm length to work well for this (e.g., B Braun 10530855). For adding inducers or cultures, 21 gauge needles are also used, but the needle for this does not need to be long enough to reach the liquid level.
7. The 96-well plate used in the calibrant measurement step should be the same type of 96-well plate as used in all downstream steps of the protocol. That is, you might use clear flat-bottom plastic 96-well plates for all steps requiring a plate reader. Ideally you should use plates not only of the same type, but from the same manufacturer at each step. In our usage of this protocol, we used the Greiner Bio-One 655161 96-well plates.

8. About 10 mL of this Wash Buffer is required for each fluorescent protein. In general, more than required of each buffer should be prepared before beginning the calibration protocol.
9. For 20 OD of cells for each fluorescent protein, 2 mL of Lysis Buffer is required. You should prepare an excess of this before embarking on absolute quantification. Adjust the volume made if more than 20 OD of cells were used for any of the fluorescent proteins according to **Note 19**.
10. This buffer should be prepared before embarking on absolute quantification with about 10 mL per fluorescent protein required.
11. Cospheric offers monodisperse SiO_2 microspheres at two diameters relevant for *E. coli* dimensions (0.890 μm and 0.961 μ m). In the development of FPCountR in Csibra and Stan [5] 0.890 μ m diameter spheres are used. FlopR discusses the option of using either size of microsphere [29] but actually makes use of 0.890 μ m microspheres [23], though some previous work on OD calibration has made use of 0.961 μm microspheres [31, 32]. In Beal et al. [31] it is argued that 0.961 μm is used due to having a similar refractive index to *E. coli* — approximately 1.4 — in line with this the Cospheric technical specifications suggest that both options have a refractive index of approximately 1.4 (https://www.cospheric.com/SiO2MS_monodisperse_silica_spheres_beads_nm_microns). Given the similar refractive indexes and the fact that previous iterations of FlopR and FPCountR used 0.890 μ m microspheres, we suggest that followers of this protocol also use these. We have not tested 0.961 μm microspheres, but comparison could be of interest for validation.
12. For Chi.Bio experiments where a synthetic gene circuit with fluorescent output is to be measured, it is important to use media that has little autofluorescence. LB, which is commonly used for bacterial cultures, is strongly autofluorescent. Therefore it is desirable to use minimal media such as M9. Similarly the FPCountR [5] method also makes use of M9 for absolute quantification. Furthermore some inducible promoters are more leaky in rich media (such as LB), providing another benefit to gene circuit characterization in M9.
13. Due to variable evaporation during autoclaving we recommend that autoclaved media and chemicals are weighed before and after autoclaving and topped up with the difference with autoclaved H_2O. This ensures consistent concentrations and volumes between media bottles and is likely particularly important when inducer concentration step changes are involved in a characterization.

14. The more free parameters there are, the greater the parameter space to be explored by the fitting algorithm. This can lead not only to better fits but also to longer computation times and overfitting. Different combinations of parameters can yield similar steady states and dynamic trajectories of the output, complicating estimation of true values of parameters. Even for a single fluorescent protein gene, various pairs of transcription and translation rates can result in the same expression. The number of plausible parameter combinations steeply rises even further as a system's complexity and the number of parameters increase. Ultimately, additional measurable outputs, such as RNA concentrations, may be necessary to fully solve this identifiability issue, but this is beyond the scope of this chapter.
15. The initial protocol suggests adding inducer to the overnight culture upon preparation [27]. In cases where proteins are poorly expressed, protein expression can sometimes be enhanced by using autoinduction media (e.g., Formedium AIMLB0101) or optimizing the concentration of inducer and time at which the inducer is added. Autoinduction media will only work for constructs where expression of the fluorescent protein is induced by IPTG.
16. As per the FPCountR protocol [27], from this point and until the end of calibrant measurement (Section 3.2) lysed cultures and resulting calibrants should be kept at 4° C. Similarly all centrifugation steps should be carried out at 4° C.
17. For poorly expressed proteins, this value can be adjusted to 40 OD of cells. In doing this, volumes of buffer that each culture is lysed in should be adjusted to maintain the same ratios as discussed in **Note 19**.
18. To gently resuspend pelleted cells, the buffer being added for resuspension should be kept at the same temperature (4° C) as the cells to avoid any shock. Then the resulting solution should be gently swirled or the buffer gently pipetted up and down until the pellet is fully resuspended.
19. 2 mL of Lysis Buffer is the required volume for resuspending 20 OD of cells giving a concentration of 10 ODmL^{-1}. If, for example, 40 OD of cells were being used for a poorly expressed protein (as suggested in **Note 17**), then 4 mL of Lysis Buffer should be added to maintain the correct concentration.
20. It was reported by the FPCountR authors in their protocol [27] that lysis by sonication gives the most reliable results. In our work, we have similarly made use of sonication with good results. However, other methods of lysis (both mechanical and chemical) are available and may also give satisfactory results. Nevertheless we have not tested these methods and therefore recommend sonication.

21. "Duration" here refers to the time during which the culture is being sonicated, not the total duration of the process. Therefore, with equal times split between sonicating and not sonicating, 2 minutes of sonication will take 4 minutes. The settings we present here are for the QSonica Cl-334 sonicator. These settings differ from the settings provided in the FPCountR protocol for the QSonica Q125 sonicator [27], which are 2 minutes sonication, 50% amplitude, and 10 seconds on/off. These settings will therefore likely require optimization for your specific sonicator and specific solution to be lysed. Generally, we recommend you start off with shorter pulses of lower amplitude. You should be careful to monitor that the suspension does not get too hot during sonication and that sonication does not cause displacement of liquid out of the Falcon tube. The sonicated/lysed suspension should go from being turbid to clear. Sonication may need to be repeated to achieve this beyond the initial duration if settings are not optimized.
22. Stock solutions of $CaCl_2$ and $MgCl_2$ should be made up prior to beginning absolute quantification. A stock solution of DNase I of 1000 Kunitz Units mL^{-1} in distilled water should also be made up prior to absolute quantification.
23. DNase I is sensitive to physical denaturation. Therefore, mixing by vortex would reduce DNase I efficiency or render DNase I inactive. As per the FPCountR protocol [27], DNase I activity *may* not be necessary for the ECmax assay (untested) but is important for the A280 assay, should it be required.
24. It may or may not be necessary to concentrate fluorescent proteins for downstream steps. Generally, high starting concentration of this sample is required for accurate quantification in downstream steps. This again may require optimization to know how much (if any) of the lysate to concentrate. For green and red fluorescent proteins, it is expected that a solution should be visibly green or red to be sufficient for downstream steps [27]. Concentration proceeds in multiple steps by adding around 400 μL lysate to the column, spinning down, discarding flow through, reloading the column, and repeating until the desired volume of lysate is concentrated.
25. Generally the sample is eluted from the column by placing the concentrator column in a clean microcentrifuge tube, upside down, and spinning for 2 minutes at 1000× g. Instructions may vary, so manufacturer's documentation for the specific column should always be consulted.
26. While transporting the loaded 96-well plate around or between measurements, the plate should be topped with a plastic lid to avoid contamination. This lid should be removed prior to

measurements. For our particular protocol there are no timecourse measurements in 96-well plates taken, and therefore the 96-well plate need ***not*** be sealed with a Breathe-Easy sealing membrane at any stage during the protocol. This differs from the original FPCountR method [5], where timecourse measurements are taken in the microplate reader, and therefore sealing of the 96-well plate is necessary at the stages described in the published protocol [27].

27. The microplate reader that is used at this step should be used for all future steps, as the quantification is specific to the particular instrument being calibrated. In our usage of this protocol, we used the Tecan Spark plate reader.

28. As per the FPCountR protocol [27], wavelengths above 900 nm are required for correction of path lengths (path length correction is described in the FPCountR supplementary material [5]), while other wavelengths in the range (200–900 nm) are required for fluorescent protein concentration quantification. As per the FPCountR protocol [27], wavelengths less than or equal to 300 nm may not be required for fluorescent proteins that absorb at longer wavelengths. In our experience of this protocol, we have used a step size of 1 nm under the assumption that more data is better, but if quantification is carried out on a microplate reader without this capability, larger step sizes may be sufficient.

29. In this protocol, by "ultimate plate reader experiment" we mean the plate reader measurements taken for Chi.Bio samples. This is important because the absolute quantification relating fluorescent protein concentration to fluorescence is specific to the particular filter settings used to measure fluorescence. Once you have sample protein concentrations per cell, this can then be related to Chi.Bio fluorescence, where the excitation and emission wavelengths can take different values compared to the plate reader settings. This is important as the Chi.Bio, with base technical specifications, is limited in the wavelengths at which it can measure fluorescence. The technical specifications for your particular version of the Chi.Bio are found at https://chi.bio/tech-specs/.

30. Taking measurements at a variety of gain values enables conversion factors to be calculated in the FPCountR R package [28] at each of these gain values. This is important as for different experiments different gain values may be either too sensitive or insensitive to the fluorescence intensity being measured over the course of an experiment. For example, if GFPs were expressed at a low level, high gain values may be required to distinguish between different concentrations of GFP. Conversely if GFP is expressed strongly, high gain values

may lead to measurements outside of the range of your specific plate reader, and therefore a lower gain is necessary. By having conversion factors for a variety of gains, gain settings in the ultimate plate reader experiment (*see* **Note 29**) can be optimized.

31. Gain values are arbitrary for each microplate reader. As per the FPCountR protocol [27], with the Tecan Spark microplate reader, it is recommended to use gains between 40 and 120, in increments of 10. In our experience with the protocol, also using the Tecan Spark, these settings work well.
32. Calibration of OD using microspheres is not strictly necessary when the desired output data is concentration rather than molecules per cell. This is because in FPCountR OD specific cell volume is used to calculate the total cell volume rather than using the number of cells to do this calculation [5]. The value used here is 3.6 μL per OD600 (% m) [30] for *E. coli*. In instances where desired data is in molecules per cell, then OD microsphere calibration is required.
33. Microspheres settle very quickly after dissolving in water; therefore it is suggested that before all steps solutions should be mixed by vigorous vortexing. During serial dilutions, thorough pipetting up and down should be carried out at each step, if not mixed by vortex.
34. The published FlopR protocol [29] is stated for 125 μL cultures and samples, while the FPCountR protocol [27] is using 225 μL cultures and calibration plates; thus we carry out the FlopR OD calibration with 225 μL of microsphere calibrant in each well, to ensure that OD measurements are taken on solutions of the same path length for Chi.Bio samples and OD calibrants. A simple way to achieve this is to make two of tube 1, so two serial dilutions series can be made—hence the number of 1.1 to 11.1 and 1.2 to 11.2. This has the additional benefit of providing a replicate of the serial dilution, which may lead to better calibration should there be variation in pipetting.
35. OD can be measured at either 600 and 700 nm or both, and correction calculations for discrepancies between microsphere calibrations for OD 600 nm and OD 700 nm are detailed in the FPCountR paper [5] and implemented in the FPCountR R package [28].
36. The R package used for analysis along with example data is provided in a GitHub repository [28]. A full worked example is also provided at https://ec363.github.io/fpcountr/articles/fpcountr.html. Briefly, firstly the `parse()` function and then the `process_absorbance_spectrum()` function are called on your data to process and normalize the path length of the data to the format required for analysis. After this,

`get_conc_ecmax()` is called on the processed data to generate the concentration of your samples. Then, the fluorescence assay data is parsed using `parse_magellan()` and conversion factors generated by running the `generate_cfs()` function on the parse fluorescence data. Detailed information on each function including likely warnings, options, and outputs is provided at https://ec363.github.io/fpcountr/articles/fpcountr.html.

37. The number of sphere particles per gram is provided for each Cospheric microsphere at https://www.cospheric.com/calculations_of_spheres_per_gram.htm. For the 0.890 μ m diameter microspheres used in this protocol, this value is 1.36×10^{12} spheres per gram. Given that we take 0.1 g of microspheres to make stock solution A, there are 1.36×10^{11} spheres in stock solution A. The total volume of stock solution A is given by

$$\text{Volume} = 1300 + \frac{\text{Mass of Spheres}}{\text{Density of Spheres}} = 1300 + \frac{0.1}{2 \div 1000} = 1350\ \mu\text{L}$$

Therefore, in stock B there are $\frac{300}{1350} \times 1.36 \times 10^{11} = 3.02 \times 10^{10}$ spheres. In tube 1.1 and 1.2 there will therefore be $\frac{150}{500} \times 3.02 \times 10^{10} = 9.07 \times 10^{9}$ spheres. The number of particles of these solutions plated in wells E1, F1, G1, and H1 is given by $\frac{225}{1500} \times 9.07 \times 10^{9} = 1.36 \times 10^{9}$ spheres. Given that the serial dilution is 3:5, the number of spheres in the next column of rows can be calculated by multiplying the previous column number of spheres by 0.6. This, for example, gives 8.16×10^{8} spheres for wells E2, F2, G2, and H2. Note: In the FlopR protocol [29], a historical value of 1.5 trillion spheres per gram for 0.89 μm diameter microspheres is given and used for their calculations. This value may be more accurate for older versions of Cospheric microspheres.

38. High expression of fluorescent proteins is desirable for good calibration results. Intuitively, the high expression of fluorescent proteins leads to more reliable results — for example, low concentrations of GFPmut3 in Fig. 4c lead to more variability in the serial dilution, potentially affecting the reliability of the final quantification.

39. We recommend that media bottles are autoclaved with standard GL-45 lids rather than Chi.Bio media bottle lids. Pipetting in the various components of working M9 media in a sterile fashion is made more challenging when dealing with lids with media tubing. Additionally, after adding inducers to media, it is important to ensure that media are well mixed, and again tubing can prove an impediment to mixing the media.

Instead we recommend that media bottles are stored with GL-45 lids on until they are required. As Chi.Bio media lids have holes, GL-45 lids have an additional benefit of maintaining sterility. At this point, one should remove the GL-45 lid in a sterile environment, sterilize their gloves with 70% ethanol, and screw on the Chi.Bio media lid, using their hands to guide tubing into the top of the bottle if required.

40. By including a magnetic stir bar in the media prior to autoclaving, we enable easy and sterile mixing of media after media components and inducers are added.
41. Chi.Bio media bottle lids are GL-45 style lids typically with a small air hole and four small holes for connecting tubing to enable pumping of media. We recommend that these lids are 3D-printed using files found on https://chi.bio/hardware/. After printing these lids, the silicone tubing should be connected to the inside of the lids. The tubing should be cut to length so that it reaches the bottom of the media bottle that the lid will be attached to, so that no media is wasted. Additionally it is recommended that tubing is either cut at an angle or a small slit included at the bottom of each tube. This prevents the tubing from suctioning onto the glass of the media bottle, which would impede flow of fresh media to the cultures, impairing turbidostat or chemostat function. Once all the tubing is connected, these lids should be autoclaved in small containers.
42. The amount of media required for an experiment is variable based on the strain of *E. coli*, expression of synthetic genes, media richness, temperature of cultures, and how these interact with the culture's growth rate. One good rule of thumb is that for four reactors at 37° C with cultures in exponential phase, 2 L of LB media will last around 12 hours. M9 media should lead to slower growth rates; nevertheless we recommend that this rule of thumb be followed to provide a margin of safety so that medium is unlikely to run out. Additionally, if planning on doing step changes in inducer, a fresh bottle for each new concentration of inducer will also be needed regardless of volumetric requirements of media.
43. For initially setting up pumps, silicone tubing should be cut to appropriate lengths. The lengths of tubing will vary depending on the physical setup of your devices, but tubing needs to be long enough to thread through peristaltic pumps, connect to Chi.Bio vials when in the Chi.Bio reactors (which should be at a level higher than pumps, waste containers, and media bottle liquid levels), and connect to either a media bottle or a waste container. The cut tubing should be inserted in the peristaltic pumps by unscrewing the clear plastic cover, threading the

tubing through the pump, and replacing the plastic cover. Next, connect the control computer to your PC by USB. Now each Chi.Bio reactor should be connected to a control computer, a pump board, and then a power supply. Turn on the power supply. Start PuTTY software on the connected PC. Connect to IP `192.168.7.2` and port `22`. Log in with username and password, which are both root. Set the current date and time by running a command in the format of `date -s "Jan 30 1649 14:00:00"`. Execute command `cd chibio` to enter the Chi.Bio directory. Start up the Chi.Bio operating system by executing the command `bash cb.sh`. Navigate to the Chi.Bio GUI in a web browser by going to the IP address (http://192.168.7.2). At this stage pumps should be tested by sterilizing tubes by pumping 70% ethanol through the tubes. The PuTTY window should be left open during running, to shut down the Chi.Bio, input ctrl+c in the PuTTY terminal. Data from Chi.Bio experiments can be accessed using the File Transfer Protocol software such as Filezilla.

44. There are a number of common maintenance jobs with the Chi.Bio system. After any moisture leaks that lead to a control computer shutting down, the reactor where the leak occurred should be disassembled and cleaned with a cotton bud and 70% ethanol, particularly making sure to clean moisture sensors, light sources, spectrometers, and any visibly dirty components. The device can then be dried with a clean cotton bud, reassembled, and left to fully air dry before use. As discussed in **Note** 2, pump failure is a common cause for moisture leaks. Peristaltic pump failure can be a result of (1) the motor turning but no media being pumped and (2) the motor failing to turn. Generally, these problems can be fixed by swapping the silicone tubing for fresh tubing as it is prone to wearing out over extended use. In other cases there may be problems with the pump motor itself, in which case the pump should be replaced. Generally, it is recommended that frequent pump maintenance be carried out/tubing replaced regularly and particularly before important experiments. In instances where a particular reactor has unreliable stirring with compatible stir bars, this can often be solved by adjusting the height of the fan which drives the stirring.
45. By "initial media" we generally mean media that is purely M9 supplemented with the necessary components for cell growth and with required antibiotics, i.e., no chemical inducers. This is because in most characterization experiments of synthetic gene circuits we want to start at a point where the transcription units are not activated, so we can then observe the dynamics that result from a step change in inducer. Generally, some of this

"initial" medium should be kept refrigerated to enable its use for making solutions of inducer to add to the vials for inducer step changes.

46. Ideally, the ends of the pump tubing should be kept in the ethanol reservoir and the tubes full of 70% ethanol while not in use to maintain sterility. Nevertheless, it is good practice to resterilize the tubing before each experiment.
47. Media could be instead added to desired volume by removing the vial lid in a sterile environment and pipetting media in. However by instead pumping the media in and excess media out with the Chi.Bio pumps, we confirm that the pumps are functioning and connected properly. Volume in a vial is determined by the level of the waste pump outlet; for the standard vials with the 3D-printed lids this is 20 mL.
48. At this stage, it is sensible to check that stirring is working as expected. A common sign that stirring is not working is hearing a repetitive clinking sound as the stir bar flops around into edges of the glass vial. Otherwise, one can switch on the 6500K LED briefly and look down into the bioreactor to visually confirm stirring.
49. The Chi.Bio GUI is intuitive to use; however a visual and detailed guide to starting the experiment and controlling settings for temperature, stirring, OD set point and control, and fluorescence measurements is provided in the Chi.Bio Operation Manual https://chi.bio/operation/.
50. There is likely slight temperature dependence of OD measurements introduced by the laser/sensor and changes to physical properties of the media (e.g., refractive index). Therefore, allowing temperature to equilibrate increases the reliability of OD calibration. Similarly, stirring improves consistency of those OD measurements. After confirming that the blanking has successfully set OD to zero, we further suggest that the experiment is allowed to run for an initial 15 minutes — importantly, with OD being measured. This confirms that there is no drift in OD over time: Measurements should ideally stay at or close to zero (normally distributed around zero if it is close to zero). If significant drift discovered, it is recommended that blanking is performed again and the reactor left running for another period of time to determine whether the drift eventually stops. This level of care over Chi.Bio OD calibration is likely not necessary for most experiments or in experiments where absolute quantification will be repeated. However, if you wish to use a Chi.Bio fluorescent protein calibration curve across multiple experiments, having consistency across your blanking is likely important. Once a satisfactory OD blanking has been achieved, ideally the Chi.Bio vial should not be moved at all.

51. While the base Chi.Bio operating system measurements are taken every minute, this could be changed by the user, should other timescales be preferable.
52. In this protocol, we activate the dither function, which means OD is allowed to zigzag around the OD set point. This enables Chi.Bio's native growth rate estimation function. The growth rate is estimated somewhat naively by calculating instantaneous growth rate at each particular timepoint by calculating natural log of current OD minus natural log of previous OD. More sophisticated methods, such as Kalman filter-based growth rate estimation [33], may be desirable if accuracy of growth rate estimation is important, for example, for model fitting.
53. For the process of model fitting and quantitative characterization of synthetic gene circuits, the concentrations chosen for inducer step changes should provide meaningful data, for example, with inducible promoters a series of increasing and distinct steady states given by each inducer concentration or with sRNA circuits various levels of repression achieved by different inductions of sRNA. You may need to iterate the experiment several times before you know how to choose the concentrations of inducers that give this idealized data. The time at which inducer step changes occur within an experiment can also be important. Ideally, one wants to get as much informative data as possible from each inducer step change before moving to the next step change. For an inducible promoter, the informative data would be steady-state fluorescence after induction and the dynamics leading up to that steady state. One can generally qualitatively see when a steady state of fluorescence is reached in the Chi.Bio GUI, and ideally another inducer step change should not be carried out until this is observed to avoid losing out on informative data. A further consideration is fluorescent protein dropout where high expression of a fluorescent protein leads to selective pressure against that protein, and eventual dropout from the culture. This is particularly problematic in continuous culture due to continuous selective pressure. Therefore, care should be taken not to induce a fluorescent protein too much; what "too much" is needs to be determined empirically.
54. Only changing the media bottle and not adding inducer to each bioreactor vial would lead to gradual addition of the new media to the culture containing the old media. This means the inducer concentration will increase exponentially to approach the reservoir concentration of inducer in the media bottle at a rate determined by the growth rate of the culture (approximately following the relation $= x(1 - e^{-\mu t})$, where x is the concentration of inducer in media bottle, t is the time since

media bottle connected, and μ is the growth rate). This type of gradual input of inducer over time complicates characterization and model fitting compared to inducer step changes.

55. Inducers are typically kept as stock solutions at something like 1000× working concentration. This means that to do normal inducer step changes, a very small amount of inducer would be added to a 20 mL vial. To reduce the impact of imprecise volume measurements, a larger volume of 1 mL of diluted inducer should be added to each vial. The concentration of these solutions is determined by the following formula: $z = \frac{21 \times 10^{-3}y - 20 \times 10^{-3}x}{1 \times 10^{-3}}$, where z is the required concentration of inducer solution, x is the current concentration of inducer in Chi.Bio experiment, and y is the new concentration of inducer in the Chi.Bio experiment. This calculation is carried out for all inducers that will be undergoing a step change, and for each vial 1.2 mL of solution is then made by diluting the inducer stock solution in the initial Chi.Bio experiment media (which should have been stored refrigerated as per **Note** 45).

56. The calculated concentration of the 1 mL inducer solution that is added to the vial is based on the culture volume being 20 mL. Given that in a particular experiment, samples may be taken from the vial, it is important that inducer step changes occur at a timepoint far enough away from sampling events that culture volume has renormalized to 20 mL. A further consideration is that, while OD control has been deactivated, the experiment continues with stirring, temperature control, and measurements being taken. Injecting 1 mL of liquid in at the same time when a measurement is taking place leads to a significant error in OD and/or fluorescence readings at that timepoint. One good practice is to ensure stirring is active before injecting the inducer solution. This ensures that the inducer is mixed into the culture by stirring, while the Chi. Bio software ensures no measurements are taking place while stirring is active and for at least 5 seconds after stirring is switched off.

57. One particularly convenient way to record the timepoints at which an inducer step change occurs is to use unused Chi.Bio light intensity settings to record metadata. For example, in an experiment measuring a GFP, the 395 nm LED will not be required. However, the Chi.Bio software outputs the intensity value of all LEDs at each timepoint as data regardless of whether the LED is switched on. Therefore, for example, one could say that a light intensity of 0.100 is the initial inducer concentration, and when the step change occurs, we increase this to 0.101.

58. Timepoints at which sampling occurs should be chosen with the aim of achieving wide coverage of the expected values of fluorescence that will be observed during an experiment. Like all interventions in Chi.Bio experiments of the form described in this chapter, these should only take place after the culture has reached the OD set point and fluorescence measurements have settled to a steady state. Otherwise, a good way of doing this is to center sampling on induction timepoints (taking into account **Note** 56). For example, a sample should be taken prior to any induction when in theory fluorescence due to fluorescent protein expression should be close to zero. Then around each induction step, you might take a sample 1 hour before induction, 1 hour after induction, 3 hours after induction, and 6 hours after induction. Such sampling would likely give for each promoter induction step: (1) baseline prior to induction step, (2) a small increase in fluorescence as a result of initial induction, (3) a larger steep increase in fluorescence, and (4) something approximating steady-state fluorescence as a result of induction. These precise timepoints may need optimization for specific gene circuits or parts, but you should be aiming to achieve baseline, small change, intermediate change, and steady-state change. Of course, more timepoints would provide more data and potentially more reliable calibrations. Finally, it is a good idea to take a sample at the final time of each experiment.
59. Most microplate readers are calibrated for cell culture OD measurements using a 200 μL sample, which has become a de facto standard volume for microbiological growth assays in 96-well plates. In this protocol, we use 225 μL to maintain consistency with the ECmax and fluorescence assays described in FPCountR [5, 27] and in Section 3.2. This choice is justified as follows: (1) The ECmax assay applies path length correction and yields absolute calibrant concentrations, so the use of 225 μL here does not constrain us to use the same volume elsewhere, (2) however, using 225 μL throughout reduces labor, as the same dilution plate can be reused for the fluorescence assay, (3) as we use 225 μL for the fluorescence assay, the plates for measurement of culture OD and fluorescence should also use 225 μL to ensure comparability, (4) OD calibration using microspheres should be performed at the same volume to match these measurement conditions, and (5) while using 225 μL may deviate from a plate reader's default calibration (giving different values of "true" OD), this has no impact on absolute quantification as long as consistent volumes are maintained across calibration and measurement steps.
60. The purpose of taking a mean of the last 30 minutes of readings prior to any fluorescent protein induction is to calculate a value

of mean background fluorescence. Given that in many cases you will wait until a steady state is reached before inducing a fluorescent protein, a period of 30 minutes prior to induction confirming flat fluorescence should exist to enable this calculation. Furthermore, due to shot-to-shot variability in the fluorescence signal, taking a mean of multiple measurements is required to get the background signal. However, the specific number of 30 minutes is to a large degree arbitrary and can be adjusted as needed. Nevertheless, 30 minutes works well in our analyses, and the times used should be kept consistent.

61. The calculation of mean fluorescence prior to fluorescent protein induction to give background fluorescence is not feasible in the cases where fluorescent proteins are expressed constitutively or significantly leaky inducible promoters are used. All inducible promoters are leaky to some extent; however the leakiness from "non-leaky" inducible promoters will have negligible impact on the fluorescence signal. In cases where this approach is not feasible, separate characterization of background fluorescence is required with nonfluorescent negative control strains. However, this presents challenges in bioreactor setups like the Chi.Bio, where swapping out cultures may lead to small changes in measurements.

62. We make use of a 21-point symmetrical moving average to smooth shot-to-shot variability in fluorescence signals. A symmetrical moving average considers points both preceding and succeeding the centered point, providing alignment with underlying trends in the data. Other approaches include trailing moving averages, where only a certain number of preceding points are considered. Each method may be more appropriate for different experimental contexts; for instance, a trailing moving average (or other approach such as exponential smoothing) is necessary for live implementation of cybergenetic control or real-time data output but introduces a phase shift that may be undesirable for post-experiment analysis. Choosing the window size is a trade-off between small windows preserving temporal resolution but under-smoothing noise, and larger windows smoothing noise well but obscuring dynamics. The window size should be chosen empirically based on circuit dynamics and time between measurements; however, we found a 21-point window to work well with our data (21 minutes being a reasonable timescale to smooth out gene expression measurements). Various other approaches for smoothing include exponential smoothing and various filters, which may be used instead of moving averages as necessary.

63. Shot-to-shot variability is the random fluctuations in measurement values between successive readings taken under approximately identical conditions. This can arise from culture

disturbances and instrument noise. Chi.Bio fluorescence measurements, growth rate estimation, and to a lesser extent OD measurements are affected by shot-to-shot variability. In the Chi.Bio, shot-to-shot variability is largely caused by culture inhomogeneities and culture stirring rather than sensor noise but is also affected by factors like the fluorescence filters, absolute levels of fluorescence, and culture density [3]. Smoothing (as described in 62) is necessary to smooth the shot-to-shot variability as, for example, in later model fitting you do not want to overfit the model to the shot-to-shot noise. However, we also do not want to smooth to an extent that biologically meaningful changes are obscured.

64. This is the final Chi.Bio fluorescence time series that remains in arbitrary units. By following the rest of the protocol, this signal can be calibrated into nM per cell or other desired units. If one is only interested in the parts of this protocol that relate to using the Chi.Bio to test synthetic gene circuits, then other forms of normalization may be desirable. One such example is that when comparing several versions of a gene circuit, it may be useful to normalize this fluorescence signal to some time series feature (i.e., a steady-state fluorescence signal); this can enable comparison of gene circuit features such as response times or efficiency of sRNA-mediated knockdown. This also facilitates some comparison between experiments and reactors, although still not quantitative as absolute quantification.

65. A Chi.Bio culture for one vial is 20 mL, compared to volumes of a single well in a 96-well plate being around 200 μL (225 μL for FPCountR [27]). This represents a 100× scale-up, meaning that in situ fluorescein-based methods and OD calibration methods using microspheres costs will be approximately 100× higher in terms of the costs of calibrants—given that calibration is likely a relatively regular event, this limits the uptake of such methods in bioreactor-based platforms. Even with FPCountR [5], a relatively cheap and low-labor workflow, scaling up to bioreactors would incur increased labor and costs. Even if such methods were cheap and low effort to achieve in situ in bioreactors, many of these methods, FPCountR included, rely on microsphere-based calibration of OD, which presents an additional cost. To carry out absolute quantification with FPCountR for quantifying 96-well plate growth assays, a 50 mL initial culture for *each* fluorescent protein is required [27]. Therefore, it is reasonable to assume that a 5 L culture for protein production may be required for a 20 mL bioreactor. Many academic microbiology labs may not have the capabilities to grow 5 L cultures routinely, while the additional labor in purifying, lysing, or processing the 5 mL culture is also a barrier to implementation.

66. FPCountR relies on experimental measurement of absorbance spectra for each fluorescent protein to calculate concentration using the ECmax assay [5]. However, with base specifications the Chi.Bio platform is limited in the number and distinctness of wavelengths of light it is able to measure absorbance at (https://chi.bio/tech-specs/). This makes it nontrivial to simply use the FPCountR protocol in situ for the Chi.Bio platform.
67. The FPCountR R package [28] at this stage of the pipeline is designed for analysis of microplate time series growth assays rather than single timepoints of samples. By collating each sampling point into a dataset in a single document, we effectively create a time series document as desired by the FPCountR R package [28]. It is possible, but untested by us, that the package works with single timepoint data, but regardless the outcome is the same. Additionally one could input the actual times of sampling from the Chi.Bio experiment for each sample into the collated dataset, again the data would be the same, but the time series plate reader curve would be more accurate.
68. The FPCountR R package [28] is then used to analyze this dataset. Briefly: An example code (https://ec363.github.io/fpcountr/articles/fpcountr.html) is followed to assemble the previously calculated conversion factors (fluorescence and OD) into the required format; the plate reader data file is parsed into the required format using `parse_magellan()`; then you run the `process_plate()` function on the parsed data which processes the data and produces a number of plots; the fluorescent protein molecules per cell is estimated using `calc_fppercell()` producing a new data file with calibrated molecules per cell and associated plots; and finally, to go to cellular concentration the `calc_fpconc()` function is called on the processed data producing a data file with molar concentration per cell and associated plots. A full tutorial for this analysis is provided at https://ec363.github.io/fpcountr/articles/fpcountr.html, reference documentation for each function can be found at https://ec363.github.io/fpcountr/reference/index.html, and the package and installation instructions/dependencies can be found at the GitHub repository for the R package [28].
69. In our workflow, we use calibrated concentration values of fluorescent proteins per cell (in nM) as our final data type. Concentration data is particularly useful for fitting deterministic models and interpreting gene circuit behavior in terms of biochemical kinetics or relating to other deterministic models. Alternatively, expressing the same data in molecules per cell may offer a more intuitive representation and can be advantageous when applying stochastic modeling approaches.

70. Generally, in bacterial growth assays fluorescence values (in AU) will to a great extent be related to the OD of a culture, and therefore fluorescence values are generally normalized by dividing by OD, giving a proxy for fluorescence per cell. In this case, we can directly relate molar concentration per cell from the plate reader to Chi.Bio fluorescence because the Chi.Bio (in turbidostat mode) is maintaining OD approximately constantly at the set point. Even when dithering is active, the changes in OD are minimal and have little to no impact on fluorescence.
71. We are not aware if there are particular timescales on which the Chi.Bio (and indeed microplate readers) should be recalibrated after carrying out this protocol. Generally, we would suggest that this protocol should be carried out before any particularly important experiments where absolute quantification is necessary. Otherwise, on the scale of a few times a year seems reasonable. There are specific concerns with Chi.Bio when it comes to recalibration. If any maintenance is carried out on the Chi.Bio reactor, i.e., it is disassembled for cleaning or other reasons, then the calibration should be repeated as sensors may have moved slightly in the reassembly. Bioreactor culture vials are also likely to be in slightly different positions during each experiment, while the reusable nature of glass vials may mean the glass is in different states during each experiment. These Chi.Bio-specific issues may relate to different measurements and invalidate calibration, hence our suggestion that before each *important* Chi.Bio experiment where absolute quantification is needed, this protocol or at least the Chi.Bio specific elements of it should be repeated.
72. Defining which molecular species and processes are necessary to include within your model will depend on the specific circuit you are investigating. For example, a circuit such as our sRNA circuit, modeling the relevant mRNA and sRNA species is important. In other cases where regulation is protein-based, models can be simplified under certain assumptions. In modeling there is an inherent trade-off between complexity and interpretability, and the aim should be to develop a model which makes useful predictions and/or tests understanding of the system being modeled.
73. Model structure should be chosen based on a priori understanding of how a system will behave. In cases where parameter fitting for a particular model structure is unable to adequately fit the data, this can suggest that the chosen model structure is not accounting for important aspects of circuit behavior. For example, in our model, we model sRNA–mRNA binding as an annihilation reaction using simple mass action kinetics; however one could imagine more complex models of sRNA–

mRNA binding which include resource competition and explicit modeling of complex association and dissociation. Furthermore, due to the messy nature of biology, even a model that explains some data well may not generalize to other similar datasets. To this end, it is good practice to test one's model against data unseen by the fitting process. Ultimately sometimes scaling factors, while reducing the quantitative usefulness of models, may be required to produce qualitatively useful models.

74. Even in some cases where a particular value is not known exactly, it may be reasonable to fix parameters to certain values. This could be done when a particular parameter is not necessarily important for overall understanding of the system, when there is redundancy between parameters on output signals, when additional experiments may be required to properly identify a value, but it would be unfeasible to do these, or when very similar values are known. A key question must be will fitting an additional parameter lead to increased understanding or a more accurate value of the parameter in question. One such example in our fitting is the degradation rater of sRNA, δ_s. For the natural version of the sRNA we use, published experimental data has been used to calculate half-life of the sRNA which can then be used to calculate δ_s. However, when making out synthetic sRNA, it is possible that sequence changes modify the true value of δ_s. Without additional experimentation, in our estimation, it is better to use this known value for a very similar molecule than fit a new value. These kinds of decisions will need to be made in the context of the particular genetic circuit being studied.

75. In cases where specific values of parameters are completely unknown, we will need to carry out our parameter estimation. However, even in these cases we can make some reasonable decisions on starting values, lower bounds, and upper bounds of these unknown values. Similar parts that have been characterized in the past can be looked at to give a rough range. For example, one could look at some well-known and well-characterized inducible promoters to see the range of possible values that the parameters describing their activity take. One could also consider processes like transcription and translation, looking at maximal, minimal, and typical rates in *E. coli*. Resources such as BioNumbers (https://bionumbers.hms.harvard.edu/) [34] can be helpful as repositories of this information. Finally, it is also a good idea to do back-of-the-envelope calculations, for example, thinking about particular rates of transcription and mRNA degradation, what steady-state concentration of mRNA would they lead to, how does this

compare to typical endogenous mRNA concentrations, to synthetic mRNA concentrations, to maximum mRNA concentrations, to total mRNA content of the cell, etc.

76. More complex cost functions may be defined as necessary to account for multiple model outputs or differences in experimental data scale. For example, when fitting models to multiple fluorescent proteins, it is often helpful to construct a composite cost function that sums or averages the errors across all outputs. In such cases, applying weightings to each term enables prioritization of particular signals—useful if one channel is noisier or less biologically important. Signal normalization, such as dividing each signal by its maximum experimental value, can prevent a composite cost from being dominated by higher magnitude signals. In situations where multiple time series of different lengths are being fit, raw SSE will bias the fit toward longer datasets; using cost functions such as the mean squared error (MSE) can help balance their influence. Other approaches, such as scaling each error term by the corresponding model-predicted value (relative error), can balance the impact of errors across points in a signal with differing magnitudes. Ultimately, the choice of cost function should be tailored to the specific problem. In addition to selecting a cost function, the optimization approach itself can be chosen; for example, Bayesian optimization may be preferable to `fmincon` when quantifying uncertainty in parameter estimates is required.

77. Optimization algorithms minimize the chosen cost function rather than deviations from "truth," so results are constrained by that choice. Even a global minimum of SSE/MSE may not be biologically correct if the cost function is inappropriate, the model is misspecified, the data is weakly informative, or parameters are non-identifiable. For nonlinear gene-circuit models, the cost landscape is typically non-convex and multimodal; approaches such as `fmincon` can converge to different local minima from different initial guesses, and multiple distinct parameter sets may attain the same minimum. Fitted parameters should therefore be viewed as one compatible explanation of the data rather than a uniquely true set; post-fitting, well-designed validation or falsification experiments can help to make sense of this. Some of these issues can be mitigated by multi-start strategies and tools such as `GlobalSearch`.

5 Discussion

In this chapter, we presented a method leveraging absolute quantification and continuous culture in a bioreactor for evaluating genetic circuits that present an effective and accessible approach to system characterization in engineering biology.

A key feature underlying our method's reproducibility is absolute quantification of genetic circuit outputs [18, 19]. Typically, synthetic biologists employ standards within experiments to compare their results in arbitrary units by scaling the outputs of novel engineered parts based on a standard transcription unit [22]. Fluorescent standards have also played an important role in connecting fluorescence to concentration. However, lacking the capacity or the instrumentation for measuring precise molecular counts should not be an impediment to reporting results in absolute rather than arbitrary units. Translating research into applications requires biological systems to be designed with specific and real-value targets [35]. To this end, implementing absolute quantification should become standard practice for characterization. The FPCountR method [5] already lowers the cost and labor threshold required for implementation of absolute quantification for microplate reader assays. The methodology we discuss extends FPCountR for obtaining absolute quantification measurements in bioreactors. The additional effort needed to collect the appropriate measurements is far outweighed by the benefits of making research instantly comparable and shareable. Making synthetic biology a true engineering discipline starts with reporting physically meaningful values.

Another important advantage of the outlined protocol is continuous culturing of bacteria in a bioreactor. Unlike microplate readers, bioreactors acting as turbidostats can maintain a constant optical density of a culture, decoupling circuit functionality from the effects of changing cell counts and nutrient availability [3, 14]. In our own experiments with sRNA circuits, significant functional differences arose when measuring circuit behavior in a plate reader as opposed to the bioreactor. The ability to control the exact concentration of bacteria and timing of adding or diluting inducers becomes invaluable for troubleshooting and finding explanations for glitches in genetic circuits. However, it is important to acknowledge the bioreactor setup's limitations. Namely, bioreactor experiments exhibit unique biological challenges, such as positive selection for biofilm formation [36]. Experiments involving many constructs or replicates are less suitable since bioreactors typically have a much lower throughput than plate readers or other smaller scale setups—but still serve as a better starting point for completely new designs thanks to providing richer insights into novel circuit behavior. Finally, operating bioreactors is associated with significant costs and technical complexities compared to microplate readers. However, open-source small-scale bioreactors, such as the Chi.Bio platform discussed in this chapter, present a workaround for this issue [3].

Although we describe a rather manual, beginner-friendly actuation approach in our methods, bioreactors can engage with microbial cultures in much more sophisticated ways. Custom software can control engineered biological systems by applying signals calculated in silico via inducer changes, optogenetic control [3, 14], media composition, or other mechanisms of control such as temperature [36, 37]. Chi.Bios and other open-source and commercial bioreactors [15, 33] can readily perform such actuation. The cumbersome sampling step can be carried out by a low-cost liquid handler (such as an Opentrons) with the ReacSight system that can be integrated with most setups [38]. For actuation, we have described model fitting and parameter estimations that can be drawn from data that then enable the deployment of more complex cybergenetics. Computer algorithms leveraging machine learning, reinforcement learning, and control theory that would be difficult to implement in a manual setup can be deployed to direct the cell state [39], which may also help steer a biosystem to states of interest, producing more informative experimental measurements [40, 41]. This layer of external control helps to relegate some of the overall computational burden away from internal cell circuits and still elicit the desired cell response. Cybergenetics makes experimenting with genetic circuit function cheaper and less cumbersome since not all of the biological actuation has to be engineered into the cell directly. The cell state can thus be controlled more carefully and reproducibly, with control protocols easily shareable in literature through computer code [42].

Outside of academic research, bioreactor-based characterization has the potential to revolutionize commercial applications. The scale-up of synthetic biology products from the lab to an industrial scale is notoriously difficult, as cells are exposed to different conditions in large cultures than in the environments that the genetic circuit was developed in (e.g., dissolved oxygen levels, shear stress, and culture homogeneity). Some biological parts are not characterized in exponential phase at all before they are assembled into a circuit. Therefore, being able to quantitatively characterize parts and circuits in a bioreactor provides better design tools along with a greater range of experimental data conditions. Bioreactor-based development in academia arguably highlights genetic circuits that are viable for commercialization with fewer false positives. Additionally, perturbing the designed genetic circuit with computer algorithms opens up opportunities for cybergenetic control in the form of model predictive control (MPC) or other control flows. Overall, this approach enables more sample- and time-efficient development and removes some of the uncertainty from the translational research process.

References

1. El Karoui M, Hoyos-Flight M, Fletcher L (2019) Future trends in synthetic biology—a report. Front Bioeng Biotechnol 7:471916. https://doi.org/10.3389/fbioe.2019.00175
2. McBride CD, Del Vecchio D (2021) Predicting composition of genetic circuits with resource competition: demand and sensitivity. ACS Synth Biol 10(12):3330–3342. https://doi.org/10.1021/acssynbio.1c00281
3. Steel H, Habgood R, Kelly CL et al (2020) In situ characterisation and manipulation of biological systems with Chi.Bio. PLOS Biol 18:1–12. https://doi.org/10.1371/journal.pbio.3000794
4. Banga JR, Villaverde AF (2025) Mechanistic dynamic modelling of biological systems: the road ahead. Curr Opin Syst Biol 42:100553. https://doi.org/10.1016/j.coisb.2025.100553
5. Csibra E, Stan GB (2022) Absolute protein quantification using fluorescence measurements with FPCountR. Nat Commun 13: 6600 (2022). https://doi.org/10.1038/s41467-022-34232-6
6. Chandran D, Copeland WB, Sleight SC et al (2008) Mathematical modeling and synthetic biology. Drug Discov Today Dis Models 5: 299–309. https://doi.org/10.1016/j.ddmod.2009.07.002
7. Zhang R, Goetz H, Melendez-Alvarez J et al (2021) Winner-takes-all resource competition redirects cascading cell fate transitions. Nat Commun 12:853. https://doi.org/10.1038/s41467-021-21125-3
8. Park JH, Holló G, Schaerli Y (2024) From resonance to chaos by modulating spatiotemporal patterns through a synthetic optogenetic oscillator. Nat Commun 15:7284. https://doi.org/10.1038/s41467-024-51626-w
9. Wang YH, Wei KY, Smolke CD. (2013) Synthetic biology: advancing the design of diverse genetic systems. Annu Rev Chem Biomol Eng 4:69. https://doi.org/10.1146/annurev-chembioeng-061312-103351
10. Melendez-Alvarez J, He C, Zhang R et al (2021) Emergent damped oscillation induced by nutrient-modulating growth feedback. ACS Synth Biol 10:1227–1236. https://doi.org/10.1021/acssynbio.1c00041
11. Monod J (1949) The growth of bacterial cultures. Annu Rev Microbiol 3:371–394. https://doi.org/10.1146/annurev.mi.03.100149.002103
12. Lin WH, Kussell E, Young LS et al (2020) Origin of exponential growth in nonlinear reaction networks. Proc Natl Acad Sci USA 117:27795. https://doi.org/10.1073/pnas.2013061117
13. Anderson PA. (1953) Automatic recording of the growth rates of continuously cultured microorganisms. J Gen Physiol 36:733–737. https://doi.org/10.1085/jgp.36.6.733
14. Gutiérrez Mena J, Kumar S, Khammash M (2022) Dynamic cybergenetic control of bacterial co-culture composition via optogenetic feedback. Nat Commun 13:4808–4808. https://doi.org/10.1038/s41467-022-32392-z
15. Wong BG, Mancuso CP, Kiriakov S et al (2018) Precise, automated control of conditions for high-throughput growth of yeast and bacteria with eVOLVER. NatBiotechnol 36:614–623. https://doi.org/10.1038/nbt.4151
16. Kim J, Darlington APS, Muñoz-Montero S et al (2025) Gene expression depends on the interplay among growth, resource biogenesis, and nutrient quality. ACS Synth Biol 14:2012–2029. https://doi.org/10.1021/acssynbio.4c00828
17. Ziv N, Brandt NJ, Gresham D. (2013) The use of chemostats in microbial systems biology. J Vis Exp 80:50168. https://doi.org/10.3791/50168
18. Beal J, Haddock-Angelli T, Baldwin G et al (2018) Quantification of bacterial fluorescence using independent calibrants. PLoS One 13: e0199432. https://doi.org/10.1371/journal.pone.0199432
19. Beal J, Haddock-Angelli T, Gershater M et al (2016) Reproducibility of fluorescent expression from engineered biological constructs in E. coli. PLoS One 11:e0150182. https://doi.org/10.1371/journal.pone.0150182
20. Mira P, Yeh P, Hall B G. (2022) Estimating microbial population data from optical density. PLoS One 17:e0276040. https://doi.org/10.1371/journal.pone.0276040
21. Beal J (2015) Bridging the gap: a roadmap to breaking the biological design barrier Front Bioeng Biotechnol 2:123819. https://doi.org/10.3389/fbioe.2014.00087
22. Kelly JR, Rubin AJ, Davis JH et al (2009) Measuring the activity of BioBrick promoters using an in vivo reference standard. J Biol Eng 3:1–13. https://doi.org/10.1186/1754-1611-3-4
23. Fedorec AJH, Robinson CM, Wen KY et al (2020) FlopR: an open source software package for calibration and normalization of plate

reader and flow cytometry data. ACS Synth Biol 9:2258–2266. https://doi.org/10.1021/acssynbio.0c00296

24. Lambert TJ (2019) FPbase: a community-editable fluorescent protein database. Nat Methods 16:277–278. https://doi.org/10.1038/s41592-019-0352-8
25. Baig H, Fontanarossa P, McLaughlin J et al (2021) Synthetic biology open language visual (SBOL visual) version 3.0. J Integr Bioinform 18. https://doi.org/10.1515/jib-2021-0013
26. Na D, Yoo SM, Chung H et al (2013) Metabolic engineering of Escherichia coli using synthetic small regulatory RNAs. Nat Biotechnol 31: 170–174. https://doi.org/10.1038/nbt.2461
27. Csibra E (2023) FPCount protocol—in-lysate (purification free) protocol V.2. 2023. Available via PROTOCOLS.IO https://www.protocols.io/view/fpcount-protocol-in-lysate-purification-free-proto-dm6gpw43plzp/v2 Accessed 8 Aug 2025. https://doi.org/10.17504/protocols.io.dm6gpw43plzp/v2
28. Csibra E (2021) FPCountR: Fluorescent protein calibration for plate readers. R package. Available via GITHUB. https://github.com/ec363/fpcountr. Accessed 8 Aug 2025. https://doi.org/10.5281/zenodo.5760028
29. Fedorec AJH, Robinson CM, Wen KY et al (2021) Correction to 'FlopR: an open source software package for calibration and normalization of plate reader and flow cytometry data ACS Synth. Biol. 10:428. https://doi.org/10.1021/acssynbio.0c00631
30. Volkmer B, Heinemann M (2011) Condition-dependent cell volume and concentration of escherichia coli to facilitate data conversion for systems biology modeling. PLoS One 6:1–6. https://doi.org/10.1371/journal.pone.0023126
31. Beal J, Farny NG, Haddock-Angelli T et al (2020) Robust estimation of bacterial cell count from optical density. Commun Biol 3: 512. https://doi.org/10.1038/s42003-020-01127-5
32. Stevenson K, McVey AF, Clark IBN et al (2016) General calibration of microbial growth in microplate readers. Sci Rep 6:38828. https://doi.org/10.1038/srep38828
33. Hoffmann SA, Wohltat C, Müller KM et al (2017) A user-friendly, low-cost turbidostat with versatile growth rate estimation based on an extended Kalman filter. PLoS One 12:1–15. https://doi.org/10.1371/journal.pone.0181923
34. Milo R, Jorgensen P, Moran U et al (2009) BioNumbers—the database of key numbers in molecular and cell biology. Nucleic Acids Res 38:D750–D753. https://doi.org/10.1093/nar/gkp889
35. Tack DS, Tonner PD, Pressman A et al (2023) Precision engineering of biological function with large-scale measurements and machine learning PLoS One 18:e0283548. https://doi.org/10.1371/journal.pone.0283548
36. Lee TA, Morlock J, Allan J et al (2025) Directing microbial co-culture composition using cybernetic control. Cell Rep Methods 5(3): 101009. https://doi.org/10.1016/j.crmeth.2025.101009
37. Lee TA, Steel H (2022) Cybergenetic control of microbial community composition. Front Bioeng Biotechnol 10:957140. https://doi.org/10.3389/fbioe.2022.957140
38. Bertaux F, Sosa-Carrillo S, Gross V et al (2022) Enhancing bioreactor arrays for automated measurements and reactive control with ReacSight. Nature Commun 13:3363–3363. https://doi.org/10.1038/s41467-022-31033-9
39. Caringella G, Bandiera L, Menolascina F (2023) Recent advances, opportunities and challenges in cybergenetic identification and control of biomolecular networks. Curr Opin Biotechnol 80:102893. https://doi.org/10.1016/j.copbio.2023.102893
40. Kryukov M, Carcano A, Batt G et al (2019) Can optimal experimental design serve as a tool to characterize highly non-linear synthetic circuits? In: 18th European control conference (ECC)
41. de Cesare I, Salzano D, di Bernardo M et al (2022) Control-based continuation: a new approach to prototype synthetic gene networks. ACS Synth Biol 11:2300–2313.
42. Perrino G, Hadjimitsis A, Ledesma-Amaro R et al (2021) Control engineering and synthetic biology: working in synergy for the analysis and control of microbial systems. Curr Opin Microbiol 62:68–75. https://doi.org/https://doi.org/10.1016/j.mib.2021.05.004

Chapter 9

Cell-Free Genetic Circuits: Extract and Template Preparation

Thales R. Spartalis, Kassidy B. Porche, Xun Tang, and Yong-Chan Kwon

Abstract

Cell-free systems have been highlighted as a transformative technology in synthetic biology, offering substantial technological advantages over traditional in vivo methods. By leveraging cellular metabolic machinery of the cells and decoupling cell growth from catalytic utilization and flexible template configuration, cell-free systems allow direct access to cell machinery and facilitate precise control over reaction conditions. Despite these advantages, standardizing DNA template preparation and cellular extract production remains a considerable challenge, often leading to variability in results. This chapter details methods for preparing cellular extracts and outlines strategies for optimizing DNA template preparation to enhance the reliability and efficiency of cell-free synthetic gene circuits.

Key words Cell-free systems, Synthetic biology, Cellular extract, Genetic circuits, DNA template preparation

1 Introduction

Synthetic gene circuits, engineered to mimic or augment natural gene regulatory networks, have become essential tools in synthetic biology. These circuits, composed of DNA, RNA, and proteins, enable precise control over gene expression, facilitating applications in diagnostics [1–3], biofuel production [4], and large-scale protein synthesis [5, 6]. However, studying and optimizing these circuits within living cells presents significant challenges due to the complex and dynamic cellular environment, where competing processes such as metabolism and replication consume energy and introduce variability [7, 8]. To overcome these limitations, the *cell-free system (CFS)* has emerged as a powerful alternative testbed, allowing researchers to isolate and manipulate transcriptional and translational machinery outside living cells [9, 10].

At the core of any CFS reaction are two essential components: *cellular extracts* (from various organisms) and *DNA templates* (either plasmids or PCR-amplified linear expression template (LET)). Cellular extracts provide the necessary protein and enzyme

Xiao-Jun Tian (ed.), *Synthetic Gene Circuits: Methods and Protocols*, Methods in Molecular Biology, vol. 3041,
https://doi.org/10.1007/978-1-0716-5304-3_9,

machinery for gene transcription and translation, while DNA templates encode the genetic information for protein production. Together, these elements not only enable rapid, tunable, and high-yield protein synthesis but also accelerate synthetic biology research without the constraints of cell growth or viability. Therefore, the open nature of CFS systems offers advantages over traditional in vivo approaches, including faster reaction kinetics, controllable reaction conditions, and the ability to monitor and adjust variables directly. These features make CFS an ideal platform for prototyping synthetic gene circuits, characterizing genetic parts, and optimizing protein expression [11–14].

The quality and composition of the cellular extract significantly affect the performance of the CFS system. For instance, cell lysates derived from *Escherichia coli* form the core reaction environment for gene expression, delivering essential enzymatic and macromolecular components, such as RNA polymerase for transcription, ribosomes for translation, and cofactor-regenerating systems, to reconstitute protein synthesis in vitro [15, 16]. By evading the complexities of cellular growth and division, this extract-based platform offers the feasibility of direct manipulation of the biochemical setting, allowing precise control over the reaction kinetics and resource allocation. However, the preparation of cellular extracts requires optimization to preserve the activity of endogenous ribosomes, polymerases, and other essential factors while removing unwanted nucleases and proteases that could degrade reaction components [17–19].

The design and preparation of the DNA templates also dictate the output of the CFS reaction in synthetic circuit research. Plasmid DNA is typically preferred in CFS due to its inherent resistance to exonuclease-mediated degradation, ensuring prolonged stability and sustained protein expression. In contrast, linear expression templates, while more susceptible to enzymatic breakdown, can be protected through the addition of exonuclease inhibitors, offering a flexible and cloning-free alternative for rapid genetic circuit prototyping [20]. Advances in DNA synthesis and assembly techniques have further streamlined the generation of high-quality templates (both plasmid and LET), enabling the rapid construction of complex gene circuits.

This chapter explores the fundamental protocols and considerations for the preparation of cellular extracts (Fig. 1) and DNA templates for CFS, highlighting the state-of-the-art practices for maximized protein yield and circuit functionality. By providing a detailed guide to these foundational techniques, we aim to facilitate researchers in leveraging CFS systems for the design, testing, and deployment of synthetic gene circuits with precision and efficiency.

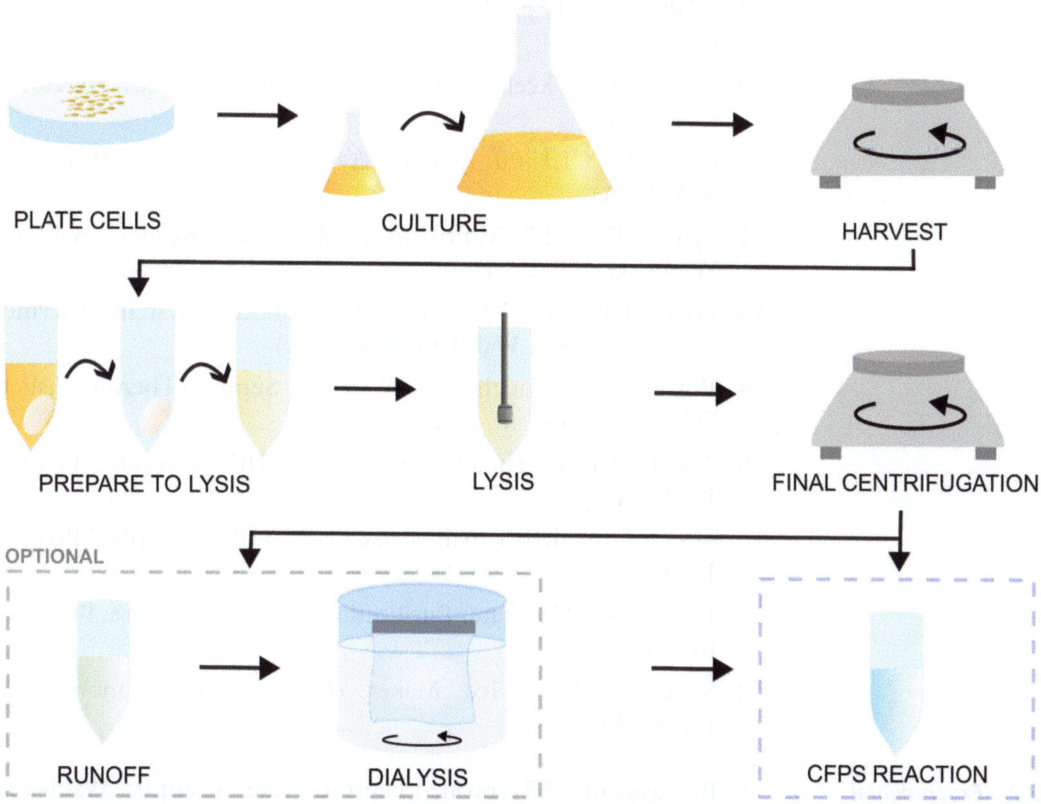

Fig. 1 Stepwise preparation of cellular extract for cell-free reaction

2 Materials

2.1 Equipment

1. New Brunswick™ Innova® 42/42R Incubator Shaker (Eppendorf, Hamburg, Germany).
2. VWR® UV-3100PC, UV/Vis Spectrophotometer (Avantor, Radnor, PA, USA).
3. Sorvall Legend X1R Centrifuge (Thermo Fisher Scientific, Waltham, MA, USA).
4. TX-400 Swinging Bucket Rotor (Thermo Fisher Scientific, Waltham, MA, USA).
5. Fiberlite™ F15-8x50cy Fixed-Angle Rotor (Thermo Fisher Scientific, Waltham, MA, USA).
6. MicroClick™ 30x2 Fixed Angle Microtube Rotor (Thermo Fisher Scientific, Waltham, MA, USA).
7. Fisherbrand™ accuSpin™ Micro 17 Microcentrifuge (Thermo Fisher Scientific, Waltham, MA, USA).
8. Q125 Sonicator with 3 mm (1/8″) diameter probe (Qsonica, Newtown, CT, USA).

9. C1000 Touch™ PCR Thermal Cycler (Bio-Rad, Hercules, CA, USA).
10. Gene Pulser Xcell™ Microbial System (Bio-Rad, Hercules, CA, USA).
11. iBright™ CL1500 Imaging System (Invitrogen, Waltham, MA, USA).
12. Synergy™ HTX Multi-Mode Microplate Reader (BioTek, Winooski, VT, USA).
13. Owl™ EasyCast™ Mini Gel Electrophoresis System (Thermo Fisher Scientific, Waltham, MA, USA).
14. PowerEase™ Touch 120 W Power Supply (Thermo Fisher Scientific, Waltham, MA, USA).
15. 2.5 L baffled Tunair shake flasks (IBI Scientific, Peosta, IA, USA).
16. 300 mL baffled Tunair shake flasks (IBI Scientific, Peosta, IA, USA).
17. Direct-Q3® UV Water Purification System (Millipore, Burlington, MA, USA).
18. Scotsman Flake Ice Maker (CurranTaylor, Canonsburg, PA, USA).

2.2 Experimental Reagents

2.2.1 *E. coli* Culture Media and Buffer Components

1. BioReagents™ Tryptone (Thermo Fisher Scientific, Waltham, MA, USA).
2. BioReagents™ Yeast Extract (Thermo Fisher Scientific, Waltham, MA, USA).
3. BioReagents™ Sodium Chloride (Thermo Fisher Scientific, Waltham, MA, USA).
4. BioReagents™ Agar (Thermo Fisher Scientific, Waltham, MA, USA).
5. Potassium Phosphate Dibasic (Sigma-Aldrich, St. Louis, MO, USA).
6. Potassium Phosphate Monobasic (Sigma-Aldrich, St. Louis, MO, USA).
7. Glucose (Thermo Fisher Scientific, Waltham, MA, USA).
8. Potassium Hydroxide (Thermo Fisher Scientific, Waltham, MA, USA).
9. Tris Base (Thermo Fisher Scientific, Waltham, MA, USA).
10. Magnesium Acetate Tetrahydrate (Thermo Fisher Scientific, Waltham, MA, USA).
11. Potassium Acetate (Thermo Fisher Scientific, Waltham, MA, USA).

12. BioReagents™ Tris-Acetate-EDTA (TAE Buffer) 50X (Thermo Fisher Scientific, Waltham, MA, USA).
13. 1,4-Dithio-DL-threitol (DTT) (Sigma-Aldrich, St. Louis, MO, USA).
14. Glacial Acetic Acid (Thermo Fisher Scientific, Waltham, MA, USA).
15. Isopropyl β-D-1-thiogalactopyranoside (Sigma-Aldrich, St. Louis, MO, USA).
16. Slide-A-Lyzer™ G2 Dialysis Cassette, 10 K MWCO, 3 mL (Thermo Fisher Scientific, Waltham, MA, USA).

2.2.2 Molecular Biology Reagents

1. BioReagents™ Agarose (Thermo Fisher Scientific, Waltham, MA, USA).
2. DpnI Restriction Enzyme (New England Biolabs, Ipswich, MA, USA).
3. GeneRuler 1 kb Plus DNA ladder (Thermo Fisher Scientific, Waltham, MA, USA).
4. GeneRuler 100 bp Plus DNA ladder (Thermo Fisher Scientific, Waltham, MA, USA).
5. Gibson Assembly® Master Mix (New England Biolabs, Ipswich, MA, USA).
6. PureLink™ HiPure Plasmid Miniprep Kit (Thermo Fisher Scientific, Waltham, MA, USA).
7. E.Z.N.A.® Gel Extraction Kit (Omega Bio-tek, Norcross, GA, USA).
8. Ampicillin sodium salt (Thermo Fisher Scientific, Waltham, MA, USA).
9. Phusion® High-Fidelity DNA Polymerase (New England Biolabs, Ipswich, MA, USA).

2.2.3 Cell-Free Reaction Components

1. Magnesium Glutamate (Sigma-Aldrich, St. Louis, MO, USA).
2. Ammonium Glutamate (MP Biomedicals, Santa Ana, CA, USA).
3. Potassium Glutamate (Sigma-Aldrich, St. Louis, MO, USA).
4. Adenosine Triphosphate (ATP) (Thermo Fisher Scientific, Waltham, MA, USA).
5. Cytidine Triphosphate (CTP) (Thermo Fisher Scientific, Waltham, MA, USA).
6. Uridine Triphosphate (UTP) (Thermo Fisher Scientific, Waltham, MA, USA).
7. Guanosine Triphosphate (GTP) (Sigma-Aldrich, St. Louis, MO, USA).

8. L-5-Formyl-5,6,7,8-tetrahydrofolic acid (Folinic acid) (Sigma-Aldrich, St. Louis, MO, USA).
9. *E. coli* total tRNA mixture (from strain MRE600) (Roche Applied Science, Indianapolis, IN, USA).
10. Phosphoenolpyruvate (PEP) (Thermo Fisher Scientific, Waltham, MA, USA).
11. 20 Amino Acids (Sigma-Aldrich, St. Louis, MO, USA).
12. Nicotinamide Adenine Dinucleotide (NAD) (Thermo Fisher Scientific, Waltham, MA, USA).
13. Coenzyme-A (CoA) (Sigma-Aldrich, St. Louis, MO, USA).
14. Potassium Oxalate (Thermo Fisher Scientific, Waltham, MA, USA).
15. Putrescine (Thermo Fisher Scientific, Waltham, MA, USA).
16. Spermidine (Thermo Fisher Scientific, Waltham, MA, USA).
17. HEPES (Thermo Fisher Scientific, Waltham, MA, USA).

2.3 E. coli *Strain*

1. BL21 Star^{TM} (DE3) (Invitrogen, Waltham, MA, USA).
2. DH5α Competent Cells (Invitrogen, Waltham, MA, USA).

3 Methods

3.1 Cellular Extract Preparation: Day 1

3.1.1 LB Media

- Dissolve 10 g/L tryptone, 5 g/L yeast extract, and 10 g/L Sodium Chloride (NaCl) in Milli-Q water.
- Sterilize the media by autoclaving at 121 ° C for 20 min.
- Cool and store at room temperature until use.

3.1.2 LB-Agar Plate

- Dissolve 10 g/L tryptone, 5 g/L yeast extract, 10 g/L Sodium Chloride (NaCl), and 15 g/L bacto-agar in Milli-Q water.
- Autoclave at 121 ° C for 30 min.
- Cool the media to 55 ° C – 65 ° C in a water bath to prevent agar degradation while ensuring the medium remains liquid.
- If required, add antibiotics once cooled to 55 ° C.
- Pour approximately 20 mL per Petri dish, ensuring even coverage.
- Allow plates to solidify at room temperature, then store at 4 ° C.

3.1.3 Bacterial Streaking

- Streak LB-agar plates with the *E. coli* strain of interest.
- Incubate overnight at 37 ° C (16 – 20 h).

3.1.4 Preparation of Buffer A Stock Solutions

- Prepare three concentrated buffer A stock solutions (100×) in separate 500 mL bottles:
- **Stock solution 1 (S1)**: 1 M Tris-Acetate
  - Dissolve 60.57 g Tris base in 400 mL Milli-Q water.
  - Adjust pH to 8.2 using glacial acetic acid.
  - Bring to a final volume 500 mL with Milli-Q water.
- **Stock solution 2 (S2)**: 1.4 M Magnesium Acetate
  - Dissolve 107.23 g magnesium acetate tetrahydrate in 500 mL Milli-Q water.
  - Mix until fully dissolved.
- **Stock solution 3 (S3)**: 6 M Potassium Acetate
  - Dissolve 294.42 g potassium acetate in 500 mL Milli-Q water.
  - Mix until fully dissolved.

(Optional) Filter-sterilize through 0.22 μm membranes. Autoclave each solution at 121 ° C for 20 min.

Store each solution at room temperature for up to 6 months.

3.2 Cellular Extract Preparation: Day 2

3.2.1 Starter Culture

- Inoculate a single colony into 30 mL LB in a 300 mL baffled Tunair flask.
- Add appropriate antibiotics if applicable.
- Incubate at 37 ° C, shaking at 250 rpm for 8 – 12 h.

3.2.2 Glucose Solution (0.4 M) Preparation

- Dissolve 18 g glucose in 250 mL Milli-Q water.

3.2.3 Preparation of 2xYTP Media

- Dissolve 16 g tryptone, 10 g yeast extract, 5 g Sodium Chloride (NaCl), 7 g Potassium Phosphate Dibasic (K_2HPO_4), and 3 g Potassium Phosphate Monobasic (KH_2PO_4) in 500 mL Milli-Q water.
- Adjust pH to 7.2 using 5 N Potassium Hydroxide (KOH).
- Bring volume to 750 mL.
- Transfer 750 mL 2xYTP media to a 2.5 L baffled Tunair flask and close the cap. Extra cover with aluminum foil is recommended.
- Sterilize the 0.4 M Glucose solution and 2xYTP media separately at 121 ° C for 20 min.

3.2.4 Main Culture Media and S30 Buffer A

- 2xYPTG:
  - Add sterilized 250 mL 0.4 M Glucose to 750 mL 2xYTP media in the biosafety cabinet. Final volume of 2xYTPG media: 1 L.

- Mix thoroughly to ensure homogeneous distribution.
- Antibiotics are not included.

- S30 Buffer A (wash buffer):
  - Combine equal volume (10 mL) of each stock solution (**S1**, **S2,** and **S3**) and 970 mL chilled Milli-Q water.
  - Aliquot and store 10 mL of mixed solution as S30 Buffer B (lysis buffer).
  - Add 2 mL 1 M DTT to the remaining 990 mL S30 Buffer A and store both buffers at 4 ° C.

3.3 Cellular Extract Preparation: Day 3

3.3.1 Main Culture Inoculation and Growth Monitoring

- Warm 2×YPTG media to 37 ° C (30 – 60 min before inoculation).
- Turn on the spectrophotometer and allow 20 – 30 min for stabilization.
- Measure OD_{600} of overnight culture from **Step 3.2.1** (*see* **Note 1**):
- Blank the spectrophotometer using 1 mL of clean LB media.
- Get a microcentrifuge tube and add 990 μL clean LB.
- Add 10 μL of overnight culture to the microcentrifuge tube (1 : 100 dilution). Mix thoroughly.
- Add the entire 1 mL mixture to a clean cuvette and measure OD_{600}. Record this value as the initial OD_{600} value for the main culture (*see* **Note 2**).
- Transfer 10 mL of the overnight culture to 1 L (1 : 100 ratio) main culture media (2xYTPG) in the biosafety cabinet.
- Transfer the inoculated main culture media to the incubator and start the main culture at 37 ° C, shaking at 250 rpm.
- Monitor the cell growth rate (OD_{600}) of the main culture:
- Use a 2 mL serological pipette to quickly collect a small sample (~ 1.5 mL) of the main culture and transfer it to a clean microcentrifuge tube. A new sample of the main culture must be collected for each OD_{600} measurement.
- Transfer 1 mL of the sample to a clean cuvette and measure OD_{600}. Once the OD_{600} value reaches 1.0, dilute the collected culture sample in fresh LB media using a 1:10 dilution ratio (Mix 900 μL of fresh LB and 100 μL of the sample). Multiply the result by 10 and plot the growth curve to monitor the cell growth until OD_{600} reaches the mid-exponential growth phase.
- Calculate doubling time (t_d) with the following Eq. 1:

$$t_d = \frac{\Delta \text{time (min)} \times \log 2}{\log[late\ OD] - \log[\text{early}\ OD]} \quad (1)$$

- (Optional) In order to carry out the cell induction, add the inducing agent at OD_{600} of around 0.5 (*see* **Notes 3** and **4**).

3.3.2 Cell Harvesting

- All subsequent cellular extract preparation steps must remain at 4 ° C or below.
- Place the centrifuge bottles on ice and turn on the centrifuge to prechill (4 ° C). Weigh empty conical tubes and mark the weight (g) on the tube, put on ice.
- Harvest cells at the mid-exponential phase using the centrifuge bottles by centrifugation at 5000 × *g*, 4 ° C for 15 min.
- Discard supernatant and immediately place the centrifuge bottles on ice.

3.3.3 Cell Washing

- Resuspend pellet in 25 mL of chilled S30 Buffer A. Transfer resuspended cells to 50 mL prechilled conical tubes (pre-weighted) and bring the chilled S30 Buffer A volume to 40 mL. Resuspend the cells by vortexing until homogeneous resuspension of the cells.
- Centrifuge resuspension at 5000 × *g*, 4 ° C for 10 min (Fiberlite™ F15-8x50cy Fixed Angle Rotor).
- Discard the supernatant and repeat three times.
- Determine the wet cell weight (g) and record.
- Flash-freeze the pelleted cells in liquid nitrogen and store at −80 ° C.

3.4 Cellular Extract Preparation: Day 4

3.4.1 Thawing and Resuspension

- Thaw the frozen cells on ice for 30 − 60 min.
- Add 1 mL of S30 Buffer B (no DTT) per 1 g wet cell mass and resuspend gently.
- Transfer the cell suspension into clean microtubes in an ice-water container (1 mL per tube).

3.4.2 Cell Lysis by Sonication Method

- Lyse the cells using the Q125 Sonicator at 20 kHz frequency and 50% amplitude [18] reaching energy (*Joules*) given by the following Eq. 2 [17] (*see* **Note 5**):

$$\text{Energy} = \frac{\text{Volume } (\mu\text{L}) - 33.6}{1.8} \qquad (2)$$

  - Process one tube at a time to maintain temperature control.
  - Immerse the tube in an ice-water bath.
  - Insert the sonication probe to approximately 2/3 of the depth of the sample.
  - Sonicate in short pulses (2–3 s on, 5–10 s off) until target energy is reached.

- Quickly after reaching the required energy (*Joules*), add 3 μL of 1 M DTT per mL of the lysate and invert the tube several times quickly.

3.4.3 Clarification

- Centrifuge the lysates at 12,000 × *g*, 10 min, 4 ° C.
- Carefully collect supernatants in fresh microtubes (*see* **Note 6**).
- Flash-freeze the cellular extract in liquid nitrogen and store at −80 ° C.

3.5 Optional Runoff Reaction and Dialysis

3.5.1 Runoff Reaction

- Seal the tube lids with parafilm and wrap in aluminum foil.
- Tape the covered and wrapped tubes to the Incubator tube holder.
- Incubate at 37 ° C, shaking at 250 rpm for 80 min.
- After the runoff reaction, immediately place the tubes on ice.
- Centrifuge at 10,000 × *g*, 10 min, 4 ° C for clarification and take supernatant as the final cellular extract.

3.5.2 Dialysis

- Prepare Dialysis Buffer C by mixing 5 mL stock solution **S1,** 10 mL each stock solution **S2** and **S3,** 975 mL chilled Milli-Q water, and 1 mL 1 M DTT.
- Pre-soak a Slide-A-Lyzer™ G2 dialysis cassette (10 K MWCO) in Dialysis Buffer C (300 mL) for 5 min, by fully submerging the cassette inside the beaker, and then place the beaker in the 4 ° C.
- Transfer 3 mL of cellular extract to the dialysis cassette and submerge the cassette in the beaker with the Dialysis Buffer C.
- Dialyze for 1 h with gentle stirring, repeating three times and exchanging the dialysis buffer each times.
- After completing dialysis, transfer the dialyzed cellular extract to clean microtubes and centrifuge for the final clarification at 12,000 × *g*, 4 ° C for 10 min.
- Carefully collect supernatants (*see* **Note 6**) in fresh microtubes and flash-freeze the tubes with liquid nitrogen and store at −80 ° C.

3.6 DNA Template Preparation

DNA templates for cell-free reactions can be prepared either as plasmids or as PCR-amplified linear expression templates (LETs). Plasmid DNA provides greater stability during reactions, while LETs offer a rapid, cloning-free option that enables fast iteration during genetic circuit prototyping. The procedures below describe primer design, amplification, purification, and template preparation for both plasmids and LETs.

3.6.1 Primer Design

- Primers should be designed to amplify the appropriate region, either the insert and vector backbone for plasmid construction, or the entire expression cassette (including promoter-UTR-gene of interest-terminator) for LET production. Primer design generally follows the same guidelines:
  - Binding region length: 18 – 35 bp.
  - Melting temperature (Tm): Target ~60 ° C for consistent annealing conditions.
  - GC content: 40 – 60%.
  - Avoid secondary structures: Check the sequence using tools such as *OligoAnalyzer* or *Primer-BLAST.*
  - Include appropriate overhangs for assembly steps (20–40 bp overlaps for Gibson Assembly or 4 bp overhangs for Golden Gate).
- For LETs, adding noncoding flanking sequences (typically 50–150 bp upstream and downstream of the expression cassette) improves template stability in crude extracts. Primers may also be synthesized with phosphorothioate bonds at the 5′ ends to provide additional protection from degradation by exonuclease.

3.6.2 PCR Amplification

- PCR amplification for plasmid inserts, vector backbones, or LETs is performed using a high-fidelity DNA polymerase (e.g., Phusion® or Q5®). Plasmid templates (1 – 10 ng) can be used directly for PCR. A standard 50 μL reaction includes high-fidelity buffer, dNTPs, primers, and polymerase.
- Typical PCR cycle conditions include:
- Initial denaturation: 98 ° C for 30 s.
- 30 cycles of: 98 ° C for 10 s; Annealing at primer-specific Tm for 20 – 30 s; Extension at 72 ° C for 20 – 30 s/kb.
- Final extension: 72 ° C for 5 min.

3.6.3 DpnI Digestion

- When PCR products are generated from Plasmid templates, DpnI digestion should be performed to remove methylated parental DNA.
- Add 1 μL DpnI to 50 μL PCR product and incubate at 37 ° C for 1 h. Digestion should occur immediately prior to the purification step (*see* **Note** 7).

3.6.4 Purification of PCR Products

- PCR products for both plasmid assembly and LET preparation are purified using a DNA gel extraction kit or spin-column purification kit.
- Gel extraction is recommended when multiple bands or primer dimers are present.

- DNA should be eluted in nuclease-free water. EDTA-containing elution buffers may interfere with transcription-translation reactions.
- DNA concentration and purity should be confirmed before use. Quantify DNA using spectrophotometry (NanoDrop®) or fluorometry (Qubit®). Assess purity using A_{260}/A_{280} ratio (should be 1.9 – 2.0).

3.6.5 Gibson Assembly (Plasmid Construction Only)

- For plasmid preparation, insert and vector fragments are assembled using Gibson Assembly Master Mix.
- Prepare DNA fragments:
- Use 50 – 100 ng linearized vector.
- Use insert(s) in a 2:1 or 3:1 molar ratio relative to the vector.
- Assembly reaction:
- 5 μL Gibson Assembly Master Mix.
- DNA fragments (*total volume* $\leq$ 20μL).
- Incubate at 50 ° C for 60 min.

3.6.6 Transformation

- Thaw competent cells on ice for 10 – 15 min before use.
- Add 2 – 5 μL assembled plasmid to 50 μL competent cells and incubate for 30 min on ice.
- Heat shock at 42 ° C for 30 – 45 s.
- Return to ice for 2 min.
- Recover in 250 μL SOC media at 37 ° C for 1 h.
- Plate on LB-Agar with appropriate antibiotic(s).
- Incubate overnight at 37 °C.
- Successful assemblies typically yield 10–100 colonies.

3.6.7 Plasmid Purification

·Pick a colony and grow in 3 – 10 mL LB media with antibiotic overnight at 37 ° C.
·Harvest cells at 5,000 rpm for 5 min.
·Use a commercial kit for the plasmid purification (*see* **Note 8**).

- Quantify DNA using spectrophotometry (NanoDrop®) or fluorometry (Qubit®).
- Assess purity using A_{260}/A_{280} ratio (should be 1.9 – 2.0).

4 Notes

1. The OD_{600} of the overnight culture should be measured after diluting it using the same ratio intended for inoculating the main culture. This diluted measurement directly reflects the starting OD_{600} of the main culture. Therefore, no additional multiplication by the dilution factor is needed.

2. Commercial BL21(DE3) strains typically yield an OD_{600} of 0.03–0.05 when inoculated at 1% (v/v). Cultures harboring plasmids may show slightly lower OD_{600} growth due to the growth burden imposed by plasmid maintenance.
3. BL21(DE3) strains express T7 RNA Polymerase under the control of the lacUV5 promoter and can be induced by IPTG. Induction should occur when the culture reaches around 0.5 OD_{600}, using final IPTG concentrations between 0.1 mM – 1 mM, depending on the expression load and toxicity of the target protein.
4. For targeted protein overexpression in cells containing inducible expression plasmids, alternative induction systems beyond standard protocols may be employed. Arabinose and tetracycline represent two commonly utilized inducers, both of which are typically administered during the early exponential growth phase (OD_{600} 0.1 – 0.5) to maximize protein production. Arabinose induction is generally performed at final concentrations ranging from 0.5 to 4 mg/mL in either 2xYT or 2xYTP culture media. However, it is important to note that arabinose-mediated induction is subject to catabolite repression by glucose. Therefore, optimal induction requires the use of glucose-free media (2xYTP) when working with arabinose-inducible expression systems. Tetracycline induction, by contrast, operates through a distinct mechanism and is effective at significantly lower concentrations, typically between 1 and 10 ng/mL.
5. In the lysis equation, "volume" refers to the amount of resuspended wet cell mass contained within a microtube. This relationship varies across *E. coli* strains and should be optimized.
6. The clarification step removes residual cell debris that can adversely affect cell-free performance. A properly clarified cellular extract should appear clear and yellow-brown. Aliquot the cellular extract in small volumes (60 – 140 μL) or a mixture of small aliquots (30, 50, 100 μL) to minimize freeze-thaw cycles, as the cellular extract activity typically decreases after 3–4 freeze-thaw cycles.
7. To preserve PCR product integrity, DpnI digestion should be performed immediately prior to gel extraction, ideally on the same day. Storing PCR products with DpnI at 4 ° C is discouraged. The digestion can conveniently proceed while the agarose gel is solidifying.
8. DNA templates intended for cell-free reactions must be eluted in nuclease-free water. Standard elution buffers contain EDTA, which chelates Mg^{2+}, an essential cofactor for both DNA polymerases and ribosomes, and can inhibit transcription-translation reactions.

References

1. Pardee K, Green AA, Ferrante T, Cameron DE, Daleykeyser A, Yin P et al (2014) Paper-based synthetic gene networks. Cell 159:940–954
2. Pardee K, Green AA, Takahashi MK, Braff D, Lambert G, Lee JW et al (2016) Rapid, low-cost detection of Zika virus using programmable biomolecular components. Cell 165:1255–1266
3. Murphy TW, Sheng J, Naler LB, Feng X, Lu C (2019) On-chip manufacturing of synthetic proteins for point-of-care therapeutics. Microsyst Nanoeng 5:13
4. Zhang F, Carothers JM, Keasling JD (2012) Design of a dynamic sensor-regulator system for production of chemicals and fuels derived from fatty acids. Nat Biotechnol 30:354–359
5. Liu DV, Zawada JF, Swartz JR (2005) Streamlining *Escherichia coli* S30 extract preparation for economical cell-free protein synthesis. Biotechnol Prog 21:460–465
6. Zawada JF, Yin G, Steiner AR, Yang J, Naresh A, Roy SM et al (2011) Microscale to manufacturing scale-up of cell-free cytokine production-a new approach for shortening protein production development timelines. Biotechnol Bioeng 108:1570–1578
7. Noireaux V, Liu AP (2020) The new age of cell-free biology. Annu Rev Biomed Eng 22: 51–77
8. Jewett MC, Swartz JR (2004) Mimicking the *Escherichia coli* cytoplasmic environment activates long-lived and efficient cell-free protein synthesis. Biotechnol Bioeng 86:19–26
9. Noireaux V, Bar-Ziv R, Libchaber A (2003) Principles of cell-free genetic circuit assembly. Proc Natl Acad Sci USA 100:12672–12677
10. Silverman AD, Karim AS, Jewett MC (2020) Cell-free gene expression: an expanded repertoire of applications. Nat Rev Genet 21:151–170
11. Takahashi MK, Chappell J, Hayes CA, Sun ZZ, Kim J, Singhal V et al (2015) Rapidly characterizing the fast dynamics of RNA genetic circuitry with cell-free transcription-translation (TX-TL) systems. ACS Synth Biol 4:503–515
12. Perez JG, Stark JC, Jewett MC (2016) Cell-free synthetic biology: engineering beyond the cell. Cold Spring Harb Perspect Biol 8: a023853
13. Hodgman CE, Jewett MC (2012) Cell-free synthetic biology: thinking outside the cell. Metab Eng 14:261–269
14. Katzen F, Chang G, Kudlicki W (2005) The past, present and future of cell-free protein synthesis. Trends Biotechnol 23:150–156
15. Garamella J, Marshall R, Rustad M, Noireaux V (2016) The all *E. Coli* TX-TL toolbox 2.0: a platform for cell-free synthetic biology. ACS Synth Biol 5:344–355
16. Takahashi MK, Hayes CA, Chappell J, Sun ZZ, Murray RM, Noireaux V et al (2015) Characterizing and prototyping genetic networks with cell-free transcription-translation reactions. Methods 86:60–72
17. Kwon YC, Jewett MC (2015) High-throughput preparation methods of crude extract for robust cell-free protein synthesis. Sci Rep 5:8663
18. Kim J, Copeland CE, Padumane SR, Kwon YC (2019) A crude extract preparation and optimization from a genomically engineered *Escherichia coli* for the cell-free protein synthesis system: practical laboratory guideline. Methods Protoc 2:1–15
19. Wagner L, Jules M, Borkowski O (2023) What remains from living cells in bacterial lysate-based cell-free systems. Comput Struct Biotechnol J 21:3173–3182
20. Sun ZZ, Yeung E, Hayes CA, Noireaux V, Murray RM (2014) Linear DNA for rapid prototyping of synthetic biological circuits in an *Escherichia coli* based TX-TL cell-free system. ACS Synth Biol 3:387–397

Part II

Control

Chapter 10

Programming RNA Trans-Splicing for Versatile Gene Regulation and Complex Cellular Logic Computation

Yuanli Gao and Baojun Wang

Abstract

Synthetic biology aims to customize biological systems by programming genetic circuits. However, the scale and complexity of genetic circuits are limited by a shortage of composable, programmable, and orthogonal regulatory parts. Recently, split-intron-enabled trans-splicing riboregulators (SENTRs) have emerged as a promising tool for regulating gene expression at the post-transcriptional level, featuring low leakage expression, wide dynamic range (> 1000-fold), high predictability using machine learning, and orthogonality across multiple intronic components. SENTR's composability with other regulatory modalities, such as split inteins, enables the establishment of a novel split-biomolecule-enabled circuit design paradigm, unleashing the information-processing capacity of a single regulator gene for complex cellular logic computation with up to six inputs. Here, we describe the methods and strategies to implement SENTRs for efficient gene regulation, including computational design, experimental assessment, and statistical analysis of SENTR-based genetic circuits. Validated on ten genes encoding fluorescent proteins, transcription factors, and ncRNAs, this method can be readily utilized to develop SENTRs for diverse applications, such as stringent gene regulation, complex logic computation, and biosensing for intracellular RNAs or physiological biomarkers.

Key words Riboregulator, RNA regulation, RNA splicing, Group I intron, Synthetic gene circuit, Biocomputing

1 Introduction

Synthetic biology applies engineering principles to customize biological systems, constructing complex synthetic circuits from modular genetic parts. However, the scaling of genetic circuits faces challenges due to the scarcity of composable parts, idiosyncratic behavior, and heavy cellular burden imposed by large genetic footprints [1, 2].

RNA-based regulators, also known as riboregulators, have emerged as appealing solutions to these challenges. The highly programmable nature of RNA molecules, with their predictable base-pairing mechanism and well-studied thermodynamics, enables

Xiao-Jun Tian (ed.), *Synthetic Gene Circuits: Methods and Protocols*, Methods in Molecular Biology, vol. 3041,
https://doi.org/10.1007/978-1-0716-5304-3_10,

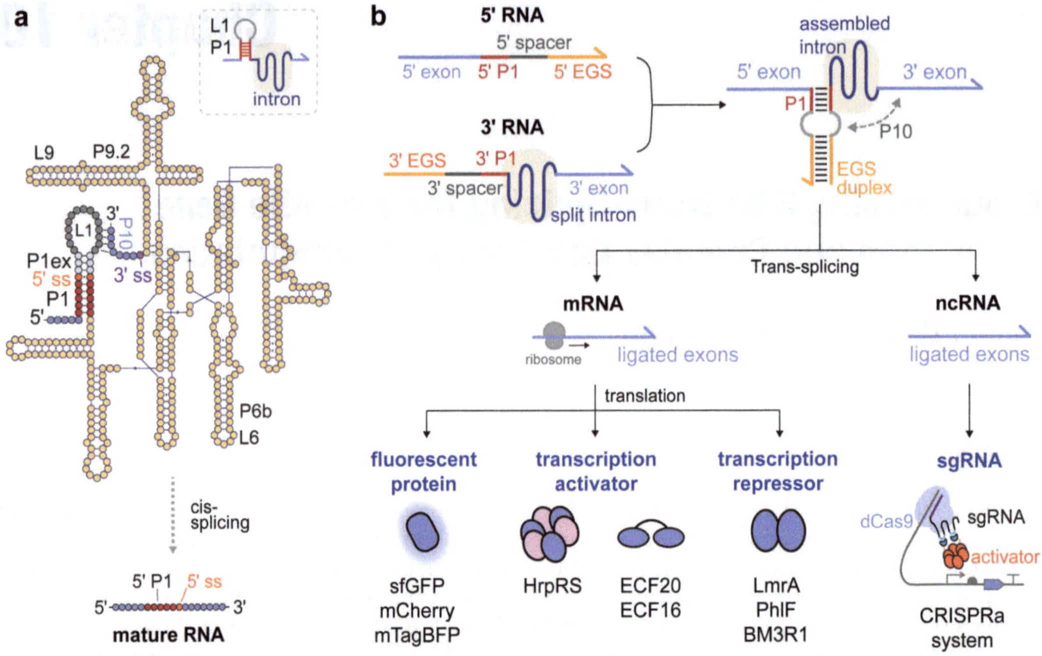

Fig. 1 Design schematics of the split-intron-enabled trans-splicing riboregulators (SENTRs). (**a**) RNA secondary structure and cis-splicing process of *Tetrahymena* rRNA intron. Colored sequence features: blue, exons; yellow, intron body; red, P1 helix; grey, loop 1 (L1); orange, 5′ splice site (5′ ss); purple, 3′ splice site (3′ ss). P1 helix formed by 5′ P1 and internal guide sequence (IGS) decides 5′ ss. P10 helix formed by L1 and 3′ exon contributes to 3′ ss recognition. Inset, simplified representation of intron secondary structure. (**b**) Design schematic of SENTRs. The split introns separate exons into 5′ RNA and 3′ RNA. The formation of the EGS duplex (colored orange) facilitates the assembly of split introns, which triggers RNA trans-splicing to reconstitute the intact RNAs. SENTRs have been demonstrated to regulate mRNAs (encoding fluorescent proteins, transcription activators, and repressors) and ncRNAs (sgRNA in the CRISPR activation system)

computational sequence design and in silico prediction of circuit behavior. This has led to the development of diverse riboregulators, including toehold switches [3], small transcription-activating RNAs (STARs) [4], and loop-initiated RNA activators (LIRAs) [5]. Furthermore, riboregulators operate at faster timescales and possess smaller genetic footprints compared to transcription factor-based circuits [6]. However, promising mechanisms for designing riboregulators are still lacking, with most synthetic riboregulators based on two predominant mechanisms, transcription and translation regulation.

Group I introns are self-splicing ribozymes that catalyze their own excision from precursor RNAs and ligate the flanking exons (Fig. 1a). Since their initial discovery as catalytic RNAs in 1982, group I introns have revolutionized our understanding of RNA functions by revealing that RNAs can catalyze chemical reactions alongside carrying genetic information [7]. Recently, group I introns have garnered increasing interest as a powerful enabling tool in synthetic biology [8, 9], due to several unique properties: (1) their autocatalytic mechanism is largely insulated from the endogenous machinery in host cells, minimizing unintended interactions between synthetic circuits and host contexts; (2) the stem-loop structures can be reprogrammed or split, enabling conditional control of splicing [10]; (3) their widespread distribution in all domains of life (bacteria [11], archaea [12], and eukaryotes [7]), as well as viruses [13], suggest that regulatory modalities derived from group I introns could be portable across diverse host organisms.

Building on group I intron-mediated splicing, we developed a class of split-intron-enabled trans-splicing riboregulators (SENTRs) [14]. SENTRs separate target RNAs into two fragments (5′ RNA and 3′ RNA) and rejoin them through RNA trans-splicing, directed by the de novo designed external guide sequences (EGSs) (Fig. 1b). SENTRs exhibited low leakage expression, wide dynamic range (>1000-fold), high predictability with machine learning, and orthogonality at multiple intronic component levels. SENTRs can sense RNA inputs, process signals through logic operations, and produce various outputs, either mRNAs or non-coding RNAs (ncRNAs). SENTRs were then integrated with the split intein-mediated protein trans-splicing to realize digital logic operation with up to six inputs. More recently, we incorporated SENTRs with transcriptional factor-based biosensors to devise trans-splicing denoiser circuits for low-leakage, on-site detection of physiological biomarkers [15].

Here, we describe the methods and strategies to design and characterize SENTRs for efficient gene regulation, including (1) in silico design of SENTRs to regulate a gene of interest; (2) experimental protocols to characterize the SENTR-based circuits, illustrated using the split-intron and intein-enabled four-input AND gate; and (3) statistical approaches for analyzing circuit performance, with emphasis on data normalization methods to improve cross-laboratory reproducibility.

2 Materials

2.1 Strains, Media, and Chemicals

1. *Escherichia coli* TOP10 for cloning experiments and circuit characterization. Bacterial strains transformed with circuit plasmids or control plasmids are used for flow cytometry assays. The circuit plasmids are designed following the procedures in Subheading 3.1 and constructed using standard Gibson assembly protocols. Note that the internal guide sequences (IGSs) in SENTRs should be modified to accommodate different exon sequences and can be introduced into circuit plasmids by PCR primers. The control plasmids include empty plasmids

(negative control) and plasmids constitutively expressing the reporter gene (positive control) (*see* Subheading 3.2).

2. Lysogeny broth (LB) Miller medium:10 g/L tryptone, 10 g/L NaCl, 5 g/L yeast extract.
3. Antibiotic stock solutions: 100 mg/mL ampicillin; 50 mg/mL kanamycin. Both antibiotics are sterilized by filtration through 0.22 μm syringe filters. The antibiotic concentrations used in the final bacterial culture are 50 μg/mL for kanamycin and 100 μg/mL for ampicillin.
4. Inducer stock solutions: 512 μM N-(3-oxohexanoyl)-L-homoserine lactone (AHL); 33.2 mM arabinose; 548 mM rhamnose; and 940 nM anhydrotetracycline (aTc). The inducer stock solutions are prepared 40× concentrated of their final concentration in the culture and sterilized by filtration through 0.22 μm syringe filters. The inducer concentrations in the final cell culture are: 12.8 μM for AHL, 830 μM for arabinose, 13.7 mM for rhamnose, and 23.5 nM for aTc.
5. Phosphate buffered saline (PBS) solution for flow cytometry assay: 10× PBS diluted tenfold in ddH_2O, supplemented with 1 mg/mL kanamycin, and sterilized by filtration through 0.22 μm bottle-top filters.

2.2 Equipment and Software

1. Transparent 96-well microplates for inoculation of bacterial strains and flow cytometry assays.
2. Plate shaker (BMG Labtech THERMOstar) for incubating the 96-well plates with continuous shaking at 37 °C and 1000 rpm.
3. Flow cytometer (Attune NxT) equipped with an autosampler for green fluorescence measurements (excitation laser: 488 nm, emission filter: 530/30 nM).
4. Multi-channel pipettes for the fast loading of samples, inducers, and media into 96-well plates. Two twelve-channel pipettes with 100 and 300 μL capacity are used here.
5. Nucleic Acids Package (NUPACK) online tool (https://nupack.org/) [16] for designing EGSs, local NUPACK software for designing orthogonal EGSs. The local NUPACK software can be installed following the NUPACK 4.0 User Guide (https://docs.nupack.org/start/).
6. Attune NxT software for acquiring flow cytometry data from the Attune NxT flow cytometer.
7. FlowJo software for analyzing flow cytometry data.

3 Methods

3.1 In Silico Design of SENTRs

SENTRs offer a powerful tool for modulating the activities of both mRNAs and ncRNAs. This section outlines the computational design process for engineering SENTRs to regulate a target gene.

1. Identify the exon sequence: Determine the exon sequence associated with the gene to be regulated, which can encode either mRNAs or ncRNAs (e.g., sgRNAs in the CRISPR-Cas system).
2. Determine the 5′ splice site: Several criteria should be considered for identifying the optimal 5′ splice site. First, the base at the 5′ splice site should ideally be uracil, as it enhances splicing efficiency compared to adenine, guanine, or cytosine. However, all four bases can serve as acceptable 5′ splice sites if necessary (*see* **Note 1**). Second, the insertion of a split intron after the 5′ splice site should effectively disrupt the normal function of the target mRNA or ncRNA, which can be restored through intron splicing. If applicable, in silico prediction algorithms, such as the RBS calculator [17], should be employed to check for any cryptic translation initiation sites downstream of the 5′ splice site, to minimize the risk of leaky gene expression (*see* **Note 2**). Third, ensure that the sequence context surrounding the 5′ splice site does not interfere with intron splicing. It is suggested to try multiple splice sites in parallel to identify the optimal 5′ splice site with the highest splicing efficiency.
3. Insert the split intron after the 5′ splice site and design the internal guide sequence (IGS) (*see* **Note 3**): Typically, IGS contains a 6-nt 5′ P10 sequence complementary to the 3′ exon (forming the P10 helix) (*see* **Note 4**), a 3-nt P1 extension sequence (comprised of triple uracil), a guanine that forms the U·G wobble pair at the 5′ splice site, and a 5-nt sequence that base pairs with the 5′ exon to form the P1 helix (Fig. 2a). P1 and P10 helices are crucial for the accurate recognition of splice sites and significantly influence splicing efficiency (Fig. 2b) [18].
4. Design external guide sequence (EGS): The EGS consists of a pair of RNA hybridization domains to facilitate split intron assembly. EGSs are typically 20-nt in length, though lengths from 10 to 50-nt are also functional (*see* **Note 5**). EGSs can be selected from the pre-validated library containing 56 EGSs with the highest dynamic ranges exceeding 1000 (Fig. 3a), or de novo designed using the NUPACK online tool (https://nupack.org/). For the de novo design of EGS duplex, a 26-nt 5′ spacer–5′ EGS chimeric RNA and 20-nt 3′ EGS are designed to be unstructured on their own and form a 20-bp

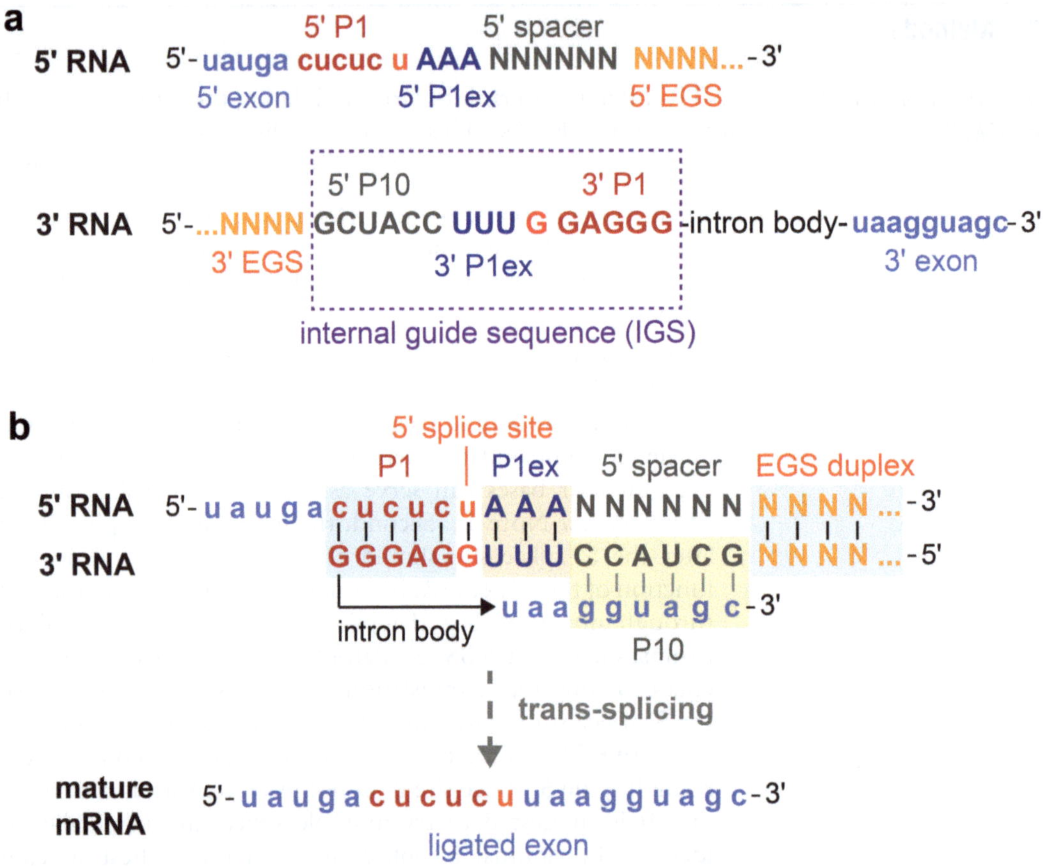

Fig. 2 Design schematics of 5′ RNA and 3′ RNA in SENTRs. **(a)** Sequence features in 5′ RNA and 3′ RNA of SENTRs and **(b)** their functions during RNA trans-splicing. Intronic sequences are denoted in uppercase, and exonic sequences are denoted in lowercase. The intronic internal guide sequence (IGS), which contains 5′ P10, 3′ P1 extension (P1ex), guanosine, and 3′ P1, is crucial for the accurate recognition of splice sites and affects splicing efficiency. The 5′ splice site is determined by the U·G wobble pair (orange). The 3′ splice site is determined by the P10 (shaded yellow) formed by 5′ P10 and a 6-nt sequence in the downstream 3′ exon. The extension of P1 by adding 3-bp P1ex (dark blue) could increase the splicing efficiency

heteroduplex when present together. The NUPACK scripts are as follows:

```
material = rna
temperature = 37.0
trials = 10
sodium[M] = 1.0
dangles = some
allowmismatch = true
# Sequence S1 contains a 6-nt 5'spacer and a 20-nt 5'EGS
# Sequence S2 contains a 20-nt 3'EGS
structure S1 = U26
structure S2 = U20
```

```
# Once co-expressed, 5' EGS binds to 3' EGS, and the 5' spacer
remains unstructured
structure Complex = U6 D20 +
domain a = N6
domain b = N20
S1.seq = a b
S2.seq = b*
Complex.seq = a b b*
prevent = AAAA, CCCC, GGGG, UUUU, KKKKK, MMMMM, RRRRR, SSSSS,
WWWWW, YYYYY
```

5. Design orthogonal EGS: To exert orthogonal regulation on multiple target genes, orthogonal EGSs with low cross-reactivity with each other are needed. One can select orthogonal EGSs from the previously established library containing 24 orthogonal EGSs with dynamic ranges above 380-fold and crosstalk below 16% (Fig. 3b). Alternatively, novel orthogonal EGSs can be designed using the multitube design program in the local NUPACK software, using the following scripts:

```
# Import Python NUPACK module
from nupack import *
# Define physical model
my_model = Model (material='rna', celsius=37)
# Set number of orthogonal systems
systems = 10
# Initialize working variables
crosstalk_targets = {}
crosstalk_excludes = []
step_tubes = []
#Loop over orthogonal systems
for i in range(systems):
# Define sequence domains
da = Domain ('N6', name= ['da', i])
db = Domain ('N20', name = ['db', i])
# Define strands containing these domains
sX = TargetStrand ([da, db], name = ['sX', i])
sY = TargetStrand ([~db], name = ['sY', i])
# ~db is the reverse complement of db
# Define the sequence diversity
hard = [Diversity (word=4, types=2), Diversity (word=6,
types=3)]
# Define target complexes
cX = TargetComplex ([sX], 'U26', name = ['cX', i])
cY = TargetComplex ([sY], 'U20', name = ['cY', i])
cXY = TargetComplex ([sX, sY], 'U6D20+', name = ['cXY', i])
# Define 2 elementary step tubes for each orthogonal system
step0_tube=TargetTube (on_targets = {cX: 1e-08, cY: 1e-08},
```

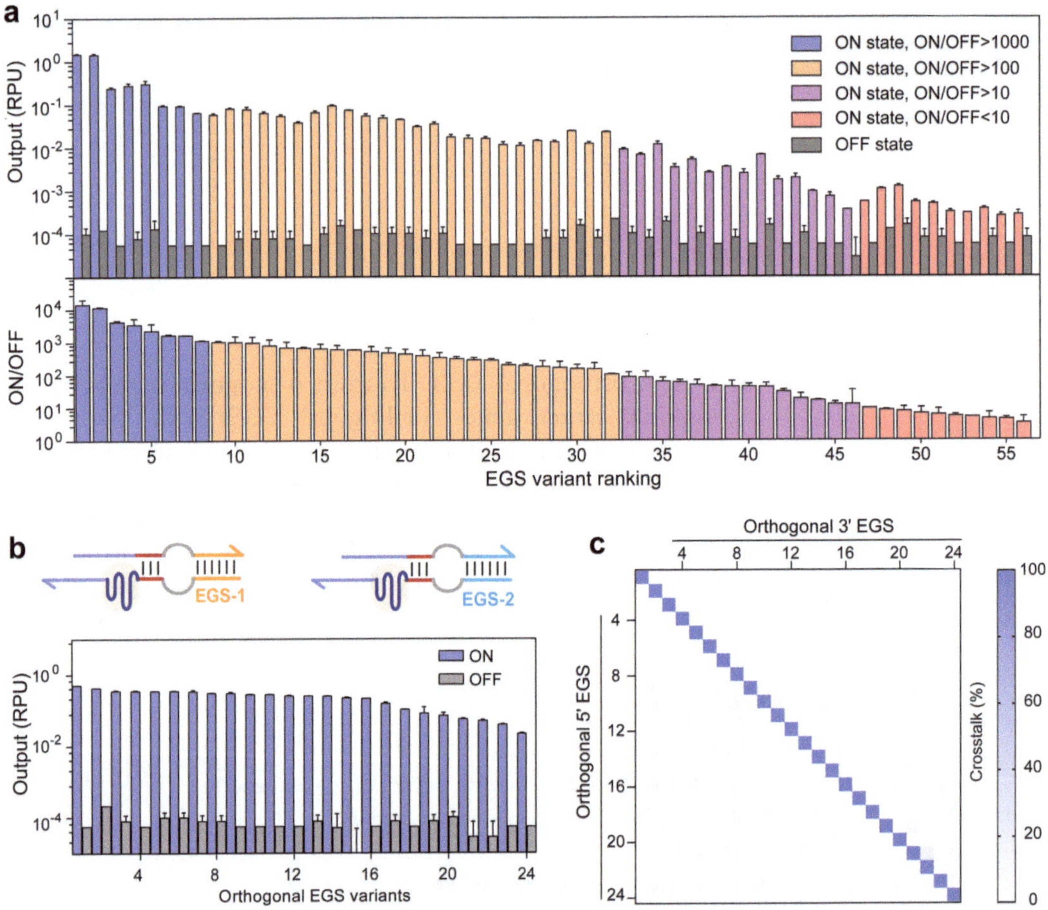

Fig. 3 Trans-splicing performance and orthogonality of SENTR libraries. **(a)** Fluorescence (upper panel) and ON/OFF ratios (lower panel) of SENTRs with 56 EGS variants. **(b)** Design schematic (upper panel) and trans-splicing performance (lower panel) for SENTRs with 24 orthogonal EGS variants. For (**a, b**), the ON-state fluorescence was measured from cells expressing 5′ RNA and 3′ RNA, and the OFF-state fluorescence was determined from cells expressing only 3′ RNA. Bars show the mean values, and error bars represent the standard deviation of $n = 3$ biological replicates. **(c)** The orthogonality matrix for 24 orthogonal EGSs. The heat maps show the mean values of $n = 3$ biological replicates. RPU, relative promoter units

```
off_targets = SetSpec (max_size = 2, exclude=[cXY]), name =
['reactants', i])
step1_tube = TargetTube (on_targets = {cXY: 1e-08},
off_targets = SetSpec (max_size = 2), name = ['products', i])
step_tubes += [step0_tube, step1_tube]
# Store contributions to the global crosstalk tube
crosstalk_targets.update({cX: 1e-08, cY: 1e-08})
crosstalk_excludes += [cXY]
# Define 1 global crosstalk tube
crosstalk_tube = TargetTube (on_targets = crosstalk_targets,
off_targets = SetSpec (max_size=2, exclude = crosstalk_ex-
cludes), name = 'crosstalk')
```

```
# Define tube set
my_tubes = step_tubes + [crosstalk_tube]
# Define defect weight of N for global crosstalk tube
my_weights = Weights(my_tubes)
my_weights[:,:,:,crosstalk_tube] *= systems
# Define and run the test tube design job
my_design = tube_design (tubes = my_tubes, defect_weights =
my_weights, model = my_model, hard_constraints = hard)
my_results = my_design.run(trials=1) [0]
# Display the design results
my_results
```

6. Compose the 5′ RNA and 3′ RNA generator sequence by assembling the intronic and exonic components designed above. In our study, the 5′ RNA and 3′ RNA are expressed on medium-copy plasmid pSB3K3 and high-copy plasmid pSB1A3, respectively (*see* **Note 6**). Note that insulators such as RiboJ [19] are used to eliminate the potential interference from upstream genetic context, especially for 3′ RNA generators.
7. Construct the plasmids following standard protocols of Gibson assembly.

3.2 Characterization of the Split-Intron and Intein-Enabled Four-Input AND Gate

1. Construct the plasmids for the four-input AND gate: In the design of the four-input AND gate, two pairs of split introns fused with two orthogonal EGSs (EGS-O8 and O25) are inserted into the CDS of the N-lobe and C-lobe of the split ECF20–M86 chimeric protein, respectively. Once co-expressed, the four RNA strands undergo two trans-splicing reactions to produce two mature mRNAs translated into two peptides ($ECF20^N$–$M86^N$ and $M86^C$–$ECF20^C$) for protein trans-splicing, to produce the functional ECF20 protein to activate downstream GFP expression (Fig. 4). The 5′ RNA-1, 3′ RNA-1, 5′ RNA-2, and 3′ RNA-2 are expressed from the arabinose, rhamnose, AHL, and aTc-inducible promoters, respectively. The 5′ RNA-1 is expressed on medium-copy plasmid pSB3K3 (pYG-G45 in Fig. 5a), and the other RNAs are expressed on high-copy plasmid pSB1A3 (pYG-G48 in Fig. 5a). The ribosome binding sites (RBSs) are selected from the iGEM RBS collection, with RBS30 (BBa_B0030) for 5′ RNA-1 and RBS32 (BBa_B0032) for 5′ RNA-2 (*see* **Note 7**).
2. Prepare the negative control and positive control strain. Each characterization experiment includes two sets of controls: the negative control is bacteria transformed with empty plasmid vectors (NC-pSB3K3 and NC-pSB1A3, Fig. 5b) and thus not expressing reporter genes; the positive control is bacteria

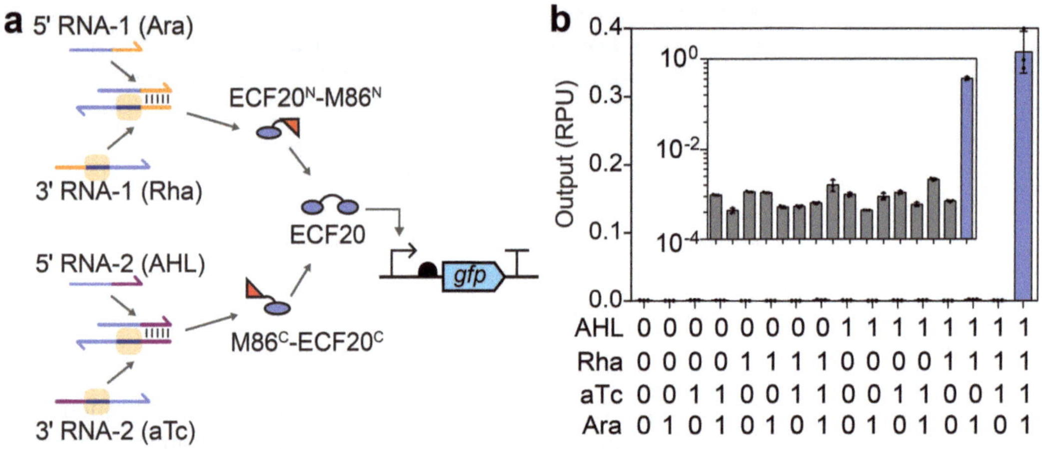

Fig. 4 Design and characterization of the split-intron and intein-enabled four-input AND gate. **(a)** Design schematic of the four-input AND gate. Two pairs of split introns fused with two orthogonal EGSs (orange and purple) trans-splice to form the mRNAs encoding the N-lobe and C-lobe of the split ECF20-M86 chimeric protein ($ECF20^N$-$M86^N$ and $M86^C$-$ECF20^C$). The two lobes of the chimeric protein trans-splice into functional ECF20, activating downstream GFP expression. The input inducers controlling the expression of certain RNA strands are shown in parentheses. **(b)** Output fluorescence for each input combination of the four-input AND gate. Inset: fluorescence levels on a logarithmic scale. Input combinations were achieved in the presence (1) or absence (0) of a certain inducer. Bars show the mean values and error bars represent the standard deviation of $n = 3$ biological replicates. RPU, relative promoter units

expressing reporter genes from a constitutive promoter J23101 (PC-J101-GFP, Fig. 5c).

3. Prepare inducer solutions for each logic state: To characterize the four-input AND gate, we need to prepare 16 inducer solutions ([0, 0, 0, 0], [0, 0, 0, 1], [0, 0, 1, 0], [0, 0, 1, 1], [0, 1, 0, 0], [0, 1, 0, 1], [0, 1, 1, 0], [0, 1, 1, 1], [1, 0, 0, 0], [1, 0, 0, 1], [1, 0, 1, 0], [1, 0, 1, 1], [1, 1, 0, 0], [1, 1, 0, 1], [1, 1, 1, 0], [1]), where 0 or 1 corresponds to the absence or presence of the inducer AHL, rhamnose, aTc, and arabinose. If a certain inducer is absent in the solution, an equal volume of ddH_2O will be added instead. For instance, 1 mL [1, 1, 1, 0] solution can be obtained by mixing 250 μL 512 μM AHL, 250 μL 548 mM rhamnose, 250 μL 23.5 nM aTc, and 250 μL ddH_2O. Keep inducer solutions at −20 °C to prevent degradation or contamination, and make sure to mix the inducer solutions thoroughly every time before use.
4. Day 0: Re-streak the bacterial strains needed (positive control, negative control, and the strains to characterize) on fresh LB agar plates containing the appropriate antibiotics.
5. Day 1: Inoculate single colonies (at least 3 colonies for each bacterial strain) into 180 μL of LB media with antibiotics in a 96-well plate. Incubate the 96-well plate overnight (16 h) in a plate shaker with continuous shaking at 37 °C and 1000 rpm.

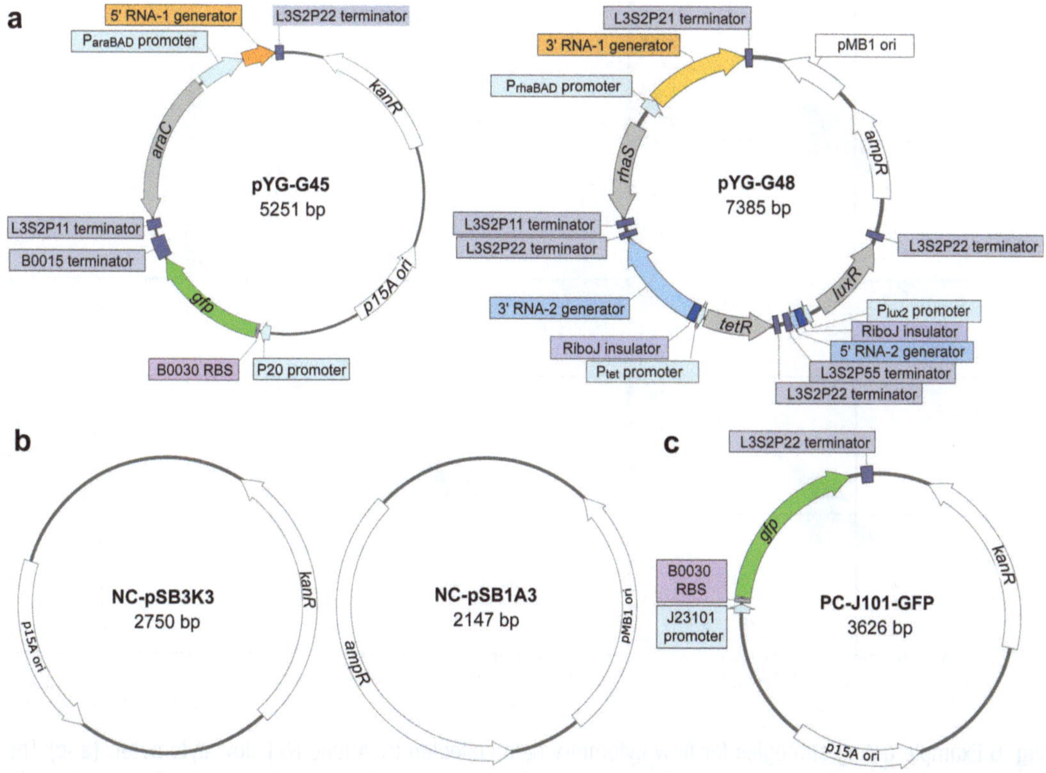

Fig. 5 Plasmid maps showing the representative circuit constructs used in this chapter. (**a**) The plasmids (pYG-G45 and pYG-G48) for the four-input AND gate in Fig. 4. (**b**) The empty plasmids (pSB3K3 and pSB1A3) for negative control. (**c**) The plasmid constitutively expressing GFP on pSB3K3 for positive control

6. Day 2: Dilute the overnight culture (usually by 100-fold) into 4 mL of fresh medium containing the appropriate antibiotic and adjust the final cell density to $OD_{600} = 0.025$. Dispense 180 μL of the appropriate culture in each well of the 96-well plate.
7. Load 20 μL of the inducer solutions into wells containing the corresponding bacterial culture (if appropriate, using a multichannel pipette) to induce synthetic gene expression. Grow the bacterial cultures containing inducers in a plate shaker at 37 °C and 1000 rpm for 6 h (*see* **Note 8**).
8. Aliquot 2 μL of bacterial culture into 198 μL of 1× PBS with 1 mg/mL kanamycin. Incubate the mixture at 4 °C for at least 1 h.
9. Acquire fluorescence data using a flow cytometer, recording at least 10,000 events per sample.

3.3 Statistical Analysis of Circuit Performance

1. Calculate fluorescence intensity using FlowJo. The detailed gating strategies, cell counts, and fluorescence histograms for negative and positive controls are shown in Fig. 6. To

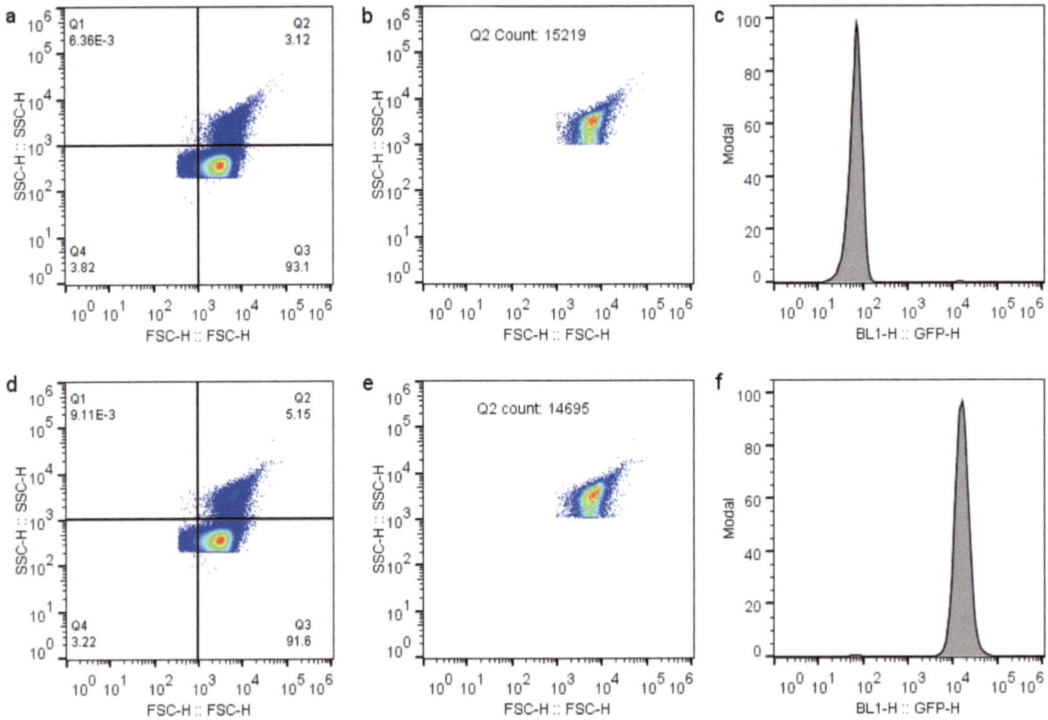

Fig. 6 Example gating strategies for flow cytometry data collected by Attune NxT flow cytometer. **(a–c)** The gating **(a)**, cell counts **(b)** and fluorescence histogram **(c)** for the negative control group. **(d–f)** The gating **(d)**, cell counts **(e)**, and fluorescence histogram **(f)** for the positive control group. Adapted with permission from [14]

differentiate cells from non-cellular particles, all events were gated on SSC-H and FSC-H, and events in Q2 with both values above 1000 a.u. were used for data analysis. Fluorescence intensity was calculated as the median fluorescence of a gated population.

2. Normalize the fluorescence and convert fluorescence values into relative promoter units (RPU) (*see* **Note 9**). First, calculate the mean fluorescence of three biological repeats of the negative control (NC) and positive control (PC) group. Second, subtract the mean fluorescence of the NC group from each sample's fluorescence value. Third, divide the NC-normalized fluorescence value by the NC-normalized mean fluorescence of the PC group. These normalization steps can be described by the following equation:

$$Fluo.(sample, RPU) = \frac{Fluo.(sample) - mean\ Fluo.(NC)}{mean\ Fluo.(PC) - mean\ Fluo.(NC)} \quad (1)$$

3. Calculate the dynamic ranges of logic gates. The output states of a genetic logic gate include TRUE states (when the output is 1) or FALSE states (when the output is 0). The dynamic range

(TRUE/FALSE ratio) of a logic gate is calculated by dividing the lowest TRUE-state fluorescence values by the highest FALSE-state fluorescence values.

4. Calculate the standard deviations (SDs). To calculate the SD of the TRUE/FALSE ratio of a genetic logic gate, we use the mean and SD of the leakiest FALSE states and the lowest TRUE states of a genetic logic gate. The equation is as follows:

$$SD\left(\frac{TRUE}{FALSE}\right) = \sqrt{\left(\frac{SD\,(TRUE)}{Mean\,(TRUE)}\right)^2 + \left(\frac{SD\,(FLASE)}{Mean\,(FALSE)}\right)^2} \times Mean\left(\frac{TRUE}{FALSE}\right) \tag{2}$$

4 Notes

1. The U·G base pair at the 5′ splice site (5′ ss) is highly conserved among group I introns, but not absolutely invariant. For instance, two introns from *Aspergillus nidulans* and *Podospora anserina* utilize a C·G base pair for 5′ ss recognition [20]. Our work further demonstrates that the *Tetrahymena* rRNA intron (TT intron) can accommodate various base pairs (including U·G, C·A, A·U, or G·C) at the 5′ ss, any of which can be used to achieve efficient trans-splicing.
2. Cryptic translation initiation downstream of the 5′ ss can generate truncated peptides with potential biological activities. For instance, in modular SENTR designs where the uracil inside the start codon serves as the 5′ ss, we observed internal translation initiation within the 3′ exon of the *mCherry* gene, which produced N-terminal truncated peptides leading to strong leaky fluorescence. Introducing synonymous mutations to disrupt this translation initiation site successfully eliminated the leaky expression of mCherry.
3. While the TT intron split at loop 1 (L1) is typically used in SENTR design for its high programmability and trans-splicing efficiency, other split introns are also applicable for SENTR, including BnrdE intron split at L6, BnrdF intron split at L6, Thy intron split at L8, Td intron split at L6, NrdB intron split at L6, SunY intron split at L9, NrdE2 intron split at L6, and TT intron alternatively split at L6 or L9.
4. In the SENTR IGS design, a 6-nt 5′ P10 sequence is recommended to maximize trans-splicing activity. In contrast, a shorter 4-nt P10 sequence is advised for designing IGS in cis-splicing circuits.
5. A 20-nt EGS generally provides optimal trans-splicing efficiency compared to shorter or longer variants (e.g., 10, 30,

40, or 50-nt). However, 10-nt EGSs can already trigger trans-splicing reactions, and 50-nt EGSs can also generate detectable trans-splicing fluorescence. It is noteworthy that the optimal lengths of EGSs are not absolute, as their sequences also significantly affect the trans-splicing performances.

6. Efficient intron splicing generally requires higher expression levels of 3′ RNA than 5′ RNA [21]. Therefore, we recommend expressing 3′ RNAs from plasmid vectors with higher copy numbers than those used for the 5′ RNAs.
7. In the split-intron and intein-enabled four-input AND gate, the four RNA inputs trans-splice to produce two mature mRNAs translated into two peptides for protein trans-splicing. Therefore, fine-tuning the translational activities of the mRNAs is crucial for maximizing the dynamic ranges of the logic gates. In particular, the RBS strength controlling 5′ RNA-2 (expressed from the high-copy plasmid) strongly influences leakiness levels, as even minor alterations can significantly affect output protein expression.
8. The optimal induction time for the end-point measurement of fluorescence can vary depending on different circuit designs. To determine the optimal induction time, measure the fluorescence using a plate reader every hour after 5 h of induction. If appropriate, use black-wall 96-well plates and a plate reader with incubation function (e.g., BMG Labtech FLUOstar fluorometer) for repeated reading (20 min/cycle) of the fluorescence and OD_{600} with continuous shaking at 37 °C and 200 rpm.
9. We normalize the data to account for variations in fluorescence (in arbitrary units, a.u.) caused by differences in experimental conditions (e.g., instrumentation, growth medium, and cellular states), which may affect reproducibility across laboratories [22]. Therefore, we employ two control groups, negative control and positive control, as biological references to mitigate such contextual effects. For the four-input AND gate, the negative control is the bacterial strain transformed with empty plasmids pSB3K3 and pSB1A3, and the positive control is the strain transformed with PC-J101-GFP on the pSB3K3 vector and empty plasmid pSB1A3.

Acknowledgments

This work was supported by the National Key R&D Program of China (2023YFF1204500), "Pioneer" and "Leading Goose" R&D Program of Zhejiang (2024C03011, 2025C01097), and National Natural Science Foundation of China (32271475, 32320103001, 32501287).

References

1. Liu Q, Schumacher J, Wan X et al (2018) Orthogonality and burdens of heterologous AND gate gene circuits in *E. coli*. ACS Synth Biol 7:553–564. https://doi.org/10.1021/acssynbio.7b00328
2. Gao Y, Wang L, Wang B (2023) Customizing cellular signal processing by synthetic multi-level regulatory circuits. Nat Commun 14: 8415. https://doi.org/10.1038/s41467-023-44256-1
3. Green AA, Silver PA, Collins JJ, Yin P (2014) Toehold switches: de-novo-designed regulators of gene expression. Cell 159:925–939. https://doi.org/10.1016/j.cell.2014.10.002
4. Chappell J, Westbrook A, Verosloff M, Lucks JB (2017) Computational design of small transcription activating RNAs for versatile and dynamic gene regulation. Nat Commun 8: 1051. https://doi.org/10.1038/s41467-017-01082-6
5. Ma D, Li Y, Wu K et al (2022) Multi-arm RNA junctions encoding molecular logic unconstrained by input sequence for versatile cell-free diagnostics. Nat Biomed Eng 6:298–309. https://doi.org/10.1038/s41551-022-00857-7
6. Green AA, Kim J, Ma D et al (2017) Complex cellular logic computation using ribocomputing devices. Nature 548:117–121. https://doi.org/10.1038/nature23271
7. Kruger K, Grabowski PJ, Zaug AJ et al (1982) Self-splicing RNA: autoexcision and autocyclization of the ribosomal RNA intervening sequence of *tetrahymena*. Cell 31:147–157. https://doi.org/10.1016/0092-8674(82)90414-7
8. Kalvapalle PB, Staubus A, Dysart MJ et al (2025) Information storage across a microbial community using universal RNA barcoding. Nat Biotechnol. https://doi.org/10.1038/s41587-025-02593-0
9. Gambill L, Staubus A, Mo K et al (2023) A split ribozyme that links detection of a native RNA to orthogonal protein outputs. Nat Commun 14:543. https://doi.org/10.1038/s41467-023-36073-3
10. Che AJ, Knight TF (2010) Engineering a family of synthetic splicing ribozymes. Nucleic Acids Res 38:2748–2755. https://doi.org/10.1093/nar/gkq186
11. Reinhold-Hurek B, Shub DA (1992) Self-splicing introns in tRNA genes of widely divergent bacteria. Nature 357:173–176. https://doi.org/10.1038/357173a0
12. Nawrocki EP, Jones TA, Eddy SR (2018) Group I introns are widespread in archaea. Nucleic Acids Res 46:7970–7976. https://doi.org/10.1093/nar/gky414
13. Landthaler M, Shub DA (1999) Unexpected abundance of self-splicing introns in the genome of bacteriophage Twort: introns in multiple genes, a single gene with three introns, and exon skipping by group I ribozymes. Proc Natl Acad Sci 96:7005–7010. https://doi.org/10.1073/pnas.96.12.7005
14. Gao Y, Mardian R, Ma J et al (2025) Programmable trans-splicing riboregulators for complex cellular logic computation. Nat Chem Biol 21:758–766. https://doi.org/10.1038/s41589-024-01781-4
15. Chen S-Y, Gao Y, Fan Y et al (2025) Trans-splicing denoiser circuits enable highly sensitive cellular sensors for physiological biomarkers. Biosens Bioelectron 287:117709. https://doi.org/10.1016/j.bios.2025.117709
16. Zadeh JN, Steenberg CD, Bois JS et al (2011) NUPACK: analysis and design of nucleic acid systems. J Comput Chem 32:170–173. https://doi.org/10.1002/jcc.21596
17. Reis AC, Salis HM (2020) An automated model test system for systematic development and improvement of gene expression models. ACS Synth Biol 9:3145–3156. https://doi.org/10.1021/acssynbio.0c00394
18. Doudna JA, Cormack BP, Szostak JW (1989) RNA structure, not sequence, determines the 5′ splice-site specificity of a group I intron. Proc Natl Acad Sci 86:7402–7406. https://doi.org/10.1073/pnas.86.19.7402
19. Lou C, Stanton B, Chen YJ et al (2012) Ribozyme-based insulator parts buffer synthetic circuits from genetic context. Nat Biotechnol 30:1137–1142. https://doi.org/10.1038/nbt.2401
20. Hur M, Waring RB (1995) Two group I introns with a C·G base pair at the 5′ splice-site instead of the very highly conserved U·G base pair: is selection post-translational? Nucleic Acids Res 23:4466–4470. https://doi.org/10.1093/nar/23.21.4466
21. Jones JT, Sullenger BA (1997) Evaluating and enhancing ribozyme reaction efficiency in mammalian cells. Nat Biotechnol 15:902–905. https://doi.org/10.1038/nbt0997-902
22. Buson F, Gao Y, Wang B (2024) Genetic parts and enabling tools for biocircuit design. ACS Synth Biol 13:697–713. https://doi.org/10.1021/acssynbio.3c00691

Chapter 11

Intercellular CRISPRi for Distributed Genetic Circuits

Abhinav Pujar, Anchita Sharma, and Manish Kushwaha

Abstract

Microbial communities and multicellular organisms employ diverse strategies for allocation of available resources, achieved through task distribution among specialized cells. Drawing inspiration from nature, several synthetic multicellular circuits have been recently constructed where a larger circuit is distributed into several cells in order to reduce the burden on individual cells. Here, we describe the implementation of multicellular logic-gate circuits in bacterial co-cultures that combine DNA messaging with CRISPRi regulation. Leveraging the easily programmable and information-dense DNA molecules, our system is composed of sender bacteria that transmit DNA messages encoding guide RNAs and receiver bacteria that receive them and express the guide RNAs to regulate transcription by CRISPR interference. We demonstrate several functional multicellular circuits representing digital logic gates that operate on timescales comparable to small molecule signaling: NOT, YES, AND, and AND-AND-NOT. The receiver cells process the inputs received to perform computations and generate a logical output.

Key words M13 bacteriophage, Distributed computation, Multicellular circuits, CRISPR interference, Combinational logic

1 Introduction

The ability of living systems to communicate and coordinate is fundamental to their function, ranging from simple bacterial quorum sensing to complex multicellular interactions [1]. In synthetic biology, multicellular circuits have several advantages: reduced metabolic burden, division of labor, specialized sub-functions, and parallel information processing [2, 3]. They have enabled distributed computation [4], pattern formation [5], and coordinated population-level behaviors [6]. However, engineered multicellular systems frequently rely on small-molecule signaling, which is often limited by signal crosstalk between different channels, constrained diversity, slow diffusion dynamics, and a low information-carrying capacity [7]. To overcome these constraints,

Authors Abhinav Pujar and Anchita Sharma have equally contributed to this chapter.

Xiao-Jun Tian (ed.), *Synthetic Gene Circuits: Methods and Protocols*, Methods in Molecular Biology, vol. 3041, https://doi.org/10.1007/978-1-0716-5304-3_11,

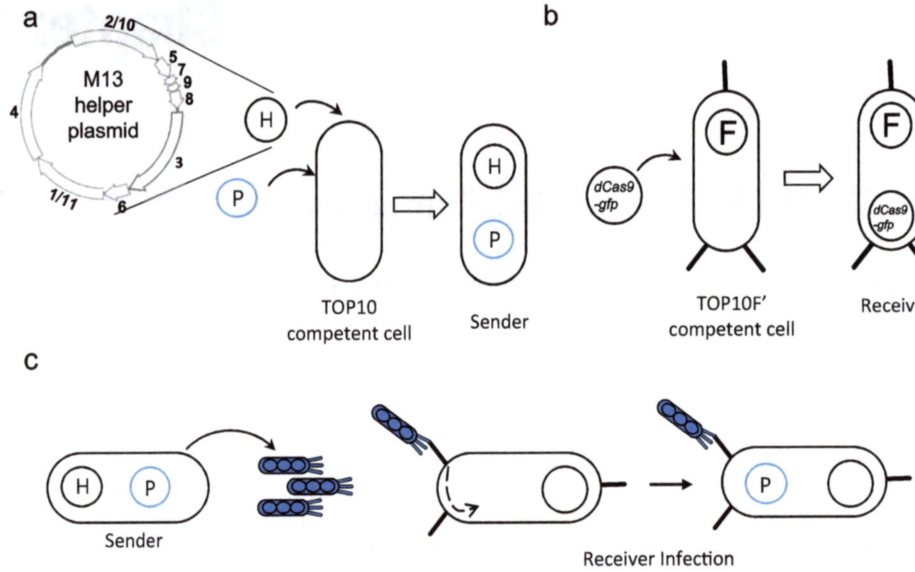

Fig. 1 Sender and receiver cells for intercellular communication. (**a**) Sender cells are made by co-transforming *E. coli* cells with a helper plasmid H (that contains all the M13 bacteriophage machinery except the phage packaging signal) and a phagemid P (a plasmid encoding a guide RNA gene and containing the M13 packaging signal). (**b**) Receiver cells are made by transforming *E. coli* F′ cells with the circuit plasmid encoding the CRISPRi logic circuitry, together with the dCas9 and the GFP gene. (**c**) When sender and receiver cells are co-incubated, M13 phages (with packaged phagemid P) are secreted by the sender cells. These infect the receiver cells via their F-pilus to deliver the phagemid

DNA has emerged as a promising alternative messaging molecule, offering high information density (200 petabytes per gram [8]), high programmability, and the ability of the message to introduce new cellular functions [9]. In order to transmit DNA from cell to cell, cellular processes of horizontal gene transfer can be re-purposed, notably conjugation and transduction [7, 10]. Among these, the non-lytic bacteriophage M13 is particularly attractive for synthetic biology applications because it can continuously produce and export single-stranded DNA (ssDNA) without lysing its host cell [7]. Here we demonstrate that M13 phagemids encoding guide RNA signals, and transmitted by sender cells, can be used to regulate gene expression by CRISPR interference in the receiver cells, resulting in several logic gate behaviors: NOT, YES, AND, AND-AND-NOT [11].

The intercellular CRISPRi (i-CRISPRi) method presented in this chapter consists of (1) sender cells engineered to constitutively secrete M13 phage particles that encode sgRNAs (Fig. 1a), and (2) receiver cells equipped with the CRISPRi machinery (i.e., a nuclease-deficient dCas9) and the CRISPR-repressible promoters that control internal logic and the final fluorescence output

(Fig. 1b). We describe how to construct the sender and receiver cells for multicellular i-CRISPRi circuits, how to grow and co-culture them, and how to analyze the fluorescence data.

2 Materials

1. Parental *E. coli* strains (details below in Table 1): TOP10, TOP10F′, ER2738F′.
2. Plasmid DNA (details below in Table 2): pSEVA63_Helper, pSEVA19_sgRNA-1_M13ps, pSEVA47_dCas9-GFP_NOT_i1, SEVA26_sgRNA-2_M13ps, pSEVA47_dCas9-GFP_YES_i2, pSEVA47_Helper, pSEVA26_sgRNA-A1_M13ps, pSEVA26_sgRNA-A0_M13ps, pSEVA63_sgRNA-B1_M13ps, pSEVA63_sgRNA-B0_M13ps, pSEVA47_dCas9-GFP_AAN_4, pSEVA19_sgRNA-C1_M13ps, pSEVA19_sgRNA-C0_M13ps.
3. LB liquid media: Luria Bertani broth prepared in de-ionized water, adjusted to pH 7.0 using NaOH.
4. LB agar: Luria Bertani broth prepared in deionized water, adjusted to pH 7.0 using NaOH + Agar added to end concentration of 1.5 g/L.
5. Kanamycin sulfate.
6. Gentamycin sulfate.
7. Tetracycline hydrochloride.
8. Spectinomycine dihydrochloride pentahydrate.
9. Ampicillin sodium salt.
10. Streptomycin sulfate.
11. 96-well plate (black)
12. 96-well plate (clear)
13. Plate-reader (Biotek Synergy HTX) 37 °C, 205 cpm, for recording OD_{600} and GFP fluorescence at 15 min intervals.
14. Spectrophotometer UVisco, for measuring OD_{600}.
15. PBS 10×.
16. Attune NxT Flow Cytometer with Cytkick autosampler.
17. Attune Focusing Fluid (1×).
18. Attune Cytometric v5.3 software.

3 Methods

3.1 Single Sender and Single Receiver i-CRISPRi with NOT Gate Circuit

1. Transform chemically competent *E. coli* Top10 cells with pSEVA19_sgRNA-1_M13ps phagemid and pSEVA63_Helper plasmid to make the sender strain (Fig. 1a). Select for positive clones by spreading cells on 50 μg/mL Amp + 5 μg/mL Gent

Table 1
Parental strains

Parental strain	Genotype	Plasmid	Antibiotic resistance/s
TOP10	K-12 *F*⁻ *mcr*A Δ(*mrr-hsd*RMS-*mcr*BC) φ80*lac*ZΔM15 Δ*lac*X74 *rec*A1 *ara*D139 Δ(*ara-leu*)7697 *gal*U *gal*K λ⁻ *rps*L(Str^R) *end*A1 *nup*G	–	Streptomycin (Str^R)
TOP10F’	K-12 *mcr*A Δ(*mrr-hsd*RMS-*mcr*BC) φ80*lac*ZΔM15 Δ*lac*X74 *rec*A1 *ara*D139 Δ(*ara-leu*)7697 *gal*U *gal*K *rps*L(StrR) *end*A1 *nup*G	F′ [*lac*Iq, *Tn*10(TetR)]	Streptomycin[a] (Str^R) / Tetracycline (Tet^R)
ER2738F’	*fhuA2 glnV Δ(lac-proAB) thi-1 Δ(hsdS-mcrB)5*	F′ [*proA⁺B⁺ lacI^q Δ(lacZ)M15 zzf::Tn*10(TetR)]	Tetracycline (Tet^R)

[a]Although this strain is resistant to streptomycin, it was not used to culture it

Table 2
Plasmids

Plasmid	Antibiotic selection	Addgene reference	Protocol/s (circuit/s)
pSEVA63_Helper	Gentamycin (Gen^R)	190432	3.1 (NOT gate), 3.2 (YES gate), 3.3 (AND gate), 3.4 (AND-AND-NOT gate)
pSEVA19_sgRNA-1_M13ps	Ampicillin (Amp^R)	231409	3.1 (NOT gate)
pSEVA47_dCas9-GFP_NOT_i1	Spectinomycin (Spec^R)	231416	3.1 (NOT gate)
SEVA26_sgRNA-2_M13ps	Kanamycin (Kan^R)	231412	3.2 (YES gate)
pSEVA47_dCas9-GFP_YES_i2	Spectinomycin (Spec^R)	231419	3.2 (YES gate)
pSEVA47_Helper	Spectinomycin (Spec^R)	231420	3.3 (AND gate), 3.4 (AND-AND-NOT gate)
pSEVA26_sgRNA-A1_M13ps	Kanamycin (Kan^R)	231414	3.3 (AND gate), 3.4 (AND-AND-NOT gate)
pSEVA26_sgRNA-A0_M13ps	Kanamycin (Kan^R)	231413	3.3 (AND gate), 3.4 (AND-AND-NOT gate)
pSEVA63_sgRNA-B1_M13ps	Gentamycin (Gen^R)	231422	3.3 (AND gate), 3.4 (AND-AND-NOT gate)
pSEVA63_sgRNA-B0_M13ps	Gentamycin (Gen^R)	231421	3.3 (AND gate), 3.4 (AND-AND-NOT gate)
pSEVA47_dCas9-GFP_AAN_4	Spectinomycin (Spec^R)	231415	3.3 (AND gate), 3.4 (AND-AND-NOT gate)
pSEVA19_sgRNA-C1_M13ps	Ampicillin (Amp^R)	231411	3.4 (AND-AND-NOT gate)
pSEVA19_sgRNA-C0_M13ps	Ampicillin (Amp^R)	231410	3.4 (AND-AND-NOT gate)

LB Agar Plate. See protocol 3.5 for testing of phage production by the sender cells.

2. Transform chemically competent *E. coli* Top10F′ cells with pSEVA47_dCas9-GFP_NOT_i1 plasmid to make the receiver strain. Select for positive clones by spreading cells on a 25 μg/mL Spec +5 μg/mL Tet LB agar plate (Fig. 1b).

3. Pick a sender colony and inoculate in 5 mL of LB media supplemented with 50 μg/mL Amp + 5 μg/mL Gent. Pick a receiver colony and inoculate in 5 mL of LB media supplemented with 25 μg/mL Spec +5 μg/mL Tet. Incubate culture tubes 12–16 h at 37 °C and 180 rpm.

4. Subculture by diluting the overnight culture 1000× in 100 mL fresh LB media supplemented with appropriate antibiotics. Incubate at 37 °C and 180 rpm. Monitor OD_{600} of both cultures and stop incubation when culture densities reach ~0.3. To stop growth, immediately move cultures to ice and incubate for at least 30 min. Measure OD_{600} again.
5. Centrifuge the cultures for 10 min, at 4500× g, at 4 °C. Discard supernatant and resuspend in cold LB media without antibiotics (*see* **Note 1**), ensuring that the re-suspended OD_{600} is higher than the final density needed. Dilute the cultures to adjust the density to an OD_{600} of 0.125.
6. Serially dilute density-adjusted sender and receiver cultures by twofold in cold LB media without antibiotics to make culture dilutions of different densities of 0.125, 0.062, and 0.031.
7. Mix 90 μL of LB media with 90 μL of receiver culture dilution in a well (input 0) of a 96-well plate (black) (Fig. 1c). Separately, mix 90 μL of sender culture dilution and 90 μL of receiver culture dilution in a well (input 1) of the 96-well plate. Incubate the plate in a plate reader at 37 °C and 205 cpm, while monitoring OD_{600} and green fluorescence (Excitation: 485/20, Emission: 528/20). Pause the plate reader after 1 h. Add 20 μL LB media +500 μg/mL Spec to the control well (input 0); add 20 μL LB media +150 μg/mL Kan + 250 μg/mL Spec to the co-culture well (input 1). Resume the plate reader incubation for an additional 18 h.
8. Collect co-culture samples for flow cytometry by transferring 10 μL of the co-culture to a new 96-well plate (clear) prepared earlier with 190 μL of cold 1× PBS (+ 2 mg/mL Kan) per well.
9. Use the PBS-diluted culture samples for flow cytometry to record fluorescence of different populations in the culture. *See* **Note 2** for flow cytometer settings, and *see* **Note 3** for population gating details. In flow cytometry analysis, expect uninfected receiver cells to have high fluorescence (ON cells), and infected receiver cells to have low fluorescence (OFF cells).
10. Plot the mean fluorescence values (BL1-H) of the receiver population (Fig. 2a). To calculate the fluorescence fold-change between ON cells (uninfected receivers) and OFF cells (infected receivers), use the following formula:

$$\text{Fold-change} = \text{Mean BL1-H (ON cells)}/\text{Mean BL1-H (OFF cells)}$$

3.2 Single Sender and Single Receiver i-CRISPRi with YES Gate Circuit

Most steps in the protocol below are similar to **3.1**, except the ones marked with "*":

1. * Transform chemically competent *E. coli* Top10 cells with pSEVA26_sgRNA-2_M13ps phagemid and pSEVA63_Helper plasmid to make the sender strain (Fig. 1a). Select for positive

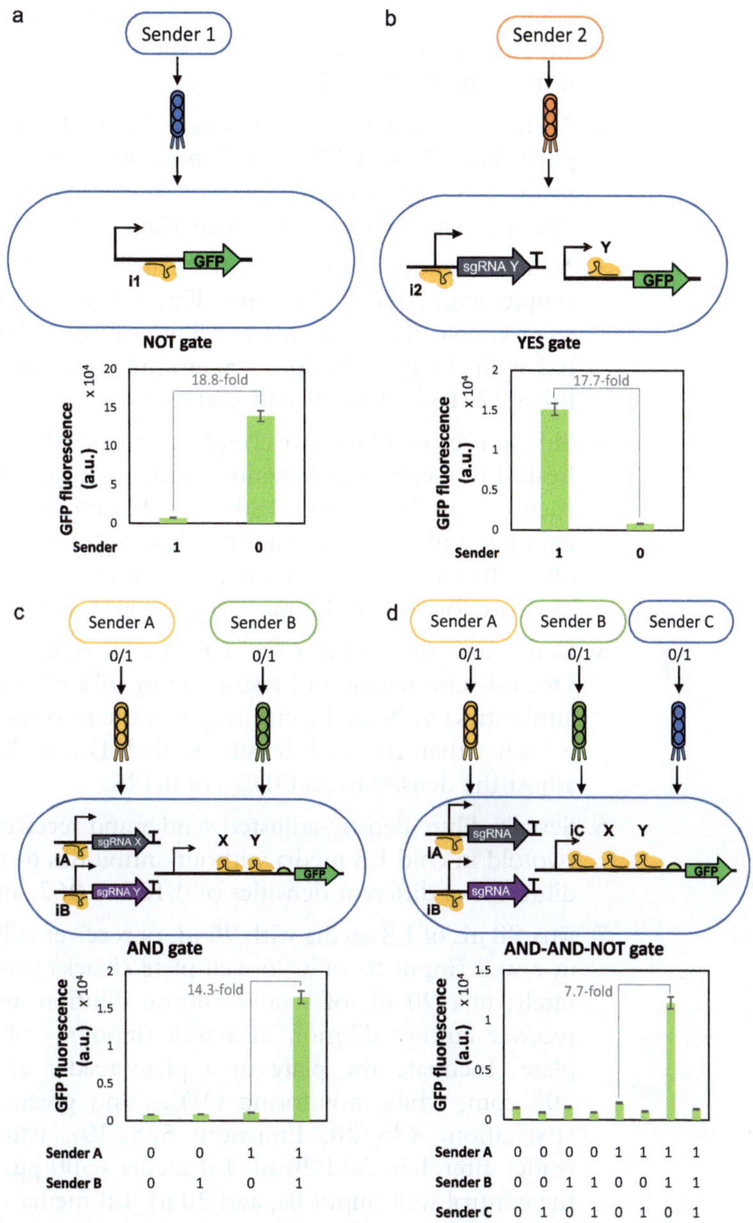

Fig. 2 Intercellular CRISPRi for digit logic gates NOT, YES, AND, and AND-AND-NOT. (**a**) NOT gate receiver cells are incubated with (input 1) or without (input 0) sender-1 cells (see methods), resulting in OFF signal in the presence and ON signal in the absence of the senders. (**b**) YES gate receiver cells are incubated with (input 1) or without (input 0) sender-2 cells (see methods), resulting in ON signal in the presence and OFF signal in the absence of the senders. (**c**) AND gate receiver cells are incubated with sender cells A and B (see methods), each of type 0 or 1, resulting in ON signal only in the presence of both type "1" senders. (**d**) AND-AND-NOT gate receiver cells are incubated with sender cells A, B, and C (see methods), each of type 0 or 1, resulting in ON signal only in the presence of type "1" senders A and B, and type "0" sender C. Data represent mean ± SD from N = 3 repeats

clones by spreading cells on 50 μg/mL Amp + 5 μg/mL Gent LB Agar Plate. See Protocol 3.5 for testing of phage production by the sender cells.

2. * Transform chemically competent *E. coli* Top10F′ cells with pSEVA47_dCas9-GFP_YES_i2 plasmid to make the receiver strain. Select for positive clones by spreading cells on a 25 μg/mL Spec +5 μg/mL Tet LB Agar Plate (Fig. 1b).
3. * Pick a sender colony and inoculate in 5 mL of LB media supplemented with 30 μg/mL Kan + 5 μg/mL Gent. Pick a receiver colony and inoculate in 5 mL of LB media supplemented with 25 μg/mL Spec +5 μg/mL Tet. Incubate culture tubes 12–16 h at 37 °C and 180 rpm.
4. Subculture by diluting overnight cultures 1000× in 100 mL fresh LB media supplemented with appropriate antibiotics. Incubate at 37 °C and 180 rpm. Monitor OD_{600} of both cultures and stop incubation when culture densities reach ~0.3. To stop growth, immediately move cultures to ice and incubate for at least 30 min. Measure OD_{600} again.
5. Centrifuge the cultures for 10 min, at 4500× g, at 4 °C. Discard supernatant and resuspend in cold LB media without antibiotics (*see* **Note 1**), ensuring that the re-suspended OD_{600} is higher than the final density needed. Dilute the cultures to adjust the density to an OD_{600} of 0.125.
6. Serially dilute density-adjusted sender and receiver cultures by twofold in cold LB media without antibiotics to make culture dilutions of different densities of 0.125, 0.062, and 0.031.
7. Mix 90 μL of LB media with 90 μL of receiver culture dilution in a well (input 0) of a 96-well plate (black) (Fig. 1c). Separately, mix 90 μL of sender culture dilution and 90 μL of receiver culture dilution in a well (input 1) of the 96-well plate. Incubate the plate in a plate reader at 37 °C and 205 cpm, while monitoring OD_{600} and green fluorescence (Excitation: 485/20, Emission: 528/20). Pause the plate reader after 1 h. Add 20 μL LB media +500 μg/mL Spec to the control well (input 0); add 20 μL LB media +150 μg/mL Kan + 250 μg/mL Spec to the co-culture well (input 1). Resume the plate reader incubation for an additional 18 h.
8. Collect co-culture samples for flow cytometry by transferring 10 μL of the co-culture to a new 96-well plate (clear) prepared earlier with 190 μL of cold 1× PBS (+ 2 mg/mL Kan) per well.
9. * Use the PBS-diluted culture samples for flow cytometry to record fluorescence of different populations in the culture. *See* **Note 2** for flow cytometer settings, and *see* **Note 3** for population gating details. In flow cytometry analysis, expect

uninfected receiver cells to have low fluorescence (OFF cells), and infected receiver cells to have high fluorescence (ON cells).

10. Plot the mean fluorescence values (BL1-H) of the receiver population (Fig. 2b). To calculate the fluorescence fold-change between ON cells (uninfected receivers) and OFF cells (infected receivers), use the following formula:

$$\text{Fold-change} = \text{Mean BL1-H (ON cells)}/\text{Mean BL1-H (OFF cells)}$$

3.3 Two Senders and Single Receiver i-CRISPRi with AND Gate Circuit

Most steps in the protocol below are similar to **3.1**, except the ones marked with "*":

1. * To make sender A1, competent *E. coli* Top10 cells must be transformed with pSEVA26_sgRNA-A1_M13ps phagemid and pSEVA63_Helper plasmid. To make sender B1, competent *E. coli* Top10 cells must be transformed with pSEVA63_sgRNA-B1_M13ps phagemid and pSEVA47_Helper plasmid. Similarly, to make sender A0 and sender B0, transform competent *E. coli* Top10 cells with pSEVA26_sgRNA-A0_M13ps phagemid + pSEVA63_Helper plasmid and pSEVA63_sgRNA-B0_M13ps phagemid + pSEVA47_Helper plasmid, respectively (Fig. 1a). See Protocol 3.5 for testing of phage production by the sender cells.
2. * To make receiver strain, competent *E. coli* Top10F′ cells must be transformed with pSEVA47_dCas9-GFP_AAN_4 (Fig. 1b).
3. * Pick a sender colony (A1, B1, A0, and B0) and inoculate in 5 mL of LB media supplemented with appropriate antibiotics (see Materials, Table 3). Pick a receiver colony and inoculate in 5 mL of LB media supplemented with 25 μg/mL Spec +5 μg/mL Tet. Incubate culture tubes 12–16 h at 37 °C and 180 rpm.
4. Subculture by diluting overnight cultures 1000× in 100 mL fresh LB media supplemented with appropriate antibiotics. Incubate at 37 °C and 180 rpm. Monitor OD_{600} of both cultures and stop incubation when culture densities reach ~0.3. To stop growth, immediately move cultures to ice and incubate for at least 30 min. Measure OD_{600} again.
5. Centrifuge the cultures for 10 min, at 4500× g, at 4 °C. Discard supernatant and resuspend in cold LB media without antibiotics (*see* **Note 1**), ensuring that the re-suspended OD_{600} is higher than the final density needed. Dilute the cultures to adjust the density to an OD_{600} of 0.125.
6. Serially dilute density-adjusted sender and receiver cultures by twofold in cold LB media without antibiotics to make culture dilutions of different densities of 0.125, 0.062, and 0.031.

7. * Mix 50 μL each of two sender culture dilutions (A0 B0, A0 B1, A1 B0, or A1 B1) and 100 μL of receiver culture dilution in a 96-well plate (black) (Fig. 1c), and incubate in a plate reader for 4 h at 37 °C and 205 cpm. Then, subculture by diluting the co-culture by 40-fold into fresh LB media supplemented with 25 μL/mL Spec +15 μL/mL Kan + 5 μL/mL Gent and incubate in a plate reader for 18 h at 37 °C and 205 cpm, while monitoring OD_{600} and green fluorescence (Excitation: 485/20, Emission: 528/20).
8. Collect co-culture samples for flow cytometry by transferring 10 μL of the culture to a new 96-well plate (clear) prepared earlier with 190 μL of cold 1× PBS (+ 2 mg/mL Kan) per well.
9. * Use the PBS-diluted culture samples for flow cytometry to record fluorescence of different populations in the culture. *See* **Note 2** for flow cytometer settings, and *see* **Note 3** for population gating details. In flow cytometry analysis, expect only the A1 B1 phage-infected receiver population to have high fluorescence (ON cells) and all other infected receiver populations to have low fluorescence (OFF cells).
10. * Plot the mean fluorescence values (BL1-H) of the receiver population (Fig. 2c). To calculate the fluorescence fold-change (conservative calculation) between ON cells (uninfected receivers) and OFF cells (infected receivers), use the following formula:

$$\text{Fold}-\text{change} = \text{Mean BL1}-\text{H (ON cells)}/\\ \text{Mean BL1}-\text{H (OFF cells with the highest fluorescence)}$$

3.4 Three Senders and Single Receiver i-CRISPRi with AND-AND-NOT Gate Circuit

Most steps in the protocol below are similar to **3.3**, except the ones marked with "*":

1. * In addition to sender A1, B1, A0, and B0, transform competent *E. coli* Top10 cells with pSEVA19_sgRNA-C1_M13ps phagemid and pSEVA63_Helper plasmid to make sender C1. Similarly, to make sender C0, transform competent *E. coli* Top10 cells with pSEVA19_sgRNA-C0_M13ps phagemid and pSEVA63_Helper plasmid (Fig. 1a). See Protocol 3.5 for testing of phage production by the sender cells.
2. To make receiver strain, competent *E. coli* Top10F′ cells must be transformed with pSEVA47_dCas9-GFP_AAN_4 (Fig. 1b).
3. * Pick a sender colony (A1, B1, A0, B0, C1, and C0) and inoculate in 5 mL of LB media supplemented with appropriate antibiotics (see Materials, Table 3). Pick a receiver colony and inoculate in 5 mL of LB media supplemented with 25 μg/mL Spec +5 μg/mL Tet. Incubate culture tubes 12–16 h at 37 °C and 180 rpm.

Table 3
Transformed strains

Protocol/circuit	Transformed strain	Parental strain	Plasmid/s transformed	Antibiotic resistance/s
Protocol 3.1 sender(NOT gate)	TOP10_H_sgRNA-1_AmpΦ	TOP10	pSEVA63_Helper + pSEVA19_sgRNA-1_M13ps	Str[Ra] / Gen[R] / Amp[R]
Protocol 3.2 sender(YES gate)	TOP10_H_sgRNA-2_KanΦ	TOP10	pSEVA63_Helper + SEVA26_sgRNA-2_M13ps	Str[Ra] / Gen[R] / Kan[R]
Protocol 3.3 and 3.4 sender (AND / AND-AND-NOT gate)	TOP10_H_sgRNA-A1_KanΦ	TOP10	pSEVA63_Helper + pSEVA26_sgRNA-A1_M13ps	Str[Ra] / Gen[R] / Kan[R]
Protocol 3.4 sender (AND-AND-NOT gate)	TOP10_H_sgRNA-C1_AmpΦ	TOP10	pSEVA63_Helper + pSEVA19_sgRNA-C1_M13ps	Str[Ra] / Gen[R] / Amp[R]
Protocol 3.3 and 3.4 sender (AND / AND-AND-NOT gate)	TOP10_H_sgRNA-B1_GentΦ	TOP10	pSEVA47_Helper + pSEVA63_sgRNA-B1_M13ps	Str[Ra] / Gen[R] / Spec[R]
Protocol 3.1 receiver (NOT gate)	TOP10F_dCas9-GFP_NOT_i1	TOP10F'	pSEVA47_dCas9-GFP_NOT_i1	Str[Ra] / Tet[R] / Spec[R]
Protocol 3.2 receiver (YES gate)	TOP10F_dCas9-GFP_YES_i2	TOP10F'	pSEVA47_dCas9-GFP_YES_i2	Str[Ra] / Tet[R] / Spec[R]
Protocol 3.3 and 3.4 receiver (AND / AND-AND-NOT gate)	TOP10F_dCas9-GFP_AAN	TOP10F'	pSEVA47_dCas9-GFP_AAN_4	Str[Ra] / Tet[R] / Spec[R]
Protocol 3.3 and 3.4 sender (AND / AND-AND-NOT gate)	TOP10_H_sgRNA-A0_KanΦ	TOP10	pSEVA63_Helper + pSEVA26_sgRNA-A0_M13ps	Str[Ra] / Gen[R] / Kan[R]
Protocol 3.4 sender (AND-AND-NOT gate)	TOP10_H_sgRNA-C0_AmpΦ	TOP10	pSEVA63_Helper + pSEVA19_sgRNA-C0_M13ps	Str[Ra] / Gen[R] / Amp[R]
Protocol 3.3 and 3.4 sender (AND / AND-AND-NOT gate)	TOP10_H_sgRNA-B0_GentΦ	TOP10	pSEVA47_Helper + pSEVA63_sgRNA-B0_M13ps	Str[Ra] / Gen[R] / Spec[R]

[a]Although these strains are resistant to streptomycin, it was not used to culture them

4. Subculture by diluting overnight cultures 1000× in 100 mL fresh LB media supplemented with appropriate antibiotics. Incubate at 37 °C and 180 rpm. Monitor OD_{600} of both cultures and stop incubation when culture densities reach ~0.3. To stop growth, immediately move cultures to ice and incubate for at least 30 min. Measure OD_{600} again.
5. Centrifuge the cultures for 10 min, at 4500× g, at 4 °C. Discard supernatant and resuspend in cold LB media without antibiotics (*see* **Note 1**), ensuring that the re-suspended OD_{600} is higher than the final density needed. Dilute the cultures to adjust the density to an OD_{600} of 0.125.
6. Serially dilute density-adjusted sender and receiver cultures by twofold in cold LB media without antibiotics to make culture dilutions of different densities of 0.125, 0.062, and 0.031.
7. * Mix 33.3 μL each of three sender culture dilutions (A0 B0 C0, A0 B0 C1, A0 B1 C0, A0 B1 C1, A1 B0 C0, A1 B0 C1, A1 B1, C0, or A1 B1 C1) and 100 μL of receiver culture dilution in a 96-well plate (black) (Fig. 1c), and incubate in a plate reader for 4 h at 37 °C and 205 cpm. Then, subculture by diluting co-culture by 40-fold into fresh LB media supplemented with 25 μL/mL Spec +15 μL/mL Kan + 5 μL/mL Gent +50 μL/mL Amp and incubate in a plate reader for 18 h at 37 °C and 205 cpm, while monitoring OD_{600} and green fluorescence (Excitation: 485/20, Emission: 528/20).
8. Collect co-culture samples for flow cytometry by transferring 10 μL of the culture to a new 96-well plate (clear) prepared earlier with 190 μL of cold 1× PBS (+ 2 mg/mL Kan) per well.
9. * Use the PBS-diluted culture samples for flow cytometry to record fluorescence of different populations in the culture. *See* **Note 2** for flow cytometer settings, and *see* **Note 3** for population gating details. In flow cytometry analysis, expect only the A1 B1 C0 phage-infected receiver population to have high fluorescence (ON cells) and all other infected receiver populations to have low fluorescence (OFF cells).
10. Plot the mean fluorescence values (BL1-H) of the receiver population (Fig. 2d). To calculate the fluorescence fold-change (conservative calculation) between ON cells and OFF cells, the following formula was used:

$$\text{Fold}-\text{change} = \text{Mean BL1}-\text{H (ON cells)}/\\ \text{Mean BL1}-\text{H (OFF cells with the highest fluorescence)}$$

3.5 Colony Forming Unit (CFU) Spot Assay

1. Pick a colony of *E. coli* ER2738F′ and inoculate it in 5 mL of LB media supplemented with 10 μg/mL Tet. Incubate culture tube for 12–16 h at 37 °C and 180 rpm.

2. Subculture by diluting the overnight culture 1000× in 10 mL fresh LB media supplemented with 10 μg/mL Tet. Incubate at 37 °C and 180 rpm. Monitor OD_{600} of the culture and stop incubation when the culture density reaches ~0.3. To stop growth, immediately move cultures to ice and incubate for at least 30 min. Measure OD_{600} again.
3. Centrifuge the culture for 10 min, at 4500× g, at 4 °C. Discard supernatant and resuspend in cold LB media without antibiotics, ensuring that the re-suspended OD_{600} is higher than the final density needed. Dilute the cultures to adjust the density to an OD_{600} of 0.125.
4. Pick one colony of the sender bacteria and resuspend it in 50 μL of sterile de-ionized water. Mix 5 μL of the resuspended sender bacteria (that produce phages encoding for kanamycin resistance) with 195 μL of resuspended *E. coli* ER2738F′ in a 96-well plate (clear). Incubate the culture in a plate reader for 5 h at 37 °C and 205 cpm.
5. Serially dilute the culture 10× five times in cold LB media without antibiotics. Use 1 μL of the culture dilutions to spot them on top of the LBA plate supplemented with 5 μL/mL Tet and 15 μL/mL Kan. Incubate the plate overnight at 37 °C for 12–16 h.
6. Bacterial growth on the spotted plates indicates that the sender colony produces viable phages. Absence of bacterial growth on the spotted region indicates that the sender colony does not produce any infective phages.

4 Notes

1. This step ensures the culture is antibiotics-free, which avoids carryover of unwanted antibiotics to the next steps.
2. The appropriate flow cytometry settings must be identified for your specific instrument. The instrument we used was a Thermofisher Attune NxT flow cytometer equipped with CytKick Autosampler. Voltages were set to: FSC (forward scatter) at 265, SSC (side scatter) at 273. To disregard smaller debris, FSC and SSC acquisition thresholds were set to 100. GFP fluorescence was measured using the 488 nM laser (excitation) and a 530/30 nM filter (BL1 channel). BL1 voltage was set to 278 volts, such that autofluorescent cells have a mean BL1-H green fluorescence of ~100.
3. In flow cytometry data, on a Cell Count against BL1-H histogram plot (Fig. 3), senders show background autofluorescence between 0 and ~500. Receivers show fluorescence between ~500 and ~50,000. The fluorescence threshold between ON

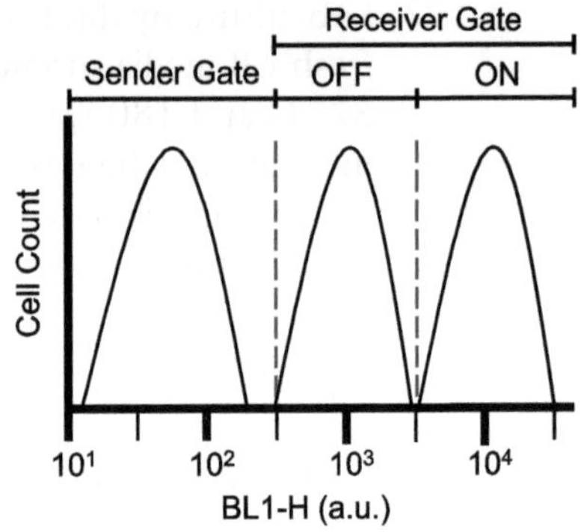

Fig. 3 Flow cytometry gating. Gating of flow cytometry data to identify bacterial cell types in a mixed population based on single-cell fluorescence. The BL1 channel corresponds to the settings used to observe green fluorescent protein (Ex λ: 488 nM, Em λ: 530/30 nM). Sender cells do not express any GFP, so they only show background autofluorescence. Receiver cells express GFP at either low (OFF) or high (ON) levels. Depending on the type of circuit, the ON or OFF states may be exhibited by infected or uninfected receivers

and OFF receiver cells was set at ~5000. These thresholds were determined by running control sender and receiver (infected or uninfected) cell populations through the flow cytometer individually, and may vary from one circuit to another. To obtain the control infected receiver population, receiver cells should be transformed with the appropriate phagemid/s.

Acknowledgments

We acknowledge support from Ile-de-France (IdF) region's DIM-RFSI (project COMBACT), Université Paris-Saclay's STIC department (project DEPEC MODE), INRAE's MICA department (starting grant and project PHEMO), and the France Excellence Eiffel Scholarship Program. This research was funded in part by the French National Research Agency (ANR) under the projects DREAMY (ANR-21-CE48-0003), and PEPR Tbox4BioProd (ANR-22-PEBB-0012).

References

1. Shapiro JA (1998) Thinking about bacterial populations as multicellular organisms. Ann Rev Microbiol 52:81–104
2. Chen YC, Destouches L, Cook A, Fedorec AJH (2024) Synthetic microbial ecology: engineering habitats for modular consortia. J Appl Microbiol 135. https://doi.org/10.1093/jambio/lxae158
3. Grozinger L, Amos M, Gorochowski TE, Carbonell P, Oyarzún DA, Stoof R, Fellermann H, Zuliani P, Tas H, Goñi-Moreno A (2019) Pathways to cellular supremacy in biocomputing. Nat Commun 10:5250

4. Macía J, Posas F, Solé RV (2012) Distributed computation: the new wave of synthetic biology devices. Trends Biotechnol 30:342–349
5. Basu S, Gerchman Y, Collins CH, Arnold FH, Weiss R (2005) A synthetic multicellular system for programmed pattern formation. Nature 434:1130–1134
6. Liao MJ, Din MO, Tsimring L, Hasty J (2019) Rock-paper-scissors: engineered population dynamics increase genetic stability. Science 365:1045–1049
7. Ortiz ME, Endy D (2012) Engineered cell-cell communication via DNA messaging. J Biol Eng 6:16
8. Church GM, Gao Y, Kosuri S (2012) Next-generation digital information storage in DNA. Science 337:1628
9. Lee IPA, Eldakar OT, Gogarten JP, Andam CP (2021) Bacterial cooperation through horizontal gene transfer. Trends Ecol Evol. https://doi.org/10.1016/j.tree.2021.11.006
10. Marken JP, Murray RM (2023) Addressable and adaptable intercellular communication via DNA messaging. Nat Commun 14:2358
11. Pujar A, Pathania A, Hopper C, Pandi A, Calderón CR, Függer M, Nowak T, Kushwaha M (2025) Phage-mediated intercellular CRISPRi for biocomputation in bacterial consortia. Nucleic Acids Res 53:gkae1256

Chapter 12

Feedback Control and Sensitivity Analysis in Synthetic Gene Circuits

Ania-Ariadna Baetica

Abstract

Synthetic gene circuits are engineered DNA sequences that control when genes turn on and how much they are expressed, alongside the regulatory machinery of cells. Many gene circuits use feedback, such that the molecules produced by a gene increase or decrease the expression of the same gene or another gene in the circuit. Feedback regulates how much of a molecule is produced and changes the timing of its production or degradation. However, how effective feedback is depends on biochemical parameters such as transcription, translation, and degradation rates, binding affinities and cooperativities, as well as cellular parameters such as plasmid copy numbers and the availability of cellular resources. All these parameters can change due to mutation, variation in the gene expression machinery, or changes in the cell's growth rate. To quantify how these changes affect the feedback regulation in a circuit, we introduce a method called sensitivity analysis, which measures how the steady-state expression level changes when each parameter is varied. Here, we explain sensitivity analysis and demonstrate its application to the negative autoregulation circuit. Sensitivity analysis can predict when a circuit's feedback will deteriorate, thus directing experiments to evaluate only the effects of the parameters predicted to be disruptive.

Key words Synthetic gene circuits, Feedback regulation, Sensitivity analysis, Parameter variation

1 Introduction

Synthetic gene circuits are engineered DNA constructs that integrate genes and regulatory sequences controlling transcription, translation, and degradation to modulate gene expression levels and timing inside cells [1, 2]. Many synthetic gene circuits employ feedback regulation, where the product of a gene either represses (negative feedback) or enhances (positive feedback) its own expression or that of other genes in the circuit [3–7]. Feedback regulation indicates that the molecules produced by a gene circuit act on the regulatory components of the same circuit, altering their rate of gene expression [2]. However, how well the feedback controls gene expression depends on many biochemical and cellular parameters such as transcription, translation, and degradation rates, promoter

Xiao-Jun Tian (ed.), *Synthetic Gene Circuits: Methods and Protocols*, Methods in Molecular Biology, vol. 3041, https://doi.org/10.1007/978-1-0716-5304-3_12,

strengths, and resource availability [8–13]. These parameters can vary due to mutation, variation in the gene expression machinery, or changes in the cell's growth rate.

Predicting how variability in these parameters affects the circuit and the ability of feedback to perform regulation requires a quantitative analysis. Sensitivity analysis provides a mathematical framework for quantifying how changes in model parameters influence a circuit's steady-state expression level [14–16]. By identifying the parameters that most strongly affect the steady-state expression, sensitivity analysis can predict whether the feedback deteriorates with parameter variation [15–18].

Sensitivity analysis uncovers which biochemical or cellular rates most strongly affect how much of a gene product is made or when it is made. Since circuits with feedback regulation are designed to control those outputs, if their parameters change too much, the circuit may no longer turn on under the correct conditions, may not respond in time, may produce an incorrect amount of gene product, or may behave inconsistently across cells. Therefore, synthetic and systems biologists can benefit from knowing which components and parameters most strongly affect the circuit's steady-state expression levels and from quantifying how much components can change before the circuit no longer functions as designed.

The theory of steady-state local sensitivity analysis is based on differentiating the model equations describing a gene circuit with respect to its parameters [2, 19, 20]. A deterministic model of a synthetic circuit can be expressed as a system of ordinary differential equations (ODEs):

$$\frac{dx}{dt} = f(x, p), \tag{1}$$

where x is the vector of molecular species concentrations such as mRNAs, proteins, and other gene products, and p is the vector of parameters describing the system, including transcription rates, degradation rates, binding affinities, and cooperativities. The function $f(x, p)$ defines the dynamics of each species. The vector x^* represents the concentrations of all species when the system is at steady state and satisfies the equation

$$f(x^*, p) = 0. \tag{2}$$

Then the local sensitivity of the steady state x^* with respect to each parameter p_i is defined as

$$S_{p_i}(x^*) = \frac{\partial x^*}{\partial p_i}\frac{p_i}{x^*}, \tag{3}$$

which represents the fractional change in steady-state expression per fractional change in the parameter [21]. An intuitive interpretation of the local steady-state sensitivity $S_{p_i}(x^*)$ is that a 1% change in the parameter p_i causes an $S_{p_i}(x^*)\%$ change in the steady-state value of x^*.

We note that sensitivity analysis is a general and interdisciplinary method that originated in control theory, chemical engineering, and applied mathematics [19, 22]. Multiple methods have been developed to quantify how system behavior depends on model parameters, including local and global sensitivity analysis, elasticity analysis, and uncertainty quantification. The steady-state local sensitivity analysis presented in this chapter is one such method. It can be applied to any biological system modeled by ordinary differential equations [23–25], and it is particularly informative for synthetic gene circuits with feedback regulation because it quantifies how parameter variations affect the feedback's ability to regulate the circuit.

In this chapter, we use the common motif of negative autoregulation to illustrate the local steady-state sensitivity analysis method for synthetic and systems biology [1]. The Method section describes the four main steps: model formulation, steady-state computation, local sensitivity calculation, and interpretation. Together, these steps provide a framework for connecting the model parameters to the circuit's steady-state behavior and for determining if changes in parameters are attenuated or amplified by the circuit.

2 Materials

1. **Software and Packages**

   Python (version $\geq$3.9) with the modules `numpy`, `scipy`, `sympy`, `matplotlib`, and `paretoset`.

   MATLAB (version $\geq$R2022a) with built-in ODE solvers (`ode45`, `ode15s`), plotting functions (`plot`, `scatter`), Pareto front plotting function `paretofront`, the SimBiology Toolbox, and the Optimization Toolbox.

   Wolfram Mathematica (version $\geq$13.0) with the functions `NDSolve`, `D`, `FindRoot`, `Eigenvalues`, and `ParetoFrontier`.

   COPASI (version $\geq$ 4.40) and iBioSim (version $\geq$ 3.0), which include integrated ODE solvers, sensitivity analysis, or multi-objective analysis tools.

   Advanced Software Option: Julia (version $\geq$ 1.9) with the packages `DifferentialEquations.jl`, `SymPy.jl`, `Plots.jl`, and `ParetoFront.jl`.

2. **Model Files**

   The ODE model describing the dynamics of the synthetic or biological circuit of interest. The model should specify all molecular species, parameters, and rate laws. Acceptable formats include `.py`, `.m`, `.cps`, `.nb`, `.jl`, or SBML (`.xml`).

3. **Computational Resources**

   Standard desktop or laptop computer (recommended ≥8 GB RAM) running Windows, macOS, or Linux. For circuits with more than five parameters, use a high-performance computing cluster or multicore workstation with parallelization, as laptop-class hardware will likely be too slow for parameter sweeps across their physiological ranges.

4. **Reference Data Sources**

   BioNumbers database [26] or equivalent literature sources for physiological parameter ranges.

5. **Visualization Tools**

   Python plotting library `matplotlib`, MATLAB's plotting functions `plot` and `scatter`, Julia's `Plots.jl` package, or the integrated visualization interfaces in COPASI, SimBiology, and iBioSim for displaying steady-state results and sensitivity functions.

3 Methods

This section describes how to perform local sensitivity analysis at steady state on synthetic gene circuits. The analysis has four main parts:

- **Model Formulation and Simulation**: Write down and simulate the deterministic ODE model that describes the synthetic gene circuit. Details are available in **Note 4.1**.
- **Steady-State Computation**: Solve for the steady-state concentrations of the molecular species of interest by setting the ODE system to zero. Details are available in **Note 4.2**.
- **Sensitivity Calculation**: Compute the local steady-state sensitivities for each parameter of interest. Details are available in **Note 4.3**.
- **Interpretation**: Analyze whether changes in parameters are amplified or reduced by the circuit. Find which parameter changes would be the most significant for the circuit's steady state. Details are available in **Note 4.3**.

Each section below provides detailed, numbered steps. Additional best practices are provided in the Notes section.

3.1 Model Formulation

1. Choose a synthetic gene circuit such as the circuits in [1, 2].
2. Write a system of ODEs that describes the dynamics of the synthetic circuit using mass-action or Hill function terms:

$$\frac{dx}{dt} = f(x, p),$$

where x is the vector of species concentrations and p is the vector of parameters.

3. Define all model parameters, including transcription and translation rates, degradation constants, Hill coefficients, and binding affinities. Ensure that parameter units are consistent: Concentration is in nM, time is in minutes, etc.
4. Nondimensionalize the model to reduce the number of parameters, if desired. Nondimensionalization means rewriting the model equations such that every term has dimension of unity. This reduces the number of parameters as only the dimensionless ratios of parameters remain [2, 27].

3.2 Steady-State Computation

1. Set the system of ODEs to zero, and find the steady-state vector of concentrations x^*, as a function of the parameters in p by solving the algebraic nonlinear equation $f(x^*, p) = 0$.
2. Choose a numerical solver that is suitable for solving nonlinear algebraic equations. A symbolic solver cannot typically solve a system of nonlinear ODEs, except in the simplest of cases [28]. The details of how to choose the numerical solver are provided in the Notes.
3. Provide the numerical solver with the system of equations f, the parameter values p, and multiple initial guesses for the steady-state values in x^*. Solve the algebraic equation numerically and obtain x^*.
4. Confirm that all solutions in x^* have real, nonnegative values. Discard all other solutions as these values cannot represent concentrations of molecular species.
5. Save the vector of solutions x^* for use in the sensitivity analysis.

3.3 Sensitivity Calculation

To compute how the steady-state concentration x^* changes with respect to a parameter p_i, we start from the steady-state condition:

3.3.1 Theory Underlying the Sensitivity Analysis Calculation

$$f(x^*, p_i) = 0,$$

where $x^* \in {}^{n}$ is the steady-state vector of state variables and $p \in {}^{m}$ is the vector of parameters. Here, p_i is one of the parameters in p.

We apply the Implicit Function Theorem to determine how the steady state x^* changes with respect to a parameter p_i, and we differentiate both sides with respect to p_i. We obtain that

$$\frac{d}{dp_i} f(x^*, p) = \frac{\partial \mathrm{f}}{\partial \mathrm{x}}(x^*, p)\frac{\partial x^*}{\partial p_i} + \frac{\partial \mathrm{f}}{\partial p_i}(x^*, p) = 0.$$

Rearranging terms gives the linear system:

$$J(x^*, p)\frac{\partial x^*}{\partial p_i} = -\frac{\partial \mathsf{f}}{\partial p_i}(x^*, p),$$

where $J(x^*, p) = \frac{\partial \mathsf{f}}{\partial \mathsf{x}}(x^*, p)$ is the Jacobian matrix evaluated at the steady state.

Solving this equation gives the unnormalized sensitivity equation:

$$\frac{\partial x^*}{\partial p_i} = -J^{-1}(x^*, p)\frac{\partial \mathsf{f}}{\partial p_i}(x^*, p).$$

The local sensitivity is then computed at steady state as

$$S_{p_i}(x^*) = \frac{p_i}{x^*}\frac{\partial x^*}{\partial p_i} = -\frac{p_i}{x^*}\left(J^{-1}(x^*, p)\frac{\partial \mathsf{f}}{\partial p_i}(x^*, p)\right).$$

3.3.2 The Computational Steps

The computational steps for the calculation of the sensitivities are as follows:

1. *Compute the Jacobian.* Symbolically compute the Jacobian matrix $J = \frac{\partial \mathsf{f}}{\partial \mathsf{x}}$, and evaluate it at the steady state (x^*, p). See the Notes section for guidance on choosing a suitable symbolic solver.
2. *Compute parameter derivatives.* For each parameter p_i, symbolically compute $\frac{\partial \mathsf{f}}{\partial p_i}$, and evaluate it at (x^*, p).
3. *Compute the inverse of the Jacobian matrix.* Numerically compute $\frac{\partial x^*}{\partial p_i} = -J^{-1}\frac{\partial \mathsf{f}}{\partial p_i}$ at the steady state (x^*, p). Check the condition number of the matrix J to ensure numerical stability [29].
4. *Calculate the local sensitivity.* For each species of interest x_j^*, numerically compute the sensitivity:

$$S_{p_i}(x_j^*) = \frac{\partial x_j^*}{\partial p_i}\frac{p_i}{x_j^*}.$$

5. *Repeat the steps.* Repeat steps 2–4 for all parameters p_i and all steady-state solutions x_j^* of molecular species of interest. Store the results as a vector or a matrix of sensitivities.

3.4 Interpretation

1. Plot histograms of the absolute values of the local sensitivities across biologically meaningful ranges of the circuit's parameters. Note whether the absolute sensitivity values are less than or greater than 1, as this indicates whether the circuit attenuates or amplifies changes in these parameters. Observe whether the absolute value of sensitivity is much greater than 1, which indicates that the circuit is highly sensitive to changes in those parameters. Conversely, observe whether the absolute value of any sensitivity is smaller than one, which indicates that the circuit is insensitive to changes in those parameters. Here, biologically meaningful parameter ranges refer to parameter

values consistent with measured or estimated kinetic rates in cells such as transcription rates, degradation rates, and binding affinities, based on available experimental data or literature.

2. Plot all pairs of the absolute values of the local sensitivities. Observe whether there are pairs in which one sensitivity can only be reduced at the expense of increasing the other. This is called a performance trade-off [30]. It represents a dependency between parameters where improving the sensitivity to one parameter makes the circuit more sensitive to another.
3. Examine the biological meaning of the parameters that the circuit is most and least sensitive to. Answer questions such as does the circuit attenuate changes in the degradation rate of the molecular species? By what percentage? In particular, identify which parameters are most likely to cause circuit failure if they vary.
4. Use these results to prioritize experimental efforts toward varying the parameters with the highest sensitivities, as changes in these are most likely to result in circuit failure.

4 Notes

4.1 Simulation Software

1. *Simulate the mathematical model of the circuit.* While this step is not necessary to perform sensitivity analysis, it is strongly recommended as it supports the interpretation of the results. Therefore, we recommend the simulation environments, grouped as follows:
   - **GUI-based tools that require no coding**: iBioSim [31], COPASI [32], the SimBiology Toolbox from MATLAB [33].
   - **Code-based simulation tools**: The Python module `scipy.integrate`, using functions `solve_ivp` and `odeint` [34]. Built-in ODE solvers in MATLAB such as `ode45`, `ode23`, and `ode15s` [35–37]. The function `NDSolve` in the Wolfram Language (Mathematica) [38].
   - **Advanced code-based simulation tools:** The `DifferentialEquations.jl` package in Julia [39].
2. *Choose kinetic parameters and initial conditions.* Choose kinetic parameters according to experimentally measured values from similar promoters, genes, or proteins in the same or a related organism. BioNumbers gives physiological ranges for a variety of organisms [26]. Since this sensitivity analysis is performed at steady state, the initial conditions of the molecular species can be chosen arbitrarily.

4.2 Numerical and Symbolic Solvers

1. Choose a computational environment that supports symbolic differentiation and numerical solvers. The following tools are recommended for beginners:
   - **The Wolfram Language (Mathematica)** supports both symbolic and numeric computation with minimal syntax overhead. Use `D[]` for derivatives and `FindRoot` for steady-state solutions. The function `D[]` can be used to find a Jacobian symbolically and `Eigenvalues[]` can compute its eigenvalues.
   - **Python**, using SymPy and SciPy: open source and widely adopted. Use SymPy for symbolic computation and SciPy's `fsolve` or `root` for numerical steady states. In SymPy, use the functions `diff()` or `jacobian()` to symbolically compute the Jacobian and `eigenvals()` to compute its eigenvalues.
   - **COPASI**: a graphical tool for biological ODE modeling and sensitivity analysis. No coding required. Suitable for users unfamiliar with programming. Enable Stability Analysis in the steady-state task to find the Jacobian matrices of the steady-state solutions and their corresponding eigenvalues.
2. Avoid tools not optimized for symbolic workflows or not appropriate for beginners. In particular:
   - **MATLAB's Symbolic Toolbox** is limited for symbolic computation.
   - **Julia's** `ModelingToolkit.jl` package offers powerful features but may be challenging to set up for beginners. `ForwardDiff.jacobian` and `LinearAlgebra.eigvals` can be used to evaluate a Jacobian matrix and its eigenvalues.
3. Any numerical solver can become stuck in local solutions [40]. Provide multiple guesses for the steady-state solutions to avoid this. Additionally, check that the solver has converged within user-specified tolerance bounds or number of iterations. Remove all solutions that are `NaN` or `Inf` from the solution vector, along with any complex values or negative real values as they cannot correspond to concentrations of molecular species.
4. Additionally, discard any solutions corresponding to unstable steady states, as these would not be observed experimentally. A steady state is unstable if small changes cause the circuit to move toward another condition. Unstable steady states can be identified by linearizing the system around the steady state (i.e., by computing the Jacobian matrix from the ODE model and evaluating it at the steady-state solution) and examining the Jacobian's eigenvalues [30]. If any eigenvalue has a positive real part, the steady state is unstable and should be discarded. If any eigenvalues are purely imaginary, the linearization is

inconclusive, and the steady state should be excluded out of caution. The tools used to compute the Jacobian matrix symbolically, evaluate it numerically at the steady-state solutions, and calculate its eigenvalues are listed above.

5. Note that synthetic circuits may have more than one steady-state solution. Provide multiple initial guesses to the numerical solver to find all steady-state solutions.
6. If the synthetic circuit has a steady-state solution at zero, as is the case in the model of positive autoregulation circuit with zero basal levels of gene expression, remove this solution from the local sensitivity analysis. The local sensitivity function cannot be defined at zero. Alternatively, consider including a term that models basal levels of gene expression.
7. There is existing computational software that calculates the local sensitivities from the ODE model of a circuit such as COPASI and SimBiology's `sbiosimulate` function [32, 33].

4.3 Local Sensitivity Analysis for the Negative Autoregulation Circuit

We derive the sensitivity analysis for the negative autoregulation circuit example.

(a) *Model Formulation:* Negative autoregulation can be modeled using the ODE:

$$\frac{dX}{dT} = \frac{\beta}{1 + \left(\frac{X}{K}\right)^n} - \gamma X,$$

where X is the concentration of the molecular species (nM), T is the time (min), β is the maximal production rate (nM min^{-1}), K is the dissociation constant (nM), n is the Hill coefficient (dimensionless), and γ is the degradation and dilution rate (min^{-1}) [1, 2]. This model has four reaction parameters: β, K, γ, and n.

To simplify the sensitivity analysis, we nondimensionalize the model to reduce its number of parameters from 4 to 2. We define the new variables

$$x = \frac{X}{K}, \qquad t = T\gamma.$$

Here, x is the dimensionless concentration of species X, and t is the dimensionless time. Therefore,

$$\frac{dX}{dT} = K\gamma \frac{dx}{dt}.$$

By substituting into the original equation, we obtain

$$K\gamma \frac{dx}{dt} = \frac{\beta}{1 + x^n} - \gamma K x.$$

We divide both sides by $K\gamma$ and define a new nondimensionalized parameter

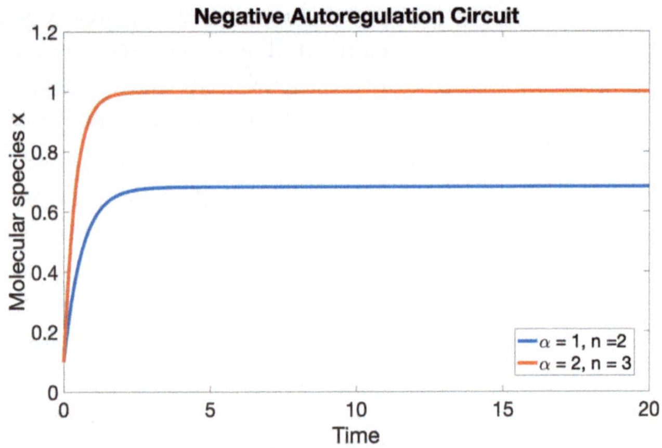

Fig. 1 Simulation of the nondimensionalized model of the negative autoregulation circuit for parameter pairs $\alpha = 1, n = 2$ and $\alpha = 2, n = 3$. The molecular species x approaches different steady-state values for each parameter pair. Simulations were performed in MATLAB using the `ode45` solver with initial condition $x(0) = 0.1$

$$\alpha = \frac{\beta}{\gamma K}.$$

Then the nondimensionalized ODE model of negative autoregulation becomes

$$\frac{dx}{dt} = \frac{\alpha}{1 + x^n} - x,$$

where α represents the ratio of the production rate to the degradation rate per unit of the dissociation constant and n represents the cooperative binding constant for species x.

(b) *Model Simulation:* Any of the tools listed in **Note 4.1** can be used to simulate the negative autoregulation circuit model. In Fig. 1, we use MATLAB's `ode45` to simulate the nondimensionalized negative autoregulation circuit.

To compute physiological ranges for the model parameters α and n, we use ranges given in the literature for yeast. Other parameter ranges can be found in the literature for other model organisms. Specifically, we assume that synthesis rates range from $[0.02, 0.1]$ nM min^{-1}, dissociation constants span $[0.01, 10]$ nM, and degradation, and dilution rates vary between $[0.01, 0.24]$ min^{-1}, as reported by [41, 42] and [43, 44], respectively. Since the dimensionless parameter α is the ratio of the production rate to the degradation rate per unit of the dissociation constant, we assume it spans the interval $[0.01, 50]$, even though we note that it can extend between values $[8 \times 10^{-3}, 1000]$. We limit its range for computational efficiency for the rest of the analysis.

Additionally, we assume that the Hill coefficient n spans values between 0.01 and 10. This wide range includes weak or negative cooperativity, where the Hill coefficient is less than 1, to very strong positive cooperativity, where the Hill coefficient could approach value 10. This range includes typical observations from enzymes, receptors, and DNA binding and allows for the modeling of systems with varying degrees of cooperativity [45–48].

(c) *Solve for the Steady-State Solution:* Mathematical models of synthetic circuits are often nonlinear due to chemical reactions such as cooperative binding, saturating reactions, feedback, and more. A nonlinear model means that changing one molecular species' concentration does not cause a proportional change in another. Nonlinear mathematical models do not typically have solutions that can be found algebraically. Instead, the steady-state concentrations of the molecular species in a circuit must be found using numerical methods and algorithms, such as the ones in **Note 4.2**.

At steady state, the molecular species in the negative autoregulation circuit, x^*, must satisfy the equation:

$$x^* = \frac{\alpha}{1 + (x^*)^n}.$$

For simplicity, we assume that the concentration of the molecular species x is at steady state for the rest of this note, and we omit the * symbol in x^*. We note that the negative autoregulation circuit has a single steady-state solution as the right-hand production term $\frac{\alpha}{1+x^n}$ decreases when x increases. Thus, the terms on the left- and right-hand sides of the equation change in opposite directions and can only equal each other once. Their intersection is at the single steady-state value. We note that the steady-state solution is always nonzero; else α would have to equal zero.

We use Python's SciPy function `fsolve` to find the steady-state value x_{ss} for different combinations of the parameters α and n. To explore how the steady state changes, we perform a simple *grid search*: We select 1,000 evenly spaced values of α between 0.01 and 50 and 1,000 values of n between 0.01 and 10. For each pair (α, n), we run `fsolve` starting from four different initial guesses to locate the steady-state solutions. This produces a fine grid of steady-state values that shows how the circuit behaves across the parameter ranges.

A steady-state solution obtained with `fsolve` is accepted when the residual norm satisfies `np.linalg.norm(fvec)` $< 10^{-6}$, indicating convergence to a true steady state, and the solver flag `intflag` = 1 confirms successful termination of `fsolve`.

We also check that the steady-state solutions in the negative autoregulation circuit are always stable. We calculate the Jacobian symbolically using SymPy's function `diff` and obtain that

$$J(x) = -\frac{\alpha n\, x^{n-1}}{(1+x^n)^2} - 1.$$

Since $\alpha > 0$, $n > 0$, and $x > 0$, it must be that $J(x_{ss}) < 0$ at steady state. Therefore, the steady-state solutions are always stable.

(d) *Steady-State Sensitivity Function Calculations.* The computation of the sensitivity functions can be performed symbolically using the Wolfram Language (Mathematica)'s function `D[]`, as described in **Note 4.2**. For simple models such as the negative autoregulation, it can also be performed by hand. We include the derivation by hand here to give a detailed example.

In this derivation, we follow the process described in the Methods section in Subheading 3.3. We define the function $f(x, \alpha, n)$ that expresses the steady-state condition for the negative autoregulation model:

$$f(x, \alpha, n) = \frac{\alpha}{1+x^n} - x = 0.$$

- *The Implicit Function Theorem.* Since x is an implicit function of the parameters α and n, then, according to the Implicit Function Theorem, it must be that

$$\frac{dx}{d\alpha} = -\frac{\partial f/\partial \alpha}{\partial f/\partial x}, \qquad \frac{dx}{dn} = -\frac{\partial f/\partial n}{\partial f/\partial x}.$$

- *Partial Derivative Computation.* Since the function f depends on $x, \alpha,$ and n, we can compute the partial derivatives of f with respect to the state and the two parameters. This step requires applying the chain rule for the steady-state condition. We obtain the following partial derivatives:

$$\frac{\partial f(x,\alpha,n)}{\partial x} = -1 - \frac{\alpha n x^{n-1}}{(1+x^n)^2}, \qquad \frac{\partial f(x,\alpha,n)}{\partial \alpha} = \frac{1}{1+x^n}, \qquad \frac{\partial f(x,\alpha,n)}{\partial n} = -\frac{\alpha x^n \ln x}{(1+x^n)^2}.$$

- *Implicit Differentiation.* Using the Implicit Function Theorem, we can compute the partial derivatives

$$\frac{dx}{d\alpha} = -\frac{\partial f(x,\alpha,n)/\partial \alpha}{\partial f(x,\alpha,n)/\partial x} = \frac{1+x^n}{\alpha n x^{n-1} + (1+x^n)^2},$$

$$\frac{dx}{dn} = -\frac{\partial \mathrm{f}(x, \alpha, n)/\partial \mathrm{n}}{\partial \mathrm{f}(x, \alpha, n)/\partial \mathrm{x}} = -\frac{\alpha x^n \ln x}{\alpha n x^{n-1} + (1 + x^n)^2}.$$

- *The Sensitivity Functions.* Using the definition of the local sensitivity functions with respect to the parameters α and n, we now calculate

$$S_\alpha(x) = \frac{dx}{d\alpha}\frac{\alpha}{x} = \frac{\alpha(1 + x^n)}{\alpha n x^n + x(1 + x^n)^2},$$

$$S_n(x) = \frac{dx}{dn}\frac{n}{x} = -\frac{\alpha n x^n \ln x}{\alpha n x^n + x(1 + x^n)^2}.$$

- *Simplified Forms of the Sensitivity Functions.* We can use the steady-state condition $\alpha = x(1 + x^n)$ to substitute α in the sensitivity function expressions and simplify them to the equivalent forms:

$$S_\alpha(x) = \frac{1 + x^n}{1 + (n + 1)x^n}, \qquad S_n(x) = -\frac{n x^n \ln x}{1 + (n + 1)x^n}.$$

(e) *Interpretation.* We use the analytical formulas of the two sensitivity functions to plot their absolute values at the steady-state molecular species values as the parameters α and n vary. First, we note that $|S_\alpha(x)|, |S_n(x)| < 1$. This means that the negative autoregulation circuit attenuates changes in its parameters.

Second, we plot the pairs of the absolute values of the sensitivities to see whether reducing the sensitivity to one parameter comes at the expense of increasing the sensitivity to another. By plotting $|S_\alpha(x)|$ against $|S_n(x)|$, we note that this is indeed the case for the negative autoregulation circuit. In Fig. 2, the curve in orange is called the Pareto front, and it shows the best possible balance between changing both parameters. Points on that curve are the designs that handle changes in both parameters as well as possible. It helps us see how much improvement in one direction must come at the expense of the other.

Additionally, we note that the circuit's design improves, that is, it approaches the orange curve when the cooperativity value n is as large as possible and the feedback strength α is at least value 2. This suggests that in an experimental implementation of negative autoregulation, both the cooperativity and the production rate of molecular species x should be maximized.

To find the Pareto set from the list of absolute sensitivity values and plot it, the Python package `paretoset` can be used. The equivalent functions are `paretofront` in MATLAB, `ParetoFrontier` in Mathematica, and

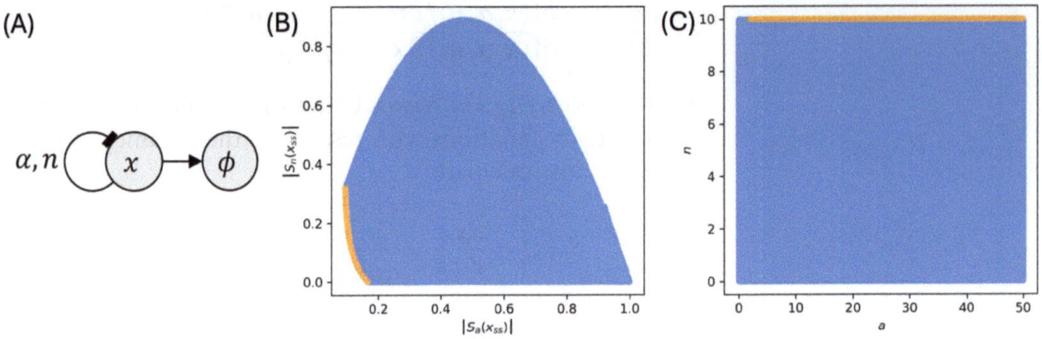

Fig. 2 Pareto analysis of the steady-state sensitivities for the negative autoregulation circuit. (**a**) Schematic of negative autoregulation of a biochemical species *x*. The blunt arrow denotes negative feedback, where *x* represses its own production. Degradation and dilution of *x* are indicated by arrows pointing to the empty-set symbol. The parameter *n* denotes binding cooperativity, and α denotes the feedback strength. (**b**) Plot of the absolute values of the steady-state sensitivities with respect to parameters α and *n*. The Pareto front corresponds to the outer boundary of the plotted points and identifies parameter choices for which reducing sensitivity to one parameter requires increasing sensitivity to the other. (**c**) Corresponding sampled parameter values and the subset of parameters that generate the Pareto front. (Figure adapted from [16])

`ParetoFront.jl` in Julia. In COPASI, the SimBiology Toolbox, and iBioSim, the Pareto front is obtained through their multi-objective optimization tools rather than a single command.

References

1. Alon U (2019) An introduction to systems biology: design principles of biological circuits. CRC Press, Boca Raton
2. Del Vecchio D, Murray RM (2015) Biomolecular feedback systems. Princeton University Press, Princeton
3. Gardner TS, Cantor CR, Collins JJ (2000) Construction of a genetic toggle switch in *Escherichia coli*. Nature 403(6767):339–342
4. Elowitz MB, Leibler S (2000) A synthetic oscillatory network of transcriptional regulators. Nature 403(6767):335–338
5. Rosenfeld N, Elowitz MB, Alon U (2002) Negative autoregulation speeds the response times of transcription networks. J Mol Biol 323(5):785–793
6. Alon U (2007) Network motifs: theory and experimental approaches. Nat Rev Genet 8(6):450–461
7. Ferrell Jr JE (1999) Xenopus oocyte maturation: new lessons from a good egg. Bioessays 21(10):833–842
8. Eldar A, Elowitz MB (2010) Functional roles for noise in genetic circuits. Nature 467(7312):167–173
9. Shakiba N, Jones RD, Weiss R, Del Vecchio D (2021) Context-aware synthetic biology by controller design: Engineering the mammalian cell. Cell Syst 12(6):561–592
10. Del Vecchio D, Dy AJ, Qian Y (2016) Control theory meets synthetic biology. J Roy Soc Interface 13(120):20160380
11. Cosentino C, Bates D (2011) Feedback Control in Systems Biology. CRC Press, Boca Raton
12. Gorochowski TE, Avcilar-Kucukgoze I, Bovenberg RA, Roubos JA, Ignatova Z (2016) Roubos, and Zoya Ignatova. A minimal model of ribosome allocation dynamics captures trade-offs in expression between endogenous and synthetic genes. ACS Synthetic Biol 5(7):710–720
13. Chu D, Barnes DJ, Von Der Haar T (2011) The role of tRNA and ribosome competition in coupling the expression of different mRNAs in

saccharomyces cerevisiae. Nucleic Acids Res 39(15):6705–6714
14. Zi Z (2011) Sensitivity analysis approaches applied to systems biology models. IET Syst Biol 5(6):336–346
15. Khammash M (2016) An engineering viewpoint on biological robustness. BMC Biol 14(1):22
16. Tran NH, Nguyen A, Rahman TW, Baetica A-A Fundamental trade-offs in the robustness of biological systems with feedback regulation. ACS Synthetic Biol 14(4):1099–1111
17. Klumpp S, Zhang Z, Hwa T (2009) Growth rate-dependent global effects on gene expression in bacteria. Cell 139(7):1366–1375
18. Klumpp S, Hwa T (2014) Bacterial growth: global effects on gene expression, growth feedback and proteome partition. Curr Opin Biotechnol 28:96–102
19. Tarantola S, Ferretti F, Piano SL, Kozlova M, Lachi A, Rosati R, Puy A, Roy P, Vannucci G, Kuc-Czarnecka M et al (2024) An annotated timeline of sensitivity analysis. Environ Model Softw 174:105977
20. Saltelli A, Tarantola S, Campolongo F, Ratto M et al (2004) Sensitivity analysis in practice: a guide to assessing scientific models, vol 1. Wiley Online Library, London
21. Cacuci DG (2018) The second-order adjoint sensitivity analysis methodology. Chapman and Hall/CRC
22. Gal T (1997) A historical sketch on sensitivity analysis and parametric programming. In: Advances in sensitivity analysis and parametric programming. Springer, Berlin, pp 1–10
23. Perumal TM, Gunawan R (2011) Understanding dynamics using sensitivity analysis: caveat and solution. BMC Syst Biol 5(1):41
24. Byrom DP, Darlington APS (2025) Genetic controllers for enhancing the evolutionary longevity of synthetic gene circuits in bacteria. Nat Commun 16(1):8590
25. Hu CY, Murray RM (2022) Layered feedback control overcomes performance trade-off in synthetic biomolecular networks. Nat Commun 13(1):5393
26. Milo R, Jorgensen P, Moran U, Weber G, Springer M (2010) Bionumbers—the database of key numbers in molecular and cell biology. Nucleic Acids Res 38(suppl_1):D750–D753
27. Bois JS (2021) justinbois/biocircuits: Version 0.1.0. https://doi.org/10.22002/D1.1615
28. Polyanin AD, Zaitsev VF (2003) Handbook of exact solutions for ordinary differential equations, 2nd edn. CRC Press, Boca Raton
29. Trefethen LN, Bau D (1997) Numerical linear algebra. SIAM.
30. Åström KJ, Murray R (2021) Feedback systems: an introduction for scientists and engineers. Princeton University Press, Princeton
31. Myers CJ, Barker N, Jones K, Kuwahara H, Madsen C, Nguyen NP (2009) iBioSim: a tool for the analysis and design of genetic circuits. Bioinformatics 25(21):2848–2849
32. Hoops S, Sahle S, Gauges R, Lee C, Pahle J, Simus N, Singhal M, Xu L, Mendes P, Kummer U. Natalia Simus, Mudita Singhal, Liang Xu, Pedro Mendes, and Ursula Kummer. COPASI—a complex pathway simulator. Bioinformatics 22(24):3067–3074
33. The MathWorks, Inc. (2024) SimBiology, version r2024b (MATLAB release 24.2). https://www.mathworks.com/products/simbiology.html
34. Virtanen P, Gommers R, Oliphant TE, Haberland M, Reddy T, Cournapeau D, Burovski E, Peterson P, Weckesser W, Bright J et al (2020) Scipy 1.0: fundamental algorithms for scientific computing in python. Nat Methods 17(3):261–272
35. Shampine LF, Reichelt MW (1997) The MATLAB ODE suite. SIAM J Sci Comput 18(1):1–22
36. Dormand JR, Prince PJ (1980) A family of embedded Runge-Kutta formulae. J Comput Appl Math 6(1):19–26
37. Bogacki P, Shampine LF (1989) A 3 (2) pair of Runge-Kutta formulas. Appl Math Lett 2(4): 321–325
38. Wolfram Research (2019) NDSolve. https://reference.wolfram.com/language/ref/NDSolve.html
39. Rackauckas C, Nie Q (2017) DifferentialEquations.jl–a performant and feature-rich ecosystem for solving differential equations in Julia. J Open Res Softw 5(1):15
40. Kharab A, Guenther R (2018) An introduction to numerical methods: a MATLAB® approach. CRC Press, Boca Raton
41. Kofahl B, Klipp E (2004) Modelling the dynamics of the yeast pheromone pathway. Yeast 21(10):831–850
42. Chen KC, Csikasz-Nagy A, Gyorffy B, Val J, Novak B, Tyson JJ Kinetic analysis of a molecular model of the budding yeast cell cycle. Mol Biol Cell 11(1):369–391
43. Helmstetter CE, Cooper S (1968) DNA synthesis during the division cycle of rapidly growing *Escherichia coli* Br. J Mol Biol 31(3): 507–518

44. Sherman F (2002) Getting started with yeast. In: Methods in enzymology, vol. 350. Elsevier, Amsterdam, pp 3–41
45. Prinz H (2010) Hill coefficients, dose–response curves and allosteric mechanisms. J Chem Biol 3(1):37–44
46. Roy M, Horovitz A (2022) Partitioning the hill coefficient into contributions from ligand-promoted conformational changes and subunit heterogeneity. Protein Sci 31(5):e4298
47. Weiss JN (1997) The hill equation revisited: uses and misuses. FASEB J 11(11):835–841
48. Huang C-Y, Ferrell Jr. JE (1996) Ultrasensitivity in the mitogen-activated protein kinase cascade. Proc Natl Acad Sci 93(19):10078–10083

Chapter 13

Dynamic Robust Control of Microbial Communities Using Cybergenetics

Ting An Lee, Scott B. Stacey, Olivia Gallup, Harrison Steel, and Kirill Sechkar

Abstract

A key enabler for the engineering of complex biological systems is the co-culturing different cell strains together in a microbial community. Co-cultures can exhibit more diverse capabilities, increased robustness, and greater efficiency than a single strain engineered for a given application. However, reliable long-term control of such intricate systems is a complex task, and achieving communities that coexist stably despite potentially fluctuating cellular environments has been difficult to achieve. This challenge can be addressed by distributing signal processing between cells and computers through cybergenetics. Specifically, cellular outputs can be measured by sensors connected to a computer, which processes these signals and adjusts the cell culture conditions to achieve a desired goal; the cells then respond to this, closing the control loop. Striking a balance between computer-based and biological control implementations allows combining the strengths of the two approaches while circumventing their limitations. Here, we describe a methodology for cybergenetic control of microbial communities, considering the associated technical requirements, possible modes of interfacing between cell consortia and computers, and the resultant scope of biological and hardware design choices. This example illustrates how combining computational modeling with direct biological implementation of gene circuits leads to more robust and controllable systems with novel functionalities and applications.

Key words Microbial community, Co-culture, Continuous culture, Cybergenetics, Bioreactor, Composition, Composition control, Quantitative, Synthetic biology

1 Introduction

Engineering a microbial organism for different applications presents significant challenges, including the often detrimental effect of new added functionalities on the cell's viability, known as burden [1, 2]. Moreover, some desirable trait combinations—such as the ability to grow on cheap feedstocks while efficiently producing a valuable compound—may be simply nonviable in a single organism due to its biological properties [3–5]. These limitations can be tackled by culturing several different cell strains together in a

Xiao-Jun Tian (ed.), *Synthetic Gene Circuits: Methods and Protocols*, Methods in Molecular Biology, vol. 3041, https://doi.org/10.1007/978-1-0716-5304-3_13,

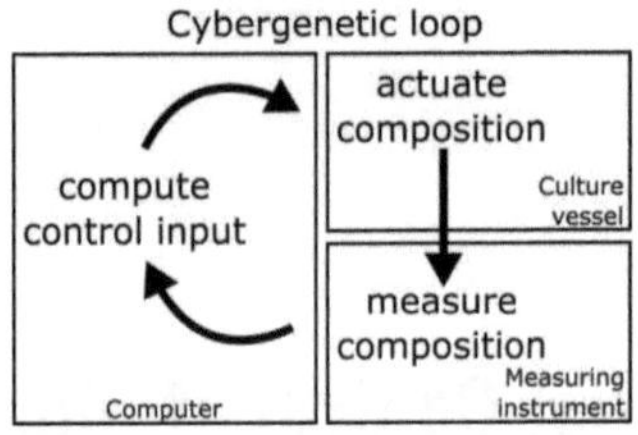

Fig. 1 Schematic of the cybergenetic control loop for community composition. Control inputs actuate the community composition in the culture vessel, which is measured by an instrument. The measurements are then passed to the computer, which calculates the next control input

bioreactor. This division of labor leverages each strain for what it does best, both expanding the range of potential functionality combinations and allowing individual cells to experience a fraction of the overall burden [3, 4]. However, microbial consortia come with their own issues, such as the difficulty of maintaining a desired co-culture composition. Without additional control, different subpopulations rarely grow at the same rate; hence the faster growing strain eventually displaces the slow-growing one leading to a non-functional monoculture [6, 7]. Auxiliary biomolecular controllers can be genetically engineered into cells to mitigate this but introduce further burden and can be subject to debilitating mutations, frequently leading to long-term breakdown of the consortia [2, 8].

Fortunately, labor can be divided not only between different microbial strains but also between cells and computers. Cybergenetics—the linking of biological and computational/robotic systems—distributes signal processing and control tasks between the culture and automatic hardware, which tracks cell populations using sensors, computes the appropriate response, and enacts it by altering the culture conditions (Fig. 1) [9]. Crucially, computations performed in silico do not directly burden the cell and are non-mutable, enabling complex actuation based on mathematical models, control algorithms, and machine learning. Moreover, computer-driven control is unconstrained by biochemical limitations, such as the impossibility of having negative values for concentrations of molecules [2]. Cybergenetic implementations are more adaptable than genetic regulation, whose properties must be hard-coded into DNA [7], enabling faster implementation and prototyping of engineered biosystems [10, 11]. Finally, tight cybergenetic regulation of the cellular environment enables more informative measurements, with higher quality data in turn improving automated control performance [12]. Although cybergenetic approaches can thus accelerate the development of advanced engineered biosystems, reliable and comprehensive measurement of a culture's state remains challenging [9], while the means of control actuation remain largely restricted to the rate of culture dilution and chemical, light, or temperature inputs [12, 13].

The two aforementioned division-of-labor paradigms can be integrated by splitting different biological functions between several co-culture strains while using a computer to cybergenetically control the microbial community's composition [7, 8]. This leverages the unique biological capabilities of different strains while benefitting from the efficiency, flexibility, and robustness of computer-based regulation [8, 14]. Automatic co-culture control can be further enhanced by biomolecular engineering of strains to render the co-culture's composition easier to measure and steer by a computer [13].

This chapter describes dynamic cybergenetic control of the composition of a community of two different strains in a bioreactor. We outline the process of selecting the hardware setup, the co-cultured strains, and the mode of control actuation. We then discuss methods for estimating population composition and measuring the ratio of component cell populations through direct approaches such as flow cytometry or fluorescence readings. Finally, we describe how to determine the control action based on these estimates, as well as approaches for iteratively improving a control loop based on experimental results. This methodological case study showcases how dividing labor between cell strains and automated hardware supports the engineering of novel living systems with expansive applications via robust control of the culture environment.

2 Materials

The first step in cybergenetic co-culture control is to identify your setup, which can be split into wetware (i.e., which microbes will constitute the co-culture), hardware (i.e., the equipment in which the culture will be grown, measured, and actuated), and software (i.e., the control algorithm that will decide the actuation strength and how machine(s) will communicate with each other). After discussing the considerations informing each of these choices, we specify the materials involved in the particular co-culturing protocol outlined in Methods (Section 3).

2.1 Wetware

The specific microbial strains that form the co-culture are typically chosen for a biological or biotechnological objective. They can be similar (e.g., two strains from the same species engineered to carry different parts of a pathway) or represent two completely different organisms (e.g., a yeast and a bacterium [15]), but the range of conditions in which they grow must overlap enough for them to coexist, because unless they are physically separated, they will share the same media and physical environment. In addition to not being too *different*, they should also not be too *similar*, so that they can be distinguished based on measurements and actuated individually. For instance, a well-chosen co-culture may consist of two microbes which can grow in the same media but utilize two different carbon sources, allowing the two strains to coexist *and* be controlled by changing the concentrations of the carbon source molecules [16]. As a counterexample, a co-culture of an obligate extremophile (i.e., a strain only growing at extreme temperatures) and a mesophile (i.e., a strain that can only grow at moderate temperatures) may be easily actuated by adjusting the culture's temperature from moderate to extreme but be unable to support simultaneous growth of both species, making the strains too different to be co-cultured. Depending on the choice of strains, means of measuring them and telling them apart, as well as ensuring distinct responses to actuating signals, can be natural or engineered. If biological engineering is required, an organism's genetic tractability and part availability should also be considered. Finally, the selection of strains is also constrained by hardware (discussed in the next section)—some organisms may require light, or controllable pH levels, or a specific gas mix (e.g., no O_2 for anaerobic conditions or CO_2 for cyanobacteria), and not all reactors equipped with these capabilities are amenable to real-time cybergenetic control.

Care should also be taken to understand how the selected strains interact. If (in the relevant environmental conditions) the species do not affect each other's growth except by competing for shared resources such as a carbon or nitrogen source, they could be considered to not have significant interactions. The effect of competition is either factored into the controller design or negated by providing it in the media in large excess. Alternatively, the strains can have positive interactions (e.g., syntrophy, inducer–producer relationships [4]), or negative interactions (e.g., from the secretion of toxins or inhibitory secondary metabolites [13]) besides mere resource competition. While positive interactions might be crucial for the co-culture and thus unavoidable, when selecting strains it can be useful to avoid negative interactions if possible, as they can have a significant effect on characterization, stability, and control inputs.

For example, amensalism—wherein one species inhibits the growth of another [17]—means that characterization cannot be carried out in monoculture and that the growth rate in a given environment is also dependent on the abundance of the other species. This can be accounted for with additional characterization but increases the system's complexity and the number of characterization experiments required. Such complicating interactions can be identified by reviewing the literature on potential strains and/or conducting a simple preliminary experiment comparing their growth rates in monoculture and in co-culture. Notably, while a quantitative understanding of the interaction is needed for an accurate model of the system, no mechanistic understanding of the interaction is required for compositional control. Additionally, even if the system has unaccounted for interactions (e.g., if it is unknown or arises during culturing from evolution), the computer's control inputs may compensate as long as measurements are accurate, though the performance may not be optimal.

In addition to the choice of microbe, one also needs to consider the method of actuating composition i.e., how to influence the growth rate of both species, such that you can influence the composition of the composition of a co-culture of those two species. This can be achieved naturally or through genetic engineering of the microbes. For example, microbes naturally have different optimal growth temperatures, and adjusting the temperature can select for one over the other. Microbes also naturally utilize different carbon/nitrogen sources, and altering the concentration of a nutrient can influence their growth rate. Alternatively, genetic engineering can be used to interface cells and computers using inducible promoters (e.g., optogenetic and chemical induction) combined with growth rate actuators (e.g., antibiotic resistance [7] and burden [18]), for example, creating a strain that grows faster/slower depending on the intensity of blue light. Design considerations may be complicated by other factors including part availability and genetic tractability, but generally, genetic engineering means that strains which are similar (e.g., two strains of *Escherichia coli* BL21) could be controlled.

Finally, one must consider how the composition of the co-culture can be measured and estimated as accurately as possible. This too can be achieved with and without genetic engineering. The most common method is to engineer a different constitutively expressed fluorescent protein into each microbe species, as fluorescent protein-based reporting is relatively reliable and easily measurable, and the genetic parts are commonly available in molecular biology labs. The reporter protein can be expressed from a plasmid or integrated into the genome for increased stability, which might be preferable if it must be highly expressed (e.g., if a given instrument requires a strong fluorescence signal for measurement). This makes cell output measurements more susceptible to deleterious mutations in continuous culture. In occasions where engineering is undesirable (e.g., for genetically intractable microbes or improved stability in longer bioprocesses) and the chosen microbes allow, natural characteristics such as pigments (e.g., chlorophyll in microalgae) or fluorescent molecules (e.g., pyoverdine in *Pseudomonas* sp.) may be exploited instead. Data that is universally available regardless of microbe choice, like optical density (OD) and bulk co-culture growth rates, can also help estimate composition [8]. However, this requires a model of the system and is not as accurate alone compared to when combined with other measurements.

Our co-culturing protocol in Methods (Section 3) is provided for two arbitrary strains actuated by changes in temperature or illumination (i.e., optogenetically) conditions, with each strain expressing a distinct fluorescent protein species for ease of measuring their cell counts. For simplicity, it is assumed that there are no resource-independent interactions between the strains. The exact solutions needed depend on the choice of strains, but below we outline M9, a minimal media routinely used for growing bacteria which does not have autofluorescence that interferes with measurements.

1. $MgSO_4$ (1 M stock solution): to prepare 250 mL, weigh out 61.62 g of $MgSO_4.7H_2O$, dissolve in 200 mL H_2O, fill to 250 mL with H_2O, and then sterilize.
2. $CaCl_2$ (50 mM stock solution): to prepare 250 mL, weigh out 1.84 g of $CaCl_2.2H_2O$, dissolve in 200 mL H_2O, fill to 250 mL with H_2O, and then sterilize.
3. Glucose (20% stock solution): to prepare 1 L, weigh out 200 g glucose, dissolve in 900 mL H_2O, fill to 1 L with H_2O, and then filter sterilize.
4. Casamino acids (10% stock solution): to prepare 1 L, weigh out 100 g casamino acids, dissolve in 900 mL H_2O, fill to 1 L with H_2O, and then filter sterilize.
5. Trace elements (100x stock solution): to prepare 1 L, first weigh out 5 g of EDTA $Na_2.2H_2O$, 0.5 g of $FeCl_3$, and 0.084 g of $ZnCl_2$ into a graduated cylinder. Then, as the concentrations of copper, cobalt, boron, and manganese salts are very low, first prepare 100,000x stocks of those salts: 13 g/L of $CuCl_2.2H_2O$, 5 g/L of $CoCl_2$, 10 g/L of H_3BO_3, and 1.6 g/L of MnCl2.4H_2O. Next, add 1 mL of each of these to the graduated cylinder. Dissolve in 900 mL of H_2O, then fill to 1L with H_2O, and sterilize. Some white precipitate will accumulate in the stock over time.
6. M9 media (minimal media supplemented with glucose, casamino acids, and trace elements): to prepare 1 L and weigh out 11.28 g of M9 salts. Fill to 900 mL and sterilize. Then, add 2 mL of 1 M $MgSO_4$ (2 mM), 2 mL of 50 mM $CaCl_2$ (0.1 mM), 20 mL of 20% glucose (0.4%), 20 mL of 10% casamino acids (0.2%), and 10 mL of 100x trace elements (1x), and then finally fill to 1 L with sterile H_2O.
7. PBS (phosphate buffered saline): to prepare 1 L, dissolve 100 mL of 10x PBS in H_2O and fill to 1 L, and then sterilize.

2.2 Hardware

The minimal requirements for the hardware setup are that it must be able to (1) culture the chosen microbes, (2) measure the co-culture composition, and (3) actuate the environment in a way that changes co-culture compositions. These can be fulfilled by a single machine, or by several machines acting in concert and integrated with liquid handlers or similar robotic tools, as long as the online measurements are fed back to the control computer in real time.

The first requirement essentially amounts to having a culture vessel where the cells may be grown, with the appropriate environment consistently maintained over time. While not a strict requirement, it is preferable for the bioreactor to be able to dilute the culture with fresh media and pump waste out, allowing it to act as a chemostat (enforcing constant cell growth rates) or, if an OD sensor is also included, a turbidostat maintaining a fixed culture OD. These setups are typical of previous examples of cybergenetically controlled cultures, as they maintain a (relatively) constant environment, avoiding the altered growth kinetics that could arise from different growth phases, nutrient limitation, or accumulation of growth-inhibiting metabolites [7, 8, 13, 16].

More constraining is the sensor requirement, which depends on the readout(s) used to determine the co-culture's composition. For example, if only OD measurements are utilized (e.g., as demonstrated to be possible in [8]), any reactor with an OD sensor is sufficient. If each strain expresses different fluorescent proteins [7, 19], reactors capable of bulk culture fluorescent measurements can be used: The Ogi3 BioReactor System or Chi.Bio bioreactors are commercially available examples. Further customizations of reactors (e.g., or those made by [19] or [20]) have been developed for specific applications. Alternatively, if a liquid handler (e.g., Opentrons) is available, bulk measurements can also be taken by integrating the culture vessel with the commonly available plate reader using the ReacSight [20] strategy, where a liquid handling robot samples from a bioreactor array, treats (e.g., dilutes) the samples, and then loads them into a measurement device. Finally, to measure fluorescence with single-cell resolution or to distinguish cells based on cell shape and size [15], the culture vessel can be integrated with a flow cytometer, also using the liquid handler-based ReacSight strategy or with a dedicated unit such as the custom built Segregostat [21], which uses peristaltic pumps to automate sampling, dilution, and loading into the cytometer.

Hardware requirements also depend on the method used for actuating the co-culture composition. Adjusting the concentration of a limiting nutrient (e.g., carbon source or amino acid) [19, 20] or a chemical inducer to control the growth rate of different strains can be done by pumping from different media reservoirs [13]. With just two media source bottles (media with and without the actuating compound), intermediate concentrations can be approximated in the bioreactor by spiking high-concentration media into the media with no actuating chemical in it [15, 16]. Alternatively, optogenetic approaches where cells respond to specific wavelengths of light [7, 22] can achieve an exact desired intensity, but these require reactors with LEDs (and typically engineered cells, unless the microbe is naturally responsive to light like cyanobacteria). Alternatively, adjusting the temperature to favor one strain over the other [8] simply requires a bioreactor with temperature control.

Overall, there are many options for hardware depending on the choice of strains (and the choice of measurement/actuation), equipment already available in the lab, and cost. Those with engineering experience looking for the cheapest setup would be best served by open-source alternatives, though building hardware from scratch typically involves a larger time investment. Parts lists and documentation for eVOLVER and Chi.Bio bioreactors are publicly available (at https://khalil-lab.gitbook.io/evolver and https://chi.bio/hardware/, respectively), with around $500 needed to assemble a single Chi.Bio reactor. Assuming the lab already has a flow cytometer, the Segregostat has documentation online [21] and, once built, could integrate the cytometer with any commercial or DIY bioreactor to expand the range of available composition measurement strategies. This approach would also be preferable for larger scale bioreactors, as pumps can sample from them more easily than a pipetting robot.

The co-culturing protocol in this chapter is outlined for the Chi.Bio platform. This is an appropriate choice for our two fluorescent strains actuated by light or temperature. Indeed, this open-source bioreactor can add fresh media and pump out waste according to OD readings, acting as a turbidostat to enforce consistent culture conditions. Moreover, it includes several LEDs for different fluorescence excitation wavelengths and a spectrometer to measure the strains' fluorescence, allowing gauge the co-culture's composition in situ without requiring sampling or flow cytometry. The same LED, as well as a controlled heat plate, allows Chi.Bio to actuate the co-culture via light and/or temperature [12].

2.3 Software and Model

Once the computer receives a readout from the sensors, it needs to (1) estimate the co-culture's composition and (2) use the estimate to compute the optimal control action that will drive the composition toward the desired value.

The methods needed to derive a composition estimate from measurements depend on the rest of the system. For instance, Kusuda et al. [19] calibrated fluorescence against OD to convert bulk culture fluorescence measurements into cell densities. Alternatively, Lee et al. [8] estimated compositions with a Kalman filter and a model with OD and bulk fluorescence measurements as inputs—the Python code for this can be found at https://github.com/Janmorlock/CompEst3000 [23]. Those using online flow cytometry require automated gating and cytometry analysis, e.g.,

MiPI Flow Cytometry Analysis toolbox for [15]. Generally, these tend to require setup or calibration with positive/negative/mixed-culture controls once and can then be used repeatedly.

Subsequently, the computer needs a control algorithm to calculate control actions from the composition estimate. Computationally simple algorithms like PID (Proportional–Integral–Derivative) provide good performance in controlling biological systems [7, 8, 19] and are straightforward to implement and tune. They do not require a model of the system, though a model can still be helpful for tuning the control algorithm or for estimation. Model-based control algorithms such as model predictive control (MPC) are also popular [20, 22, 24] and successful at compositional control, though they are more complex and computationally intensive.

When running a setup with separate machines (e.g., multiple culture vessels, an Opentrons, and a flow cytometer), software that integrates them is also required to port signals from all sensors into the computer and then distribute the actuating signal from the computer. Here, relying on previously published frameworks (e.g., ReacSight [20] and Segregostat's mFCAtoolbox [21]) allows researchers to build on established tools rather than starting from scratch.

Composition estimation in our case study is from fluorescence measurements calibrated against cell density in monoculture. Control signal calculation is performed as outlined in detail in Section 3.3.

3 Methods

The protocol outlined below describes in detail an approach to implementing cybergenetic control for a co-culture. The broad steps are generally applicable to any hardware setup and any co-culture (independent of microbes chosen or method used to measure/actuate composition), but for the purposes of having a detailed protocol with practical tips we will focus on specific control inputs and outputs in a Chi.Bio bioreactor. Nevertheless, we strongly encourage those interested in co-cultures to adopt and customize whatever hardware is best suited to their goals and circumstances, advancing the field of cybergenetics as a whole.

We focus on the case of a co-culture consisting of two arbitrary species, and assume that each stain is engineered to constitutively express a different fluorescent protein (e.g., GFP and RFP) at a high enough level to be observable. We also assume that either temperature or optogenetics is used as a control input, requiring either strains with different optimal culturing temperatures or strains engineered with optogenetic actuators of growth (as in [7]), with control inputs calculated using a straightforward PID control algorithm. More complex estimation/controllers or

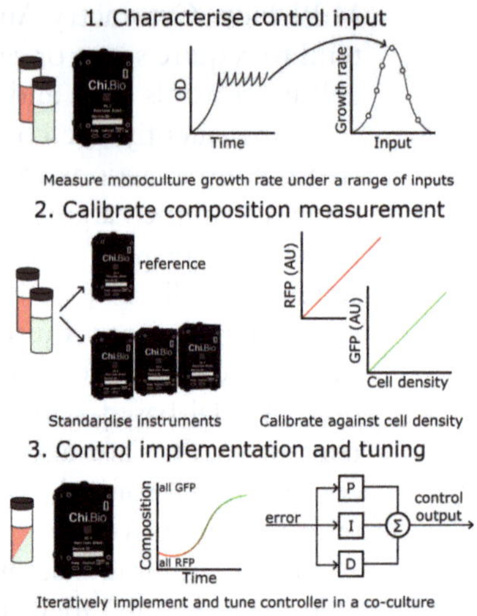

Fig. 2 Schematic: graphical overview of the entire method presented in this chapter. The method begins with characterization of the control input's influence on the growth rate of each co-cultured strain. This is then followed by calibration of the co-culture's composition measurements. Finally, the control algorithm is implemented in the bioreactor, and its parameters are iteratively tuned over several runs of the co-culture

mathematical models and simulation to tune control parameters could improve performance but are not strictly necessary and thus are not covered here.

Broadly, the steps involved are monoculture characterization of the control input, monoculture calibration of composition measurement, and then implementation and tuning of control in a co-culture (Fig. 2). A minimal amount of expertise in control engineering and coding (Python) would be helpful, but this can be feasibly achieved with just a background in molecular biology.

3.1 Monoculture Characterization of Control Inputs

The first step is to characterize the effect of the chosen control input on growth rate in your reactor of choice. Assuming negligible cross-species interactions, characterization can be carried out in monoculture, greatly simplifying the process. Below we outline how to set up a monoculture in the Chi.Bio, but these steps should be followed for any Chi.Bio setup in other sections. Documentation on how to set up the software of a Chi.Bio for the very first time is available online in official Chi.Bio guidance (https://chi.bio/software/) and thus is not covered here.

1. The day before the experiment, plate your strains, e.g., onto LB-agar, and incubate overnight.
2. The day of the experiment, inoculate 5 mL of your media with a colony from each plate, and incubate/shake in optimal conditions.

3. Autoclave glass vials (e.g., 11593532 from Fisher Scientific). We recommend capping with a 3D-printed lid (files available at https://chi.bio/hardware/) and using an O-ring between the glass vial and lid (e.g., 196-5666 from RS) (*see* **Note 1**).
4. Prepare and autoclave media on a standard liquid cycle. We recommend preparing it in standard Duran bottles and capping with 3D-printed lids (files available at https://chi.bio/hardware/) or any commercially available GL-45 lid with ports for tubing. Tubing should be attached to the underside of the lid prior to autoclaving (*see* **Note 2**).
5. Cut 50 cm and 100 cm lengths of 2.5 mm ID 4.5 mm OD tubing (e.g., 01-93-1416/20 from Altec) for pumping. Each pump board requires 1x 50 cm and 1x 100 cm. This only needs to be done once and can then be reused for several experiments until the tubing is worn out (*see* **Note 3**).
6. Unscrew the plastic cover of each pump, thread the tubing in, and screw the cover back on (*see* **Note 4**).
7. Connect the Chi.Bio control computer to your computer (*see* **Note 5**).
8. In order, connect reactors to the control computer and pump boards and then into power supply, and then turn on the power (*see* **Note 6**).
9. On Windows/Linux, open PuTTY. Connect to IP address (default is `192.168.7.2`) and port `22`, and log in (username and password are `root`). On Mac, open terminal, and then input `ssh -X root@192.168.7.2`.
10. Set it to the current date/time with command formatted like `date-s "Sep 17 1999 13:00:00"`.
11. Enter command `cd chibio` and then `bash cb.sh` to start the operating system.
12. Once it has successfully started, open the GUI on a web browser by going to the IP address (default is http://192.168.7.2:5000).
13. Place the ends of all tubing into 70% ethanol or diluted bleach to sterilize the ends (*see* **Note 7**).
14. Sterilize by setting the pump to 1 and switching it on, pumping ethanol through for around 10 seconds (*see* **Note 8**).
15. Sterilize stir bars in ethanol, and then in a sterile environment put them into the autoclaved vials (*see* **Note 9**).
16. (Optional for culture without antibiotics and longer experiments) Sterilize a short piece of tubing around 3 cm long and connect one to a port on each reactor and the media reservoirs. Then add a 0.2 um filter at the other end to sterilize all incoming air (*see* **Note 10**).

17. Flush ethanol out of the tube by connecting the start of the input tubes to media, setting Pumps to 1, and then switching on (*see* **Note 11**).
18. Wipe the outside of each vial with ethanol (to prevent aberrant OD readings), then place into reactors, and connect the ends of the input tubes to a port in each. Connect the start of each reactor's waste pump tube to its port (the one with a long protruding black tube), and then fill the reactors. This can be done by pipetting media in under sterile conditions or connecting a vial to the tubing, setting the value of Pump 1 to "1" in the GUI, and switching it on until the meniscus is roughly at the height of the black waste pump tube from the lid (*see* **Note 12**).
19. Calibrate each reactor by turning on temperature control and stirring, leaving it for around 10 minutes (to come up to the temperature set point), then stop stirring, measure the OD, and copy the "Raw" value into the "Measured" box and click set (*see* **Note 13**).
20. Inoculate the reactors, either in a sterile environment or through a port using a syringe (in the latter case, temporarily disconnect one of the other tubes) (*see* **Note 14**).
21. Set the control input in each reactor, covering a range of input values. For example, if you are using six reactors, for an optogenetic input you could set LED intensity values of 0-0.2-0.4-0.6-0.8-1, or for temperature, 26-28-30-32-34-36 ° C (*see* **Note 15**).
22. Start all reactors. Immediately switch on OD regulation and dithering to allow for online growth rate calculation (*see* **Note 16**).
23. Once the data is collected, stop all reactors, and then download the files from the control computer. They can be accessed with FTP software such as Filezilla on Windows/Linux or Cyberduck on mac, with the same IP address (i.e., 192.168.7.2), port, and username/password. Data is stored as excel files in the /root/chibio folder.
24. Plot and derive the relationship between the input and the growth rate.

This characterization step determines each species' growth rate in response to a range of input values in monoculture and is generally recommended regardless of hardware setup or method of actuation (optogenetic, temperature, and concentration of carbon source or amino acid). It is used later to identify a suitable operating range and reasonable parameters for the controller, as well as a model of the system if necessary.

3.2 Monoculture Calibration of Composition Measurement

The next step is to calibrate measurements of co-culture composition, standardizing measurements between reactors and calibrating against known quantities. Below, we assume that composition is determined from the Chi.Bio's measurements of fluorescence intensity from fluorescent proteins (e.g., GFP, RFP) engineered into cells. We outline first how to standardize measurements between reactors to account for inter-instrument variability, then how to make a calibration curve for converting fluorescence into cell density, and how to characterize this under a range of input conditions. For brevity, we avoid repeating the setup steps for the Chi.Bio, as they remain the same with the exception that there is no need to prepare any media reservoirs.

3.2.1 Standardization

For the Chi.Bio, inter-reactor variability in fluorescence measurements for a single sample can arise from slight differences in their LEDs or sensors, especially between different batches of reactors. This must be accounted for in order to prevent errors in control. The ideal way to calibrate the Chi.Bio (or indeed many spectrophotometers) is through absolute quantification against a known reference standard, e.g., as done in [25].

However, for compositional control it is sufficient to know the relative abundance of each species as a percentage rather than in absolute terms, and it is therefore possible to simply standardize between reactors so that they are consistent with each other in terms of arbitrary units (AUs) of fluorescent intensity. This can be accomplished by measuring a dilution series of a fluorescent solution (e.g., culture of cells expressing fluorescent protein) in every reactor and observing the relationship between dilution and measured fluorescence. Each reactor will have a slightly different relationship. One reactor can (arbitrarily) be delegated as the reference against which all reactors are standardized, and the gradient of every other reactor can be scaled to the reference reactor with a scaling factor. These scaling factors can be saved and applied to reactors during subsequent control experiments to adjust fluorescence measurements, accounting for hardware variations.

1. Prepare a fluorescent solution. This could be a culture of the cells, purified fluorescent protein, or highlighter fluid (the color should match the fluorescent proteins used). If using a culture, inoculate 50 mL of media in 250 mL flasks by adding colonies of your strains, and then incubate at optimal temperature and revolutions per minute (RPM) in a shaking incubator for long enough that it at least reaches the OD set point that will be used in experiments (*see* **Note 17**).
2. Record a unique, persistent identifier (i.e., name) for each reactor.

3. Set up and connect all Chi.Bio reactors to be calibrated as above, except without tubing and media reservoirs. Set the excitation/emission to those that match your fluorescent protein.
4. Create a linear dilution series of each monoculture (e.g., 33-66-100%), around 20 mL each in different Chi.Bio vials with stir bars (*see* **Note 18**).
5. Place each vial containing a certain dilution into a reactor, briefly turn on stirring to mix, then disable stirring to allow the culture to settle for approximately 30 seconds, and then measure fluorescence thrice, noting down each reading (*see* **Note 19**).
6. Repeat this for all reactors that needs to be calibrated. Then, repeat this for every vial from the different dilutions.
7. Plot the average fluorescence intensity at each dilution for each reactor. Each reactor should show a linear increase in fluorescence intensity, with a slightly different gradient.
8. Calculate the scaling factor for each reactor as the gradient divided by the gradient of the reference reactor.

Each reactor identifier can then be associated with its unique scaling factor. A fluorescent intensity measurement in a reactor can be converted into the reference reactor value by dividing by that reactor's scaling factor. This can be coded into the Chi.Bio operating system, so that during the control experiment, the fluorescence measured in each reactor is corrected by the reactor's scaling factor (see Section on closed-loop control below). This only needs to be done once for each reactor, although changes to the alignment of the laser/sensor (occurring, e.g., if the reactor is taken apart for a repair and then reassembled) can slightly alter these values.

3.2.2 Calibrating Fluorescence Against Cell Density

In order to drive the co-culture to specific compositions, we then need to characterize the relationship between the fluorescence intensity and the number of cells. It is technically only necessary to engineer fluorescence into one strain—its relative abundance can be calculated from its fluorescence and then subtracted from the overall cell count to get the relative abundance of the other strain. However, for simplicity we assume here that both strains are fluorescent.

1. Inoculate 50 mL of media in 250 mL flasks with colonies of your strains.
2. Incubate at optimum temperature and RPM for long enough that it reaches the OD set point that will be used in experiments, but not much higher.

3. Set up and connect one Chi.Bio reactor as above, except without tubing and media reservoirs. Set the excitation/emission to those that match your fluorescent protein.
4. Once the target OD has been reached, ensure that the OD of both flasks is the same with a spectrophotometer. Dilute with media otherwise.
5. Create a linear dilution series of each strain by diluting with the *other* strain. For example, for a 0-25-50-75-100% series, vials would contain 0% strain 1 and 100% strain 2, then 25% strain 1 and 75% strain 2, etc. (*see* **Note 20**).
6. Place each vial containing a certain dilution into the reactor, briefly turn on stirring, and then measure fluorescence thrice, noting down each reading.
7. Repeat this with the vials of every dilution.
8. Plot the average fluorescence intensity at every dilution of both strains. This creates a calibration curve that can relate fluorescence intensity to relative abundance, though this is only valid at that OD set point.

This simple calibration curve can then be used to estimate composition during control. The custom program containing the controller should first standardize the fluorescence according to that reactor's scaling factor and then use the calibration curve to find the equivalent density of that strain.

Depending on the source of fluorescence, this can be complicated by the fact that fluorescent output can be affected by the state of the cell, which may fluctuate depending on the present and previous inputs and environment. Bulk measurements (such as those of the Chi.Bio) lack the single-cell resolution needed to distinguish whether the cells of one species are decreasing in number but maintaining the level of fluorescence per cell or whether they are being maintained at the same number but with less fluorescence per cell. However, since the input is calculated by the controller and thus is known for every historical time point, it is possible to account for this if the relationship between input and fluorescence is understood. In this case calibrating fluorescence would require a similar setup as above, where monocultures are grown in a Chi.Bio, except that instead of a single static measurement, the fluorescence over time should be measured across a range of inputs.

3.3 Co-culture Implementation and Control

3.3.1 Open-Loop Co-culture Control

To validate and integrate the setup and characterization completed thus far, a suitable next step is to perform open-loop control of a co-culture. This entails growing cells together with preset actuation signals and observing how changes in composition conform. Although not technically necessary, this step serves to iron out bugs in the code and physical setup, compare experimental results to simulations from a model, and act as a sanity check for every part of the control loop other than the controller.

1. Set up a monoculture of each strain in the Chi.Bio as described above, i.e., inoculated from precultures at a fixed density, connected to media reservoirs (*see* **Note 21**).
2. Operate them as turbidostats with the same OD set point as that used in the control experiments (i.e., mid exponential) or slightly higher.
3. Once it has reached the OD set point, pause reactors, and bring the vials into a sterile environment (*see* **Note 22**).
4. Mix the monocultures into co-cultures with the desired ratios. (*see* **Note 23**).
5. Set the inputs on the reactors. A simple open-loop test is to start from 50:50 and test conditions that strongly favor one strain or the other or to test conditions that should enable equal growth rates and maintain the 50:50 ratio.
6. Download the data as described above, and plot the estimated composition over time.

It is unlikely that without any feedback signal a 50:50 ration will be maintained, as even with stable growth rates, random drift can cause the composition to fluctuate, but there should be a clear ability to drive one strain or the other to dominance. This step also presents a good opportunity to confirm the absence/presence/magnitude of interactions between strains by observing whether the co-culture behaves in line with monoculture characterization.

3.3.2 Closed-Loop Co-culture Control

The final step is to close the loop with controller that utilizes feedback, which in this example is a PID (Proportional–Integral–Derivative) controller. At every time point (i.e., once per minute in the Chi.Bio), the error ($e(t)$) is calculated as the difference between the observed community composition and the desired composition. The three components of the controller act to correct this error:

The P term is proportional to the error, such that if there is a large error, there is a larger correction.

$$P = K_P \times e(t) \tag{1}$$

The I term responds to the accumulated error over time (i.e., the integral of error), eliminating any offset or steady-state error.

$$I = K_I \times \int_0^t e(t)dt \tag{2}$$

Finally, the D term responds to the rate of change of the error, acting to minimize overshooting or oscillations.

$$D = K_D \times \frac{de(t)}{dt} \tag{3}$$

The correction or input at every time point is the sum of these three terms. Consequently, tuning the controller simply involves modifying the parameters K_P, K_I, and K_D. Besides tuning them in real experiments, modeling and simulating the system can help find good starting points for these parameters *in silico*.

The PID controller is straightforward to program and can be implemented modularly to result in PI or P control for further simplification. It should be implemented on the computer that collects measurements, estimates composition, and controls the input to the culture vessel. For the Chi.Bio, this is the BeagleBone control computer connected to the bioreactor. It contains the Chi. Bio operating system written in Python. A straightforward way to implement the controller is to add it as a custom program and then select it during setup, as outlined in the steps below:

1. Get the `app.py` file for the Chi.Bio operating system. It can be accessed by going to http://192.168.7.2:3000 in a web browser when the control computer is connected or from GitHub.
2. Code the PID controller into the custom program section. It involves calculating the three terms highlighted above and then setting the input (e.g., temperature and optogenetic intensity) as the sum of those three terms (*see* **Note 24**).
3. Go to http://192.168.7.2:3000 when a control computer is connected.
4. Replace `app.py` with the new version containing your custom program.
5. Save, run, and restart the reactors if they were already running (i.e., type ctrl+C in PuTTY for Windows/Linux or Terminal for mac and then `bash cb.sh`).
6. Set up co-cultures in the Chi.Bio as described above, i.e., OD adjusted precultures, used to inoculate monocultures operating as a turbidostat, mixed into co-cultures after they reach their set point.
7. Immediately after mixing, select the custom program in the GUI and switch it on. There should now be closed-loop cybergenetic control of the composition. Manual intervention now is only necessary to replace media if it runs out, to sample the culture if desired, or to replace the vials to prevent biofilms (*see* **Note 25**).
8. Once the experiment is complete, download and analyze the data, and then tune the controller parameters according to the performance.
9. Repeat the process starting from step 1 until suitable control performance is achieved.

4 Notes

1. The O-ring is useful as it helps to prevent moisture leaking from the vials during operation. Each reactor has two moisture sensors (silver metal rings) at the top and middle of the reactors. If they are short-circuited by a liquid spill, they will stop the reactor and any pumping to avoid water damage to the electronics—this is a common reason for the "Failed Pump Comms" error message. Leaks from prior runs can typically be observed as green/black color on the sensor's rings—these should be cleaned on both sides (top and bottom) of both moisture sensors, e.g., with a cotton bud and 70% ethanol. Leaks can be almost completely avoided with O-rings and careful operation. Optionally, the top moisture sensor can also be disconnected to make the system even less susceptible to crashing minor leaks by taking off the left and right sides of the reactor and then unplugging the wire coming from the top moisture sensor. The bottom moisture sensor should still detect any major leaks.
2. The bottom of the tubing on the inside of the media bottle should be cut at an angle, such that it cannot accidentally suction onto the wall/floor of the media bottle and prevent pumping, which might occur if it is cut flush. The amount of media needed depends heavily on the microbe, media richness, and how quickly they will grow in that environment. It can be estimated by assuming a requirement of approximately 13 mL per generation of growth; hence for *E. coli* growing at 37 ° C with an approximate 20 minute doubling time the system consumes around 1 liter of LB per 24-hour day.
3. The recommended Altec tubing is of generally good quality but prone to wearing out over prolonged use or lengthy experiments, particularly for the waste pump (which is used heavily). This is problematic, since if tubes are too worn out for the pump to effectively remove liquid, the culture can overflow and thus crash the reactor. To avoid this, you can either (1) frequently replace the tubing or (2) leave the sturdier tubing that comes with the peristaltic pumps inside and instead connect Altec tubing to either end (e.g., with appropriate barbed connectors). This tubing is more robust and less susceptible to wearing out.
4. The motors in these peristaltic pumps can wear down from prolonged use. Pump 2 (the waste pump) is particularly prone to this, as in a typical experiment with a fast growing microbe it is operating in most cycles to prevent overflow and help with aeration. Its failure is also particularly problematic, as it will cause an overflow that crashes all reactors on the same

control computer. Worn pumps can be replaced with other pumps on the pump board (i.e., Pump 3/4) by simply swapping their positions and plugs. Alternatively, if all pumps have already been worn out, they can be replaced with other 6V peristaltic pumps (these can be purchased separately online), requiring only very basic soldering.

5. More than one control computer can be connected to the same computer, though you will need to reconfigure the IP addresses of every additional control computer to not be the default `192.168.7.2`, as explained in the user manual. It is helpful to split reactors between control computers, so that one of them crashing would not affect others.
6. Plug the power supply into a reactor which is already connected to the control computer/pump board BEFORE turning the mains power on, as doing otherwise can damage the reactor. Additionally, place the reactors on a higher surface, so that the liquid in them is always above the level of liquid in your media reservoir. This way, a reservoir would not drain into a reactor if a pump fails.
7. There can be backflow of media from the reactor into the tubes leading to the reservoir if there is an imperfect seal in the pump, particularly if the tubing is worn down. For shorter experiments this is acceptable, but over time aerosolized microbes (from the stirring) enter the tubing and can multiply/form a biofilm. Backflow then allows them to eventually contaminate the media reservoir, preventing accurate readings and proper dilutions. For longer experiments, 1-way valves (e.g., 7757K41 from McMaster-Carr) can be placed near the end of the tube, followed by a short (e.g., 3 cm) tube going from the valve to the vial input. A fully connected length of tubing could therefore look like: media reservoir—Altec tubing—tube connector—tubing in peristaltic pump—tube connector—Altec tubing—valve—Altec tubing—vial input.
8. Pumps can fail in several ways: (1) The motor turns, but the pump does not work. In this case, it is likely that the pump does not have a tight enough seal on the tubing. Add a small strip of tape onto the white plastic of the pump walls, so that it would sit between the tubing and the wall, which should allow the pump to sufficiently squeeze the tubing. (2) The motor does not turn. Check if it turns when the tubing if removed. If it does not, then it needs to be replaced/swapped with another pump motor. If it does turn, then either the motor needs to be replaced or the tubing is unsuitable for the pump. Note that PharMed tubing (typically used for peristaltic pumps) of 2.5 mm ID and 4.5 mm OD is not compatible and will have this issue. If the peristaltic pumps pull the tubing during

operation, wrap some tape around where the tubing enters the pump to stop the tubes from being pulled in.

9. Stir bars can fail if they are too strong in terms of size/force of their internal magnet (stalling the motor and preventing it from starting) or too weak (leading to a failure to stir properly, usually accompanied by a tinkling sound as the bar hits the glass vial). If stirring is problematic, use a previously tested stir bar (e.g., 11818862 from Fisher Sci). Even then, problems can still arise occasionally. If stirring stops randomly during an experiment but is fixed by moving/shuffling the reactor, it is likely *just* strong enough to stall the motor. Then, it is possible to take off the sides of the reactor and then slightly unscrew the bottom where the motor is, so that it is *slightly* further away from the bottom of the vial. Placing some washers to extend the distance between the motor and the vial will also achieve the same effect. Alternatively, if the stirring does not work properly from the beginning, it is possible that the plastic holding the magnets has melted (*see* **Note 14**, the other note about heating). If this is the case, either 3D print a new plastic holder, add the magnets and glue it onto a replacement fan, or replace the reactor.

10. Air is "pulled" into the reactors whenever the waste pump is pumping out, and the filters ensure that only sterile air enters. Similarly, whenever media are pumped out of the media reservoirs, it needs to be replaced with air that should be sterilized. Care should be taken to ensure that all other ports in the reactor/media are plugged whenever pumping is occurring, as otherwise air will circumvent the filter. This can be done by plugging two ports with one connected tube or adding a piece of tube to a port and then plugging it (e.g., with 5463K73 from McMaster-Carr).

11. Between the ethanol in the tube and the media in the reservoir, there is a pocket of air. Once the pump is turned on, it will flow, briefly pause (once it reaches the air), and then flow again, at which point the ethanol has been flushed out, and you can stop. For sterility, ensure that (1) you do not turn pumping on until it is connected to the media reservoir, such that ambient (non-sterile) air does not enter the tubing, and (2) all other ports are already connected to their ports/plugged except the one with the filter.

12. The opening of the waste pump tube sets the volume in the reactor. You may overfill and then turn on Pump 2 until no more liquid pumps out.

13. The temperature/settling of particulates in the media can slightly affect OD readings, and they should therefore be calibrated close to the conditions they will operate in.

14. For reproducibility between experiments and reactors, precultures should be resuspended to a standard OD (e.g., 2) and inoculated at a fixed dilution (e.g., 1 in 200, i.e., 100 uL in 20 mL). The precultures therefore do not need reach any particular OD before they can be used. If the preculture used a different medium than the experiment, wash with PBS/experimental media. When removing vials from the reactor, it is *essential to ensure you switch off temperature control* (which is automatically enabled when starting an experiment in the GUI), as otherwise it will overheat and possibly damage the stirring mechanism. Temperature control should only be switched on when there is liquid in the reactor.
15. It may be necessary to start cells off with no control input until they reach the OD set point, as that may completely arrest growth in the lag phase. In that case, start all reactors at the same conditions, and only apply the control input when they have all reached the OD set point.
16. The operating system has a growth rate calculator, but it may be susceptible to noise in OD measurements, and any interventions—e.g., when removing a tube for sampling or due to a crash—cause sharp spikes/drops in OD. The growth rate can also be manually calculated from the periods of growth between each dilution (i.e., between cycles where Pump 1 activity is shown as above 0).
17. The exact amount of time depends on the growth rate in the media you are using. An overnight incubation is typical, e.g., for *E. coli* in M9 supplemented with glucose and trace elements. In order to get a calibration curve for the full range of possible compositions (i.e., 0%-100% of each strain), the OD should be at least as high as the OD set point. Going too much above this OD, particularly if it enters stationary phase, is also not desirable as that may alter the fluorescent characteristics, but precision is not necessary.
18. If using fluorescent cells, only do this once the flask culture in the incubator has reached a high enough OD (i.e., the target OD in control experiments). Wipe the outside of each vial with ethanol prior to measurement. It is also possible to dilute monocultures with the other strain in order to do this as the same time as calibration of fluorescence against cell density.
19. There is generally more inter- rather than intra-reactor noise, so readings from the same reactor are likely to be very similar.
20. Cell density impacts fluorescence measurements. In a control experiment running as a turbidostat, the cell density is always maintained even as the relative abundance of one strain changes. The Chi.Bio provides raw fluorescence intensity values and those normalized by the "Clear" filter, which can

help to account for cell density. However, by calibrating against solutions of variable composition but constant *total* cell density, this does not need to be considered. It is possible to use these solutions for standardizations as well and do them at the same time.

21. Starting as a co-culture straightaway (i.e., from a very low starting density of each strain) is not ideal because there is a period of growth before the culture reaches the OD and the cells reach the physiological state at which they were characterized. Any control applied before then would be imperfect, creating imbalances in relative abundance or exacerbating any errors in the initial inoculation.
22. This does not need to be done *immediately* when it reaches the set point, e.g., you could plate on day 1, inoculate precultures on the morning of day 2, inoculate Chi.Bios in the afternoon, and then mix into co-cultures on day 3.
23. While OD is relatively consistent between reactors, for the most accurate mixes it is best to measure the OD of each monoculture in an external spectrophotometer and dilute as necessary to make them as consistent as possible.
24. The native Chi.Bio OS https://github.com/HarrisonSteel/ChiBio/blob/master/app.py contains several examples of controllers in the custom program section. For example, "C1" (the first program) encodes an integral controller for optogenetic control of gene expression, where green light ('LEDD') activates expression of a fluorescent protein (intensity measured as "Emit 1"). It calculates the error as the difference between the desired set point ("Status," which can be set and altered at any point during the experiment in the GUI) and the fluorescent intensity (Emit 1). Depending on the input, it may also be necessary to limit it to stay within the limits of the hardware or the biological system, e.g., even if there is a large error, the temperature should not be raised high enough to kill both species. If there is very poor understanding of the system, a possible starting point is to use only the P term of the PID controller and to set K_P based on the open-loop experiment above.
25. In continuous culture systems, biofilms arise and are selected for as they persist on the inner surfaces of the bioreactor while planktonic cells are diluted out. They complicate control because (1) biofilms can affect readings (e.g., OD, fluorescence) by obscuring the sensor, and also (2) even if measurements are taken in a separate machine such as a cytometer, biofilms can impact growth kinetics, as they can contribute to the undesired persistence of one strain over the other in the system. Detergents such as Tween (polysorbate) 20 are

typically used to reduce biofilm formation, though these are not always sufficient [8]. By transferring the culture into a new vial/reactor, these issues can be avoided, but they need to be transferred before biofilms can arise.

5 Discussion

Co-cultures provide a way to combine the individual strengths of different microbial species while retaining flexibility. Currently, the way in which microbial species will grow together is difficult to predict and requires experimental characterization before stable growth can be engineered. Even small amounts of data help immensely in determining the type of relationship between chosen microbes and whether their co-culturing will be successful. Previous studies have benefited greatly from bioreactor setups for controlling co-cultures [7, 8, 15]. Above we outlined basic steps that can be taken to determine whether positive or negative interaction effects (potentially resulting in unmanageably complex growth dynamics) are present and how each species responds to media conditions. Following this, baseline hardware and software setups with control capabilities that can easily and cheaply grow co-cultures were discussed. We also recommend many useful steps for preliminary debugging, such as monoculture characterization and open-loop step-changes as sanity checks.

Overall, compared to ready-made solutions and conventional culturing techniques, a setup with customizable hardware such as the Chi.Bio platform enables more quantitative measurement of co-culture composition while offering a comprehensive set of tools to subsequently actuate changes in composition [12, 19, 20]. Unexpected behaviors can be caught early thanks to frequent collection of growth and fluorescence data. This is especially valuable for co-culturing engineered microbes, which may contain burdensome genetic circuits with fluorescent reporters or light-responsive modules and are therefore essentially guaranteed to deviate from their expected growth properties over time, resulting in poor performance of open-loop control approaches. Although the composition of species can be estimated from OD and other bulk measurements, measurements enabled extending a bioreactor with additional sensors can help to identify individual species and provide a more precise path to determining cell counts. For example, fluorescence can be related to the number of cells through experimental characterization. Further precision may be gained by integrating the culture vessel with a flow cytometer, allowing single-cell resolution of count and fluorescence level [21].

Once validated and characterized, the same small-scale setup further presents a powerful way to control co-culture composition. With a computer in the control loop, complex control approaches can be orchestrated to modulate media composition, temperature, light [7, 22], or other inputs [16] automatically, where a simple PID algorithm may be sufficient. While in single-strain non-cybergenetic approaches the entire computational and metabolic burden is typically imposed upon the cell by genetic engineering [2], the modularity of co-cultures also means a more flexible way to distribute biological computations and reactions [8, 13]. Further shifting some aspects of computation to external digital actuators can help retain optimal culture composition across time and respond nimbly to noise and drifts of the cell's properties over time while getting the most out of the engineered cells. Less burden on the cell further helps to retain engineered components in a cell population over longer evolutionary timescales [2].

We therefore argue that the customizability and responsiveness of cybergenetic systems is not just a recommended part of experimental setups, but crucial to co-culture development. Although a more customizable setup tends to be less standardized and therefore requires time and effort to implement and troubleshoot, this is well worth the investment given the importance of optimizing growth conditions for the strains involved. When the actuating signal is determined *in silico*, optimal controller parameters for maintaining a specific composition can be explored easily without any cumbersome reengineering of the strains' DNA. The results of this free exploration can then inform the design choices when genetically encoding purely biomolecular feedback circuits into the cells for downstream applications [11]. Developing more sophisticated ways to estimate composition in large bioreactors, whether from species-specific features or even from bulk measurements, would make it possible to deploy co-culture control algorithms at scale for commercial purposes and reduce the need for costly screening campaigns. Cybergenetics thus presents a powerful way to both study and engineer co-cultures and unlock entirely novel combinations of biological functions.

References

1. Tsoi R, Wu F, Zhang C et al (2018) Metabolic division of labor in microbial systems. Proc Natl Acad Sci USA 115:2526–2531. https://doi.org/10.1073/pnas.1716888115
2. Perrino G, Hadjimitsis A, Ledesma-Amaro R et al (2021) Control engineering and synthetic biology: working in synergy for the analysis and control of microbial systems. Curr Opin Microbiol 62:68–75. https://doi.org/https://doi.org/10.1016/j.mib.2021.05.004
3. McCarty NS, Ledesma-Amaro R (2019) Synthetic biology tools to engineer microbial communities for biotechnology. Trends Biotechnol 37:181–197. https://doi.org/10.1016/j.tibtech.2018.11.002
4. Gasparek M, Steel H, Papachristodoulou A (2023) Deciphering mechanisms of production of natural compounds using inducer-producer microbial consortia. Biotechnol Adv 64:108117. https://doi.org/10.1016/j.biotechadv.2023.108117
5. Johns NI, Blazejewski T, Gomes, ALC et al (2016) Principles for designing synthetic microbial communities. Curr Opin Microbiol 31:146–153. https://doi.org/10.1016/j.mib.2016.03.010
6. Hardin G (1960) The competitive exclusion principle. Science 131:1292–1297. https://doi.org/10.1126/science.131.3409.1292

7. Gutiérrez Mena J, Kumar S, Khammash M (2022) Dynamic cybergenetic control of bacterial co-culture composition via optogenetic feedback. Nat Commun 13:4808–4808. https://doi.org/10.1038/s41467-022-32392-z
8. Lee TA, Morlock J, Allan J et al (2025) Directing microbial co-culture composition using cybernetic control. Cell Rep Methods 5(3): 101009. https://doi.org/10.1016/j.crmeth.2025.101009
9. Caringella G, Bandiera L, Menolascina F (2023) Recent advances, opportunities and challenges in cybergenetic identification and control of biomolecular networks. Curr Opin Biotechnol 80:102893. https://doi.org/10.1016/j.copbio.2023.102893
10. Chait R, Ruess J, Bergmiller T et al (2017) Shaping bacterial population behavior through computer-interfaced control of individual cells. Nat Commun 8:1535 https://doi.org/10.1038/s41467-017-01683-1
11. Kumar S, Rullan M, Khammash M (2021) Rapid prototyping and design of cybergenetic single-cell controllers. Nat Commun 12:651–5651. https://doi.org/10.1038/s41467-021-25754-6
12. Steel H, Habgood R, Kelly CL et al (2020) In situ characterisation and manipulation of biological systems with Chi.Bio. PLOS Biol 18:1–12. https://doi.org/10.1371/journal.pbio.3000794
13. Lee TA, Steel H (2022) Cybergenetic control of microbial community composition. Front Bioeng Biotechnol 10:957140. https://doi.org/10.3389/fbioe.2022.957140
14. Espinel-Ríos S, Bettenbrock K, Klamt S et al (2023) Machine learning-supported cybergenetic modeling, optimization and control for synthetic microbial communities. Comput Aided Chem Eng 52:2601–2606. https://doi.org/10.1016/B978-0-443-15274-0.50413-3
15. Martinez JA, Bouchat R, Gallet de Saint Aurin T (2025) Automated adjustment of metabolic niches enables the control of natural and engineered microbial co-cultures. Trends Biotechnol 43:1116–1139. https://doi.org/10.1016/j.tibtech.2024.12.005
16. Martinez JA, Delvenne M, Henrion L et al (2022) Controlling microbial co-culture based on substrate pulsing can lead to stability through differential fitness advantages. PLoS Comp Biol 18:e1010674. https://doi.org/10.1371/journal.pcbi.1010674
17. Tshikantwa TS, Ullah MW, He F et al (2018) Current trends and potential applications of microbial interactions for human welfare. Front Microbiol 9:372532. https://doi.org/10.3389/fmicb.2018.01156
18. Boo A, Mehta H, Amaro RL et al (2023) Host-aware RNA-based control of synthetic microbial consortia. bioRxiv 2023.05.15.540816. https://doi.org/10.1101/2023.05.15.540816
19. Kusuda M, Shimizu H, Toya Y (2021) Reactor control system in bacterial co-culture based on fluorescent proteins using an Arduino-based home-made device. Biotechnol J 16: 2100169. https://doi.org/10.1002/biot.202100169
20. Bertaux F, Sosa-Carrillo S, Gross V et al (2022) Enhancing bioreactor arrays for automated measurements and reactive control with ReacSight. Nature Commun 13:3363–3363. https://doi.org/10.1038/s41467-022-31033-9
21. Sassi H, Nguyen, TM, Telek S et al (2019) Segregostat: a novel concept to control phenotypic diversification dynamics on the example of -Gram-negative bacteria. Microb Biotechnol 12:1064–1075. https://doi.org/10.1111/1751-7915.13442
22. Aditya C, Bertaux F, Batt G et al (2021) A light tunable differentiation system for the creation and control of consortia in yeast. Nat Commun 12:5829. https://doi.org/10.1038/s41467-021-26129-7
23. Morlock J (2025) CompEst3000. Version v1.0.0. Available via GITHUB. https://github.com/Janmorlock/CompEst3000. https://doi.org/10.5281/zenodo.14882855. Accessed 8 Aug 2025
24. Milias-Argeitis A, Rullan M, Aoki SK. et al (2016) Automated optogenetic feedback control for precise and robust regulation of gene expression and cell growth. Nat Commun 7: 12546. https://doi.org/10.1038/ncomms12546
25. Csibra E, Stan GB (2022) Absolute protein quantification using fluorescence measurements with FPCountR. Nat Commun 13: 6600. https://doi.org/10.1038/s41467-022-34232-6

Chapter 14

From Negative Feedback to Integrated Control: Recombinase-Based Strategies for Mitigating Resource Competition

Rixin Zhang, Rong Zhang, and Xiao-Jun Tian

Abstract

One of the key challenges in synthetic biology is the unintended interference between independently designed gene modules caused by resource competition within host cells. This interference disrupts circuit modularity and undermines the reliability of genetic systems. In this chapter, we discussed the developments of recombinase-based control strategies to address these challenges. We first describe a negative feedback controller, Re-NF controller, that leverages recombinase-mediated promoter flipping to automatically adjust circuit activity in response to resource usage, thereby stabilizing module performance under fluctuating cellular conditions. Building on this foundation, we introduce the Re-NF-FF controller, which integrates both negative feedback and feedforward regulation through a single recombinase-mediated promoter flipping. This dual strategy effectively balances resource allocation, reduces crosstalk, and improves predictability across diverse circuit contexts. We outline the design principles of both controllers, review their performance in controlling modular circuits, and discuss their implications for synthetic biology. Together, these strategies illustrate how recombinase-based regulation can be harnessed to enhance robustness, predictability, and modularity in the engineering of genetic systems.

Key words Gene circuit design, Circuit modularity, Resource decoupling, Resource controller, Promoter flipping

1 Introduction

In synthetic biology, genetic circuits are expected to function independently and reliably, with each module operating as a distinct and predictable unit [1]. However, in practice, circuit performance is often compromised by unintended interactions between genetic modules [2, 3]. These interferences arise mainly from the limited availability of cellular resources—such as RNA polymerases (RNAP) and ribosomes—that are shared by all genes expressed within the host cell for transcription and translation [2–4]. This phenomenon, known as resource competition, introduces

Xiao-Jun Tian (ed.), *Synthetic Gene Circuits: Methods and Protocols*, Methods in Molecular Biology, vol. 3041, https://doi.org/10.1007/978-1-0716-5304-3_14,

emergent behaviors that are difficult to predict when resource limitations are not accounted for during the design phase. For instance, resource competition can give rise to a Winner-Takes-All outcome in systems where coactivation was intended [5], or induce nonmonotonic dose—response relationships in activation cascades [3]. Such unintended behaviors reflect a loss of circuit modularity and robustness, complicating both the design and implementation of synthetic genetic systems.

Several strategies have been proposed to address the challenges imposed by resource competition in synthetic genetic systems. Since cellular resources—such as ribosomes and RNAP—are shared among multiple genetic modules, one approach is to employ orthogonal resources (such as orthogonal ribosomes/RNAP systems) to different modules [6, 7]. This breaks the resource bottleneck and reduces inter-module coupling. Additionally, emergent behaviors in synthetic circuits often arise from imbalanced resource allocation, wherein the activity of one module depletes resources required by another. Redirecting cellular resources toward the resource-limited module can mitigate interference and restore functional balance. This strategy of resource reallocation has been applied to improve the robustness of genetic circuits through regulatory schemes such as negative feedback [8, 9], antithetic integral feedback [10], and feedforward control [11–14].

Site-specific recombinases modify DNA by recognizing specific binding sites and catalyzing recombination events at those locations [15]. Depending on the orientation of the recombinase sites, recombination can occur in three ways: integration, excision, or inversion. While recombinase-mediated integration is commonly employed in DNA assembly, its excision and inversion capabilities enable precise logic-based control in genetic circuits. Specifically, DNA inversion occurs when a target sequence is flanked by two recombination sites arranged in a head-to-head orientation, whereas excision takes place when the sites are aligned in the same direction. Leveraging these mechanisms, recombinases have been widely applied in synthetic biology—for example, in logic computation [16], biological data storage [17–20], adaptive controllers [21], and noise reduction strategies [22].

In the following sections, we first describe our recombinase-based negative feedback controller that autonomously adjusts gene expression to stabilize circuit behavior under varying resource conditions. Building on this foundation, we then introduce the Re-NF-FF controller, which integrates negative feedback with feedforward regulation through recombinase-mediated promoter flipping. This combined strategy enables a simple, tunable, and efficient way to combine both control mechanisms within a single, compact genetic architecture. Together, these designs illustrate how recombinase-based regulation can be leveraged to construct reliable and predictable genetic systems, and offer a conceptual framework for future efforts to engineer higher-order control architectures in synthetic biology.

2 Resource Competition and Previous Control Strategies

2.1 Unintended Coupling via Resource Competition

In an ideal scenario with unlimited cellular resources, gene expression would occur independently. For example, in Fig. 1a, two modules coexist within the host cell: Module 1, comprising an inducible promoter—driven gene (Gene A), and Module 2, comprising a constitutively expressed gene (Gene B). Under ideal conditions, induction of Gene A would not affect the behavior of Gene B, and thus the expression level of Gene B would remain

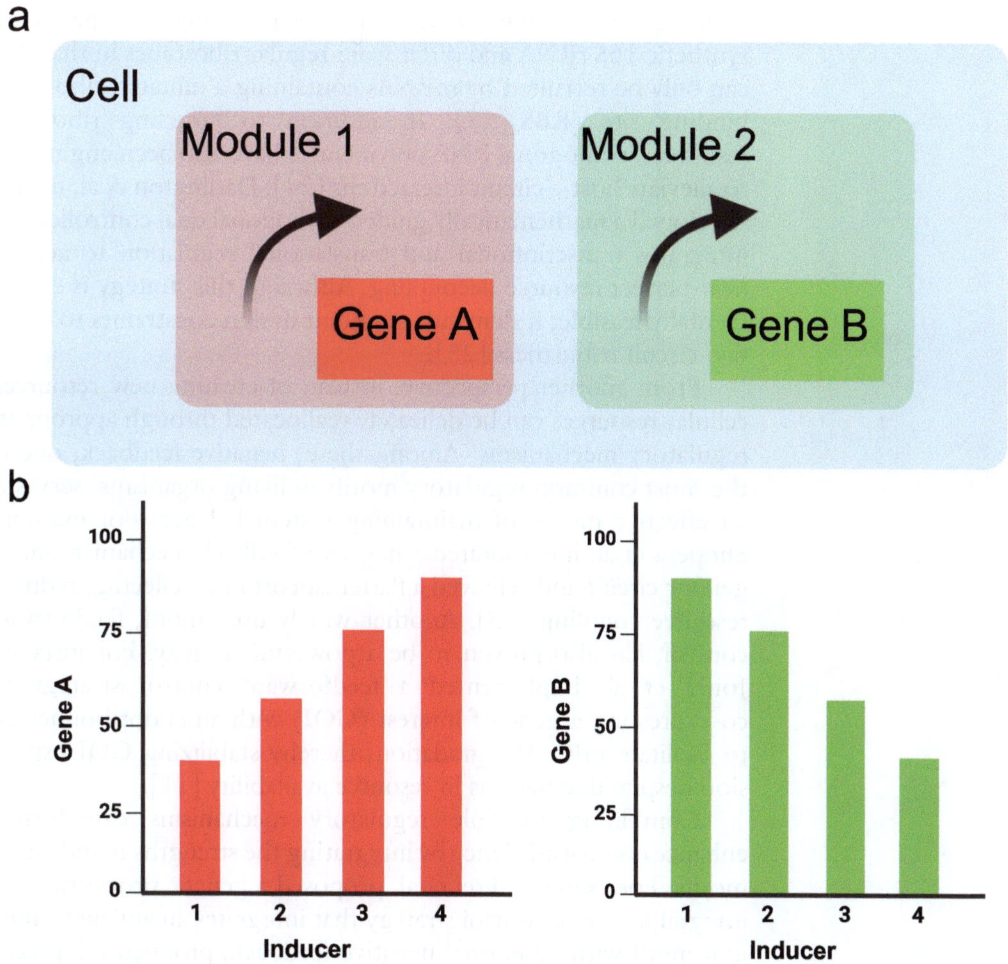

Fig. 1 Illustration of resource competition between co-expressed gene modules. (**a**) Two distinct genetic modules coexisting within the host cell: Module 1 expresses Gene A under an inducible promoter, and Module 2 constitutively expresses Gene B. (**b**) Induction of Gene A elevates its expression while concomitantly reducing the expression of Gene B due to competition for limited shared resources [2]

constant regardless of changes in Gene A. However, in reality, cellular resources are inherently limited, leading to significant competition between gene modules and resulting in unintended coupling [2, 23]. When the expression of Gene A is induced, this module sequesters a substantial fraction of the shared resources, thereby decreasing their availability for Gene B and leading to a reduction in Gene B expression (Fig. 1b).

2.2 Previous Mitigation Approaches

Many strategies have been proposed to restore the modularity of genetic circuits by decoupling resource interference. Since resource coupling arises from the sharing of common cellular resources, gene modules utilizing distinct resource pools exhibit minimal interference with each other. For example, Darlington et al. used orthogonal ribosomes (o-ribosomes) to reduce resource coupling within genetic circuits. O-ribosomes are generated by expressing a synthetic 16S rRNA and differ from regular ribosomes in that they can only be recruited by mRNAs containing a mutated ribosome-binding site (RBS) [7]. In addition to targeting ribosomal resources, orthogonal RNA polymerases have also been engineered to alleviate host—circuit interactions [24]. Darlington et al. further developed a mathematically guided orthogonal dual controller that integrates transcriptional and translational regulation to achieve near-perfect resource decoupling. Although this strategy is experimentally feasible, it demands stringent design constraints to maintain circuit robustness [25].

From another perspective, instead of creating new resources, cellular resources can be delicately reallocated through appropriate regulatory mechanisms. Among these, negative feedback, one of the most common regulatory motifs in living organisms, serves as an effective means of maintaining system balance. For example, Shopera et al. incorporated a negative feedback mechanism into a genetic circuit and achieved a flatter isocost line, reflecting reduced resource coupling [23]. Another widely used motif, feedforward control, has also proven to be a powerful strategy. For instance, Jones et al. implemented a feedforward control strategy by co-expressing a gene of interest (GOI) with an endoribonuclease to facilitate mRNA degradation, thereby stabilizing GOI expression despite fluctuations in resource availability [11].

Combining multiple regulatory mechanisms can further enhance control efficiency by integrating the strengths of individual motifs. For example, Frei et al. proposed a genetic proportional—integral feedback control strategy that integrates an antithetic integral motif with additional negative feedback, producing a precise and robust controller capable of adapting to variable cellular environments [10]. Similarly, Yang et al. developed buffering devices termed "Equalizers" to reduce copy-number variation at the single-cell level. At their core, Equalizers couple a negative feedback loop with incoherent feedforward control, outperforming either regulation alone in minimizing cell-to-cell expression variability [26].

3 Recombinase-Based Controllers

Despite the significant progress made in addressing resource interference among genetic modules, many existing strategies still suffer from low efficiency or are constrained by complex circuit designs. Therefore, developing new regulatory strategies that integrate multiple control mechanisms to enhance efficiency without increasing circuit complexity will not only expand the current synthetic biology toolbox but also pave the way for scaling up circuit complexity and functionality. To this end, we recently constructed genetic circuits capable of modulating cellular resource allocation through recombinase-mediated regulation. The rationale for employing recombinases to control resource allocation is fourfold:

1. *Proven logical control capability*: Recombinases have been extensively characterized and widely used in synthetic biology for implementing logic operations and memory functions, demonstrating strong potential for constructing resource-aware genetic circuits.
2. *Tunable recombination efficiency*: Recombinase activity can be precisely modulated by adjusting expression levels, enabling the creation of adaptive regulatory modules—particularly valuable for managing resource allocation under dynamic cellular and environmental conditions.
3. *High orthogonality and scalability*: Recombinase-mediated regulation offers orthogonal and modular control, allowing large-scale and multiplexed circuit regulation that extends beyond simple feedback architectures.
4. *Integration of multiple control strategies*: Recombinases can be readily combined with other regulatory motifs, such as feedback and feedforward control, providing a simple yet powerful framework for implementing complex multi-layered regulation within compact circuit designs.

3.1 Recombinase-Based Negative Feedback (Re-NF) Controller

3.1.1 Design Rational

Negative feedback is one of the most common regulatory motifs employed by living organisms to adapt to dynamic and fluctuating environments. In such systems, the output feeds back to modulate its own production by suppressing the strength of the input signal. The inversion-based recombination mechanism functions as a natural form of negative feedback and thus holds promise for mitigating resource competition in genetic circuits. In the resource competition model described in Fig. 1, the two competing genes, Gene A and Gene B, can be replaced with any genes of interest

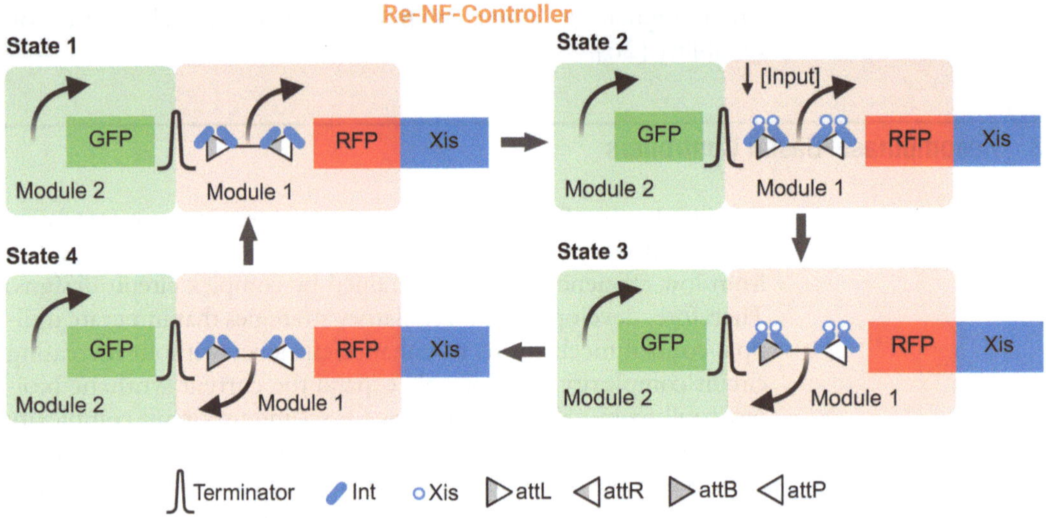

Fig. 2 Topology and operation of the Re-NF-controller. RFP is driven by an inducible promoter, with its transcriptional output modulated through bicistronic co-expression of excisionase and the constitutively expressed integrase. GFP is constitutively expressed, and its expression dynamics serve as an indicator of resource coupling within the system

depending on the application. For convenient and quantitative characterization, fluorescent proteins serve as reliable reporters for assessing the controller's performance. Therefore, Gene A and Gene B are substituted with RFP (Red Fluorescent Protein) and GFP (Green Fluorescent Protein), respectively. In our design shown in Fig. 2, Module 1 represents an inducible RFP that, when activated, draws cellular resources away from GFP in Module 2, which is constitutively expressed. A terminator is placed between the two modules to prevent transcriptional readthrough from the flipped promoter (*see* **Note 1**). To implement a recombinase-based negative feedback controller (*Re-NF-controller*), the promoter driving RFP expression is flanked by recombinase recognition sites—attL (attachment Left) and attR (attachment Right)—which are recognized and bound by an integrase dimer (Fig. 2, state 1). In this system, integrase is constitutively expressed and, by itself, is unable to mediate attL/attR-to-attB/attP recombination [27]. To establish a negative feedback loop responsive to changes in RFP expression, a recombination directionality factor, excisionase, is co-expressed with RFP in a bicistronic arrangement, thereby ensuring that its expression level is proportional to that of RFP. When RFP is induced and highly expressed, it imposes a significant burden on cellular resources. At the same time, excisionase expression also increases. Excisionase can recognize and bind to the integrase—DNA complex, forming a synaptic tetramer in which recombination occurs (Fig. 2, state 2) [28, 29]. Specifically, the integrase–excisionase complex catalyzes promoter inversion by cleaving, flipping, and re-ligating the recognition sites, thereby converting attL/attR into attB (attachment Bacteria) and attP (attachment Phage) (Fig. 2, state 3). The inversion reduces the proportion of promoters oriented to drive RFP expression, thereby automatically limiting the expression of both RFP and excisionase as resource consumption increases. As excisionase levels decline, the concentration of the excisionase–integrase complex diminishes (Fig. 2, state 4), allowing free integrase to catalyze the reverse reaction and restore the attB/attP configuration back to attL/attR. This bidirectional recombination mechanism forms the core of the Re-NF-Controller, which dynamically and reversibly tunes gene expression in response to fluctuations in resource demand from one module.

3.1.2 Modeling and Theoretical Insight

Ordinary differential equations (ODEs) can be employed to model dynamic biological processes. For example, the dynamics of constitutively expressed proteins $[P_i]$ in the system can be described as:

$$\frac{d[P_i]}{dt} = CN * v_{pi} - d_{pi} * [P_i]$$

where v_{pi} is the expression rate of protein P_i, d_{pi} is the degradation rate of protein P_i, and CN is the gene copy number.

For inducible proteins, the expression rate depends on the fraction of active promoters. Accordingly, the production term in the above function is modified to $R_\mathrm{i} * v_{pi}$, where R_i denotes the fraction of active promoters capable of driving protein expression (*see* **Note 2**).

$$\frac{d[P_i]}{dt} = R_\mathrm{i} * CN * v_{pi} - d_{pi} * [P_i]$$

To model the limitations imposed by resource competition into the equations, a partition function framework is applied to describe the distribution of resources among competing gene expression processes [5]. Specifically, the general form of the equation modeling resource competition for a single protein is:

$$\frac{d[P_i]}{dt} = CN * v_{pi} \cdot \frac{R_i/Q_i}{\mathrm{PF_Q}} - d_{pi} \cdot [P_i]$$

where $PF_Q = CN\sum R_i/Q_i + 1$ with Q_i represents the resource competing capability of the gene module. Under idealized conditions with unlimited resources, we can set all $Q_i \rightarrow \infty$, and thus PF_Q approaches 1 and the model reduces to the conventional gene expression framework that does not account for resource competition.

Based on the recombination mechanisms, the dynamics of promoter flipping mediated by serine recombinase could be modeled as follows:

$$\frac{d[\phi]}{dt} = v_{\text{on}} * \frac{[INT]^4}{K_{I1}{}^4 + [INT]^4} * (1 - \phi) - v_{\text{off}} * \frac{[XIS]^4}{K_X{}^4 + [XIS]^4} \frac{[INT]^4}{K_{I2}{}^4 + [INT]^4} * \phi$$

Here, ϕ represents the proportion of promoter copies in the attL/attR configuration, ranging from 0 to 1. The first term of ODE represents the attB/attP-to-attL/attR recombination, yielding a circuit topology in the attL/attR configuration, with v_{on} denoting the promoter flipping rate mediated by integrase. The second term captures the reverse recombination process, characterized by a flipping rate constant v_{off}. $K_{I1,\,2}$ and K_X are equilibrium constants that define the integrase and excisionase concentrations required to achieve half-maximal flipping rates. The Hill coefficients are set to 4, reflecting the tetrameric nature of the integrase–attachment site and integrase–excisionase complexes.

We can apply the above modeling framework to construct a model for the system with Re-NF-controller and analyze its performance in mitigating resource competition.

Explicitly, the ODEs of the system are formulated as follows to describe the dynamics of concentration of RFP ($[RFP]$), GFP ($[GFP]$), integrase $[INT]$, and excisionase ($[XIS]$):

$$\frac{d[INT]}{dt} = u_I{}^* \frac{CN/Q_I}{PF_Q} - d_I{}^*[INT]$$

$$\frac{d[XIS]}{dt} = u_A{}^* \frac{CN^*\phi^*R_i/Q_X}{PF_Q} - d_X{}^*[XIS]$$

$$\frac{d[GFP]}{dt} = u_{G1}{}^* \frac{CN/Q_G}{PF_Q} - d_G{}^*[GFP]$$

$$\frac{d[RFP]}{dt} = u_R{}^* \frac{CN^*\phi^*R_i/Q_R}{PF_Q} - d_R{}^*[RFP]$$

$$\frac{d[\phi]}{dt} = v_{\text{on}}{}^* \frac{[INT]^4}{K_{I1}{}^4 + [INT]^4}{}^* (1 - \phi) - v_{\text{off}} * \frac{[XIS]^4}{K_X{}^4 + [XIS]^4} \frac{[INT]^4}{K_{I2}{}^4 + [INT]^4}{}^* \phi$$

where ϕ denotes the proportion of promoters directing the expression of RFP and excisionase, $PF_Q = 1 + CN * \left(\frac{1}{QI} + \frac{1}{Q_G} + \frac{\phi * R_i}{Q_R} + \frac{\phi * R_i}{Q_X}\right)$ (*see* **Note 3**).

3.1.3 Plasmid Construction

Illustrative Design of the Re-NF Controller

To translate the conceptual framework of resource control into a practical implementation, genetic circuits incorporating the Re-NF controller must first be constructed. In the general design, the induction pattern is left unspecified to maintain conceptual clarity.

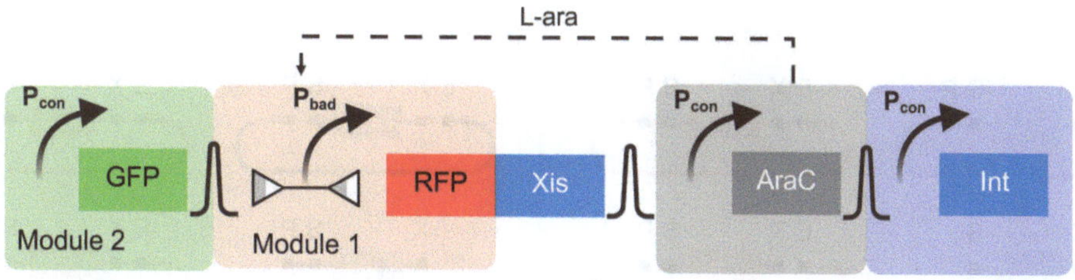

Fig. 3 A functional Re-NF circuit with all genetic components intact. The circuit is designed to maintain the modularity of constitutive GFP expression, which is affected by resource depletion resulting from AraC-induced RFP expression

Module 1 acts as an inducible component that competes with Module 2 for cellular resources. Various regulatory systems can be used to construct such an inducible module, including AraC and LuxR. In this chapter, we use the AraC system for Module 1. In the AraC-inducible system, addition of arabinose forms the AraC–arabinose complex, which activates Pbad promoter, leading to increased transcription rates with higher inducer concentrations. Based on this inducible system, a functional Re-NF controller can be designed as shown in Fig. 3.

Circuit Assembly

Plasmids are constructed using the BioBrick Standard Assembly method [30], as illustrated in Fig. 4. The restriction sites located at the prefix and suffix of each part are used for assembly.

For example, as shown in Fig. 4a, Part A can be assembled into the second circuit following these steps:

1. Part A is excised using EcoRI and SpeI.
2. The second circuit is digested with EcoRI and XbaI.
3. Equimolar amounts of the digested insert and backbone are combined in a ligation reaction (*see* **Note 4**).
4. The ligation mixture is transformed into competent cells, which are plated on agar plates containing the appropriate antibiotics.
5. Colony PCR is performed to identify positive colonies, which are then inoculated into media with suitable antibiotics.
6. Plasmids are extracted using a miniprep kit, followed by EcoRI and PstI digestion to confirm the insert size and obtain the final construct.

To swap the positions of genes during construction, alternative restriction sites can be selected for digestion as shown in Fig. 4b. In this case, Part A is excised with XbaI and PstI, and the backbone circuit is digested with SpeI and PstI. The subsequent procedures are identical to those described above.

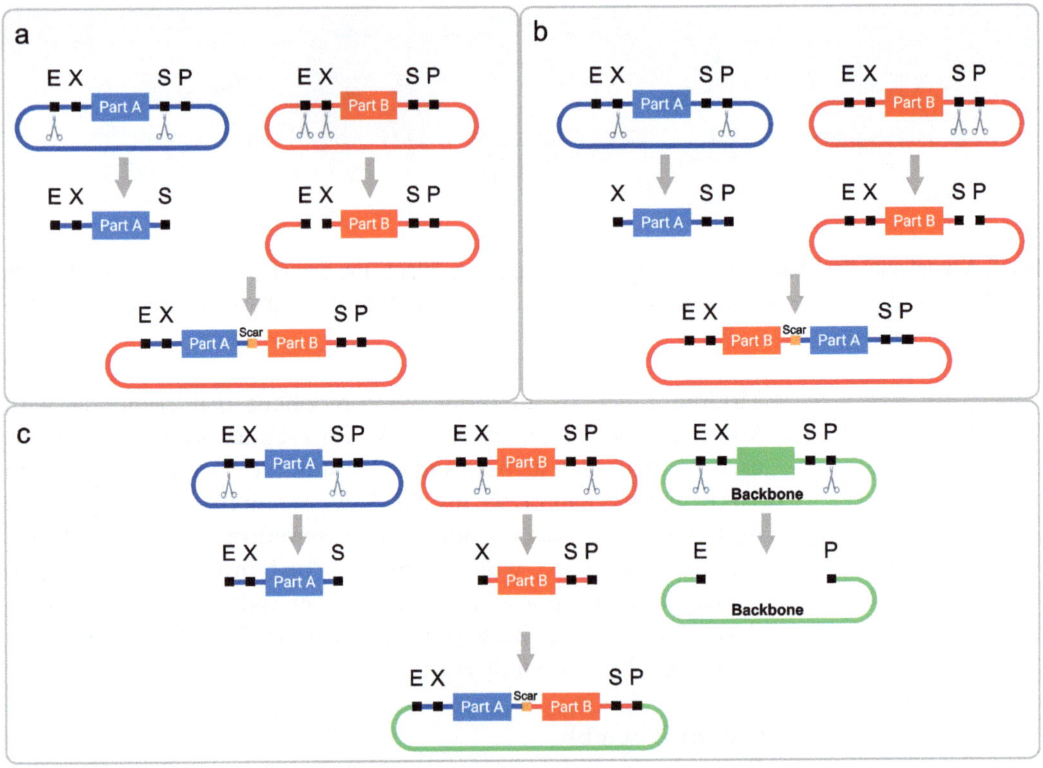

Fig. 4 BioBrick Standard Assembly method. (**a**, **b**) one-part ligation. (**c**) two-part ligation

The assembly described above represents a single-part ligation, where one insert (Part A) is incorporated into the circuit. The BioBrick Standard Assembly can also be used to perform a two-part ligation, as shown in Fig. 4c. Specifically, the left part is excised using EcoRI and SpeI, while the right part is excised using XbaI and PstI (*see* **Note 5**). To ligate these two parts into a circuit, the backbone is digested with EcoRI and PstI. The resulting construct retains these restriction sites, allowing further cloning into desired plasmids using the same approach.

To assemble the Re-NF circuit, we begin by constructing a plasmid containing both the RFP and excisionase genes arranged in a bicistronic configuration. This plasmid can be generated using a two-part ligation strategy. First, both genes must be modified with the appropriate prefixes and suffixes containing the four required restriction sites to ensure proper digestion. According to the protocol, the RFP fragment is digested with EcoRI and SpeI, while the excisionase fragment is digested with XbaI and PstI. The plasmid backbone is digested with EcoRI and PstI to allow ligation of the two fragments, yielding a plasmid carrying both inserts. This intermediate plasmid can then be used for subsequent cloning steps, following the same workflow, until the full Re-NF circuit is constructed.

3.1.4 Induction Regime for Resource Competition Assay

To examine the regulatory effects of recombinase-based control under resource competition, the resource-utilizing module is implemented as an inducible Module 1 (Fig. 3). Upon induction, RFP expression draws cellular resources away from the constitutively expressed Module 2, thereby establishing a controlled resource-competition scenario.

The general induction procedure is as follows:

1. Plasmids are transformed into *E. coli* strain K-12 MG1655*ΔlacIΔaraCBAD* (*see* **Note 6**).
2. Three colonies are picked for each test group and inoculated into 300 μL of LB medium supplemented with the appropriate antibiotics. Cultures are grown at 37 °C with shaking at 250 rpm for 5 h to allow saturation, ensuring that the cell concentrations of the three replicates are comparable for more reliable results.
3. 5 μL of each culture is transferred into each well of a 96-well plate containing 200 μL of LB medium with the corresponding antibiotic.
4. A 25% L-arabinose stock solution is serially diluted to 8.3%, 5%, 2.5%, 0.625%, 0.25%, and 0.125% to prepare working solutions. Each working solution is further diluted 1:1000 into the corresponding medium to induce the gene circuit.
5. Plates are incubated overnight at 37 °C with shaking at 250 rpm to allow gene expression to reach steady state.

Following induction as described above, 200 μL of each culture is transferred to a transparent 96-well plate for fluorescence intensity measurement using a Synergy H1 Hybrid Reader (BioTek). For GFP, excitation and emission wavelengths are set to 485 nM and 515 nM, respectively. For RFP, excitation and emission wavelengths are set to 438 nM and 485 nM, respectively. To correct background fluorescence from the medium, three wells containing 200 μL of LB medium without cells are used as blank controls. To normalize for differences in cell density, fluorescence values are calculated by dividing the background-subtracted fluorescence by the corresponding optical density at 600 nM (OD_{600}).

3.1.5 Results

To evaluate the resource-decoupling performance of the Re-NF controller, an open-loop control circuit lacking the recombination mechanism is required for direct comparison. Several strategies can be used to disable recombination. One approach is to disrupt key recombination components, such as by mutating nucleotides within the attachment sites recognized or excised by the integrase, or by introducing mutations into essential integrase domains to prevent recombination. A simpler alternative is to omit the recombinase genes entirely, although this design consumes fewer resources and therefore produces a weaker degree of resource coupling in the control circuit.

After circuit induction and plate reader measurements, fluorescence intensities from the GFP and RFP modules are collected to characterize how GFP expression responds to changes in resource availability driven by increasing RFP expression. Before comparing the Re-NF controller with the open-loop circuit, fluorescence data are normalized. GFP expression levels are normalized to their initial (non-induced) values, allowing clear visualization of induction-dependent trends. RFP levels are normalized to their maximum value, constraining all measurements to the 0–1 range and ensuring consistency across experiments. Using these normalized values, an interdependence curve between GFP and RFP can be constructed. A steeper slope indicates stronger resource coupling, whereas a flatter slope reflects reduced resource competition. In the ideal scenario without resource competition, the curve would be completely flat, indicating that GFP expression remains unaffected by RFP induction.

The experimental results showed that although the Re-NF controller partially reduced resource coupling, it did not fully eliminate resource interference. This is evident from the interdependence curve, which falls between that of the open-loop circuit and the ideal no-coupling scenario (Fig. 5a). The suboptimal decoupling arises from intrinsic limitations in regulatory capacity: once the regulatory components (e.g., repressors or activators) become saturated, they can no longer respond proportionally to increasing perturbations, preventing complete decoupling. Similar saturation effects have been observed in other negative-feedback-based strategies for mitigating resource coupling, and the same constraint applies to our recombinase-based feedback system. In the Re-NF

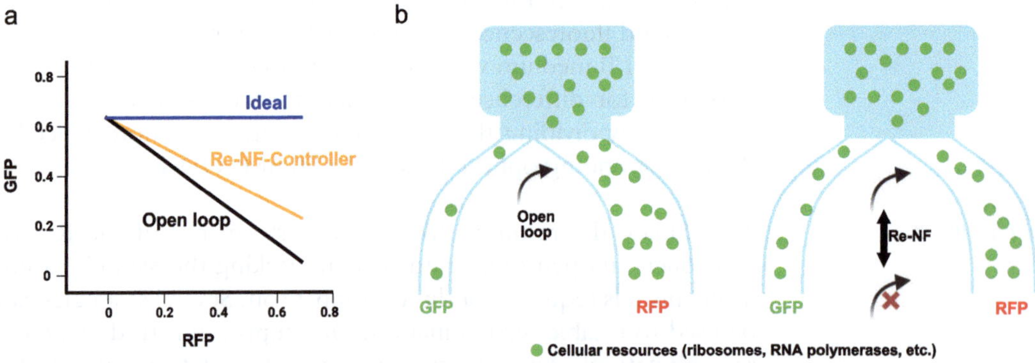

Fig. 5 Resource decoupling effect of the Re-NF controller. (**a**) Interdependence curves between GFP and RFP expression for circuits regulated by the Re-NF controller (yellow) and for the unregulated open-loop circuit (black). The flat blue line represents the ideal scenario in which GFP expression remains fully independent of RFP induction. (**b**) Schematic illustration of resource allocation in genetic circuits, comparing the unregulated open-loop system with the system regulated by the Re-NF controller

controller, decoupling occurs primarily through restricting RFP expression, thereby freeing resources that can be redirected to support GFP expression (Fig. 5b). However, this mechanism alone is insufficient to fully eliminate resource interference, underscoring the need for a more effective regulatory topology.

3.2 Combined Negative Feedback and Feedforward (Re-NF-FF) Controller

3.2.1 Design Concept

To address the limitations of the Re-NF controller, its circuit topology can be redesigned to achieve stronger resource decoupling. By leveraging the promoter-flipping event triggered by recombination, additional resources can be redirected to the GFP module by allowing it to receive transcriptional input from the reverted promoter for enhanced compensation. In this design, the GFP module is placed downstream of the flipped promoter, and the terminator used in the original Re-NF controller is removed (Fig. 6), enabling GFP expression to benefit from a dual-promoter effect that helps counteract resource depletion during competition. This modification yields a hybrid controller that integrates both feedback and feedforward regulation, referred to as the Re-NF-FF controller. The regulatory efficiency of the Re-NF-FF controller can be experimentally evaluated by assembling a circuit that includes the AraC-inducible system, analogous to the configuration described in Fig. 3.

3.2.2 Mathematical Model of the Re-NF-FF-Controller

In the Re-NF-FF system, the flipped promoter not only restricts the expression of RFP but also enhances the expression of GFP. Accordingly, building upon the modeling framework of the

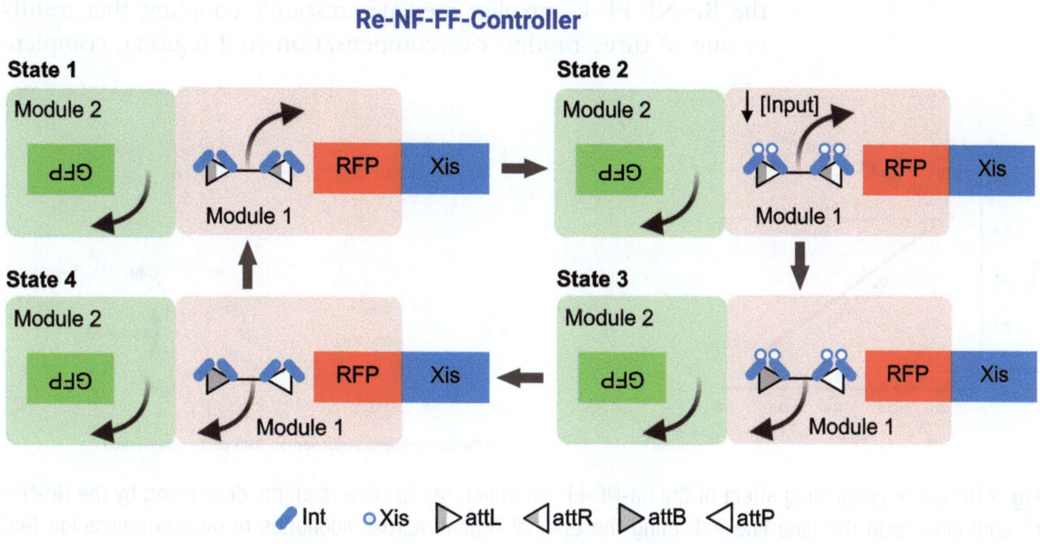

Fig. 6 Topology and operation of the Re-NF-FF-controller. RFP is driven by an inducible promoter, with its transcriptional output modulated through bicistronic co-expression of excisionase and the constitutively expressed integrase. GFP is constitutively expressed, and its expression dynamics serve as an indicator of resource coupling within the system

Re-NF-controller, the equation for GFP in the Re-NF-FF system is modified by introducing an additional term that represents the extra transcriptional input provided by the flipped promoter:

$$\frac{d[GFP]}{dt} = u_{G1} * \frac{CN/Q_G}{PF_Q} + u_{G2} * \frac{(1-\phi) * CN * R_i/Q_R}{PF_Q} - d_G * [GFP]$$

Accordingly, PF_Q should change into $1 + CN * \left(\frac{1}{QI} + \frac{1}{Q_G} + \frac{\phi * R_i}{Q_R} + \frac{\phi * R_i}{Q_X} + \frac{(1-\phi) * R_i}{Q_R}\right)$ due to additional resource distribution regime.

3.2.3 Performance and Tunability of the Re-NF-FF-Controller

The additional GFP expression driven by the flipped promoter can offset the reduction caused by resource competition. However, this compensation may overshoot, leading to overcorrection. Thus, achieving optimal resource regulation with the Re-NF-FF controller requires a tunable mechanism that balances the compensatory response and mitigates both overcompensation and residual resource coupling (Fig. 7).

The promoter inversion process is catalyzed by the integrase–excisionase complex, which assembles in a defined stoichiometric ratio, thereby conferring tunability to the Re-NF-FF-Controller. This tunability enables the controller to adapt to varying levels of cellular resource demand, such as elevated competition in high-copy plasmid systems, highly resource-intensive gene modules, or complex genetic circuits characterized by limited resource availability due to sharing among multiple modules. As shown in Fig. 7a, the Re-NF-FF-Controller mediates resource coupling that results in one of three modes: overcompensation (red region), complete

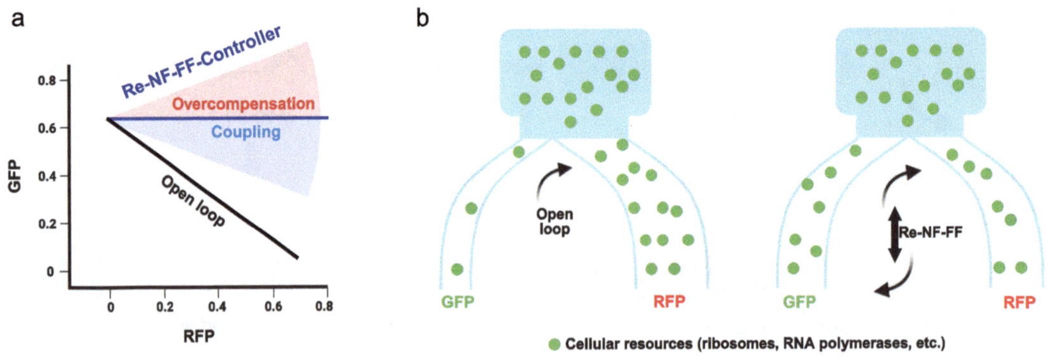

Fig. 7 Resource decoupling effect of the Re-NF-FF controller. (**a**) Tunable resource decoupling by the Re-NF-FF-controller, with the blue line indicating the optimal regime across scenarios of overcompensation (red shaded area) and resource coupling (blue shaded area). The black line shows the open-loop system for comparison. (**b**) Schematic illustration of resource allocation in genetic circuits: unregulated gene expression and gene expression regulated by the Re-NF-FF-controller

decoupling (blue line), or suboptimal decoupling (blue region). Consequently, fine-tuning of the controller is necessary under specific conditions to achieve optimal mitigation of resource-mediated interference. Such tuning can be accomplished by modulating the expression levels of integrase and excisionase to regulate recombination efficiency, thereby controlling the extent of resource reallocation to the target module. Under conditions of intensified resource competition, an increased fraction of plasmids harboring the inverted promoter is preferred to mitigate resource depletion within the affected gene module. Achieving this requires a balanced expression of excisionase and integrase to maintain elevated recombination activity. Conversely, when resource competition is reduced, attenuated recombination is desirable to prevent overcompensation. This can be achieved by perturbing the integrase–excisionase balance. For example, the excisionase expression level can be tuned by using different strengths of degradation tags. Faster turnover rate of excisionase is necessary when a weaker resource competing module is introduced to avoid overcompensation, vice versa, in a heightened resource competing environment, enhanced expression of excisionase with lower degradation rate in the Re-NF-FF controller is required to effectively perform resource reallocation. In conclusion, the operation of Re-NF-FF controller includes a feedforward mechanism inherently, without requiring additional regulatory genes that could further burden the system. By combining negative feedback and feedforward control through a single promoter inversion event (Fig. 7b), a tunable resource mitigator can be constructed. With appropriate tuning, this strategy can effectively alleviate resource competition and serve as a valuable tool for building resource-aware gene circuits.

3.3 Discussion

In this chapter, we described a design strategy that incorporates both negative feedback and feedforward control mediated by one promoter flipping recombination. This approach has been demonstrated to perform more efficiently than negative feedback alone. In addition to enhanced resource reallocation capabilities, it also offers tunability by adjusting the levels of recombinase enzymes. These features make the Re-NF-FF-Controller a promising tool for mitigating resource competition.

We demonstrated the performance of the Re-NF-FF-Controller using a two-gene system, which serves as a standard testbed for assessing controller efficacy, as demonstrated in several previous studies [2, 23]. However, the application of the Re-NF-FF strategy is not limited to this two-gene model. More than 25 putative large serine recombinases have been identified [17, 31, 32], and for each recombinase, up to six orthogonal recombination site pairs can be engineered by introducing distinct, non-symmetric 2-bp central overlap sequences [33]. The combination of these recombinases with their corresponding recombination sites enables the scalable implementation of the Re-NF-FF strategy, offering a potentially valuable approach for addressing resource competition among multiple genes or within complex genetic circuits. It is important to note that, compared to the use of orthogonal recombinases, employing recombination sites in constructing the Re-NF-FF-Controller reduces the resource burden imposed by recombinase expression, thereby alleviating additional strain on the system. While the integration of controller proteins into genetic circuits inevitably introduces burden, the critical consideration is to evaluate the trade-off between regulatory performance and the corresponding negative impact of the burden imposed.

Beyond its application in mitigating resource competition, the Re-NF-FF strategy also holds potential for optimizing production, controlling cellular growth, and regulating metabolism, particularly in scenarios where suppression of one module benefits other modules. The performance of the Re-NF-FF-Controller has been demonstrated to effectively alleviate resource competition. Future studies aimed at assessing its regulatory precision and robustness under perturbations will provide valuable insights into its functional capabilities. Additionally, investigating the feasibility of implementing the Re-NF-FF strategy in mammalian cells will expand its potential applications, including cancer research and therapeutic interventions.

4 Notes

1. The terminator used here may exhibit reduced termination efficiency, possibly due to steric hindrance introduced during the recombination process. To improve termination efficiency, a double terminator can be used in place of a single terminator.
2. For unregulated, constitutively expressed genes in the system, both ϕ and R_i are set to 1.
3. R_i denotes the level of the regulated promoter activity, the value of which depends on the specific induction system employed.
4. An 8-base pair "scar" sequence is generated during the ligation process. This scar results from sticking the compatible ends of XbaI and SpeI and is not recognized by either restriction enzyme in further cloning steps.
5. When transforming a two-part ligation mixture, a double volume of competent cells is recommended to improve transformation efficiency. Accordingly, the volume of LB medium should also be doubled to ensure optimal cell growth conditions.

6. A mutated *E. coli* K-12 MG1655 strain is used for transformation, in which the araBAD operon is deleted to prevent interference with the arabinose-inducible (AraC) system.

Acknowledgments

This work was supported by a grant from the US National Institutes of Health (R35GM142896 to X.-J.T.).

Competing Interests The authors declare no competing interests.

References

1. Lucks JB, Qi L, Whitaker WR, Arkin AP (2008) Toward scalable parts families for predictable design of biological circuits. Curr Opin Microbiol 11:567–573
2. Gyorgy A, Jiménez JI, Yazbek J, Huang H-H, Chung H, Weiss R, Del Vecchio D (2015) Isocost lines describe the cellular economy of genetic circuits. Biophys J 109:639–646
3. Qian Y, Huang H-H, Jiménez JI, Del Vecchio D (2017) Resource competition shapes the response of genetic circuits. ACS Synth Biol 6:1263–1272
4. Klumpp S, Zhang Z, Hwa T (2009) Growth rate-dependent global effects on gene expression in bacteria. Cell 139:1366–1375
5. Zhang R, Goetz H, Melendez-Alvarez J, Li J, Ding T, Wang X, Tian X-J (2021) Winner-takes-all resource competition redirects cascading cell fate transitions. Nat Commun 12:853
6. Liu CC, Jewett MC, Chin JW, Voigt CA (2018) Toward an orthogonal central dogma. Nat Chem Biol 14:103–106
7. Darlington APS, Kim J, Jiménez JI, Bates DG (2018) Dynamic allocation of orthogonal ribosomes facilitates uncoupling of co-expressed genes. Nat Commun 9:695
8. Jones RD, Qian Y, Ilia K, Wang B, Laub MT, Del Vecchio D, Weiss R (2022) Robust and tunable signal processing in mammalian cells via engineered covalent modification cycles. Nat Commun 13:1720
9. Venturelli OS, Tei M, Bauer S, Chan LJG, Petzold CJ, Arkin AP (2017) Programming mRNA decay to modulate synthetic circuit resource allocation. Nat Commun 8:15128
10. Frei T, Chang C-H, Filo M, Arampatzis A, Khammash M (2022) A genetic mammalian proportional–integral feedback control circuit for robust and precise gene regulation. Proc Natl Acad Sci 119:e2122132119
11. Jones RD, Qian Y, Siciliano V, DiAndreth B, Huh J, Weiss R, Del Vecchio D (2020) An endoribonuclease-based feedforward controller for decoupling resource-limited genetic modules in mammalian cells. Nat Commun 11:5690
12. Barajas C, Huang H-H, Gibson J, Sandoval L, Del Vecchio D (2022) Feedforward growth rate control mitigates gene activation burden. Nat Commun 13:7054
13. Stone A, Ryan J, Tang X, Tian X-J (2022) Negatively competitive incoherent feedforward loops mitigate winner-take-all resource competition. ACS Synth Biol 11:3986–3995
14. Stone A, Rijal S, Zhang R, Tian X-J (2024) Enhancing circuit stability under growth feedback with supplementary repressive regulation. Nucleic Acids Res 52:1512–1521
15. Grindley NDF, Whiteson KL, Rice PA (2006) Mechanisms of site-specific recombination*. Annu Rev Biochem 75:567–605
16. Weinberg BH, Pham NTH, Caraballo LD, Lozanoski T, Engel A, Bhatia S, Wong WW (2017) Large-scale design of robust genetic circuits with multiple inputs and outputs for mammalian cells. Nat Biotechnol 35:453–462
17. Roquet N, Soleimany AP, Ferris AC, Aaronson S, Lu TK (2016) Synthetic recombinase-based state machines in living cells. Science (1979) 353:aad8559
18. Friedland AE, Lu TK, Wang X, Shi D, Church G, Collins JJ (2009) Synthetic gene networks that count. Science (1979) 324: 1199–1202
19. Bonnet J, Subsoontorn P, Endy D (2012) Rewritable digital data storage in live cells via

engineered control of recombination directionality. Proc Natl Acad Sci 109:8884–8889

20. Zhao J, Pokhilko A, Ebenhöh O, Rosser SJ, Colloms SD (2019) A single-input binary counting module based on serine integrase site-specific recombination. Nucleic Acids Res 47:4896–4909
21. Steel H, Papachristodoulou A (2019) Low-burden biological feedback controllers for near-perfect adaptation. ACS Synth Biol 8: 2212–2219
22. Folliard T, Steel H, Prescott TP, Wadhams G, Rothschild LJ, Papachristodoulou A (2017) A synthetic recombinase-based feedback loop results in robust expression. ACS Synth Biol 6:1663–1671
23. Shopera T, He L, Oyetunde T, Tang YJ, Moon TS (2017) Decoupling resource-coupled gene expression in living cells. ACS Synth Biol 6: 1596–1604
24. An W, Chin JW (2009) Synthesis of orthogonal transcription-translation networks. Proc Natl Acad Sci 106:8477–8482
25. Darlington APS, Bates DG (2020) Architectures for combined transcriptional and translational resource allocation controllers. Cell Syst 11:382–392.e9
26. Yang J, Lee J, Land MA, Lai S, Igoshin OA, St-Pierre F (2021) A synthetic circuit for buffering gene dosage variation between individual mammalian cells. Nat Commun 12:4132
27. Ghosh P, Pannunzio NR, Hatfull GF (2005) Synapsis in phage Bxb1 integration: selection mechanism for the correct pair of recombination sites. J Mol Biol 349:331–348
28. Ghosh P, Wasil LR, Hatfull GF (2006) Control of phage Bxb1 excision by a novel recombination directionality factor. PLoS Biol 4:e186
29. Ghosh P, Bibb LA, Hatfull GF (2008) Two-step site selection for serine-integrase-mediated excision: DNA-directed integrase conformation and central dinucleotide proofreading. Proc Natl Acad Sci 105:3238–3243
30. Shetty R, Lizarazo M, Rettberg R, Knight TF (2011) Chapter thirteen – assembly of BioBrick standard biological parts using three antibiotic assembly. In: Voigt C (ed) Methods in enzymology, vol 498. Academic Press, pp 311–326
31. Brown WRA, Lee NCO, Xu Z, Smith MCM (2011) Serine recombinases as tools for genome engineering. Methods 53:372–379
32. Yang L, Nielsen AAK, Fernandez-Rodriguez J, McClune CJ, Laub MT, Lu TK, Voigt CA (2014) Permanent genetic memory with >1-byte capacity. Nat Methods 11:1261–1266
33. Colloms SD, Merrick CA, Olorunniji FJ, Stark WM, Smith MCM, Osbourn A, Keasling JD, Rosser SJ (2014) Rapid metabolic pathway assembly and modification using serine integrase site-specific recombination. Nucleic Acids Res 42:e23–e23

Chapter 15

Harnessing CRISPRi Competition to Develop Multimodule Controllers for Resource-Aware Circuit Design

Sadikshya Rijal, Rong Zhang, and Xiao-Jun Tian

Abstract

Cellular resource limitations give rise to resource competition, undermining the modularity and predictability of engineered genetic circuits. In systems containing positive feedback, such competition can drive Winner-Takes-All (WTA) dynamics, resulting in severe imbalances in resource allocation across circuit modules. In this chapter, we present an experimental implementation of a Negatively Competitive Regulatory (NCR) controller based on CRISPR interference (CRISPRi) in dual self-activation (DSA) circuits. We describe a detailed workflow for chromosomal integration of a tunable *dCas9* expression cassette, as well as the design of self-activation modules and module-specific guide RNAs that induce self-repression through competition for limiting dCas9. This architecture introduces effective negative feedback to the more active module while reallocating resources to the less active one, thereby promoting balanced module activity. Finally, we provide guidelines for quantitatively assessing the regulation of resource competition between DSA modules using the NCR strategy. Overall, these guidelines demonstrate how CRISPRi can be leveraged to implement NCR strategy in gene circuits, thereby enhancing circuit modularity and predictability.

Key words Resource decoupling, dCas9 competition, Module coactivation, Modularity, Multi-module control

1 Introduction

Cellular transcriptional, translational, and post-translational resources are inherently limited in host chassis, constraining the performance of synthetic genetic systems in both bacterial and mammalian cells [1, 2]. Such limitations restrict the achievable expression levels of engineered constructs and compromise the modularity and predictability of genetic circuits [3–8]. When two or more heterologous gene modules are expressed within a single strain, they inevitably compete for shared cellular resources, leading to substantial interference between the modules [9–11]. Such resource competition can reduce the expression of each module by up to 60%, significantly altering circuit behavior and

Xiao-Jun Tian (ed.), *Synthetic Gene Circuits: Methods and Protocols*, Methods in Molecular Biology, vol. 3041,
https://doi.org/10.1007/978-1-0716-5304-3_15,

performance [10]. To mitigate these effects, a variety of strategies, such as engineering orthogonal cellular resources (ribosomes, polymerases) and incorporating negative-feedback or feedforward controllers, have been developed to decouple synthetic modules from one another [12–16].

Under certain circuit architectures, however, resource competition manifests in a more extreme form known as "Winner-Takes-All" (WTA) behavior [9]. WTA has been observed in gene circuits driven by positive feedback loops, including dual self-activation (DSA) circuits and cascading bistable switches (CBS). Both architectures consist of two self-activation modules in which transcription factors reinforce their own expression. In these contexts, shared cellular resources become disproportionately allocated to one module, causing the other to be silenced. As a result, the two modules fail to coactivate, and overall circuit modularity is severely compromised. One experimentally verified strategy for mitigating WTA has been to distribute gene modules across different microbial strains, thereby dividing the cellular workload [9]. While effective in the short term, this approach remains vulnerable to long-term instability due to differential strain growth rates.

Previously, we developed theoretical frameworks for multiple single-strain engineering strategies that achieve multimodule regulation through negatively competitive regulation (NCR) [17–20]. NCR uses a deliberately limited controller resource to implement competitive negative feedback or feedforward regulation. In one implementation using CRISPR interference (CRISPRi) to enable NCR, module-specific guide RNAs compete for a shared pool of catalytically dead Cas9 (dCas9) to achieve the negative feedback regulation [18]. This design was recently validated experimentally [21]. In this chapter, we describe the detailed experimental procedures for implementing this system, including genomic integration of dCas9 into *E. coli*, circuit design and construction, and testing of multimodule NCR behavior.

2 Materials

2.1 Bacterial Cell Culture Reagents

1. Luria Bertani (LB) Broth (Miller) (25 g/L).
2. Luria Bertani (LB) Agar (Miller) (40 g/L).

2.2 Antibiotics

1. Kanamycin (50 μg/μL).
2. Ampicillin (100 μg/μL).

2.3 Restriction Digestion Reagents

1. FastDigest Buffer (10×).
2. rCutsmart Buffer (10×).

3. FastDigest EcoRI.
4. FastDigest XbaI.
5. FastDigest SpeI.
6. FastDigest PstI.
7. SapI 10,000 units/mL.

2.4 T7 DNA Ligation Reagents

1. StickTogether™ DNA Ligase Buffer (10×).
2. T7 DNA Ligase (3,000,000 units/mL).

2.5 Gel-Electrophoresis Reagents

1. TAE buffer: 40 mM Tris, 20 mM acetate, 1 mM EDTA.
2. 100 bp and 1 kb DNA Ladder.
3. Gel Loading Dye, Purple (6×).
4. DNA gel stain.
5. Molecular-grade agarose.

2.6 Other Materials

1. QIAprep Spin Miniprep Kit or any other plasmid extraction kit.
2. QIAquick PCR Purification Kit or any other PCR cleanup kit.
3. *Taq* PCR Kit.
4. Any high-fidelity PCR Kit.
5. Nuclease-free water.
6. Mix-and-Go! *E. coli* transformation kit.

3 Methods

In the CRISPRi-based NCR controller, dCas9 is integrated into the genome to serve as a limited controller resource. Its availability is further constrained by an inducible pTet–TetR–aTc regulatory system (Fig. 1a). Each module expresses a distinct dCas9 single-guide RNA (sgRNA) in the plasmid, enabling negative feedback through competition for the shared dCas9 pool. In this design, the dominant ("winning") module produces more sgRNA, thereby depleting the limited dCas9 resource and increasing repression of its own activity. Conversely, the less active module experiences weaker NCR-mediated inhibition and gains greater access to core cellular resources. In this chapter, we focus on implementing and characterizing this controller within the DSA circuit architecture (Fig. 1b).

3.1 dCas9 Integration into E. coli K-12 MG1655 (ΔLacI ΔAraC) Genome

1. Add BioBrick prefix and suffix sites on dCas9 using forward and reverse primers and a high-fidelity polymerase. Perform a PCR cleanup step for purification of the dCas9 PCR product (Fig. 2a and Table 1).

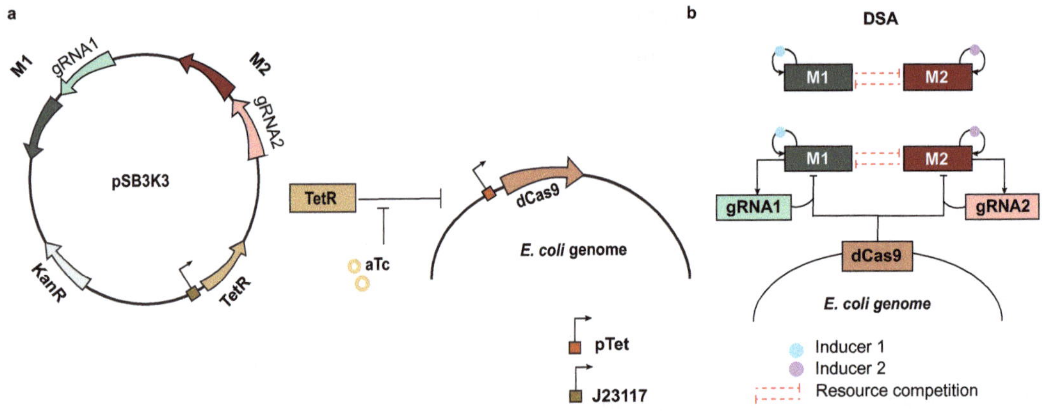

Fig. 1 Schematic of the CRISPR-linked multimodule negative feedback. (**a**) Multicopy plasmid pSB3K3 encoding two circuit modules (M1 and M2) and a constitutively expressed TetR repressor, which suppresses genome-encoded dCas9 expression in an anhydrotetracycline (aTc)–dependent manner. (**b**, top) Schematic of the Dual Self-Activation (DSA) circuit, which exhibits WTA behavior. Each module is self-activating, with its transcription factor promoting its own expression. Red dotted lines indicate unintended mutual inhibition arising from resource competition. (**b**, bottom) Implementation of the CRISPR-based NCR controller in the DSA circuit. Each module expresses a gRNA that targets its own promoter, enabling module-specific transcriptional repression through competition for a shared, limited pool of dCas9. (Adapted from our previous work [21])

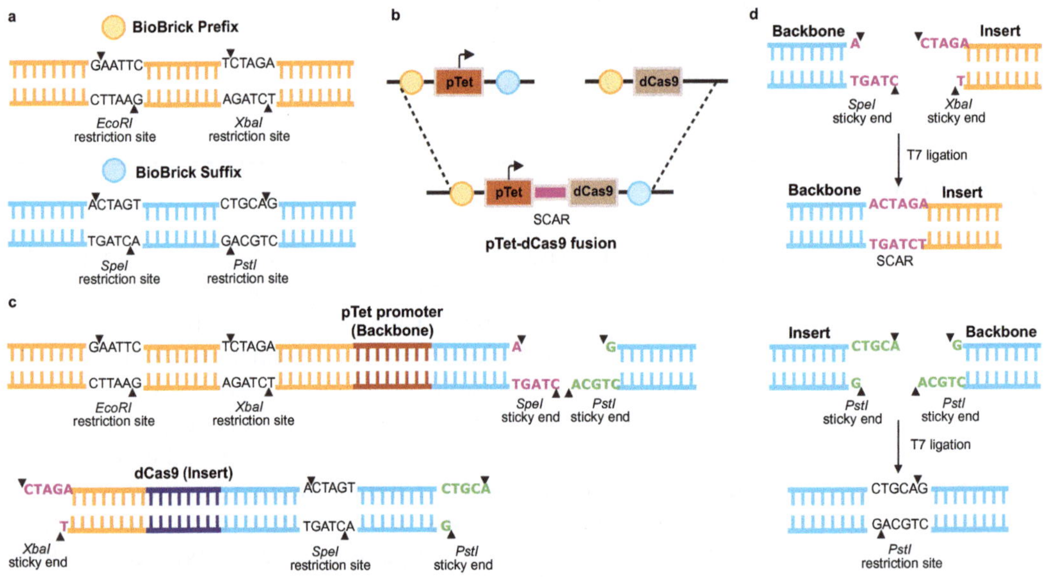

Fig. 2 BioBrick cloning of the dCas9 gene to the pTet promoter backbone. (**a**) BioBrick Prefix (yellow) containing EcoRI and XbaI restriction sites and BioBrick suffix (blue) containing SpeI and PstI restriction sites. (**b**) Schematic of BioBrick cloning in which the dCas9 gene (insert), flanked by BioBrick-compatible sites, is assembled into the pTet promoter backbone. (**c**) pTet promoter backbone digest with SpeI and PstI sticky ends (top). The dCas9 gene fragment with XbaI and PstI sticky ends (bottom). (**d**, top) The SpeI and XbaI feature complementary sticky ends and ligate to form a SCAR site. (**d**, bottom) PstI sticky ends in the backbone and insert ligate to reconstruct the PstI restriction site in the BioBrick suffix

Table 1
List of oligos used in the experiments

Sequence (5′→3′)	Oligonucleotide description
agagttgcgataaaaagcgt**g**	sgRNA targeting pBAD (c > **g** at 3′)
tgacacctgtaggatcgtac	sgRNA targeting pLux9
agaagagcacaggctcttct	Non-targeting sgRNA (NT)
tctggaattcgcggccgcttctagagaaagaggagaaaggatctat	Forward addition of BioBrick restriction sites to dCas9 (EcoRI and XbaI)
ggactgcagcggccgctactagtacgcagaaaggcccacccgaa	Reverse addition of BioBrick restriction sites to dCas9 (SpeI and PstI)
ggactgcagcggccgctactagtacgctcttcttatggagaaacagtagagag	Reverse deletion of pBAD bases after TSS site (SapI, SpeI, and PstI)
ctggaattcgcggccgcttctaga	BioBrick prefix primer
ggactgcagcggccgctactagt	BioBrick suffix primer

2. Perform SpeI and PstI digestion on pTet promoter backbone and XbaI and PstI digestion on the dCas9 PCR product. Wash digested fragments using a PCR cleanup kit (Fig. 2b, c and Table 2).
3. Ligate the digested fragments from **step 2** to build a pTet-driven inducible dCas9 module. The SpeI and XbaI sticky ends will ligate to form a SCAR site, unrecognized by either enzyme. Likewise, PstI sticky ends in the backbone and insert ligate to reconstruct the PstI restriction site (Figs. 2d and 3a and Table 2).
4. Transform the ligation reaction mixture from **step 3** into competent *E. coli* DH10B cells.
5. Spread transformed cells on an LB plate with appropriate antibiotic for overnight induction at 37 °C.
6. After 16 h, perform PCR on 5–10 colonies using insert sequence-specific primers, followed by verification via gel-electrophoresis in a 1% agarose gel in TAE with DNA stain. Culture the verified colonies overnight in antibiotic-enriched LB broth at 37 °C in a shaking incubator.
7. Extract plasmids from overnight cell cultures and verify the addition of the dCas9 gene to the pTet promoter by Sanger sequencing or whole plasmid sequencing.
8. pOSIP-KH is a BioBrick-compatible integration plasmid with kanamycin resistance. pOSIP-KH expresses tyrosine integrase from the phase HK022 to integrate the gene of interest into the HK022 integration site in *E. coli K-12 MG1655* (*ΔLacI*

Table 2
Reagents required for restriction digestions and ligations

Reagents	Amount
Reagents for restriction digestion for BioBrick cloning	
FastDigest buffer (10×)	5 μL
Enzyme I	2.5 μL
Enzyme II	2.5 μL
PCR product/backbone/insert	400 ng/1000 ng/2000 ng
Nucleate-free water	Total to 50 μL
Reagents for ligation using T7 DNA ligase	
StickTogether™ DNA ligase buffer (10×)	5 μL
Insert	2 μL
Backbone	2 μL
T7 DNA ligase	1 μL
Nucleate-free water	Total to 10 μL
Reagents for restriction digestion using SapI for sgRNA cloning	
rCutSmart buffer (10×)	5 μL
SapI	1 μL
PCR product/backbone/insert	400 ng/1000 ng/2000 ng
Nucleate-free water	Total to 50 μL

ΔAraC). Digest pOSIP-KH with SpeI and PstI and the dCas9 module from **step 7** with XbaI and PstI.

9. Clean the digested fragments from **step 8** with a PCR cleanup kit and ligate them with T7 DNA ligase.
10. Transform the pOSIP-KH expressing the integrase and dCas9 into the competent *E. coli K-12 MG1655 (ΔLacI ΔAraC)* host cells.
11. Perform the selection for positive colonies using overnight induction at 37 °C in LBK plates, followed by colony PCR with insert-specific primers and verification by gel-electrophoresis in a 1% agarose gel in TAE with DNA stain.
12. Culture verified colonies in 5 mL LBK broth at 37 °C overnight in a shaking incubator.
13. Inoculate 400 μL overnight inoculum into fresh 5 mL LBK broth and incubate the culture for 2–3 h (OD ~0.4) at 37 °C in a shaking incubator.

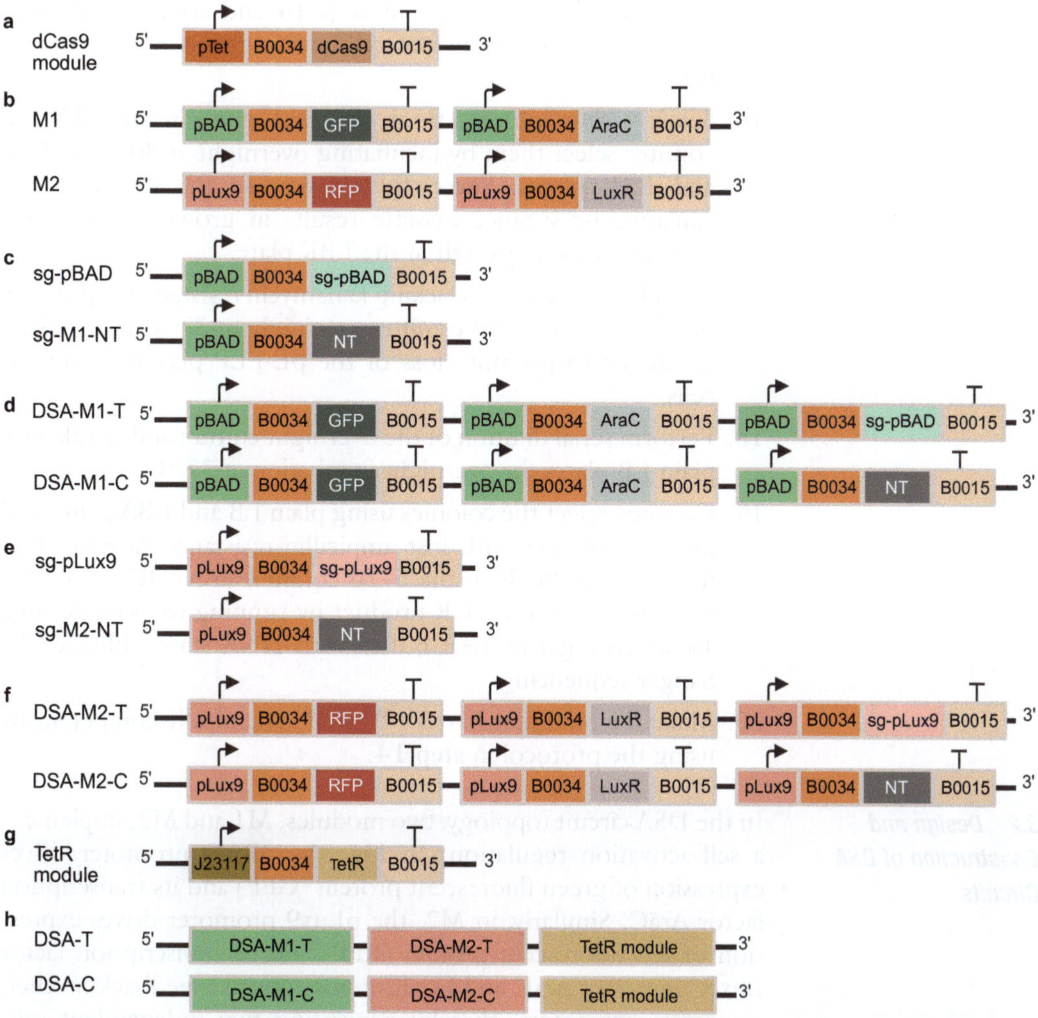

Fig. 3 Circuit schematics of all constructs built for implementing NCR in the DSA circuit. (**a**) dCas9 gene expression driven by an inducible pTet promoter. (**b**) M1 and M2 self-activation modules. (**c**) pBAD promoter-driven sg-pBAD and non-targeting gRNA (sg-M1-NT) modules. (**d**) DSA-M1-T and DSA-M1-C comprising M1 module with sg-pBAD and sg-M1-NT sgRNA modules for M1, respectively. (**e**) pLux9 promoter-driven sg-pLux9 and non-targeting gRNA (sg-M2-NT) modules. (**f**) DSA-M2-T and DSA-M2-C comprising M2 module with sg-pLux9 and sg-M2-NT sgRNA modules for M2, respectively. (**g**) Constitutive promoter driving the gene expression of the transcription factor, TetR. (**h**) Final test (DSA-T) and control (DSA-C) constructs for the implementation of the NCR strategy in DSA. B0034 and B0015 correspond to the iGEM Registry parts for a strong ribosome binding site (RBS) and a double terminator, respectively. Sequences for each component are provided in the supplementary information of the original publication [21]. All guide RNAs used in this study are listed in Table 1

14. Prepare chemically competent *E. coli* cells with genome-integrated pOSIP-KH using Mix-and-Go! *E. coli* transformation kit.
15. pE-FLP is an ampicillin-resistant plasmid to perform the FLP recombinase-mediated excision. Transform pE-FLP into the

competent cells prepared in **step 14** and select the positive colonies by incubating the cells overnight at 30 °C in LBA plates.

16. Select a few colonies from the LBA plate in **step 15** and counter-select them by incubating overnight at 30 °C in LBA and LBK plates. Successful excision of the integrase and kanamycin-resistance cassette results in growth in the LBA plate and loss of growth in the LBK plate.
17. Inoculate the colonies lacking kanamycin resistance in plain LB broth and incubate the culture overnight at 37 °C in a shaking incubator to promote loss of the pE-FLP plasmid from the cells.
18. Perform serial dilution of the overnight culture and streak it on plain LB plates for overnight incubation at 37 °C.
19. Counter-select the colonies using plain LB and LBA plates and identify colonies with lost ampicillin-resistance cassette. Perform colony PCR on the 5–10 colonies using dCas9-specific primers. Verify the PCR product by running the samples in a 1% agarose gel in TAE buffer with DNA stain, followed by Sanger sequencing.
20. Make the modified *E. coli*^{dCas9} strain chemically competent by using the protocol in **step 14**.

3.2 Design and Construction of DSA Circuits

In the DSA circuit topology, two modules, M1 and M2, implement a self-activation regulation. In M1, the pBAD promoter drives expression of green fluorescent protein (GFP) and its transcription factor AraC. Similarly, in M2, the pLux9 promoter drives expression of red fluorescent protein (RFP) and its transcription factor LuxR (Fig. 3b). AraC and LuxR impose positive feedback on their respective promoters, thereby generating two independent self-activating modules.

Construct M1 and M2 by performing sequential BioBrick restriction digestion, PCR cleanup, ligation, transformation, and verification. Refer to Table 2 for the reaction reagents and volumes. Digest the backbone with SapI and PstI, and the inserts are digested with XbaI and PstI (Fig. 2b–d). Gel extractions of backbone or insert fragments may be performed using standard protocol, if necessary. Transform the ligation reaction mixture into *E. coli* DH10B cells. Perform PCR verification after each ligation cycle with at least one insert-specific primer pair. *See* **Notes 1–3** for the necessary mutations made to the standard BioBrick parts for the experiments.

3.3 Guide RNA Design and Test

1. The pSG4K5 plasmid expresses the sgRNA scaffold under the constitutive J23119 promoter. The region between the promoter and the scaffold is flanked by two SapI sites. Search for

20 bp sequences with 'CC' PAM site in the 5′ end in the promoter driving M1 gene expression, pBAD.

2. Synthesize complementary forward and reverse oligo sequences for selected sgRNAs with sticky ends matching the digested SapI sites in the pSG4K5 plasmid.
3. Resuspend the oligos in nuclease-free water to make the stock concentration of 100 μM. Dilute 10 μL of the oligos in 90 μL of nuclease-free water to achieve a final concentration of 1 μM.
4. Mix 2.5 μL forward and reverse oligos and 45 μL nuclease-free water to a total volume of 50 μL in a PCR tube.
5. Heat the tube for 2 min at 95 °C in a temperature block and cool the reaction mixture by turning off the heat block.
6. Digest pSG4K5 with SapI, followed by PCR cleanup, and ligate the sgRNA insert using T7 DNA ligase (Table 2).
7. Transform the ligation reaction mixture to chemically competent *E. coli* DH10B cells.
8. Perform overnight incubation in an LBK plate at 37 °C. Perform PCR on 5–10 colonies using forward/reverse sgRNA oligo and a backbone-specific primer. Verify PCR products using gel-electrophoresis in a 1% agarose gel in TAE with DNA stain.
9. Repeat **steps 2–8** for all the possible CRISPRi sgRNAs for the M1 promoter.
10. To test the oligos, co-transform the sgRNAs along with M1 into chemically competent *E. coli*^{dCas9}. The undigested pSG4K5 can be co-transformed with M1 as the control (*see* **Note 4**). Spread the transformed cells into LB plates enriched with antibiotics specific to both sgRNA and M1 plasmids. Incubate the plate overnight at 37 °C.
11. Inoculate 300 μL LB broth enriched with appropriate antibiotics with colonies from **step 10** in triplicate. Incubate the cultures for 5 h at 37 °C in a shaking incubator. Ensure three colonies from the control plate are also incubated.
12. Prepare a 96-well plate (200 μL per well) with antibiotic-enriched LB broths containing multiple inducer doses for M1 (pick 2–3 concentrations ranging from high to low).
13. Inoculate 5 μL of the cells from **step 11** into the 96-well plate in **step 12**. Include the control triplicates, where no down-regulation of M1 reporter fluorescence is expected.
14. Incubate the plate at 37 °C in a shaking incubator for 16 h.
15. Read the absorbance at 600 nm (OD_{600}) and fluorescence at appropriate wavelengths (excitation/emission filters 485/515 nm for GFP and 546/607 nm for RFP) using the plate reader.

16. Analyze the fluorescence/OD_{600} data for different sgRNAs compared to the control and select a few that exhibit CRISPRi-mediated fluorescence downregulation for further experiments.
17. Repeat **steps 1–16** for the promoter driving gene expression (pLux9) in M2.
18. Find the pre-validated sgRNAs for M1 (sg-pBAD) and M2 (sg-pLux9) in Table 1.

3.4 Modules DSA-T and DSA-C Design

1. Synthesize the gBlock gene fragment spanning the BioBrick prefix, pBAD promoter, RBS, sgRNA targeting pBAD (sg-pBAD), sgRNA scaffold, terminator, and the BioBrick suffix. Replace sg-pBAD with the non-target sgRNA sequence (sg-NT) to construct the sg-M1-NT control (Fig. 3c, *see* **Notes 5** and **6**).
2. Resuspend gBlocks in nuclease-free water to a final concentration of 1 ng/mL.
3. Perform PCR using BioBrick prefix and suffix-specific forward and reverse primers and a high-fidelity polymerase to amplify the fragment. Wash the PCR product using a PCR cleanup kit.
4. Digest the M1 plasmid with SpeI and PstI, and the PCR product from **step 3** using XbaI and PstI. Perform PCR cleanup on digested fragments.
5. Ligate the backbone and insert fragments from **step 4** using T7 DNA ligase to construct DSA-M1-T and DSA-M1-C, comprising the sg-pBAD module and the sg-M1-NT module, respectively (Fig. 3d).
6. Transform 10 μL of the ligation reaction mixture from **step 5** into chemically competent *E. coli* DH10B and spread the cells in an LB plate enriched with backbone-specific antibiotic.
7. Incubate the plate at 37 °C overnight.
8. Pick 5–10 colonies from the plate in **step 7** to perform colony PCR using the insert-specific primer pair. Verify the PCR products in a 1% agarose gel in TAE with DNA stain.
9. Inoculate the selected colonies for overnight culture at 37 °C for plasmid extraction and verify the sequences using Sanger sequencing or whole plasmid sequencing.
10. Repeat **steps 1–9** to build a plasmid comprising M2 and pLux9-driven sgRNAs to make DSA-M2-T and DSA-M2-C plasmids (Fig. 3e, f).
11. Digest the DSA-M1-T plasmid with SpeI and PstI, and DSA-M2-T using XbaI and PstI.

12. Clean digested fragments from **step 11** using a PCR cleanup kit and ligate them using T7 DNA ligase. If DSA-M1-T and DSA-M2-T are under plasmids with the same antibiotic resistance, perform gel extraction of digests using a gel-extraction kit instead of PCR cleanup.
13. Repeat the transformation and verification cycle (**steps 6–9**).
14. Repeat **steps 11–13** for DSA-M1-C and DSA-M2-C to construct a control for the NCR.
15. Build a constitutive TetR module by cloning iGEM parts J23117 to TetR (P0440) by performing XbaI and PstI digestion for TetR and SpeI and PstI digestion for J23117, followed by PCR cleanup and ligation (Fig. 3g).
16. Transform the ligation reaction mixture from **step 15** to chemically competent *E. coli* DH10B and spread the cells on an LB plate containing ampicillin for overnight growth at 37 °C.
17. Verify 5–10 colonies using colony PCR with TetR-specific primers, followed by gel-electrophoresis in a 1% agarose gel in TAE with DNA stain.
18. Inoculate and incubate the selected colonies at 37 °C in a shaking incubator overnight for plasmid extraction, followed by verification via Sanger sequencing.
19. Finally, construct DSA-T and DSA-C by ligating XbaI and PstI-digested TetR module from **step 18** to SpeI and PstI-digested test and control plasmids verified in **steps 13** and **14**, respectively (Fig. 3h and *see* **Note** 7).
20. Repeat the transformation and verification cycle (**steps 6–9**).

3.5 Single-Module Regulation

1. Transform the DSA-T and DSA-C plasmids into chemically competent *E. coli*^{dcas9} and culture the cells in LB plates with kanamycin overnight at 37 °C.
2. Inoculate three colonies each for DSA-T and DSA-C plates from **step 1** in 300 μL LB broth with kanamycin at 37 °C for 5 h.
3. Prepare a 96-well plate with antibiotic-enriched LB media (200 μL per well) containing a fixed aTc concentration and multiple Lara concentrations for M1 induction.
4. Inoculate the 96-well plate from **step 3** with cell cultures (5 μL per well) from **step 2**.
5. Incubate the plate from **step 4** at 37 °C in a plate reader under shaking conditions for 16 h. Read the absorbance at 600 nm (OD_{600}) and fluorescence at wavelength 485/515 nm for GFP excitation/emission every 15–30 min.

6. Analyze the fluorescence/OD_{600} data for the sgRNA compared to the control by plotting a dose-response curve at different L-ara concentrations for DSA-T and DSA-C.
7. Repeat **steps 1–6** at multiple aTc concentrations to investigate CRISPRi on M1 at different dCas9 expression strengths (*see* **Note 8**).
8. Repeat **steps 1–7** for M2 single-module regulation. Use C6 for pLux9 induction and set the plate reader wavelengths to 546/607 nm for RFP excitation/emission.

3.6 NCR Implementation in DSA

1. Transform the DSA-T and DSA-C plasmids into chemically competent *E. coli*^{dcas9} and culture the cells in LB plates with kanamycin overnight at 37 °C.
2. Inoculate three colonies each for DSA-T and DSA-C plates from **step 1** in 300 μL LB broth with kanamycin at 37 °C for 5 h.
3. Refer to the single-module regulation data and choose an aTc dose for dCas9 strength regulation. At the effective aTc dose, dCas9 should downregulate gene expression on both M1 and M2 and must not be deleterious on either module.
4. Prepare a 96-well plate with antibiotic-enriched LB media containing a fixed aTc (15 ng/mL) and Lara concentration (1.25×10^{-3}%) for a constant dCas9 and M1 induction, and multiple C6 concentrations (0, 9×10^{-9} M, 1×10^{-8} M, 3×10^{-8} M, 5×10^{-8} M) across the plate for M2 induction.
5. Inoculate 5 μL of cells from **step** 2 into the wells.
6. Incubate the 96-well plate for 16 h at 37 °C under shaking conditions.
7. For flow cytometry, use excitation/emission filters 480 nm/530 nm for GFP and 480 nm/>670 nm for RFP detection.
8. Use FSC-A/FSC-H gating strategy to eliminate doublets and non-cellular small particles. Use the plain LB medium without any cells as a negative control. Include unstained and single-stained cells as controls for fluorescence compensation during flow cytometry.
9. Prepare the samples from **step 6** for flow cytometry by diluting the cells ~1000-fold in 1× PBS.
10. Perform flow cytometry for samples across the plate and analyze the coactivation population in test and control samples. Reference the cell population in the sample with 0 C6 for DSA-C and DSA-T to set the quadrant threshold during analysis.

4 Notes

1. pLux9 used in the study is a mutant of the wild-type pLux promoter (A → T at −10 and G → T − 11 relative to the TSS).
2. A LuxR variant, LuxRG2C, was used in our study to reduce leakiness from the pLux9 promoter. LuxRG2C features S116A and M135I mutations.
3. The fluorescent reporter proteins RFP and GFP include lva-degradation tags at the C-terminal.
4. The control should not target promoter, gene, or terminator sequences in M1 or M2. Here, we use the sequence flanked by SapI sites in pSG4K5 as a non-targeting sgRNA control.
5. Ensure that there is no scar site between the promoter and the sgRNA targeting it. As BioBrick ligation results in scar site formation, we recommend a scarless cloning strategy or promoter-sgRNA gBlocks.
6. Some promoters might contain sequences downstream of the transcription start site (TSS), for instance, the pBAD promoter (iGEM part K206000). These bases should be removed to prevent unnecessary bases upstream of sgRNAs, as they might interfere with dCas9 binding and activity (Table 1).
7. The NCR performed the best when DSA-C and DSA-T were cloned to the medium-copy pSB3K3 plasmid with kanamycin resistance. The WTA behavior and regulation in DSA-C and DSA-T were difficult to tune when expressed under a high-copy expression vector.
8. If the dCas9-mediated transcriptional downregulation is disproportionately asymmetrical between two modules, sgRNAs can be mutated for further tuning.

Acknowledgments

This work was supported by a grant from the US National Institutes of Health (R35GM142896 to X.-J.T.).

Competing Interests The authors declare no competing interests.

References

1. Ceroni F, Algar R, Stan G-B, Ellis T (2015) Quantifying cellular capacity identifies gene expression designs with reduced burden. Nat Methods 12:415–418. https://doi.org/10.1038/nmeth.3339
2. Carbonell-Ballestero M, Garcia-Ramallo E, Montañez R, Rodriguez-Caso C, Macía J (2016) Dealing with the genetic load in bacterial synthetic biology circuits: convergences with the ohm's law. Nucleic Acids Res 44:

496–507. https://doi.org/10.1093/nar/gkv1280

3. Di Blasi R, Pisani M, Tedeschi F, Marbiah MM, Polizzi K, Furini SMN, Siciliano V, Ceroni F (2023) Resource-aware construct design in mammalian cells. Nat Commun 14:3576. https://doi.org/10.1038/s41467-023-39252-4
4. Stone A, Youssef A, Rijal S, Zhang R, Tian X-J (2024) Context-dependent redesign of robust synthetic gene circuits. Trends Biotechnol 42: 895–909. https://doi.org/10.1016/j.tibtech.2024.01.003
5. Boo ALC, Ellis T, Stan G-B (2019) Host-aware synthetic biology. Curr Opin Syst Biol 14:66–72. https://doi.org/10.1016/j.coisb.2019.03.001
6. Di Blasi R, Gabrielli J, Shabestary K, Ziarti I, Ellis T, Kontoravdi C, Ledesma-Amaro R, Ceroni F (2024) Understanding resource competition to achieve predictable synthetic gene expression in eukaryotes. Nat Rev Bioeng 2: 721–732. https://doi.org/10.1038/s44222-024-00206-0
7. Şimşek E, Yao Y, Lee D, You LC (2022) Toward predictive engineering of gene circuits. Trends Biotechnol 41:760–768. https://doi.org/10.1016/j.tibtech.2022.11.001
8. Kumar S, Hasty J (2023) Stability, robustness, and containment: preparing synthetic biology for real-world deployment. Curr Opin Biotechnol 79:102880. https://doi.org/10.1016/j.copbio.2022.102880
9. Zhang R, Goetz H, Melendez-Alvarez J, Li J, Ding T, Wang X, Tian X-J (2021) Winner-takes-all resource competition redirects cascading cell fate transitions. Nat Commun 12:853. https://doi.org/10.1038/s41467-021-21125-3
10. Gyorgy A, Jiménez JI, Yazbek J, Huang H-H, Chung H, Weiss R, Del Vecchio D (2015) Isocost lines describe the cellular economy of genetic circuits. Biophys J 109:639–646. https://doi.org/10.1016/j.bpj.2015.06.034
11. Qian Y, Huang HH, Jimenez JI, Del Vecchio D (2017) Resource competition shapes the response of genetic circuits. ACS Synth Biol 6:1263–1272. https://doi.org/10.1021/acssynbio.6b00361
12. Darlington APS, Kim J, Jiménez JI, Bates DG (2018) Dynamic allocation of orthogonal ribosomes facilitates uncoupling of co-expressed genes. Nat Commun 9:695. https://doi.org/10.1038/s41467-018-02898-6
13. Jones RD, Qian Y, Siciliano V, DiAndreth B, Huh J, Weiss R, Del Vecchio D (2020) An endoribonuclease-based feedforward controller for decoupling resource-limited genetic modules in mammalian cells. Nat Commun 11:5690. https://doi.org/10.1038/s41467-020-19126-9
14. Zhang R, Zhang R, Tian X-J (2025) Integrating recombinase-based feedback and feedforward control for optimal resource decoupling. Nucleic Acids Res 53:gkaf913. https://doi.org/10.1093/nar/gkaf913
15. Huang H-H, Qian YL, Del Vecchio D (2018) A quasi-integral controller for adaptation of genetic modules to variable ribosome demand. Nat Commun 9:5415. https://doi.org/10.1038/s41467-018-07899-z
16. Frei T, Cella F, Tedeschi F, Gutierrez J, Stan GB, Khammash M, Siciliano V (2020) Characterization and mitigation of gene expression burden in mammalian cells. Nat Commun 11: 4641. https://doi.org/10.1038/s41467-020-18392-x
17. Stone A, Ryan J, Tang X, Tian X-J (2022) Negatively competitive incoherent feedforward loops mitigate winner-take-all resource competition. ACS Synth Biol 11:3986–3995. https://doi.org/10.1021/acssynbio.2c00318
18. Stone A, Zhang R, Tian X-J (2021) Coupling shared and tunable negative competition against winner-take-all resource competition via CRISPRi moieties. In: 2021 American control conference (ACC), New Orleans, LA, USA. IEEE, Piscataway, pp 1–6
19. Chakravarty S, Guttal R, Zhang R, Tian X-J (2024) Mitigating winner-take-all resource competition through antithetic control mechanism. ACS Synth Biol 13:4050–4060. https://doi.org/10.1021/acssynbio.4c00476
20. Chakravarty S, Zhang R, Tian X-J (2024) Noise reduction in resource-coupled multi-module gene circuits through antithetic feedback control. Proc IEEE Conf Decis Control 2024:5566–5571. https://doi.org/10.1109/cdc56724.2024.10886586
21. Rijal S, Standage-Beier K, Zhang R, Stone A, Youssef A, Wang X, Tian X-J (2025) CRISPRi-linked multimodule negative feedback loops to address winner-take-all resource competition. ACS Synth Biol 14:3646–3654. https://doi.org/10.1021/acssynbio.5c00394

Part III

Dynamics

Chapter 16

Modeling Growth-Mediated Dilution in Synthetic Circuits: Effects of Circuit Topology and Phase Separation

Xiao-Jun Tian and Rong Zhang

Abstract

Growth-mediated dilution is a major source of variability and instability in synthetic gene circuits, influencing both expression dynamics and circuit memory. Our previous work demonstrated that circuit topology plays a critical role in determining how robustly a circuit maintains its function under changing growth conditions, and that phase separation can serve as a physical mechanism to buffer these perturbations. In this chapter, we present a comprehensive modeling framework that integrates deterministic gene regulation, stochastic gene expression, cell-volume growth and division, and condensate formation. We introduce two complementary stochastic simulation approaches. The first is a hybrid stochastic–deterministic population-level model that treats intracellular gene expression deterministically while modeling cell growth, division, and burden-dependent feedback stochastically to capture population-level dilution effects. The second is a single-cell, fully stochastic simulation that uses the Gillespie algorithm to model transcription, translation, and degradation reactions while continuously updating cell growth, dilution, division, and condensate size through a thermodynamic phase-separation model. The modeling frameworks presented here provide generalizable tools for analyzing growth–circuit interactions, evaluating robustness across parameter regimes, and guiding the design of synthetic circuits resilient to physiological fluctuations.

Key words Transcriptional condensates, Stochastic simulation, Hybrid stochastic–deterministic modeling, Cell growth and division, Condensate buffering, Single-cell modeling

1 Introduction

Synthetic gene circuits are engineered to perform programmable cellular functions, from biosensing and signal processing to metabolic pathway regulation and therapeutic gene control. These circuits enable cells to make decisions, respond to environmental cues, and coordinate complex behaviors in a manner analogous to electronic systems. However, their reliable operation depends not only on the regulatory design of the circuit itself but also on the physiological state of the host cell in which the circuit resides [1–5]. In actively growing cells, growth directly influences circuit behavior by

Xiao-Jun Tian (ed.), *Synthetic Gene Circuits: Methods and Protocols*, Methods in Molecular Biology, vol. 3041,
https://doi.org/10.1007/978-1-0716-5304-3_16,

determining the rates of resource allocation and dilution of circuit components [6–8]. While a synthetic circuit can often be tuned to function as intended under constant laboratory conditions, its performance frequently deteriorates in realistic or applied environments where cell growth and resource availability fluctuate dynamically. Such variations can distort gene expression patterns, alter feedback strengths, and shift circuit operating regimes, ultimately compromising the robustness and stability of engineered cellular functions [9–11].

Our previous work demonstrated a topology-dependent mechanism by which synthetic circuits lose their intended function under growth-mediated dilution [10]. Specifically, we found that a self-activation switch rapidly lost its memory after dilution into fresh growth medium, whereas a toggle switch maintained stable bistability across growth conditions. These findings revealed that circuit topology critically determines susceptibility to growth feedback and must be carefully considered during circuit design to enhance robustness under variable growth conditions.

Aiming to develop generalizable control strategies that can protect diverse circuit architectures from growth-induced perturbations, we subsequently demonstrated that engineering phase-separated transcriptional regulators can stabilize synthetic circuit function against growth-mediated dilution [11]. By forming transcriptional condensates at target promoters, phase-separated activators maintain high local concentrations even as cells grow rapidly or undergo dilution, thereby buffering gene expression from host-dependent perturbations and enhancing circuit robustness. This work introduced a physical layer of control, showing that molecular condensation can act as an intrinsic noise filter for growth-coupled systems.

In this chapter, we present a detailed stochastic simulation workflow for modeling growth-mediated dilution in synthetic circuits, emphasizing the roles of circuit topology and phase separation. We describe how growth feedback can be incorporated into stochastic gene expression models, how circuit topology alters system stability under dilution, and how phase separation can buffer stochastic fluctuations. The modeling framework integrates stochastic kinetics, cell growth dynamics, and compartmentalized molecular states to capture the dynamic behavior of growth-coupled circuits. The methods outlined here provide a general computational approach for analyzing and designing synthetic circuits that remain robust under dynamically changing growth environments.

2 Materials

2.1 Software and Programming Environment

MATLAB Implementation

1. **MATLAB (R2020a or later)** for implementing ODE-based deterministic models and Gillespie-style stochastic simulations.
2. Required MATLAB toolboxes:
   - MATLAB ODE Suite (built-in) for numerical integration of ordinary differential equations (e.g., ode45, ode15s, ode23t).
   - Statistics and Machine Learning Toolbox for random number generation and probability distributions used in stochastic simulations.
   - Optimization Toolbox (optional) for free-energy minimization during condensate size calculations.
3. For parallel or large-scale simulations:
   - Parallel Computing Toolbox for running computationally intensive parameter sweeps or ensemble simulations across cores.

2.2 Computational Resources

1. A standard scientific workstation with:
   - $\geq$ 8 GB RAM.
   - $\geq$ 2.0 GHz multi-core CPU.
   - Operating system compatible with MATLAB environments (Windows, macOS, or Linux).
2. For large simulations or extensive parameter sweeps (optional):
   - Access to high-performance computing (HPC) resources or cloud computing platforms.
   - MATLAB Parallel Computing Toolbox for parallelization.

3 Methods

The methods presented in this chapter combine deterministic modeling, stochastic simulation, cell growth dynamics, and phase-separation thermodynamics to analyze growth-mediated dilution in synthetic gene circuits. We begin by formulating baseline and growth-coupled deterministic models that describe molecular expression and dilution within self-activation and toggle switch topologies. We then introduce two stochastic simulation approaches: a hybrid stochastic–deterministic population-level model that incorporates growth-mediated dilution through probabilistic cell division, and a fully stochastic single-cell model that uses the Gillespie algorithm to simulate transcription, translation, degradation, cell-volume growth, and condensate formation. Together, these methods provide a unified framework for quantifying how circuit topology and phase separation influence circuit stability under dynamic growth conditions.

3.1 Define the Growth-Dilution Model Framework

3.1.1 Baseline Deterministic Model: No Growth, No Dilution, No Burden

Under conditions of no growth or constant growth, the dynamics of molecular concentrations in a gene circuit can be described by ordinary differential equations (ODEs) in which dilution effects are incorporated into a constant degradation term, without explicit modeling of cell growth. The general form of the ODE for a molecular species $[X]$ is

$$\frac{d[X]}{dt} = f([X]) - d \cdot [X]$$

where $f([X])$ represents the production rate, determined by the designed regulatory interactions within the circuit, and d is the constant degradation rate. In this formulation, degradation is assumed to be a first-order process that depends linearly on the molecular concentration. This baseline model captures the intrinsic regulatory dynamics of the circuit in isolation from cellular growth effects. It is particularly useful for analyzing steady-state behavior, where the equilibrium concentration is determined solely by the balance between production and degradation. Under these assumptions, the steady-state level of each component reflects the internal topology and regulatory logic of the circuit itself, without influence from host physiology or growth-mediated feedback.

3.1.2 Inclusion of Growth and Dilution

When the cell growth rate varies over time, it is necessary to explicitly model the dynamics of cell growth and the resulting dilution of intracellular components (Fig. 1a). In this case, molecular concentrations change not only due to synthesis and degradation but also because of growth-mediated volume expansion, which effectively reduces the concentration of all cellular components.

The general form of the concentration dynamics for a molecular species $[X]$ is

$$\frac{d[X]}{dt} = f([X]) - d \cdot [X] - \mu(t) \cdot [X]$$

where $f([X])$ is the regulated production rate, d is the intrinsic degradation constant, and $\mu(t)$ is the instantaneous specific growth rate (units: time^{-1}). The term $\mu(t) \cdot [X]$ explicitly accounts for dilution due to cell volume increase during growth.

The instantaneous specific growth rate is defined as:

$$\mu(t) = \frac{dN}{dt} \cdot \frac{1}{N}$$

where N represents the cell density (or cell number) at time t. To describe the time evolution of N, a separate growth equation is

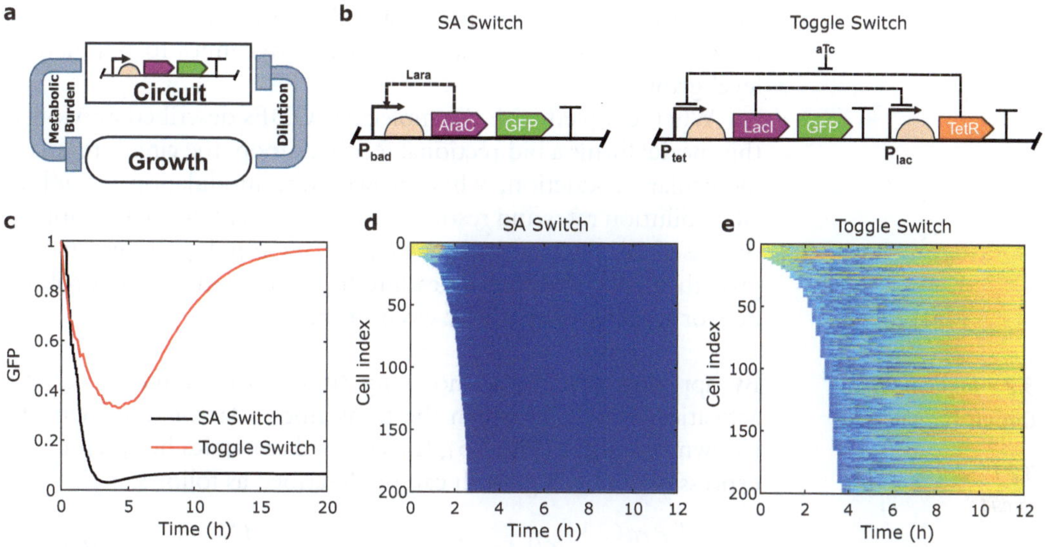

Fig. 1. Growth feedback and circuit topologies. (**a**) Schematic diagram of general growth feedback in synthetic gene circuits, illustrating how gene expression imposes metabolic burden that reduces growth rate, which in turn modulates dilution of circuit components and feeds back onto gene expression dynamics. (**b**) Network diagrams of the self-activation (SA) circuit and the toggle switch circuit. The SA circuit consists of a single positive-feedback loop in which a transcription factor (e.g., AraC) activates its own promoter, whereas the toggle switch comprises two mutually repressive transcription factors (e.g., LacI and TetR) forming a bistable regulatory architecture. (**c**) Comparison of normalized, averaged molecule-number trajectories for cells initialized in the ON state in the SA circuit and the toggle switch during cell growth. (**d**, **e**) Representative normalized molecule-number trajectories for simulations initiated with 10 seeding cells in the ON state for the SA circuit (**d**) and the toggle switch (**e**). Here only the first 200 cells from a total simulated population of 1000 cells are shown

introduced. A common choice is a logistic growth model, which captures both exponential and saturation phases:

$$\frac{dN}{dt} = \mu_{max} \cdot g([X]) \cdot \left(1 - \frac{N}{N_0}\right) \cdot N$$

where μ_{max} is the maximum specific growth rate, N_0 is the carrying capacity of the system, and $g([X])$ is a dimensionless function representing the growth attenuation caused by gene circuit expression.

The function $g([X])$ captures the dependency of growth rate on the circuit-induced metabolic or translational burden (Fig. 1a). A simple and widely used form is

$$g([X]) = \frac{1}{[X]/J + 1}$$

where $[X]$ represents the concentration of a burden-imposing circuit species), J is a constant defining the half-saturation level at which growth rate is reduced to 50% of its maximum. In this formulation, $g([X])$ monotonically decreases as circuit expression increases, linking molecular-level circuit dynamics to population-level growth.

When coupled with the molecular ODEs described previously, this model forms a bidirectional feedback loop: the circuit regulates molecular production, while growth rate modulation via $g([X])$ alters dilution rates and resource availability. This feedback captures the essential mechanism by which circuit topology interacts with growth-mediated dilution, leading to topology-dependent robustness or fragility of synthetic circuit function.

3.2 Implement Circuit Topologies

3.2.1 Self-activation Circuit

By applying the above modeling framework to the AraC self-activation switch, in which the transcription factor AraC activates its own promoter (Fig. 1b), the coupled dynamics between circuit expression and cell growth can be described as follows:

$$\frac{d[AraC]}{dt} = f([AraC]) - d \cdot [AraC] - \frac{dN}{dt} \cdot \frac{1}{N} \cdot [AraC]$$

$$\frac{dN}{dt} = \mu_{max} \cdot g([AraC]) \cdot \left(1 - \frac{N}{N_0}\right) \cdot N$$

where [AraC] is the concentration of AraC protein. The production rate $f([AraC])$ is defined by the autoregulatory activation kinetics

$$f([AraC]) = k_0 + k_1 \cdot \frac{S_a \cdot [AraC]^2}{S_a \cdot [AraC]^2 + 1},$$

where k_0 and k_1 are the basal and maximal synthesis rates, respectively, and S_a represents the activation strength that depends on the presence of the inducer L-arabinose. The activation parameter S_a can be further described as:

$$S_a = C_{min} + (C_{max} - C_{min}) \cdot \frac{Lara^n}{Lara^n + K^n} \cdot$$

where C_{min} and C_{max} are the minimum and maximum activation coefficients, K is the half-saturation constant, and n is the Hill coefficient.

The burden function $g([AraC])$ captures how AraC expression reduces the effective cell growth rate

$$g([AraC]) = \frac{1}{[AraC]/J + 1}$$

where J is the characteristic expression level at which the growth rate is reduced by 50%.

In this formulation, AraC self-activation generates a positive feedback loop that enables bistability and memory in the circuit. However, as growth rate and dilution vary dynamically, the effective intracellular concentration of AraC can fluctuate, leading to loss of circuit memory or state switching.

3.2.2 Toggle Switch Circuit

The toggle switch consists of two mutually repressive transcription factors (LacI and TetR) that inhibit each other's expression, forming a bistable regulatory circuit capable of maintaining one of two distinct steady states (Fig. 1b). Incorporating growth-mediated dilution and feedback, the coupled dynamics of the system are described as

$$\frac{d[LacI]}{dt} = f1([LacI],[TetR]) - d\cdot[LacI] - \frac{dN}{dt}\cdot\frac{1}{N}\cdot[LacI]$$

$$\frac{d[TetR]}{dt} = f2([LacI],[TetR]) - d\cdot[TetR] - \frac{dN}{dt}\cdot\frac{1}{N}\cdot[TetR]$$

$$\frac{dN}{dt} = \mu_{max}\cdot g([LacI]+[TetR])\cdot\left(1-\frac{N}{N_0}\right)\cdot N$$

where $[LacI]$ and $[TetR]$ are the protein concentrations of the two repressors. The regulated production rates for each gene are defined as

$$f1 = crl + \frac{1}{1+\left[\left(\frac{[TetR]}{k_t}\right)\cdot\left(1+\frac{ATc}{k_{atc}}*\frac{kt}{[TetR]}\right)^{-m}\right]^{nt}}\cdot(cil-crl)$$

$$f2 = crt + \frac{1}{1+\left(\frac{[LacI]}{k_l}\right)^{nl}}\cdot(cit-crt)$$

where crl and crt are basal synthesis rates for LacI and TetR, respectively, cil and cit are maximal induced synthesis rates, k_t and k_l are repression thresholds, nt and nl are Hill coefficients.

The growth attenuation function is defined as:

$$g = \frac{1}{([LacI]+[TetR])/J+1}$$

where J represents the effective burden threshold at which total circuit expression reduces the growth rate by half.

This toggle-switch model captures the mutual repression topology and its interaction with growth feedback. The dynamic balance between LacI and TetR expression defines the circuit's bistable states, while growth-mediated dilution introduces additional coupling between molecular fluctuations and cellular physiology.

3.3 Hybrid Stochastic–Deterministic Simulation of Growth Feedback

This section describes a hybrid stochastic–deterministic simulation framework that integrates deterministic gene circuit dynamics with stochastic cell-level growth and division events. In this approach, intracellular molecular dynamics are modeled deterministically using ordinary differential equations (ODEs), while stochasticity arises from discrete, probabilistic population processes such as cell division, asymmetric partitioning, and growth feedback. This hybrid method efficiently captures the interplay between gene circuit regulation and growth-mediated dilution, enabling quantitative analysis of circuit robustness and heterogeneity at the population level without the computational burden of fully stochastic molecular simulations.

Step 1. Initialize the system

Set the initial conditions for all simulation variables:

- Circuit type: SA or toggle switch.
- Cell number N.
- Simulation time $t = 0$.
- mRNA and protein copy numbers for each cell: M_i and P_i, initialized according to the Initial OFF or Initial ON condition (*see* **Note 1**).

Step 2. Advance time

Increase simulation time by a small time increment Δt,

$$t = t + \Delta \mathrm{t}.$$

Step 3. Compute division probability

For each cell, calculate the probability of undergoing division within the current time step cell

$$P_i = \frac{\log(2) \cdot dt}{DT} \cdot g(x_i)\left(1 - \frac{N}{N_0}\right) \cdot H\left(t_i^{postdiv} - DT\right) \cdot \frac{N}{N_{div}}$$

where DT is the cell doubling time, $t_i^{postdiv}$ is the elapsed time since the last division for cell, N_{div} is the number of ready-for-division cells based on the condition $t_{postdiv} > DT$, H is the Heaviside step function, and $g(x)$ represents the reduction in growth rate due to circuit-induced burden, $\frac{\log(2)}{DT}$ gives the idealized per-step division rate in the absence of burden (*see* **Note 2**).

Step 4. Determine division events

For each cell, generate a uniformly distributed random number $r_i \in [0, 1]$.

If $r_i > P_i$, the cell divides.

- Update intracellular expression after division, and reset post-division timers

$$x_i = x_i \cdot \lambda \cdot \alpha, t_i^{postdiv} = 0$$

- Create a new daughter cell:

$$x_{new} = x_i \cdot \lambda \cdot (1 - \alpha),$$

- Set post-division timers for new cells:

$$t_{new}^{postdiv} = 0,$$

- Increment the total cell number:

$$N = N + 1.$$

Here, λ represents the fold change of x_i (0.5~0.7) for each cell division by considering the expression changes before the cell division. α (0~1) is used for simulation of unsymmetrical cell division (*see* **Note 3**).

If $r_i \le P_i$, the cell does not divide:

$$t_i^{postdiv} = t_i^{postdiv} + dt.$$

Step 5. Update circuit dynamics.

Between division events, simulate the gene circuit dynamics for each cell using the ODEs defined for the molecular network without explicit growth feedback.

Step 6. Iterate until termination

Repeat **steps 2–5** until the total simulation time t exceeds the predefined maximum simulation time T_{max}.

This hybrid simulation approach focuses on modeling stochastic growth dynamics at the population level, rather than intrinsic expression noise within individual cells. By treating intracellular gene regulation deterministically and introducing stochasticity only through cell growth and division processes, this method could track the expression profiles of all cell populations. The simulations indicate that the SA circuit rapidly loses its ON state following dilution into fresh medium, reflecting its high sensitivity to growth-mediated dilution (Fig. 1c). By contrast, the toggle switch recovers and returns to the ON state after dilution, as observed at both the population-averaged level (Fig. 1c) and the single-cell level (Fig. 1d, e). It provides an efficient and mechanistically grounded framework for exploring how population-level stochasticity shapes the stability, memory, and robustness of synthetic gene circuits under dynamic growth conditions.

3.4 Stochastic Simulation of Growth Feedback and Condensate-Buffered Circuits

In this section, we model the stochastic dynamics of gene circuits that incorporate a phase separation feature, in which transcriptional condensate formation modulates gene expression by fusing with intrinsic disordered region (IDR) (Fig. 2a). To capture these dynamics in both the self-activation (SA) and condensate-forming (Drop-SA) circuits at the single-cell level, we developed a stochastic simulation algorithm that integrates stochastic gene expression, cell-volume growth, cell division, and condensate formation with dynamic size adjustment. This framework extends the standard Gillespie algorithm by coupling transcriptional reactions with cellular growth and phase-separation thermodynamics. Unlike the

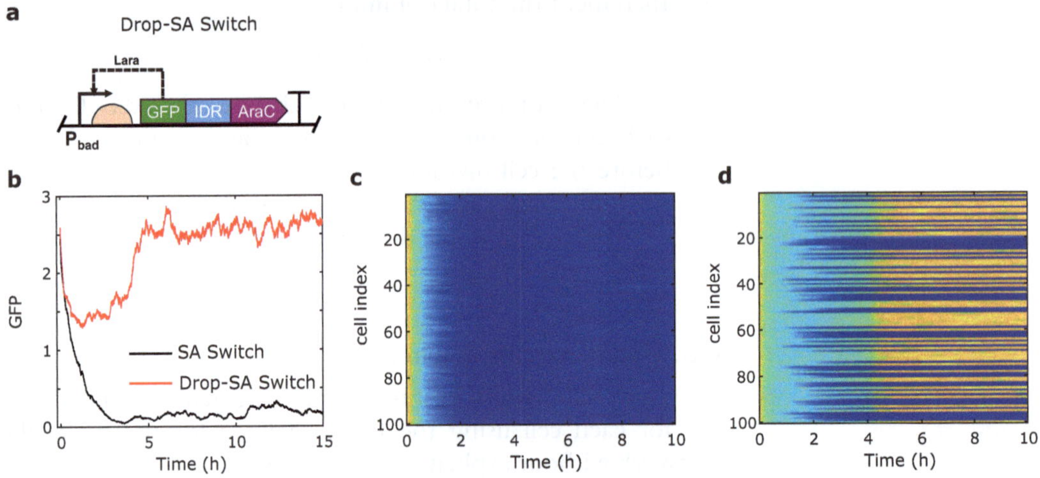

Fig. 2. Condensate-buffered circuit and stochastic simulation workflow. (**a**) Schematic diagram of the Drop-SA circuit, in which the transcription factor AraC is fused to an intrinsically disordered region (IDR) to promote condensate formation. (**b**) Example single-cell stochastic protein trajectory for cells initialized in the ON state in the SA circuit and the Drop-SA circuit. (**c, d**) Representative normalized molecule-number trajectories from 100 simulation replicates initialized in the ON state for the SA circuit (**c**) and the Drop-SA circuit (**d**)

population-level hybrid model described earlier, this approach focuses on a single representative cell, tracking its molecular fluctuations, condensate dynamics, and growth over time.

1. **Initialize the system:**
   Set the initial parameters and variables
   - **Time**: $t = 0$.
   - **Circuit type**: SA or Drop-SA.
   - **Inducer level**: [L-ara] (L-arabinose concentration).
   - **mRNA and protein copy numbers**: M and P, initialized according to the Initial OFF or Initial ON condition.
   - **Cell volume:** $V_t = V_o$.
   - **Droplet size:** $N_{droplet} = 0$.
   - **post-division timer**: $t_{new}^{postdiv} = 0$.
2. **Define reaction propensities:**
   The system includes four stochastic reactions
   - Transcription M → M + 1 with a propensity v_m.
   - mRNA degradation M → M-1 with a propensity $d_m M$.
   - Translation P → P + 1 with a propensity $v_p M$.
   - Protein degradation P → P-1 with a propensity $d_p P$.

The transcription propensity v_m is defined as

$$v_m = \left(k_m \frac{S_a * C_{TF}^2}{Sa * C_{TF}^2 + 1} + k_{m0} \right) * \Omega * V_t$$

where k_m and k_{m0} are basal and regulated transcription constants, and S_a is the activation parameter dependent on the L-arabinose inducer

$$S_a = \left(C_{min} + (C_{max} - C_{min}) * \frac{L_{ara}^3}{L_{ara}^3 + K_m^3} \right)$$

The effective transcription-factor concentration C_{TF} depends on,

- No condensate formation: C_{TF} is the average TF concentration, calculated as

$$C_{TF} = C_{ave} = \frac{P}{V_t} \frac{1}{\Omega}.$$

- Condensate formation in Drop-SA system: C_{TF} is the concentration of TF in the droplet phase ($C_{droplet}$), expressed as

$$C_{TF} = C_{droplet} = \frac{1}{V_{TF}} \frac{1}{\Omega} * \varepsilon_{drop}.$$

where parameter ε_{drop} quantifies TF enrichment within the condensate.

Factor Ω is used to convert the variable concentration (C) in micromolar (μM) and number (N) based equation $N = C * V * \Omega$. Here, $\Omega = N_A * \frac{\text{micromolar}}{\text{mol/m}^3} = 6.0 * 10^{23} * 10^{-3} = 6.0 * 10^{20}$ with the Avogadro constant $N_A = 6.0 * 10^{23}$.

3. **Simulate stochastic reactions:**
   Apply the Gillespie algorithm to determine the next reaction and time step (Δt). Update the time ($t = t + \Delta t$) and copy numbers (M and P) by stoichiometry accordingly based on the chosen reaction. This captures intrinsic transcriptional and translational fluctuations.

4. **Update growth and division:**
   Between reaction events, increase the cell volume (V_t) linearly according to Doubling Time (DT),

$$V_t = V_0 \left(\frac{t_{new}^{postdiv}}{DT} + 1 \right),$$

where DT is calculated based on the growth rate μ as $DT = log(2)/\mu$ and the growth rate is calculated as $\mu = \mu_{max}(1 - \text{N/Nmax})$ for batch culture (*see* **Note 4**).

if $V_t = 2V_0$, the cell divides, and the mRNA and protein molecules are randomly partitioned between the two daughter cells (*see* **Note 5**). For this single-cell model, the simulation continues with one representative daughter cell ($V_t = V_0$) while updating cell number N to reflect total population size in the logistic term and post-division timer $t_{new}^{postdiv} = 0$ (*see* **Note 6**).

5. **Condensate formation and size calculation:**

   For Drop-SA, determine condensate formation at each time step by minimizing the total system free energy

$$F = F_{dilute} + F_{droplet} + F_{SurfaceTension}$$
$$= V_{dilute} * f_{dilute} + V_{droplet} * f_{droplet} + \gamma * A$$

   where droplet volume ($V_{droplet}$) is calculated as $V_{droplet} = N_{droplet} * V_{TF}$, where $N_{droplet}$ is the number of *TF* in the droplet and V_{TF} is the volume of a single *TF* molecule. Consequently, the volume of the dilution phase is given by $V_{dilute} = V_t - V_{droplet}$. The free energy density in the droplet phase ($f_{droplet}$) is defined as $f_{droplet} = -\mu / V_{TF}$, with μ representing the relative chemical potential of the single *TF* molecule. In the dilute phase, the free energy density (f_{dilute}) is expressed as $f_{dilute} = k_B T / V_{TF} * (\varnothing_{dilute} * log(\varnothing_{dilute}) - \varnothing_{dilute})$, where $\varnothing_{dilute} = N_{dilute} * V_{TF}/V_{dilute}$ is the volume fraction of TF in the dilution phase. The last term accounts for the surface tension γ and the droplet surface area $A = 4 * \pi * \left(\frac{3}{4\pi} * V_{TF} * N_{droplet}\right)^{\frac{2}{3}}$.

   Minimization of *F* yields the equilibrium $N_{droplet}$, defining condensate size and effective TF sequestration (*see* **Note 7**).

6. **Iterate cell population:**

   Repeat **steps 2–5** until time exceeds the maximum simulation time Tmax, corresponding to several hours of real culture time. Record trajectories of M, P, V_T, and $N_{droplet}$ to analyze growth, gene expression, and condensate behavior over time for further analysis and interpretation (*see* **Note 8**).

This single-cell stochastic modeling framework simulates gene expression noise and growth-mediated dilution within one representative cell. It focuses on stochastic gene expression rather than population variability, allowing direct comparison between conventional self-activation (SA) and condensate-forming (Drop-SA) circuits. The simulations show that the SA circuit rapidly loses its ON state after dilution into fresh medium, in agreement with the simulation in Subheading 3.3. By contrast, the Drop-SA circuit successfully reestablishes and maintains the ON state following dilution (Fig. 2b–d). The inclusion of phase-separation thermodynamics demonstrates how transcriptional condensates buffer stochastic fluctuations, stabilize expression states, and enhance circuit robustness under dynamic growth conditions. The sensitivity of circuit performance to model parameters can be examined through parameter sweeps, enabling identification of stability regimes for SA and Drop-SA circuits (*see* **Note 9**).

3.5 Discussion and Expected Results

The modeling frameworks presented in this chapter provide a unified approach to analyzing how growth-mediated dilution interacts with circuit topology and phase separation to shape synthetic gene-circuit behavior. By combining deterministic circuit equations, stochastic gene expression, and explicit treatment of cell growth and division, these models capture the essential mechanisms that give rise to the robustness or fragility of gene circuits under dynamically changing physiological conditions.

For the self-activation (SA) circuit, stochastic simulations show that rapid cell growth and frequent dilution events significantly destabilize the activated state. As cell volume increases and protein concentration decreases, transcriptional activation weakens, leading to spontaneous switching to the OFF state. This behavior reflects the underlying topology: a single positive-feedback loop is inherently sensitive to multiplicative dilution, causing the system to lose memory in high-growth environments.

In contrast, the condensate-forming (Drop-SA) circuit demonstrates substantially enhanced robustness. Phase separation concentrates transcription factors within condensates, maintaining high local transcriptional activity and buffering the promoter from fluctuations in bulk concentration. Stochastic simulations reveal that this buffering mechanism reduces noise, increases ON-state residence time, and supports bistability across a wider range of growth conditions. The dynamic adjustment of condensate size stabilizes expression by compensating for dilution during growth.

These results highlight two generalizable principles:

1. **Circuit topology determines susceptibility to growth-mediated perturbation.** Self-activation circuit alone is insufficient to maintain memory when dilution is strong.
2. **Physical compartmentalization can serve as an effective buffering strategy.** Condensate formation provides a robust, topology-independent means of stabilizing expression under variable growth rates.

Together, these models illustrate how integrating biochemical regulation, cell physiology, and phase separation into a single simulation framework yields mechanistic insight into circuit failure modes and suggests new design strategies for growth-robust synthetic systems.

4 Notes

1. *ON/OFF Initialization Conditions.* The initial molecular counts for the ON and OFF states should be selected based on the steady-state solutions of the deterministic model under the corresponding induction conditions, typically no inducer for the OFF state and sufficiently high inducer for the ON state. Using steady-state values ensures that simulations begin from biologically meaningful expression levels and avoids artificial switching caused by initializing near unstable or transitional states.
2. *Growth Rate and Carrying Capacity.* In logistic growth modeling, the maximum growth rate (μ_{max}) and the carrying capacity (N_{max}) are key parameters that determine the magnitude and timing of growth-mediated dilution. Both parameters depend on culture conditions such as temperature, nutrient availability, and medium composition. High values of μ_{max} result in frequent dilution events that can destabilize gene expression, whereas very high N_{max} values extend the exponential growth phase.
3. *Partitioning Noise.* Random partitioning of mRNA and protein molecules during cell division can substantially influence gene expression noise, particularly when molecule copy numbers are low. For simulations aimed at investigating noise generation or propagation, explicit stochastic segregation should be modeled to capture biologically relevant variability. In contrast, for mean-field or computationally simplified simulations, symmetric partitioning may be employed to minimize variability and focus on average system behavior.
4. *Volume Dynamics and Numerical Stability.* Updating cell volume through $V_t = V_0\left(\frac{t_{new}^{postdiv}}{DT} + 1\right)$ requires sufficiently small time steps within the Gillespie loop to maintain numerical stability. When reaction propensities span several orders of magnitude or when the system exhibits stiff dynamics, using a modified next-reaction method or a tau-leaping approximation can substantially improve computational efficiency and stability. However, it might lead to large volume changes occurring between stochastic reaction events that can introduce errors in concentration calculations and reaction propensities.
5. *Cell Division Timing.* Cell division is triggered when the cell volume reaches $2\,V_0$. In simulations molecular reaction rates are assumed much faster than growth, many stochastic reaction events may occur between divisions.
6. *Choice of Representative Cell for Single-Cell Stochastic Model.* The single-cell stochastic model approximates population-level

behavior by updating the total cell number N only within the logistic growth term, while explicitly tracking gene expression in a single representative lineage. This approximation is suitable for studying growth-mediated dilution and condensate buffering without modeling full population heterogeneity. To capture true cell-to-cell variability, the simulation can be run for many independent cells followed by statistical analysis of the resulting ensemble. It should also be noted that, due to intrinsic stochasticity, individual cells may lose circuit memory even when initialized in the same ON state, highlighting the role of noise in determining cell-fate outcomes.

7. *Condensate Free-Energy Minimization.* The free-energy minimization used to determine condensate size can be implemented efficiently by directly scanning over all plausible transcription factor (TF) counts. If this step becomes computationally expensive, precomputing a lookup table of N_{droplet} values or restricting the search space to biologically feasible droplet sizes can substantially accelerate the simulation without compromising accuracy.
8. *Interpreting Stochastic Trajectories.* Short-term fluctuations in mRNA levels are expected due to burst-like transcription and rapid degradation, whereas protein trajectories are typically smoother because of longer molecular lifetimes and slower turnover. Condensate size changes occur on intermediate timescales determined by transcription factor abundance and phase-separation dynamics. Together, these multilayered fluctuations provide insight into how transcriptional noise, protein stability, and condensate buffering shape overall circuit behavior.
9. *Parameter Sweeps and Robustness Analysis.* To assess the generality and robustness of the model predictions, perform parameter sweeps over key variables such as the burden constant J, the droplet enrichment factor $\varepsilon_{\text{drop}}$, and the maximal growth rate μ_{max}. Visualizing the results using heat maps or phase diagrams can highlight parameter regimes in which condensate formation markedly enhances circuit stability and mitigates growth-mediated dilution effects.

Acknowledgments

This work was supported by a grant from the US National Institutes of Health (R35GM142896 to X.-J.T.).

Competing Interests The authors declare no competing interests.

References

1. Borkowski O, Ceroni F, Stan G-B, Ellis T (2016) Overloaded and stressed: whole-cell considerations for bacterial synthetic biology. Curr Opin Microbiol 33:123–130. https://doi.org/10.1016/j.mib.2016.07.009
2. Boo ALC, Ellis T, Stan G-B (2019) Host-aware synthetic biology. Curr Opin Syst Biol 14:66–72. https://doi.org/10.1016/j.coisb.2019.03.001
3. Şimşek E, Yao Y, Lee D, You LC (2022) Toward predictive engineering of gene circuits. Trends Biotechnol 41:760–768. https://doi.org/10.1016/j.tibtech.2022.11.001
4. Kumar S, Hasty J (2023) Stability, robustness, and containment: preparing synthetic biology for real-world deployment. Curr Opin Biotechnol 79:102880. https://doi.org/10.1016/j.copbio.2022.102880
5. Stone A, Youssef A, Rijal S, Zhang R, Tian X-J (2024) Context-dependent redesign of robust synthetic gene circuits. Trends Biotechnol 42: 895–909. https://doi.org/10.1016/j.tibtech.2024.01.003
6. Klumpp S, Zhang ZG, Hwa T (2009) Growth rate-dependent global effects on gene expression in bacteria. Cell 139:1366–1375. https://doi.org/10.1016/j.cell.2009.12.001
7. Scott M, Gunderson CW, Mateescu EM, Zhang ZG, Hwa T (2010) Interdependence of cell growth and gene expression: origins and consequences. Science 330:1099–1102. https://doi.org/10.1126/science.1192588
8. Klumpp S, Hwa T (2014) Bacterial growth: global effects on gene expression, growth feedback and proteome partition. Curr Opin Biotechnol 28:96–102. https://doi.org/10.1016/j.copbio.2014.01.001
9. Stone A, Rijal S, Zhang R, Tian X-J (2024) Enhancing circuit stability under growth feedback with supplementary repressive regulation. Nucleic Acids Res 9:1512–1521
10. Zhang R, Li J, Melendez-Alvarez J, Chen XW, Sochor P, Goetz H, Zhang Q, Ding T, Wang X, Tian X-J (2020) Topology-dependent interference of synthetic gene circuit function by growth feedback. Nat Chem Biol 16:695–701. https://doi.org/10.1038/s41589-020-0509-x
11. Zhang R, Yang WF, Zhang RX, Godar A, Rijal S, Youssef A, Nielsen DR, Zheng WW, Tian X-J (2025) Phase separation to buffer growth-mediated dilution in synthetic circuits. Cell 188:1–15. https://doi.org/10.1101/2024.11.01.621586

Chapter 17

Quantitative Coarse-Grained Modeling of Microbial Metabolism: Uncovering Growth Laws, Environmental Responses, and Host-Circuit Coupling

Yi-Hui Lai and Ting Lu

Abstract

Microbial cell metabolism is highly complex, involving a vast number of molecular species and multi-level regulatory processes. To address this complexity and decode cellular behavior, coarse-grained modeling provides a powerful quantitative framework by grouping similar biological entities while preserving essential network properties. This approach offers insights into intracellular regulation and resource allocation, enables systematic analysis of natural biological systems, and generates predictions for synthetic engineering outcomes. In this chapter, we present a comprehensive overview of coarse-grained modeling, from model construction to applications, and demonstrate its utility in uncovering microbial metabolic behavior.

Key words Mathematical modeling, Kinetic modeling, Coarse-grained modeling, Proteome allocation, Resource allocation, Metabolic response, Nutrient limitation, Host-circuit coupling, Metabolic burden

1 Introduction

Microorganisms are widely adopted as synthetic platforms for metabolic engineering and have been extensively studied for their capacity to produce valuable compounds. This synthetic capacity fundamentally relies on how cells allocate their limited resources, particularly proteins, which can account for more than half of the cellular dry biomass—up to ~55% in *E. coli* [1]. Proteins are the primary functional components that sustain cellular activities: enzymes drive the catabolism and anabolism of essential nutrients, while RNA polymerases and ribosomes enable the synthesis of these enzymes. Therefore, understanding how cells allocate protein resources across different functions is critical for explaining how they adapt and thrive under diverse growth conditions.

To understand how protein allocation is controlled, it is necessary to examine the underlying layers of cellular regulation.

Xiao-Jun Tian (ed.), *Synthetic Gene Circuits: Methods and Protocols*, Methods in Molecular Biology, vol. 3041,
https://doi.org/10.1007/978-1-0716-5304-3_17,

However, cellular metabolism is inherently complex, involving multiple regulatory mechanisms such as allosteric control, signaling transduction, transcriptional regulation, and protein translation (Fig. 1). These layers do not act independently; instead, they communicate and coordinate to provide the cell with optimal strategies in response to environmental cues. As a result, experimental approaches alone often struggle to disentangle these overlapping layers, making mathematical modeling an essential tool for generating quantitative insights into how regulation shapes cellular behavior.

Here, we present a coarse-grained mathematical approach to capture the essence of different regulatory layers and their interconnections, which together shape intracellular dynamics. This method follows a top-down analysis, where essential behaviors of metabolic networks and regulatory systems are represented by selecting key metabolites, proteins, and regulatory factors. The term "coarse-grained" originates from grouping proteins into functional categories, providing an intuitive view of how cells allocate resources [2, 3]. This principle can be extended beyond proteins to other cellular components, such as metabolites and regulatory factors [4]. By simplifying complex cellular networks, coarse-grained modeling enables systematic analysis of how different regulatory layers coordinate to drive cellular dynamics and adaptation.

In the following sections, we provide detailed instructions on the construction and application of coarse-grained models. Subheading 2 describes the step-by-step design of the modeling framework, including the selection of representative sectors and units, the organization of reactions, and the derivation of kinetic equations for the essential network. Subheading 3 illustrates how this framework can be applied to study cellular resource allocation under different contexts, such as changes in growth rate, nutrient composition, environmental conditions, and the expression of synthetic gene constructs. The presented modeling examples demonstrate how coarse-grained modeling serves as a useful platform for quantitatively understanding and systematically exploring cellular resource investment strategies across diverse conditions. By revealing how cells manage internal resources and adapt to environmental pressures, the coarse-grained modeling framework provides a systems-level perspective on resource allocation and metabolic flux distribution. Such insights are valuable for guiding strategies to control cellular metabolism, improve the robustness of engineered systems, and accelerate progress in metabolic engineering and synthetic biology.

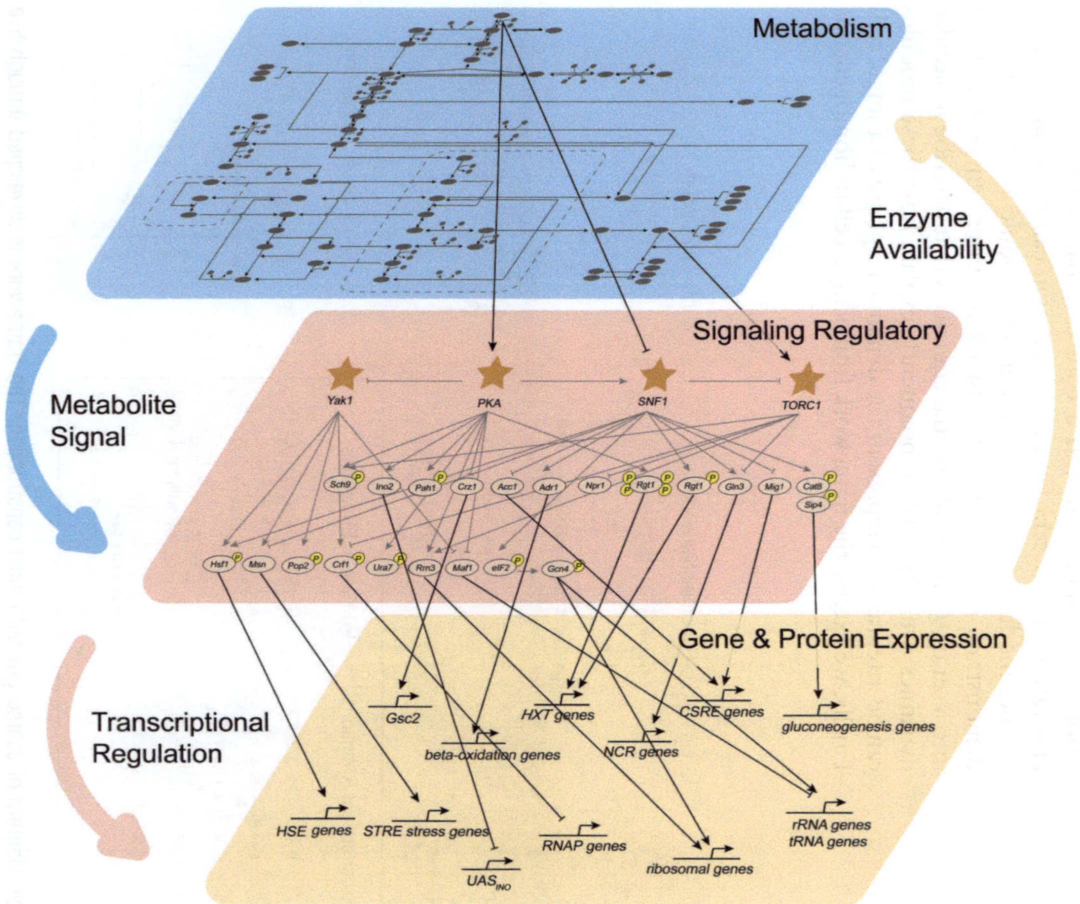

Fig. 1 System-level overview of multi-level regulation coordination. Cellular metabolic responses are coordinated through three interconnected layers: the metabolic layer, the signaling regulatory layer, and the gene and protein expression layer. This coordination begins with the metabolic layer, which senses changes in fluxes. These changes trigger signaling pathways that regulate transcription factors and subsequent gene expression. The resulting gene and protein expression determines the cellular protein and enzyme distribution, which redirect the metabolic fluxes

2 Methods

In this section, we provide step-by-step instructions for constructing a coarse-grained kinetic modeling framework that captures the essential regulatory coordination within the cellular system (Fig. 2). The framework begins with the description of single-cell behavior. We first demonstrate how the complex whole-cell regulatory network can be simplified into a core essential network. Next, we illustrate how different layers of regulation can be organized into distinct modules. Then, we present one module as an example to demonstrate how a series of network reactions can be translated into a set of ordinary differential equations that describe molecular dynamics over time. We subsequently explain how these modules can be integrated to represent the dynamic response of a single cell. Finally, we extend the framework from single-cell to homogeneous

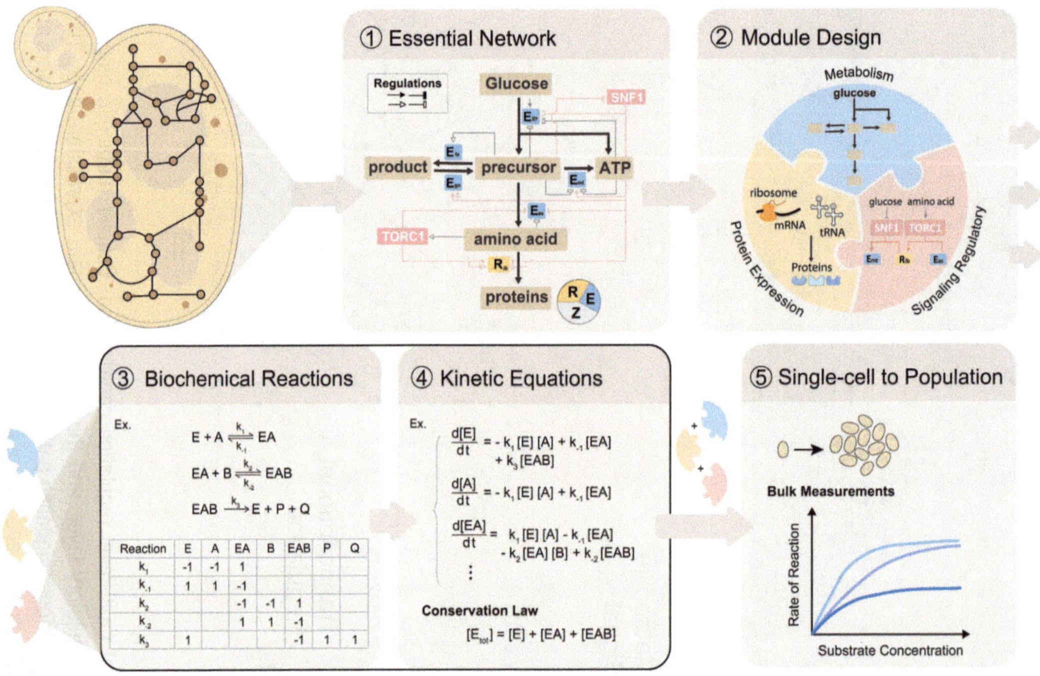

Fig. 2 Schematic diagram of coarse-grained model construction. The framework is developed through five main steps: (i) selecting coarse-grained representative units for metabolites, proteins, and regulations to construct an essential network of the whole system; (ii) separating different layers of dynamics and regulation into distinct modules, where the output of one module serves as the input for another; (iii) listing all biochemical reactions within each module and compiling them into reaction tables; (iv) formulating rate equations to describe molecular changes, applying the protein conservation law, and deriving kinetic equations; and (v) combining all modules into integrated equation sets to simulate overall system behavior under homogeneous population dynamics. The resulting simulations can then be directly compared with bulk experimental measurements

population dynamics, thereby enabling direct comparison with bulk experimental measurements.

2.1 Essential Network Construction

The main idea of coarse graining is to group components into broader units, which reduces modeling complexity while preserving the system behaviors of interest. Therefore, before simplification, we should always ask: *What specific question are we addressing, and how detailed are the measurements available to us?* Incorporating unmeasured metabolites or overly fine detail can overcomplicate the framework and hinder parameter estimation. Since this is a top-down approach, the first step is to decide the lowest level of detail we want the model to represent. The guiding principle is to keep the framework as simple as possible.

In the following, we illustrate how to construct a minimal essential network that captures our interest in understanding central carbon metabolism in baker's yeast (*S. cerevisiae*). Despite its simplicity, the network represents the major fluxes from glucose uptake to biomass formation, bioproduct synthesis, and energy generation. To capture these fluxes, we first identify the key components of the network, with changes in their abundance enabling quantification of flux distributions. We then define relative proteome sectors that determine reaction activities. Finally, we incorporate regulators and signaling factors that govern flux sensing and the expression of enzymes and proteins. Together, this streamlined design enables the representation of core carbon flux–related reactions and their regulatory mechanisms.

2.1.1 Representative Metabolites

In this step, we select representative units that most effectively capture the flux distribution of interest. Because the focus is on how carbon influx is partitioned into energy, biomass, and bioproducts, six core metabolites are identified. Glucose represents the primary carbon source entering the system. Precursors mark the key intermediate branching points where carbon is directed toward bioproduct, energy, or biomass formation (*see* **Note 1**). ATP functions as the universal energy currency, driving diverse reactions and supporting cellular growth. Amino acids serve as the building blocks required for biomass synthesis (*see* **Note 2**). Bioproducts, such as ethanol, represent carbon sinks that accumulate as metabolites not directly utilized by the cell. Finally, proteins reflect the accumulation of cellular biomass (*see* **Note 3**). To ensure that these representatives are accurately defined, it is useful to consult curated metabolic maps such as KEGG (https://www.genome.jp/kegg/pathway.html) to verify stoichiometry and identify essential pathways.

2.1.2 Representative Proteome Sectors

Once the main components of the network are defined, the next step is to assign representative reactions that govern the conversion between them. Note that the rate of each reaction depends not only on substrate and product concentrations but also on enzyme availability. For this reason, it is important to quantify both metabolite levels and enzyme concentrations. To account for how enzyme's availability changes the rate of representative reactions, proteins are organized into representative sectors that regulate these processes.

It is reported that proteins belonging to the same functional pathways often display similar dynamic patterns [5]. Because yeast typically expresses around 6000 proteins, grouping them into functional sectors is critical for reducing complexity. Gene Ontology (GO) annotations provide a systematic way to classify proteins according to their functions. The Gene Ontology database (https://current.geneontology.org/products/pages/downloads.html) offers annotation files (.gaf) for commonly studied organisms. By combining these annotations with GO Slim term mappings (*see* **Note 4**), proteins can be grouped into representative sectors that capture their functional roles. Visualization tools such as Proteomaps (https://www.proteomaps.net/) further help illustrate this grouping process [2].

In the illustrated example, six representative reactions are considered: glucose to precursor, glucose to ATP, precursor to product, product to precursor, precursor to amino acids, and amino acids to proteins. Accordingly, proteins are categorized into six functional sectors, each governing one of these reactions, plus an additional sector that encompasses proteins responsible for other essential cellular functions (Z sector in the illustration). Using GO terms, protein functions associated with the representative reactions can be grouped into these sectors. Within the six functional sectors, we further distinguish between enzyme-based proteins (E sector) and ribosomal-based proteins (R sector). The E sector supports carbon metabolism and the synthesis of essential building blocks, while the R sector governs transcriptional and translational processes, enabling protein production and contributing to cellular growth.

To summarize, this step shows that proteins can be grouped into representative sectors that govern specific reactions in the essential network, while further classification into R, E, and Z sectors provides a higher-level understanding of their functional roles.

2.1.3 Representative Regulators

Moving forward, this step is to understand how metabolite signals influence the regulation of gene expression. To achieve this, we examine the major regulatory pathways and signaling networks. The BioCyc database (https://biocyc.org/) provides curated regulatory networks for many common organisms, including transcription factors and their regulons. For instance, the *E. coli* regulatory overview maps are available at https://ecocyc.org/overviewsWeb/regOv.shtml?orgid=ECOLI. Nevertheless, literature searches are often necessary to capture the full spectrum of metabolic signals and the pathways they regulate.

After constructing a comprehensive map of the regulatory and signaling network, we identify the key regulators that respond to the representative metabolites. Using the baker's yeast example, cAMP–PKA and SNF1 are the primary regulators of glucose availability [6], while amino acid concentrations influence TORC1 signaling [7, 8], which plays a central role in biomass formation. We then map how these signaling factors exert control, either through direct regulation of enzyme activity [9] or through transcriptional [10] and translational control [10, 11, 12] of specific functional proteome sectors. Furthermore, cAMP–PKA and SNF1 can be grouped together, as they often produce opposing regulatory effects on the same representative pathways [13].

Altogether, this mapping provides a framework for understanding how representative metabolites trigger regulatory signals, which in turn modulate protein activity and abundance. At this stage, we have constructed a coarse-grained essential network that captures the topology of the original complex whole-cell system.

2.2 Module Design and Integration

From the construction of the essential network, it becomes clear that cellular responses can be viewed across three layers: changes in metabolic flux, the buildup of signaling regulators, and the resulting alterations in protein abundance and activity. These layers can then be separated into individual modules for a clearer understanding of their interactions.

Each module represents a certain level of regulation and dynamics. The metabolism module captures the biochemical reactions and how changes in reaction rates alter the concentrations of representative metabolites. The signaling regulatory module responds to metabolite signals, where the active fraction of signaling molecules builds up [13, 14]. The protein expression module reflects the downstream outcomes of signaling, altering gene expression and thereby reallocating proteome sectors. In addition, signaling molecules can directly modify protein activity through post-translational regulation and, in turn, affects metabolic reaction rates.

This modular design provides a systematic view of the cellular interconnections. Each module represents a specific level of regulation, with the output of one module serving as the input to the next. Taken together, the modules capture how the entire system responds to environmental and nutritional signals and adjusts cellular fluxes accordingly.

2.3 Biochemical Network Reactions

Once the system has been divided into modules, the next step is to specify the reactions within each module. These reactions describe how molecular concentrations change over time and offer the fundamentals for establishing the corresponding rate equations.

In practice, this certain step consists of two main tasks: identifying the reactions and organizing them into a reaction table. Because reaction specification and rate equation formulation are closely connected, they are always performed together to construct a set of Ordinary Differential Equations (ODE-based) for each module. Repeating these two-steps process across all modules yields a complete set of ODEs for the entire cellular system.

Here, we provide an example of the protein expression module to illustrate how a set of mechanistic reactions (*see* **Step 3**) can be transformed into ODEs (*see* **Step 4**) to describe molecular dynamics within the module. Specifically, we demonstrate how the tRNA charging process and the activity of signaling molecules influence ribosomal protein partitioning and, in turn, shape the translational outcomes of different proteome sectors.

2.3.1 List of Reactions

The process of protein translation is driven by a few key reactions. The first is tRNA charging, which determines the supply of charged tRNAs available for translation. The second concerns the role of charged and uncharged tRNAs during elongation: charged tRNAs support peptide chain elongation, whereas uncharged tRNAs can bind to ribosomes and cause translation stalling. These reactions are closely coupled to ribosome state transitions and directly influence the translational output of mRNAs. In the following, we outline the reactions underlying tRNA charging and ribosome partitioning in translation.

The tRNA aminoacylation process requires amino acids, ATP, and uncharged tRNA. Catalyzed by aminoacyl-tRNA synthetases, the reaction proceeds in two steps: first, the amino acid is activated by ATP to form an aminoacyl-adenylate (AA-AMP) intermediate; second, the activated amino acid is transferred to the corresponding tRNA, producing charged tRNA (aminoacyl-tRNA) and releasing AMP. This process can be described as a sequential series of four reactions.

1. Amino acid and ATP binding/unbinding to the synthetase

$$E + Aa + ATP \underset{k_{-1}}{\overset{k_1}{\rightleftharpoons}} E \cdot Aa \cdot ATP$$

2. Formation of aminoacyl adenylate (AA–AMP) and release of pyrophosphate

$$E \cdot Aa \cdot ATP \xrightarrow{k_2} E \cdot Aa \cdot AMP + PP$$

3. tRNA binding/unbinding to the synthetase

$$E \cdot Aa \cdot AMP + t_u \underset{k_{-3}}{\overset{k_3}{\rightleftharpoons}} E \cdot Aa \cdot AMP \cdot t_u$$

4. Transfer of the amino acid to the tRNA molecule, releasing charged tRNA and AMP

$$E \cdot Aa \cdot AMP \cdot t_u \overset{k_4}{\longrightarrow} E + t_c + AMP$$

Here, E denotes the aminoacyl-tRNA synthetase, Aa the amino acid, and t_u and t_c the uncharged and charged tRNAs, respectively.

Next, we consider ribosome state transitions, which encompass translation initiation, elongation, and stalling. Translation initiation within a given proteome sector depends on both the concentration of the corresponding mRNA and the translational activity of ribosomes, both of which are regulated by TORC1 and Snf1. Ribosome states are further influenced by the binding of charged or uncharged tRNAs. Binding of a charged tRNA allows the ribosome to elongate the peptide chain; once elongation is completed, the ribosome is ready to bind another tRNA molecule. In contrast, binding of an uncharged tRNA can lead to ribosome stalling, thereby interrupting translation. Further details are listed in the following reactions.

5. Ribosome binding/unbinding to the mRNA, preparing the ribosome for tRNA recruitment

$$R_f + M_j \underset{k_{-5}(s^*)}{\overset{k_5(\tau^*)}{\rightleftharpoons}} R_{o,j}$$

6. Charged tRNA binding/unbinding to the ribosome to enable peptide chain elongation

$$R_{o,j} + t_c \underset{k_{-6}}{\overset{k_6}{\rightleftharpoons}} R_{e,j}$$

7. Peptide elongation followed by the release of uncharged tRNA

$$R_{e,j} + (peptide) \overset{k_7(ATP)}{\longrightarrow} R_{o,j} + t_u + (peptide^{+1})$$

8. Uncharged tRNA binding/unbinding to the ribosome, which leads to the ribosome stalling

$$R_{o,j} + t_u \underset{k_{-8}}{\overset{k_8}{\rightleftharpoons}} R_{s,j}$$

where R_f stands for the free ribosome, while $R_{o,\ j}$, $R_{e,\ j}$, $R_{s,\ j}$ represent the empty, elongating, and stalled ribosomes, respectively, associated with mRNA of protein sector type j. M_j refers to the mRNA corresponding to protein sector type j. The symbol $(\cdot)$ indicates functional dependence; τ^* and s^* stands for the active states of the signaling transduction molecule TORC1 and Snf1, respectively.

2.3.2 Reactions Table

Based on the reactions outlined above, we can now organize them into reaction tables to systematically track the consumption and formation of each molecule. This representation will facilitate the subsequent analysis of molecular changes. In the tables that follow, each reaction is listed alongside its corresponding molecular changes and reaction rate. For simplicity, we assume that each molecule contributes in a first-order relationship to the reaction rate. The tRNA charging reactions (1–4) are summarized in Table 1, while the ribosome partitioning reactions (5–8) are presented in Table 2.

2.4 Step 4: Equations for Biochemical Reactions

With the reaction tables in place, we can now track the changes for each molecule. However, a model that includes every reaction rate and molecule becomes computationally intensive and requires numerous parameters. Therefore, we further simplify the model by applying time-scale domain knowledge, using the rapid equilibrium assumption or the quasi-steady-state assumption. This approach allows us to derive Michaelis-Menten-type rate laws from the detailed reaction networks. We will demonstrate this simplification process by applying it to both the tRNA charging and ribosome partitioning processes.

2.4.1 Assumptions and Derivations

For the tRNA charging process, the following assumptions can be made:

1. The binding and unbinding of molecules occur much faster than the other reactions.
2. The molecules dynamics are in equilibrium, meaning that the rate of aminoacyl adenylate formation is equal to the rate of charged tRNA synthesis.

Table 1
tRNA charging reactions and rates

Reaction	Molecule										Rate
	E	Aa	ATP	$E \cdot Aa \cdot ATP$	$E \cdot Aa \cdot AMP$	PP	$E \cdot Aa \cdot AMP \cdot t_u$	t_u	t_c	AMP	
$E + Aa + ATP \xrightarrow{k_1} E \cdot Aa \cdot ATP$	−1	−1	−1	1							$k_1[E][Aa][ATP]$
$E \cdot Aa \cdot ATP \xrightarrow{k_{-1}} E + Aa + ATP$	1	1	1	−1							$k_{-1}[E \cdot Aa \cdot ATP]$
$E \cdot Aa \cdot ATP \xrightarrow{k_2} E \cdot Aa \cdot AMP + PP$				−1	1	1					$k_2[E \cdot Aa \cdot ATP]$
$E \cdot Aa \cdot AMP + t_u \xrightarrow{k_3} E \cdot Aa \cdot AMP \cdot t_u$					−1		1	−1			$k_3[E \cdot Aa \cdot AMP][t_u]$
$E \cdot Aa \cdot AMP \cdot t_u \xrightarrow{k_{-3}} E \cdot Aa \cdot AMP + t_u$					1		−1	1			$k_{-3}[E \cdot Aa \cdot AMP \cdot t_u]$
$E \cdot Aa \cdot AMP \cdot t_u \xrightarrow{k_4} E + t_c + AMP$	1						−1		1	1	$k_4[E \cdot Aa \cdot AMP \cdot t_u]$

Table 2
Ribosome partitioning reactions and rates

Reaction	Molecule							Rate
	R_f	M_j	$R_{o,j}$	$R_{e,j}$	t_c	t_u	$R_{s,j}$	
$R_f + M_j \overset{k_5(\tau^*)}{\rightarrow} R_{o,j}$	−1	−1	1					$k_5(\tau^*)[R_f][M_j]$
$R_{o,j} \overset{k_{-5}(s^*)}{\rightarrow} R_f + M_j$	1	1	−1					$k_{-5}(s^*)[R_{o,j}]$
$R_{o,j} + t_c \overset{k_6}{\rightarrow} R_{e,j}$			−1	1	−1			$k_6[R_{o,j}][t_c]$
$R_{e,j} \overset{k_{-6}}{\rightarrow} R_{o,j} + t_c$			1	−1	1			$k_{-6}[R_{e,j}]$
$R_{e,j} + (peptide) \overset{k_7(ATP)}{\rightarrow} R_{o,j} + t_u + (peptide^{+1})$			1	−1		1		$k_7(ATP)[R_{e,j}]$
$R_{o,j} + t_u \overset{k_8}{\rightarrow} R_{s,j}$			−1			−1	1	$k_8[R_{o,j}][t_u]$
$R_{s,j} \overset{k_{-8}}{\rightarrow} R_{o,j} + t_u$			1			1	-1	$k_{-8}[R_{s,j}]$

From assumption (1), we apply the rapid equilibrium approximation to reactions (1) and (3), leading to the association and dissociation reaction relationships as:

$$k_1[E][Aa][ATP] = k_{-1}[E \cdot Aa \cdot ATP] \tag{1}$$

$$k_3[E \cdot Aa \cdot AMP][t_u] = k_{-3}[E \cdot Aa \cdot AMP \cdot t_u] \tag{2}$$

These can be rearranged as:

$$[E \cdot Aa \cdot ATP] = K_1[E][Aa][ATP], \text{where } K_1 = \frac{k_1}{k_{-1}} \tag{3}$$

$$[E \cdot Aa \cdot AMP \cdot t_u] = K_3[E \cdot Aa \cdot AMP][t_u], \text{where } K_3 = \frac{k_3}{k_{-3}} \tag{4}$$

From assumption (2), we relate reactions (2) and (4) as:

$$k_2[E \cdot Aa \cdot ATP] = k_4[E \cdot Aa \cdot AMP \cdot t_u] \tag{5}$$

which simplifies to:

$$[E \cdot Aa \cdot AMP \cdot t_u] = K_2[E \cdot Aa \cdot ATP], \text{where } K_2 = \frac{k_2}{k_4} \tag{6}$$

Applying the conservation law for the total aminoacyl-tRNA synthetase:

$$[E_{tot}] = [E] + [E \cdot Aa \cdot ATP] + [E \cdot Aa \cdot AMP] + [E \cdot Aa \cdot AMP \cdot t_u] \tag{7}$$

Substituting Eqs. (3), (4), and (6) into (7), we derive the pseudo-steady-state concentrations of each complex:

$$[E] = \frac{1}{1 + K_1[Aa][ATP] + \frac{K_2K_1[Aa][ATP]}{K_3[t_u]} + K_2K_1[Aa][ATP]}[E_{tot}] \tag{8}$$

$$[E\cdot Aa\cdot ATP] = \frac{K_1[Aa][ATP]}{1 + K_1[Aa][ATP] + \frac{K_2K_1[Aa][ATP]}{K_3[t_u]} + K_2K_1[Aa][ATP]}[E_{tot}] \tag{9}$$

$$[E\cdot Aa\cdot AMP] = \frac{\frac{K_2K_1[Aa][ATP]}{K_3[t_u]}}{1 + K_1[Aa][ATP] + \frac{K_2K_1[Aa][ATP]}{K_3[t_u]} + K_2K_1[Aa][ATP]}[E_{tot}] \tag{10}$$

$$[E\cdot Aa\cdot AMP\cdot t_u] = \frac{K_2K_1[Aa][ATP]}{1 + K_1[Aa][ATP] + \frac{K_2K_1[Aa][ATP]}{K_3[t_u]} + K_2K_1[Aa][ATP]}[E_{tot}] \tag{11}$$

Finally, since our primary interest is the formation of charged tRNA, which directly impacts ribosome partitioning and peptide elongation, we obtain the synthesis rate as a Michaelis–Menten–type expression derived from reaction (4) and eq. (11):

$$\begin{aligned} J_{t_c,syn} &= k_4[E\cdot Aa\cdot AMP\cdot t_u] \\ &= \frac{k_4[E_{tot}]}{1 + \frac{1}{K_2} + \frac{1}{K_2K_1[Aa][ATP]} + \frac{1}{K_3[t_u]}} \end{aligned} \tag{12}$$

This rate law explicitly demonstrates the dependence of charged tRNA synthesis on the availability of three substrates, amino acids, ATP, and uncharged tRNA, as well as on enzyme concentration (also *see* **Note 5**).

For the ribosome partitioning process, we assume that ribosome dynamics are balanced across all states, meaning each ribosome type can be treated under a quasi–steady-state approximation. Accordingly, the dynamics of ribosomes can be written as:

$$\frac{d[R_f]}{dt} = -k_5(\tau^*)[R_f][M_j] + k_{-5}(s^*)[R_{o,j}] = 0 \tag{13}$$

$$\begin{aligned} \frac{d[R_{o,j}]}{dt} &= k_5(\tau^*)[R_f][M_j] - k_{-5}(s^*)[R_{o,j}] - k_6[R_{o,j}][t_c] \\ &\quad + k_{-6}[R_{e,j}] + k_7(ATP)[R_{e,j}] - k_8[R_{o,j}][t_u] \\ &\quad + k_{-8}[R_{s,j}] \\ &= 0 \end{aligned} \tag{14}$$

$$\frac{d[R_{e,j}]}{dt} = k_6[R_{o,j}][t_c] - k_{-6}[R_{e,j}] - k_7(ATP)[R_{e,j}] = 0 \tag{15}$$

$$\frac{d[R_{s,j}]}{dt} = k_8[R_{o,j}][t_u] - k_{-8}[R_{s,j}] = 0 \tag{16}$$

Rearranging Eqs. (13–16), we obtain:

$$[R_{o,j}] = K_5(\tau^*, s^*)[R_f][M_j], \text{where } K_5(\tau^*, s^*) = \frac{k_5(\tau^*)}{k_{-5}(s^*)} \quad (17)$$

$$[R_{e,j}] = K_5(\tau^*, s^*)K_6(ATP)[R_f][M_j][t_c], \text{where } K_6(ATP) = \frac{k_6}{k_{-6} + k_7(ATP)} \quad (18)$$

$$[R_{s,j}] = K_5(\tau^*, s^*)K_8[R_f][M_j][t_u], \text{where } K_8 = \frac{k_8}{k_{-8}} \quad (19)$$

From the ribosome conservation law:

$$[R_f] = [R_{tot}] - \sum_j [R_{o,j}] - \sum_j [R_{e,j}] - \sum_j [R_{s,j}] \quad (20)$$

The protein synthesis rate of a type-j protein can then be defined as the peptide elongation rate divided by the amino acid length l_j of the protein:

$$J_{P_j,syn} = \frac{k_7(ATP)[R_{e,j}]}{l_j} \quad (21)$$

The utilization rate of charged tRNA during peptide elongation is:

$$J_{t_c,deg} = -\sum_j k_7(ATP)[R_{e,j}] \quad (22)$$

From Eqs. (18), (21), and (22), the results indicate that protein translation rates are regulated by τ^*, s^*, and ATP. These factors influence the equilibrium constants (K_5, K_6) as well as the rate constant (k_7). In addition, the concentration of $[M_j]$ for each protein type j reflects transcriptional output, which is also regulated by τ^*, s^* (*see* **Note 6**).

2.4.2 Differential Equations for Module Dynamics

We successfully derived the rates for two key processes: tRNA charging and its utilization during peptide elongation, and the overall protein synthesis rate. By focusing on the formation of the final products (charged tRNAs and proteins), we implicitly integrated all other reactions into the rate functions of these core processes. Based on these derived rates and considering growth-induced dilution (λ), we can now formulate the following differential equations to describe the dynamics of the final products within the protein expression module, where the λ stands for the growth rate:

$$\frac{d[t_c]}{dt} = J_{t_c,syn} - J_{t_c,deg} - \lambda[t_c] \quad (23)$$

$$\frac{d[P_j]}{dt} = J_{P_j,syn} - \lambda[P_j] \quad (24)$$

2.5 From Single Cell to Population Growth

By following the procedures outlined in **Steps 3** and **4**, each module can be expressed as a set of differential equations. To obtain a system-level view, these module equations are then integrated such that the output of one module serves as the input to another. For instance, in the protein expression module, the dynamics of different protein sectors are described by Eq. (24). The resulting protein sector concentrations can then feed into the metabolism module, where they regulate the corresponding reaction rates.

Up to this point, our discussion has focused on the single-cell level and its intracellular biochemical reactions. From an experimental perspective, however, cellular growth is typically monitored at the population level, such as through optical density measurements. This raises a key question: how can protein synthesis rate (biomass formation at the single-cell level) be connected to the observed population growth rate? To enable such comparisons, the single-cell framework must be extended to describe population-level dynamics. In this step, we demonstrate how to derive the overall population growth rate from biomass synthesis rate at the single-cell level.

First, we define the cellular biomass as:

$$M_{cell} = (a[P_{tot}])V_{cell} = const. \tag{25}$$

where biomass is assumed to be proportional to the total protein concentration $[P_{tot}]$ within the cell. In addition, the cellular biomass M_{cell}, the total protein concentration $[P_{tot}]$, and the cell volume V_{cell} are set as constants.

From Eqs. (24, 25), the rate of change of cellular biomass can be written as:

$$\frac{dM_{cell}}{dt} = aV_{cell}\frac{d[P_{tot}]}{dt} = aV_{cell}\left(\sum_j J_{P_j,syn} - \lambda\sum_j [P_j]\right) = 0 \tag{26}$$

Rearranging Eq. (26), the cellular growth rate is derived as:

$$\lambda = \frac{\sum_j J_{P_j,syn}}{\sum_j [P_j]} = \frac{\sum_j J_{P_j,syn}}{[P_{tot}]} \tag{27}$$

Thus, cellular growth is directly proportional to the total protein synthesis rate, while the total protein concentration $[P_{tot}]$ remains constant.

At the system level, the change in total biomass M_{all} can be written as:

$$\frac{dM_{all}}{dt} = \lambda \cdot M_{all} \tag{28}$$

where the system biomass is the product of cellular biomass and cell number N,

$$M_{all} = M_{cell} \cdot N \tag{29}$$

Taking the derivative on both side of eq. (29) gives:

$$\frac{dM_{all}}{dt} = M_{cell} \cdot \frac{dN}{dt} \tag{30}$$

By comparing Eqs. (28, 29) with Eq. (30), we arrive at the population growth equation (*see* **Note 7**):

$$\frac{dN}{dt} = \lambda N \tag{31}$$

3 Models and Extensions

3.1 Growth Laws: Steady-State and Dynamic Controls

With the essential network constructed, the coarse-grained model helps elucidate how different levels of regulation coordinate under varying conditions. By grouping proteins according to their functions, the model reveals how cells allocate resources to adapt to environmental cues. A central focus is the relationship between the abundance of ribosomal and non-ribosomal proteins and the cellular growth rate. Ribosomal protein concentration directly determines the overall protein synthesis rate, which in turn is proportional to the cellular growth rate. As described in Subheading 2.1.2, proteins can be grouped based on functional annotations, and in Subheading 2.4.1, we showed how ribosome abundance and ribosome states determine protein synthesis. Building on this, Subheading 2.5 demonstrated that protein synthesis is explicitly proportional to growth rate. Together, the coarse-grained model provides a framework to investigate trade-offs between ribosomal and enzymatic proteins (Fig. 3, left panel).

It is widely studied that cellular proteome allocation shifts in response to changes in growth rate, nutritional conditions, and stress. Under steady-state conditions, where a cell grows at a constant exponential rate, its proteome composition is carefully adapted to reach maximal growth. In *E. coli*, two empirical growth laws have been discovered, showing a linear relationship between the growth rate and ribosomal abundance under both varying nutritional conditions and antibiotic exposure [3]. Modeling analyses have further shown that this relationship emerges from a robust regulatory strategy that tunes fluxes to achieve optimal growth [15, 16]. Similarly, a linear growth law relating growth to ribosomal proteins has been discovered for *S. cerevisiae* under carbon-limited nutritional conditions [17]. Modeling results for *S. cerevisiae* have also confirmed that the system achieves maximal

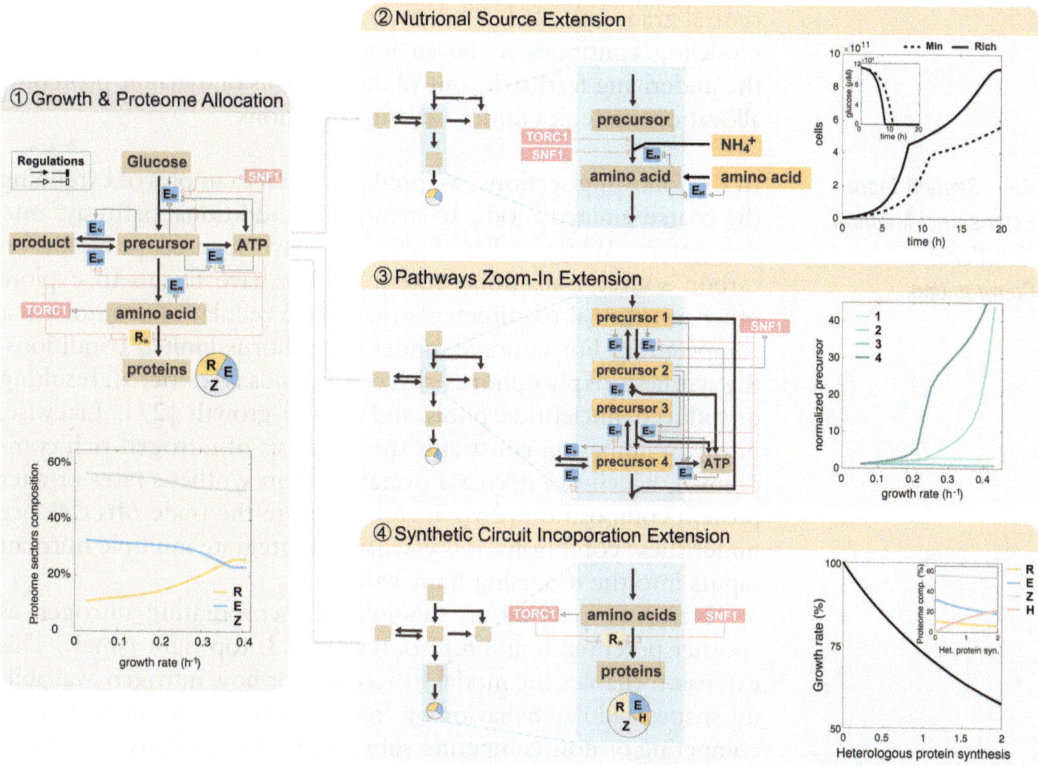

Fig. 3 Model applications and extensions. (i) The constructed coarse-grained modeling framework can be applied to investigate cellular trade-offs, such as proteome allocation across different growth rates and environmental conditions. In addition, the framework can be extended at selected breakpoints to provide deeper insights into specific pathways. For example, (ii) integrating essential nutrients other than glucose, such as ammonium and amino acids, allows the model to capture how medium composition influences cellular growth. (iii) Zooming into glycolysis enables the framework to distinguish metabolic breakpoints between the upper and lower glycolytic profiles and to identify key rate-limiting steps. (iv) Incorporating external gene expression helps quantify the metabolic burden imposed by heterologous circuits, improve understanding of host–circuit interactions, and guide strategies for mitigating these coupled effects

growth allocation, as perturbations to regulatory factors led to lower growth rate [18].

Moving beyond steady-state conditions, recent studies have begun to explore how microorganisms navigate fluctuating environments. Nutritional upshifts and downshifts are used to identify the dynamic allocation strategies of the proteome, which reveal how cells partition their resources in non-constant conditions [19–21]. Key phenomena during these transitions include the hierarchical utilization of carbon sources [22, 23] and the diauxic shifting events [24]. A common metric for evaluating a cell's adaptability is the duration of the lag phase during these transitions [24, 25]. A longer transition time indicates a lower adaptation ability, and researchers have identified that the allocation of protein resources for ribosomal synthesis or for stress responses becomes a central trade-off [26]. This is an area of ongoing research, and modeling continues to be an invaluable tool for understanding the underlying feedback control mechanisms of dynamic proteome allocation strategies under varying conditions.

3.2 System Input Extensions: Medium Nutritional Composition.

In the following sections, we present several examples of extending the coarse-grained model to incorporate additional pathways into the system. While earlier discussions have focused primarily on carbon source transitions, recent studies have begun to explore other nutritional requirements that shape cellular macromolecular composition. For example, under phosphorus-limited conditions, the synthesis of phosphorus-rich compounds is restricted, resulting in reduced nucleotide pools and slower growth [27]. Likewise, nitrogen limitation constrains the synthesis of nitrogen-rich compounds, which can decrease overall protein synthesis rates or alter protein composition [28]. To fully capture the trade-offs cell face under these conditions, it is essential to integrate multiple nutrient inputs into the modeling framework.

Here, we provide an example of incorporating nitrogen as another potential limiting resource (Fig. 3, top right panel). This extension enables the model to account for how nitrogen availability shapes cellular behavior by treating nitrogen as an additional competing or non-competing substrate in the reactions. For example, under nutrient-rich conditions, the flux from precursors to amino acids is often modeled without considering ammonium concentration. However, when ammonium becomes limiting, its concentration directly constrains amino acid synthesis rates [27]. To capture this effect, two modifications are required: introducing a differential equation to track ammonium concentration over time, and adding an ammonium-dependent term to the relevant Michaelis–Menten rate equations. By incorporating nitrogen-related entries in this way, the model can systematically explore the cellular behavioral landscape across varying carbon and nitrogen compositions in the culture medium [29].

In addition, to compare rich and minimal media, the model can be expanded to include amino acid uptake pathways from the culture, allowing cells to acquire amino acids directly from the environment rather than relying solely on de novo synthesis. As shown in Fig. 3 (top right panel), modeling results demonstrate that rich media lead to faster glucose uptake and greater biomass accumulation compared to minimal media. This modeling framework not only enables the dissection of proteome allocation but also allows for tracking changes in cellular composition and growth rates under different nutritional conditions [17, 24, 30]. Such insights provide valuable guidance for medium optimization and dynamic control in bioreactors [31].

3.3 Pathway Extensions: One Precursor to Four Precursors

Another area of study has focused on understanding how carbon flux is partitioned between assimilation for biomass formation and catabolism for energy production during environmental fluctuations. To better quantify this distribution, the coarse-grained model can be "zoomed in" to expand the one-step reaction from glucose to precursor into a more detailed, four-step pathway (Fig. 3, middle right panel). In this refinement, representative upper-glycolysis precursors (fructose-6-phosphate and fructose–1,6-bisphosphate) and lower-glycolysis precursors (phosphoenolpyruvate and pyruvate) are selected, and the proteome sectors in the original one-step model are subdivided to govern distinct sets of reactions [24]. In addition, regulatory effects on protein activities and the signaling roles of these precursor metabolites are re-analyzed. The newly constructed essential network can then be described by a corresponding set of equations, following the same steps outlined in Subheading 2.

This approach quantitatively identifies rate-limiting steps and clarifies how key carbon-based precursors are distributed toward amino acid, ATP, and other product synthesis [24]. As shown in Fig. 3, the significant concentration shifts of precursors 2 and 4 across different growth rates indicate their sensitivity for growth-related control. Quantifying these precursors pinpoints the most limited precursor pools and rate-limiting proteome sectors, which enhance the understanding of cellular resource allocation strategies [25]. Furthermore, flux distributions can be validated by externally supplementing precursor inputs, either through isotope tracing [30] or intermediate supplementation [32]. These analyses can ultimately guide enzyme overexpression strategies to overcome bottlenecks and improve bioproduction [33].

3.4 Module Extensions: Synthetic Circuits Incorporation

Although simulations can help us understand and optimize native cellular investment strategies, introducing external gene constructs often disrupts normal fluxes and produces unexpected results [34, 35]. One major challenge is the host–circuit coupling effect, where the expression of integrated genes is constrained by the limited capacity of the host's transcriptional and translational machinery [36, 37].

Coarse-grained modeling helps analyze how heterologous gene expression affects cells. In this framework, translation of external genes can be represented as the formation of a new protein sector, which diverts amino acid resources away from native protein synthesis (Fig. 3, lower right panel) [38, 39]. As the rate of additional protein synthesis increases, the fractions of ribosomal and enzymatic proteins decline, leading to reduced glycolysis and slower overall growth. This phenomenon, known as metabolic burden [40], also impacts synthetic constructs, which result in unexpected behaviors such as bistability or oscillations [38, 41]. Current research efforts focus on developing context-aware feedback controllers to mitigate these undesirable outcomes and improve the robustness of synthetic designs [42, 43].

4 Notes

1. Representative precursors are often selected as pyruvate or acetyl-CoA, since they function as key metabolic intermediates and central branching points in cellular metabolism. As the end product of glycolysis, pyruvate serves as a hub that connects multiple pathways. In the mitochondria, pyruvate can be converted to acetyl-CoA to fuel the TCA cycle, generating ATP and providing precursors such as oxaloacetate and α-ketoglutarate for amino acid biosynthesis. In the cytosol, pyruvate can also be directed toward fermentation, producing compounds such as ethanol and acetate. Acetyl-CoA itself represents another crucial branching point. In mitochondria, it supports the TCA cycle and amino acid synthesis, while in the cytosol it provides the building blocks for lipid and acetate production.
2. Glutamine is selected as a representative amino acid to reflect growth-limiting conditions. Glutamine concentration has been shown to closely correlate with cellular growth rate under nitrogen limitation in both *E. coli* and *S. cerevisiae*, where its synthesis via glutamine synthetase is tightly regulated by external ammonia availability [8, 44]. Moreover, under nitrogen limitation, translation elongation is slowed at glutamine codons [45].
3. The overall protein composition of a cell is typically assumed to be constant; therefore, a higher protein accumulation rate reflects a faster cellular growth rate (see Subheading 2.5 for detailed derivations).
4. GO Slim is a curated subset of the Gene Ontology (GO) that provides a simplified view by grouping detailed terms into broader, high-level categories. These terms are drawn from the three main GO domains: Biological Process, Molecular Function, and Cellular Component. GO Slim terms are selected to represent major branches of the ontology, allowing researchers to summarize large gene annotation sets. For example, the *go_slim_mapping.tab* file from the Saccharomyces Genome Database provides a widely used mapping for yeasts (http://sgd-archive.yeastgenome.org/curation/literature/) [46].

5. Based on the conservation of tRNA, the relationship between uncharged tRNA (t_u) and charged tRNA (t_c) can be written as:

$$[t_u] = [t_{tot}] - [t_c] \quad (32)$$

where $[t_{tot}]$ stands for the total concentration of tRNA and is assumed to be proportional to the concentration of ribosomal components.

6. In Eqs. (17–22), protein translation rates are expressed as functions of τ^*, s^*, and ATP. The equilibrium constants (K_5, K_6) and the rate constant (k_7) are regulated parameters, denoted as $(\cdot)$, and can be represented explicitly with Hill functions. Similarly, $[M_j]$, the mRNA concentration for protein sector j, reflects transcriptional expression and is also regulated by τ^* and s^*. Thus, $[M_j]$ should be seen as $[M_j(\tau^*, s^*)]$, with its transcriptional regulation also described by Hill functions.
7. Note that λ is not a constant; it is a time-dependent variable that changes with the protein translation rate. This reflects the fact that cellular growth is dynamically coupled to intracellular processes. When λ is assumed constant, cellular population growth can be approximated by the classical exponential form:

$$N(t) = N_0 e^{\lambda t} \quad (33)$$

This formulation has been used in microbial physiology to describe exponential growth under balanced conditions (e.g., in batch culture during the exponential phase). In our framework, however, λ varies over time, as it is determined by protein synthesis dynamics.

Acknowledgments

This work was funded by the DOE Center for Advanced Bioenergy and Bioproducts Innovation (U.S. Department of Energy, Office of Science, Biological and Environmental Research Program under Award Number DE-SC0018420) and the National Institutes of Health (GM133579). Any opinions, findings, and conclusions or recommendations expressed in this publication are those of the authors and do not necessarily reflect the views of the U.-S. Department of Energy.

References

1. Milo R, Jorgensen P, Moran U, Weber G, Springer M (2010) BioNumbers-the database of key numbers in molecular and cell biology. Nucleic Acids Res 38:D750–D753. https://doi.org/10.1093/nar/gkp889
2. Liebermeister W, Noor E, Flamholz A, Davidi D, Bernhardt J, Milo R (2014) Visual account of protein investment in cellular functions. Proc Natl Acad Sci USA 111(23): 8488–8493. https://doi.org/10.1073/pnas.1314810111
3. Scott M, Gunderson CW, Mateescu EM, Zhang ZG, Hwa T (2010) Interdependence of cell growth and gene expression: origins and consequences. Science 330(6007): 1099–1102. https://doi.org/10.1126/science.1192588
4. Chubukov V, Gerosa L, Kochanowski K, Sauer U (2014) Coordination of microbial metabolism. Nat Rev Microbiol 12(5):327–340. https://doi.org/10.1038/nrmicro3238
5. Murphy JP, Stepanova E, Everley RA, Paulo JA, Gygi SP (2015) Comprehensive temporal protein dynamics during the diauxic shift in. Mol Cell Proteomics 14(9):2454–2465. https://doi.org/10.1074/mcp.M114.045849
6. Zaman S, Lippman SI, Schneper L, Slonim N, Broach JR (2009) Glucose regulates transcription in yeast through a network of signaling pathways (vol 5, pg 245, 2009). Mol Syst Biol 5:257. https://doi.org/10.1038/msb.2009.20
7. González A, Hall MN (2017) Nutrient sensing and TOR signaling in yeast and mammals. EMBO J 36(4):397–408. https://doi.org/10.15252/embj.201696010
8. Ukai H, Araki Y, Kira S, Oikawa Y, May AI, Noda T (2018) Gtr/ego-independent TORC1 activation is achieved through a glutamine-sensitive interaction with Pib2 on the vacuolar membrane. PLoS Genet 14(4):e1007334. https://doi.org/10.1371/journal.pgen.1007334
9. Tripodi F, Nicastro R, Reghellin V, Coccetti P (2015) post-translational modifications on yeast carbon metabolism: regulatory mechanisms beyond transcriptional control. Biochim Biophys Acta (BBA) 1850(4):620–627. https://doi.org/10.1016/j.bbagen.2014.12.010
10. Schüller HJ (2003) Transcriptional control of nonfermentative metabolism in the yeast. Curr Genet 43(3):139–160. https://doi.org/10.1007/s00294-003-0381-8
11. Janapala Y, Preiss T, Shirokikh NE (2019) Control of translation at the initiation phase during glucose starvation in yeast. Int J Mol Sci 20(16):4043. https://doi.org/10.3390/ijms20164043
12. Cherkasova VA, Hinnebusch AG (2003) Translational control by TOR and TAP42 through dephosphorylation of eIF2α kinase GCN2. Genes Dev 17(7):859–872. https://doi.org/10.1101/gad.1069003
13. Milanesi R, Tripodi F, Vertemara J, Tisi R, Coccetti P (2021) AMPK phosphorylation is controlled by glucose transport rate in a PKA-independent manner. Int J Mol Sci 22(17):9483. https://doi.org/10.3390/ijms22179483
14. Urban J, Soulard A, Huber A, Lippman S, Mukhopadhyay D, Deloche O, Wanke V, Anrather D, Ammerer G, Riezman H, Broach JR, De Virgilio C, Hall MN, Loewith R (2007) Sch9 is a major target of TORC1 in. Mol Cell 26(5):663–674. https://doi.org/10.1016/j.molcel.2007.04.020
15. Scott M, Klumpp S, Mateescu EM, Hwa T (2014) Emergence of robust growth laws from optimal regulation of ribosome synthesis. Mol Syst Biol 10(8). https://doi.org/10.15252/msb.20145379
16. Bosdriesz E, Molenaar D, Teusink B, Bruggeman FJ (2015) How fast-growing bacteria robustly tune their ribosome concentration to approximate growth-rate maximization. FEBS J 282(10):2029–2044. https://doi.org/10.1111/febs.13258
17. Björkeroth J, Campbell K, Malina C, Yu R, Di Bartolomeo F, Nielsen J (2020) Proteome reallocation from amino acid biosynthesis to ribosomes enables yeast to grow faster in rich media. Proc Natl Acad Sci USA 117(35): 21804–21812. https://doi.org/10.1073/pnas.1921890117
18. Nguyen V, Xue P, Li YF, Zhao HM, Lu T (2023) Controlling circuitry underlies the growth optimization of Saccharomyces cerevisiae. Metab Eng 80:173–183. https://doi.org/10.1016/j.ymben.2023.09.013
19. Liao C, Priyanka P, Lai YH, Rao CV, Lu T (2024) How does allocate proteome? ACS Synth Biol 13(9):2718–2732. https://doi.org/10.1021/acssynbio.3c00537
20. Kohanim YK, Levi D, Jona G, Towbin BD, Bren A, Alon U (2018) A bacterial growth law out of steady state. Cell Rep 23(10): 2891–2900. https://doi.org/10.1016/j.celrep.2018.05.007

21. Mori M, Schink S, Erickson DW, Gerland U, Hwa T (2017) Quantifying the benefit of a proteome reserve in fluctuating environments. Nat Commun 8:1225. https://doi.org/10.1038/s41467-017-01242-8
22. Erickson DW, Schink SJ, Patsalo V, Williamson JR, Gerland U, Hwa T (2017) A global resource allocation strategy governs growth transition kinetics of. Nature 551(7678):119. https://doi.org/10.1038/nature24299
23. Okano H, Hermsen R, Kochanowski K, Hwa T (2020) Regulation underlying hierarchical and simultaneous utilization of carbon substrates by flux sensors in. Nat Microbiol 5(1):206. https://doi.org/10.1038/s41564-019-0610-7
24. Nguyen V, Li YF, Lu T (2024) Emergence of orchestrated and dynamic metabolism of Saccharomyces cerevisiae. ACS Synth Biol 13(5): 1442–1453. https://doi.org/10.1021/acssynbio.3c00542
25. Basan M, Honda T, Christodoulou D, Hörl M, Chang YF, Leoncini E, Mukherjee A, Okano H, Taylor BR, Silverman JM, Sanchez C, Williamson JR, Paulsson J, Hwa T, Sauer U (2020) A universal trade-off between growth and lag in fluctuating environments. Nature 584(7821):470. https://doi.org/10.1038/s41586-020-2505-4
26. Zhu ML, Wang Q, Mu HY, Han F, Wang YL, Dai XF (2023) A fitness trade-off between growth and survival governed by Spo0A-mediated proteome allocation constraints in. Sci Adv 9(39):eadg9733. https://doi.org/10.1126/sciadv.adg9733
27. Boer VM, Crutchfield CA, Bradley PH, Botstein D, Rabinowitz JD (2010) Growth-limiting intracellular metabolites in yeast growing under diverse nutrient limitations. Mol Biol Cell 21(1):198–211. https://doi.org/10.1091/mbc.E09-07-0597
28. Hui S, Silverman JM, Chen SS, Erickson DW, Basan M, Wang JL, Hwa T, Williamson JR (2015) Quantitative proteomic analysis reveals a simple strategy of global resource allocation in bacteria. Mol Syst Biol 11(2). https://doi.org/10.15252/msb.20145697
29. Li YF, Lai YH, Lu T (2025) Coarse-grained modeling elucidates differential metabolism of Saccharomyces cerevisiae under varied nutrient limitations. ACS Synth Biol 14(5):1523–1532. https://doi.org/10.1021/acssynbio.4c00803
30. Jalihal AP, Kraikivski P, Murali TM, Tyson JJ (2021) Modeling and analysis of the macronutrient signaling network in budding yeast. Mol Biol Cell 32(21):ar20. https://doi.org/10.1091/mbc.E20-02-0117
31. Xia JY, Sánchez BJ, Chen Y, Campbell K, Kasvandik S, Nielsen J (2022) Proteome allocations change linearly with the specific growth rate of under glucose limitation. Nat Commun 13(1):2819. https://doi.org/10.1038/s41467-022-30513-2
32. You CH, Okano H, Hui S, Zhang ZG, Kim M, Gunderson CW, Wang YP, Lenz P, Yan DL, Hwa T (2013) Coordination of bacterial proteome with metabolism by cyclic AMP signalling. Nature 500(7462):301–306. https://doi.org/10.1038/nature12446
33. Boecker S, Slaviero G, Schramm T, Szymanski W, Steuer R, Link H, Klamt S (2021) Deciphering the physiological response of under high ATP demand. Mol Syst Biol 17(12):e10504. https://doi.org/10.15252/msb.202110504
34. Zhang R, Goetz H, Melendez-Alvarez J, Li J, Ding T, Wang X, Tian XJ (2021) Winner-takes-all resource competition redirects cascading cell fate transitions. Nat Commun 12(1): 853. https://doi.org/10.1038/s41467-021-21125-3
35. Gyorgy A, Jimenez JI, Yazbek J, Huang HH, Chung H, Weiss R, Del Vecchio D (2015) Isocost lines describe the cellular economy of genetic circuits. Biophys J 109(3):639–646. https://doi.org/10.1016/j.bpj.2015.06.034
36. Gorochowski TE, Aycilar-Kucukgoze I, Bovenberg RAL, Roubos JA, Ignatova Z (2016) A minimal model of ribosome allocation dynamics captures trade-offs in expression between endogenous and synthetic genes. ACS Synth Biol 5(7):710–720. https://doi.org/10.1021/acssynbio.6b00040
37. Klumpp S, Zhang Z, Hwa T (2009) Growth rate-dependent global effects on gene expression in bacteria. Cell 139(7):1366–1375. https://doi.org/10.1016/j.cell.2009.12.001
38. Liao C, Blanchard AE, Lu T (2017) An integrative circuit-host modelling framework for predicting synthetic gene network behaviours. Nat Microbiol 2(12):1658–1666. https://doi.org/10.1038/s41564-017-0022-5
39. Weisse AY, Oyarzun DA, Danos V, Swain PS (2015) Mechanistic links between cellular trade-offs, gene expression, and growth. Proc Natl Acad Sci USA 112(9):E1038–E1047. https://doi.org/10.1073/pnas.1416533112
40. Borujeni AE, Zhang J, Doosthosseini H, Nielsen AAK, Voigt CA (2020) Genetic circuit characterization by inferring RNA polymerase movement and ribosome usage. Nat Commun 11(1):5001. https://doi.org/10.1038/s41467-020-18630-2

41. Melendez-Alvarez J, He CH, Zhang R, Kuang Y, Tian XJ (2021) Emergent damped oscillation induced by nutrient-modulating growth feedback. ACS Synth Biol 10(5): 1227–1236. https://doi.org/10.1021/acssynbio.1c00041
42. Stone A, Youssef A, Rijal S, Zhang R, Tian XJ (2024) Context-dependent redesign of robust synthetic gene circuits. Trends Biotechnol 42(7):895–909. https://doi.org/10.1016/j.tibtech.2024.01.003
43. Sechkar K, Steel H, Perrino G, Stan GB (2024) A coarse-grained bacterial cell model for resource-aware analysis and design of synthetic gene circuits. Nat Commun 15(1):1981. https://doi.org/10.1038/s41467-024-46410-9
44. Doucette CD, Schwab DJ, Wingreen NS, Rabinowitz JD (2011) α-ketoglutarate coordinates carbon and nitrogen utilization via enzyme I inhibition. Nat Chem Biol 7(12):894–901. https://doi.org/10.1038/Nchembio.685
45. Li SHJ, Li ZY, Park JO, King CG, Rabinowitz JD, Wingreen NS, Gitai Z (2018) Translation strategies differ across carbon, nitrogen and phosphorus limitation conditions. Nat Microbiol 3(8):939–947. https://doi.org/10.1038/s41564-018-0199-2
46. Engel SR, Aleksander S, Nash RS, Wong ED, Weng S, Miyasato SR, Sherlock G, Cherry JM (2024) Saccharomyces genome database: advances in genome annotation, expanded biochemical pathways, and other key enhancements. Genetics 229(3). https://doi.org/10.1093/genetics/iyae185

Chapter 18

Gene Expression Across Growth Stages (GEAGS): Modeling Circuit Dynamics in Batch Cultures

Chelsea Y. Hu and Hari R. Namboothiri

Abstract

The dynamics of recombinant gene expression in bacterial systems are often modeled under the assumption of constant growth. However, most synthetic gene circuit assays are performed in enclosed batch cultures, where nutrient depletion and physiological transitions from exponential to stationary phase substantially influence transcription, translation, mRNA degradation, and protein dilution. This chapter introduces the Gene Expression Across Growth Phases (GEAGS) framework—a dual-scale modeling approach that couples molecular reaction networks with logistic population growth through growth-dependent rate modifier functions (RMFs). The framework enables accurate simulation of gene expression dynamics under time-varying growth rate and resource competition. We provide step-by-step methods to construct GEAGS models, starting from a minimal constitutive reporter system and extending to complex applications involving optogenetic regulation and layered feedback control. This protocol establishes a generalizable computational workflow for synthetic biologists aiming to model the effects of growth, molecular crowding, and resource limitation on gene expression dynamics in batch culture.

Key words Dynamic modeling, Growth-aware model, Feedback control, Batch culture

1 Introduction

Gene expression in living cells is tightly coupled to cellular physiology and environmental conditions. In bacterial systems, population growth dynamics imposes time-varying constraints on transcription, translation, mRNA degradation, and protein dilution by altering metabolic activity and resource allocation [1]. Conventional ordinary differential equation (ODE) models of gene circuits often assume steady-state growth and constant rate parameters, which limits their predictive power in batch cultures where growth rate varies substantially over time.

To bridge this gap, the Gene Expression Across Growth Stages (GEAGS) framework extends minimal gene expression models by explicitly coupling molecular kinetics with population growth dynamics. The approach introduces *rate modifier functions*

Xiao-Jun Tian (ed.), *Synthetic Gene Circuits: Methods and Protocols*, Methods in Molecular Biology, vol. 3041,
https://doi.org/10.1007/978-1-0716-5304-3_18,

(*RMFs*) as empirical scaling terms that map the logistic growth state of the culture to key molecular processes, such as transcription, translation, folding, and degradation. These RMFs allow a single model to represent different physiological regimes (lag, exponential, and stationary) without requiring separate parameter sets.

The GEAGS framework can be implemented modularly in BioCRNpyler [2], a Python tool for programmatic construction of chemical-reaction-network (CRN) models and simulated using Bioscrape [3] for deterministic or stochastic analyses. This integration provides an extensible workflow for exploring how growth-phase dynamics influence gene expression and how resource competition and protein crowding shapes circuit behavior.

This chapter provides practical guidance for building GEAGS models of increasing complexity. Beginning with a minimal transcription-translation model, readers will learn how to (1) couple molecular-level kinetics to logistic growth, (2) apply RMFs to growth-dependent processes, (3) decompose translation into explicit resource-binding reactions, and (4) extend the model to optogenetic and layered-feedback systems. The protocol concludes with example simulations demonstrating how the GEAGS framework accurately reproduces experimental gene-expression trajectories in batch cultures.

2 Methods

2.1 Modeling Gene Expression Dynamics

In synthetic biology, systems design frequently uses simplified, minimal models that represent the coarse-grained dynamics of gene expression using lumped rate equations. A common example is shown here (Eqs. 1 and 2), where the transcription–translation of a single gene expression cassette (GEC) with a constitutively expressed gene P coding for a protein is modeled using two ODEs and two state variables, namely, mRNA (M) and protein (P) (Fig. 1). Here, P is the observable output, typically a fluorescent protein [4].

$$\frac{dM}{dt} = \beta_m - d_m \cdot M \tag{1}$$

$$\frac{dP}{dt} = k_{tl} \cdot M - d_p \cdot P \tag{2}$$

Here, β_m and k_{tl} denote the rates of transcription and translation, respectively, while d_m represents the degradation rate of mRNA. The parameter d_p accounts for the combined effects of protein degradation (d_{deg}) and dilution (d_{dil}) due to cell growth, such that $d_p = d_{deg} + d_{dil}$. All gene expression models discussed in this chapter build on this fundamental two-equation framework. Any extensions to this system of ODEs depend on the specific synthetic

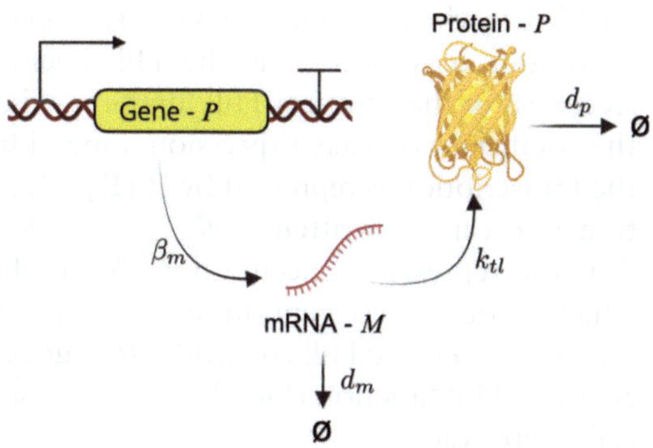

Fig. 1 Schematic representation of the two-equation dynamic model for gene expression

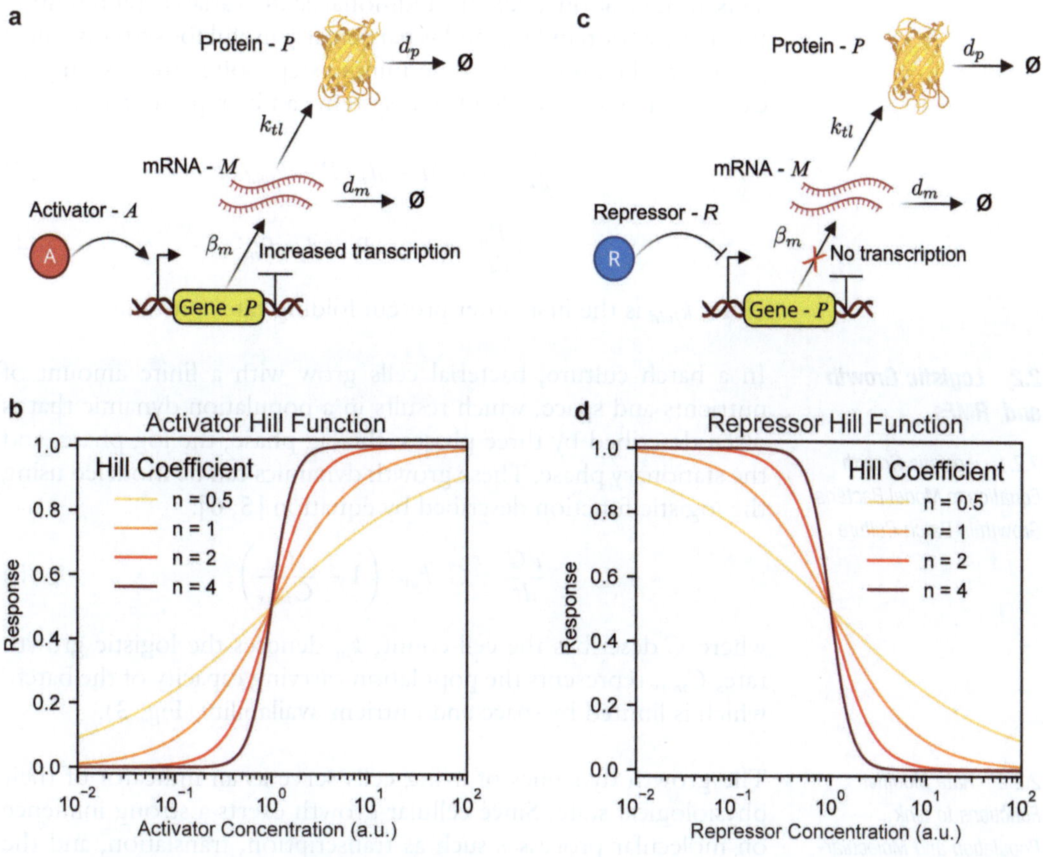

Fig. 2 Activator and repressor Hill functions

gene circuit being modeled. For instance, if the transcription is activated by A (Fig. 2a), the dynamics of the inducible promoter can be described using an activator Hill function. In this case, the transcription rate can be written as $\beta'_{m,activator} = \beta_m \cdot \left(\frac{A^n}{A^n + K_A^n} + l_0\right)$, where A is the activator concentration, K_A is the activation coefficient which is equal to concentration of A required for 50% of maximum activation, n is the Hill coefficient that governs the steepness of the activator Hill function (Fig. 2b), and l_0 represents the coefficient of leaky expression, ranged from 0 to 1. Similarly, if the transcription is repressed by R (Fig. 2c), the effective transcription rate can be written as $\beta'_{m,repressor} = \beta_m \cdot \left(\frac{K_R^n}{R^n + K_R^n} + l_0\right)$, where R is the repressor concentration, K_R is the activation coefficient which is equal to concentration of R required for 50% of maximum repression, n is the Hill coefficient that governs the steepness of the activator Hill function (Fig. 2d), and l_0 represents the coefficient of leaky expression.

Additionally, a simple yet biologically relevant extension of the minimal model involves introducing a protein maturation step, which captures folding and maturation as a first-order process. This modification adds an additional state variable representing the mature protein (P_m) and is particularly useful for systems where protein folding is the rate-determining step—often the case in gene expression. The extended ODE system can be represented as:

$$\frac{dP}{dt} = k_{tl} \cdot M - d_p \cdot P - k_{fold} \cdot P \tag{3}$$

$$\frac{dP_m}{dt} = k_{fold} \cdot P - d_p \cdot P_m \tag{4}$$

Here, k_{fold} is the first-order protein folding rate constant.

2.2 Logistic Growth and RMFs

2.2.1 Logistic Growth Equation to Model Bacterial Growth in Batch Culture

In a batch culture, bacterial cells grow with a finite amount of nutrients and space, which results in a population dynamic that is often described by three phases: the lag phase, the log phase, and the stationary phase. These growth dynamics can be modeled using the logistic function described by equation [5, 6]:

$$\frac{dC}{dt} = C \cdot k_{gr} \cdot \left(1 - \frac{C}{C_{max}}\right), \tag{5}$$

where C describes the cell count, k_{gr} denotes the logistic growth rate, C_{max} represents the population carrying capacity of the batch, which is limited by space and nutrient availability (Fig. 3).

2.2.2 Rate Modifier Functions to Link Population and Molecular-Scale Dynamics

The growth dynamics of living cells serve as an indicator of their physiological state. Since cellular growth exerts a strong influence on molecular processes such as transcription, translation, and the degradation of mRNA and proteins [1], the dynamics of gene expression at the molecular scale are inherently intertwined with growth dynamics at the population scale. The molecular-scale dynamics can be coupled with cell growth, using the rate modifier functions (RMFs) [7]. These functions establish empirical relationships between growth-dependent reactions in the CRN and the bacterial growth phases. RMFs estimate the growth phase of the population using the following equation:

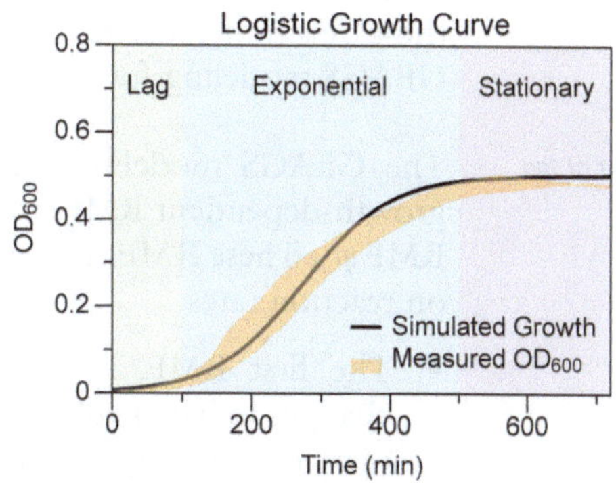

Fig. 3 Population dynamics in batch culture modeled using the logistic growth equation. The shaded background indicates the lag, log, and stationary phases. The experimentally measured OD_{600} is shown using the orange shaded region, representing mean ± standard deviation across replicates; data adapted from Namboothiri et al. [7]. The solid black line represents the simulated growth using the logistic growth equation

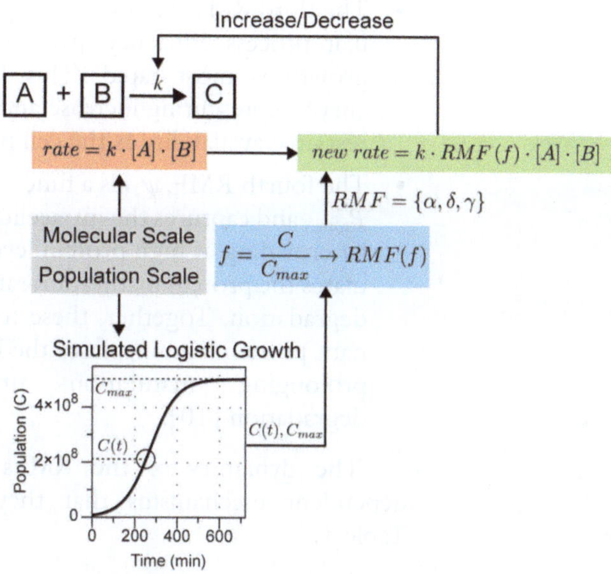

Fig. 4 Schematic showing how the growth-dependent RMFs scale the molecular-scale reactions

$$f = \frac{C}{C_{max}}, \tag{6}$$

where C represents the current population size, and C_{max} is the population carrying capacity of the batch culture. The value of f, bound by 0 and 1, indicates the proximity of the population to its carrying capacity. The RMFs are functions of f and are used to scale the rate constants of elementary reactions, dynamically modifying rates based on the growth phase of the population (Fig. 4). Linking these reaction rates to growth-dynamics forms the basis of the GEAGS modeling framework.

2.2.3 Description of the RMFs and Their Applications

The GEAGS modeling framework incorporates 4 RMFs: three growth-dependent RMFs (α, γ, δ) and one non-growth-dependent RMF ψ_P. These RMFs represents a distinct effect of cellular context on reaction rates.

- The first RMF, α, quantifies how the logistic growth rate changes in batch culture as the bacterial population approaches carrying capacity. This RMF decreases monotonically with growth and modulates the first-order dilution rates of molecular species.
- The second RMF, δ, quantifies the impact of growth-phase transitions on key intracellular processes, increaseing sharply upon entry into stationary phase. Described by a Hill function, δ increases monotonically in a sigmoidal manner, capturing the upregulation of molecular rates during this transition.
- The third RMF, γ, models the growth phase-dependent molecular process efficiency, peaking during mid-log phase when growth is most rapid. This RMF resembles a bell-shaped curve, representing increase and decrease in the process rate or resource availability as the cell population grows.
- The fourth RMF, ψ_P, is a function of total recombinant protein P_{total}, and captures the intracellular effects arising from molecular crowding at high protein levels [8, 9]. Specifically, ψ_P modulates the protein maturation rate, while $(1 - \psi_P)$ scales protein degradation. Together, these terms reflect that high recombinant protein load increases the likelihood of misfolding [8, 9], prolonging maturation times [9] and enhancing degradation [10].

The dynamics of the RMFs and the respective growth-dependent mechanisms that they modify are summarized in Table 1.

2.3 Application Example 1: Building a minimal GEAGS Model

In this section, we extend the three-equation model described by Eqs. 7, 8, and 9 to a minimal four-equation GEAGS model, demonstrating how to build a simple growth-dependent model. We use BioCRNpyler [2], a Python-based CRN construction tool, to model the system in a modular and extensible way. The complete BioCRNpyler implementation can be found in the accompanying GitHub repository at: https://github.com/synbiosystems/Minimal-GEAGS-model-in-BioCRNpyler.git

In BioCRNpyler, the model and its associated RMFs are encoded in a composable structure. The GEAGS framework is constructed by defining system components (see the accompanying GitHub repository), specifying their interaction mechanisms, and

Table 1
Table summarizing the different RMFs, their dynamics, and which growth-dependent mechanisms are modified by them

RMF	RMF dynamics	Mechanisms modulated by RMF
$\alpha = 1 - f$	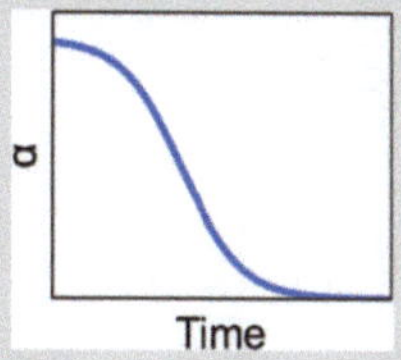	Modifies the first-order dilution (d_{dil}) of molecular species and the coarse-grained mRNA degradation rate
$\delta = \frac{f^{n_\delta}}{1+f^{n_\delta}}$ Here, n_δ is the Hill coefficient that governs the steepness of δ	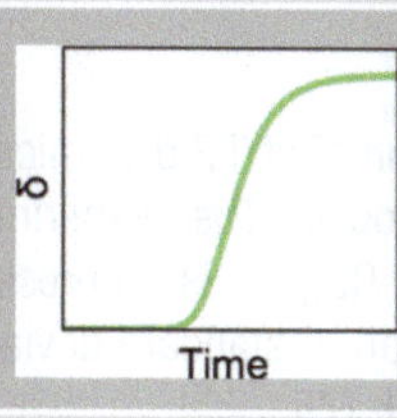	Modifies the endogenous protein degradation; if mRNA dynamics is critical, δ also modifies the dissociation rates of open-promoter and nuclease-mRNA complexes
$\gamma = (f \cdot (1 - f))^m$ Here, m modulates the fold change in γ	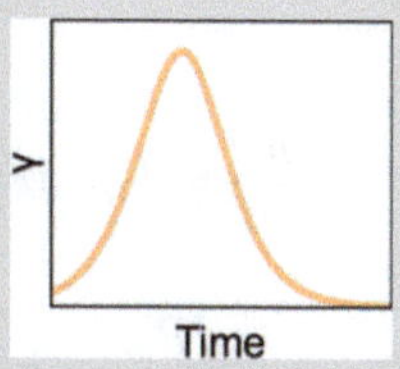	Modifies the rates that peak in mid-log phase, including transcription, translation, protein maturation, amino acid biosynthesis, and degron-mediated protein degradation. In a CRN formulation, γ scales the abundances of resource species involved in gene expression
$\psi_P = \frac{K_P^{n_\psi}}{K_P^{n_\psi} + P_{total}^{n_\psi}}$ Here, n_ψ is the Hill coefficient that governs the steepness of ψ_P. P_{total} is the total recombinant protein in the system	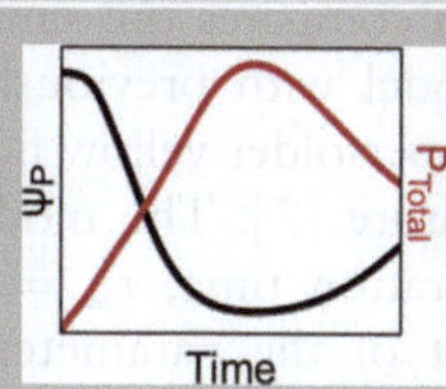	Captures the effects of molecular crowding when recombinant proteins are overexpressed. The RMF ψ_p modifies the maturation rate and the RMF $1 - \psi_p$ modifies the endogenous protein degradation

assembling these into a mixture object that hosts all components and provides shared resources. Growth phase-dependent modeling is achieved by extending BioCRNpyler's internal libraries with additional mechanisms, introducing RMFs through custom reaction propensities, and defining the associated parameters.

2.3.1 Constructing Minimal Single-Scale Model

Synthetic gene circuits are often characterized using fluorescent proteins, whose slow maturation can become a rate-determining step. Under these conditions, explicitly modeling the maturation process is critical, as described in Subheading 2.1 (Eqs. 3 and 4). Here, we present a minimal single-scale model, built from Eqs. 7, 8, and 9, represented by the following system of ODEs:

$$\frac{dM}{dt} = \beta_m - d_m \cdot M \tag{7}$$

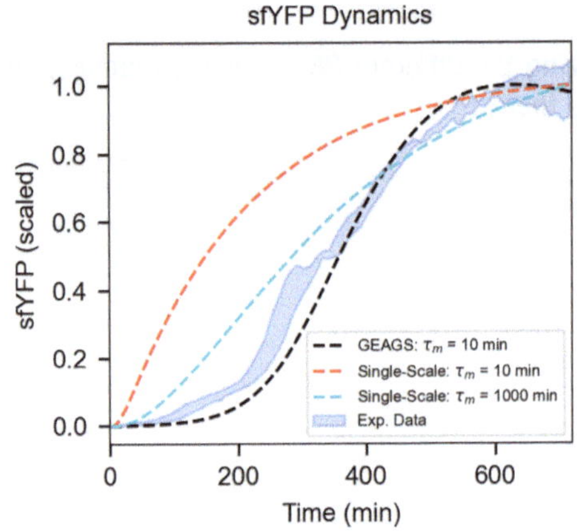

Fig. 5 Comparison of sfYFP dynamics prediction using the minimal single-scale and GEAGS model. The experimentally measured fluorescence signal, normalized by OD_{600}, is represented using the blue shaded region, representing mean ± standard deviation across replicates. (Data adapted from Namboothiri et al. [7])

$$\frac{dP}{dt} = k_{tl} \cdot M - d_p \cdot P - k_{fold} \cdot P \tag{8}$$

$$\frac{dP_m}{dt} = k_{fold} \cdot P - d_p \cdot P_m \tag{9}$$

Figure 5 compares the simulation of P_m from the single-scale minimal model with previously published data of constitutively expressed superfolder yellow fluorescent protein (sfYFP) dynamics in batch culture [7]. The model was simulated using the typical sfYFP maturation time, $\tau_m = 10$ min [11] ($k_{fold} = 0.1\ \text{min}^{-1}$) and the rest of the parameters were taken from the literature: $\beta_m = 3\ \text{nM} \cdot \text{min}^{-1}$ [12], $d_m = 0.15\ \text{min}^{-1}$ [13], $k_{tl} = 0.1\ \text{min}^{-1}$ [12] and d_p= 5e-5 min^{-1} [14]. Under these conditions, the model predicts a substantially faster initial rise in sfYFP fluorescence than observed experimentally, where sfYFP shows a prolonged early lag. To test whether maturation alone could account for this lag, we increased τ_m to 1000 min ($k_{fold} = 0.001\ \text{min}^{-1}$). Even with this unrealistically slow maturation rate for a fast-folding protein, the single-scale model still fails to reproduce the observed lag in sfYFP accumulation.

2.3.2 Constructing a minimal GEAGS Model

The discrepancy highlighted in Subheading 2.3.1 can be resolved using a minimal GEAGS model. This model is built by adding the logistic growth equation and applying the RMFs to the minimal single-scale model as described in Eqs. 7, 8, and 9. The steps to construct the model are outlined below:

1. Include the logistic growth equation (Eq. 5) in the system of ODEs (Eqs. 7, 8, and 9).
2. Since the mRNA (M), the unfolded protein (P), and the mature protein (P_m) are diluted by cell growth, add a dilution rate d_{dil} to their respective degradation terms.
3. Apply α to the dilution rate (d_{dil}). Additionally, apply α to the active mRNA degradation rate (d_m) to capture the increase in mRNA stability as the culture enters the stationary phase [15].
4. Apply γ to the transcription rate (β_m), translation rate (k_{tl}), and the protein folding rate (k_{fold}). These rates peak during the mid-log phase, due to the increased abundance of resources. Additionally, add b_{fold} to account for the basal folding rate as shown in Eq. 19.
5. Apply δ to the active protein degradation rate (d_p) to model the increase in endogenous proteolysis as the cells enter stationary phase, reflecting the physiological shift to utilize recycled amino acids as an adaptation strategy under nutrient-limited conditions [16].
6. The minimal GEAGS model ODEs are shown below:

$$\frac{dM}{dt} = \boldsymbol{\beta_m} - (\boldsymbol{d_m} + \boldsymbol{d_{dil}}) \cdot M \tag{10}$$

$$\frac{dP}{dt} = \boldsymbol{k_{tl}} \cdot M - \left(\boldsymbol{d_p} + \boldsymbol{d_{dil}}\right) \cdot P - \boldsymbol{k_{fold}} \cdot P \tag{11}$$

$$\frac{dP_m}{dt} = \boldsymbol{k_{fold}} \cdot P - \left(\boldsymbol{d_p} + \boldsymbol{d_{dil}}\right) \cdot P_m \tag{12}$$

$$\frac{dC}{dt} = k_{gr} \cdot C \cdot \left(1 - \frac{C}{C_{max}}\right) \tag{13}$$

The rates modified by RMFs are highlighted in bold.

7. The modified growth-dependent rate equations are:

$$\boldsymbol{\beta_m} = \beta_m \cdot \gamma \tag{14}$$

$$\boldsymbol{k_{tl}} = k_{tl} \cdot \gamma \tag{15}$$

$$\boldsymbol{d_m} = d_m \cdot \alpha \tag{16}$$

$$\boldsymbol{d_p} = d_p \cdot \delta \tag{17}$$

$$\boldsymbol{d_{dil}} = d_{dil} \cdot \alpha \tag{18}$$

$$\boldsymbol{k_{fold}} = k_{fold} \cdot (\gamma + b_{fold}) \tag{19}$$

The model was built using BioCRNpyler and simulated in Bioscrape. As shown in Fig. 5 the GEAGS model reproduces the dynamics observed experimentally [7], particularly capturing the characteristic slow initial rise of the sfYFP signal.

This framework readily extends to systems with varying complexities, as demonstrated in the following sections.

2.3.3 Constructing a minimal GEAGS Model in BioCRNpyler: A Step-by-Step Tutorial

1. Define Model Species and Parameters:
   Represent the three species, mRNA (M), protein (P), and the cell population (C), as *Species* object. Define the model parameters: transcription (β_m), degradation (d_m, d_p), translation (k_{tl}), and growth-related parameters (k_{gr}, C_{max}), as *ParameterEntry* objects to enable symbolic referencing within rate expressions.
2. Define Growth-Dependent RMFs:
   Define the RMFs as algebraic functions of the variable f as shown in Table 1. Write them as BioCRNpyler compatible string expressions, and use the *GeneralPropensity* function to specify custom growth-dependent propensities.
3. Reaction Network Formulation:
   Represent each biological process as a *Reaction* object, with an associated general propensity function that defines its rate law.
4. Network Assembly and Simulation:
   Combine all species and reactions into a *ChemicalReactionNetwork* object. This network can be automatically exported into an SBML format file and simulated using Bioscrape. Deterministic trajectories are obtained by specifying initial conditions and integrating over the desired time span.

2.4 Application Example 2: Optogenetic Control in Batch Culture

In this section, we apply the the GEAGS framework to model the dynamics of an optogenetically controlled two-component system in a bacterial batch culture. As shown in Fig. 6a, the CcaSR optogenetic system regulates the expression of a superfolder green fluorescent protein (sfGFP). The two-component system consists of CcaS, a membrane-bound histidine kinase that senses green and red light, and CcaR, the corresponding response regulator [17] (Fig. 6a). Upon green light exposure, CcaS autophosphorylates to form CcaSp, which transfers the phosphate to CcaR. The phosphorylated CcaR dimerizes to form the activation complex Ac that activates the PcpcG2 promoter and induces sfGFP transcription. Red light reverses this process by promoting CcaS dephosphorylation, thereby repressing expression. Figure 6a shows the schematic representation of all the mechanistic steps occurring in the optogenetic control system.

In the optogenetic control experiment, the cells are either exposed to red (repression) or green (activation) light continuously over a time of 16 h [18]. The dynamics of the system are shown in Fig. 6c. Readers are encouraged to refer to Namboothiri et al. [18] for the detailed system description and the python codes used to simulate the GEAGS model.

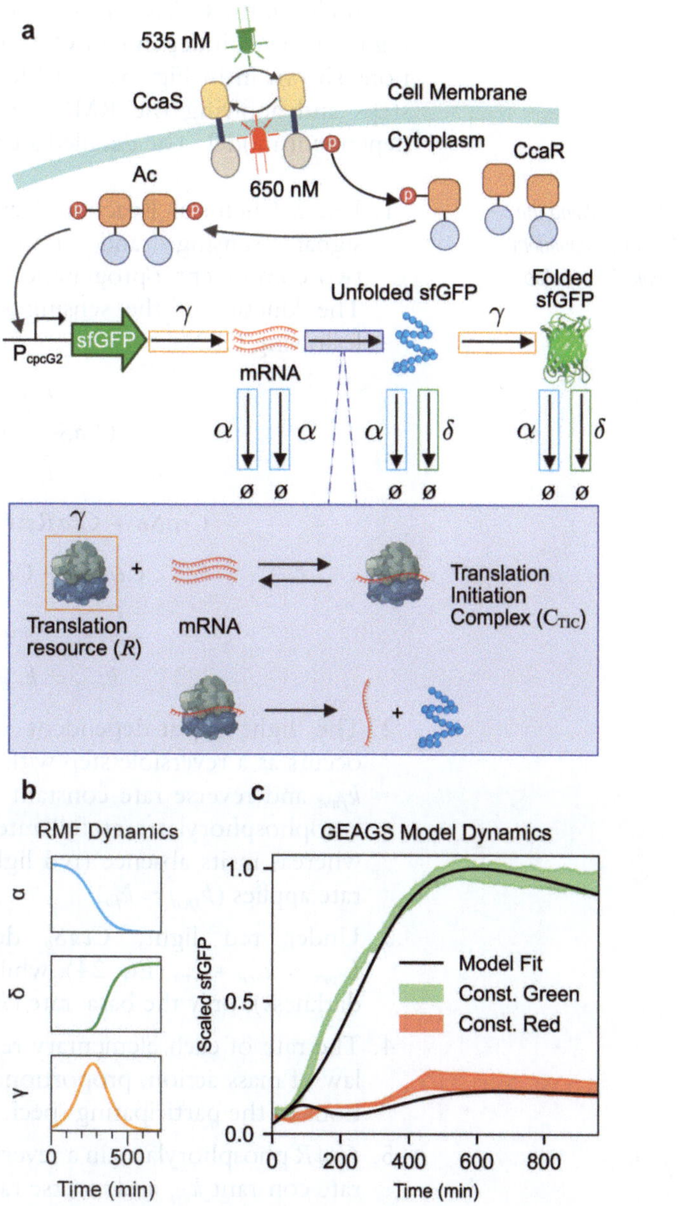

Fig. 6 Applying the GEAGS framework to model the CcaSR optogenetic system. (**a**) Schematic of the CcaSR two-component system. Reactions rates and molecular species modified by the RMFs are highlighted with colored boxes, corresponding to the RMFs defined in (**b**). (**b**) The temporal profiles of the three RMFs. (**c**) Comparison of model simulations and experimental data. Solid lines denote model simulations, while the shaded regions represent the experimental mean $\pm$ standard deviation across replicates. (Data adapted from Namboothiri et al. [18])

Building the GEAGS model involves two steps: (1) Constructing the growth-independent ODE system from the species interactions shown in in Fig. 6a. (2) Identifying the growth-dependent steps and applying the RMFs in Fig 6b to form the growth-dependent model. The detailed steps are shown below.

2.4.1 Building the Growth-Independent (Single-Scale) Model

1. Use a Chemical Reaction Network (CRN) to represent the signal sensing and transduction dynamics of the two-component optogenetic system (as shown in Fig. 6a). The kinetics of the sensing system are described with the following CRNs:

$$CcaS \underset{k_{rvrs}}{\overset{k_{fwd}}{\rightleftarrows}} CcaS_P \tag{20}$$

$$CcaS_P + CcaR \rightleftarrows CcaS + CcaR_P \tag{21}$$

$$CcaR_p + CcaR_p \rightleftarrows Ac \tag{22}$$

$$k_{fwd} = k_{green} + k_{fb} \tag{23}$$

$$k_{rvrs} = k_{red} + k_{rb} \tag{24}$$

2. The light input-dependent autophosphorylation of $CcaS$ occurs as a reversible step with forward reaction rate constant k_{fwd} and reverse rate constant k_{rvrs}. Under green light, $CcaS$ autophosphorylates at full rate $k_{fwd} = k_{green} + k_{fb}$ (Eq. 23), whereas in its absence (red light or darkness), only the basal rate applies ($k_{fwd} = k_{fb}$).
3. Under red light, $CcaS_p$ dephosphorylates at full rate $k_{rvrs} = k_{red} + k_{rb}$ (Eq. 24), while in its absence (green light or darkness), only the basal rate ($k_{rvrs} = k_{rb}$) is applied.
4. The rate of each elementary reaction in the CRN follows the law of mass action, proportional to the product of concentrations of the participating species.
5. $CcaR$ phosphorylates in a reversible step with forward reaction rate constant k_{R_b} and reverse rate constant k_{R_u}.
6. Phosphorylated $CcaR$ ($CcaR_p$) forms the activating complex dimer Ac reversibly with forward reaction rate constant k_{Rp_b} and reverse rate constant k_{Rp_u}.
7. The dynamics of the phosphorylation described in the CRNs (Eqs. 20, 21, and 22) are modeled using the ODEs (Eqs. 25, 26, 27, 28, and 29), representing the mass balances of the species.
8. The transcription rate of mRNA (M) is $\beta_m \cdot \left(\frac{Ac^n}{Ac^n+Kc^n} + l_0\right)$, where Ac is the concentration of activating complex, Kc is the activation coefficient, n is the Hill coefficient that governs the

steepness of the activator Hill function, and l_0 represents the coefficient of leaky expression.

9. The steps of mRNA degradation and sfGFP translation, maturation, and degradation are modeled as described in Subheading 2.1.
10. In the final set of ODEs, the first 5 ODEs (Eqs. 25, 26, 27, 28, and 29) describe the phosphorylation dynamics specified by the CRN and the remaining 3 ODEs (Eqs. 3, 4, and 30) describe the gene expression dynamics of sfGFP.
11. The optogenetic system dynamics can be represented by the following ODEs:

$$\frac{dCcaS}{dt} = -(k_{green} + k_{fb}) \cdot CcaS + (k_{red} + k_{rb}) \cdot CcaS_P + k_{R_b} \cdot CcaS_p \cdot CcaR - k_{R_u} \cdot CcaS \cdot CcaR_P \tag{25}$$

$$\frac{dCcaS_p}{dt} = (k_{green} + k_{fb}) \cdot CcaS - (k_{red} + k_{rb}) \cdot CcaS_P - k_{R_b} \cdot CcaS_p \cdot CcaR + k_{R_u} \cdot CcaS \cdot CcaR_P \tag{26}$$

$$\frac{dCcaR}{dt} = -k_{R_b} \cdot CcaS_p \cdot CcaR + k_{R_u} \cdot CcaS \cdot CcaR_P \tag{27}$$

$$\frac{dCcaR_p}{dt} = k_{R_b} \cdot CcaS_p \cdot CcaR - k_{R_u} \cdot CcaS \cdot CcaR_P - k_{Rp_b} \cdot CcaR_p^2 + k_{Rp_u} \cdot Ac \tag{28}$$

$$\frac{dAc}{dt} = k_{Rp_b} \cdot CcaR_p^2 - k_{Rp_u} \cdot Ac \tag{29}$$

$$\frac{dM}{dt} = \beta_m \cdot \left(\frac{Ac^n}{Kc^n + Ac^n} + l_0 \right) - d_m \cdot M \tag{30}$$

$$\frac{dP}{dt} = k_{tl} \cdot M - (d_p + k_{fold}) \cdot P \tag{31}$$

$$\frac{dP_m}{dt} = k_{fold} \cdot P - d_p \cdot P_m \tag{32}$$

2.4.2 Constructing the GEAGS Model

Gene translation is inherently a higher-order, growth-dependent process, yet it is often simplified as a first-order reaction dependent solely on mRNA concentration, and assumes translational resources to be abundant. In Example 1, we modified this rate to explicitly incorporate growth dependence. Here, we further decompose the model to account for resource competition. This refinement is essential because mRNA concentrations under green and red light differ substantially; as a result, translational rates could differ significantly, as the translational resources are not equally abundant under the two conditions. We therefore explicitly model the growth-dependent availability of a coarse-grained translational resource (R) within a CRN, and incorporate it into the GEAGS modeling framework.

To construct the GEAGS model, follow the steps:

1. Include the logistic growth equation (Eq. 5) in the system of ODEs.
2. Decompose the first-order translation rate into a CRN (Eqs. 33 and 34) as shown in Fig. 6a (violet shaded box). The mRNA binds with the freely available growth-dependent translational resource R (R_{free}) to form the translation initiation complex (C_{TIC}) with forward reaction rate constant $k_{tli\,b}$ and reverse rate constant $k_{tli\,u}$. C_{TIC} then undergoes translation elongation with a first-order rate constant of k_{tl} to form free mRNA (M) and unfolded sfGFP (P).

$$M + R_{free} \underset{k_{tli_u}}{\overset{k_{tli_b}}{\rightleftarrows}} C_{tic} \tag{33}$$

$$C_{tic} \xrightarrow{k_{tl}} M + R_{free} + P \tag{34}$$

3. Since the mRNA (M), the unfolded sfGFP (P), and the mature sfGFP (P_m) get diluted due to cell growth, add a dilution rate d_{dil} to their respective degradation terms.
4. Apply α to the dilution rate (d_{dil}). Additionally, apply α to the active mRNA degradation rate (d_m) to capture the increase in mRNA stability as the culture enters the stationary phase [15].
5. Apply γ to the transcription rate (β_m), translation resource availability (R_{Total}), and the protein folding rate (k_{fold}). These rates peak during the mid-log phase, due to the increased abundance of resources. Additionally, add b_{fold} to account for the basal folding rate as shown in Eq. 19.
6. Apply δ to the protein degradation rate (d_p), to model the increase in endogenous proteolysis as the cells enter stationary phase, reflecting the physiological shift to utilize recycled amino acids as an adaptation strategy under nutrient-limited conditions [16].
7. The new set of equations are:

$$\frac{dM}{dt} = \boldsymbol{\beta}_{\boldsymbol{m}} \cdot \left(\frac{Ac}{Kc + Ac} + l_0\right) - (\boldsymbol{d}_{\boldsymbol{m}} + \boldsymbol{d}_{\boldsymbol{dil}}) \cdot M - k_{tli_b} \cdot R_{free} \cdot M + k_{tli_u} \cdot C_{tic} + k_{tl} \cdot C_{tic} \tag{35}$$

$$\frac{dC_{tic}}{dt} = k_{tli_b} \cdot R_{free} \cdot M - k_{tli_u} \cdot C_{tic} - k_{tl} \cdot C_{tic} \tag{36}$$

$$\frac{dP}{dt} = k_{tl} \cdot C_{tic} - (\boldsymbol{d}_{\boldsymbol{p}} + \boldsymbol{d}_{\boldsymbol{dil}}) \cdot P - \boldsymbol{k}_{\boldsymbol{fold}} \cdot P \tag{37}$$

$$\frac{dP_m}{dt} = \boldsymbol{k_{fold}} \cdot P - \left(\boldsymbol{d_p} + \boldsymbol{d_{dil}}\right) \cdot P_m \tag{38}$$

$$\frac{dC}{dt} = k_{gr} \cdot C \cdot \left(1 - \frac{C}{C_{max}}\right) \tag{39}$$

8. The rates in bold are modified by RMFs, as described in Eqs. 14, 15, 16, 17, 18, and 19.
9. Model the free resource concentration using the mass-balance equation for R_{free}, that is $R_{free} = R_{max} \cdot \gamma - C_{TIC}$, where R_{max} is the maximum cap on total available R.

The simulation results obtained using the GEAGS framework (Fig. 6c) show a strong agreement with the experimentally observed sfGFP dynamics, demonstrating its suitability for modeling gene expression dynamics for complex circuitry with dynamic inputs, in batch cultures. Note: Species associated with the CcaSR system are assumed to be constitutively expressed, and therefore continuously available at abundant levels.

2.5 Application Example: Layered Feedback Control Circuit

In this section, we apply the GEAGS framework to a layered feedback circuit [12], demonstrating its utility in while modeling complex systems. These constructs consist of a two-stage activation cascade, where the first GEC encodes CinR, an activator that drives expression of sfYFP in the second GEC. The four key circuit configurations are shown in Fig. 7a include: (i) an open-loop system with no feedback, (ii) a *cis* feedback system where an antisense RNA (sRNA) [19] expressed from the second GEC represses its own, (iii) a *trans* feedback system where LacI expressed from the second GEC represses the combinatorial promoter of the first GEC, and (iv) a layered feedback system that combines both layers of feedback.

To build the GEAGS model, follow two steps: (1) Construct the growth-independent ODE system from species interactions shown in the schematic in Fig. 7b. Readers are encouraged to refer to Hu et al. [12] for more details on the model derivation and construction. (2) Identify the growth-dependent steps and apply the RMFs in Fig 7c to relevant reactions or species to form the growth-dependent model. The detailed steps are written below.

2.5.1 Building the Growth-Independent (Single-Scale) Model

1. The first GEC, the inducer cassette, is induced by Rhl and drives the expression of CinR. Rhl and CinR are types of acyl-homoserine-lactone (AHL) inducers, which are signaling molecules involved in quorum sensing.
2. The inducer cassette contains the combinatorial promoter $P_{Rhl/LacO}$, which is activated by the AHL inducer Rhl [12] (x), and is repressed by LacI.

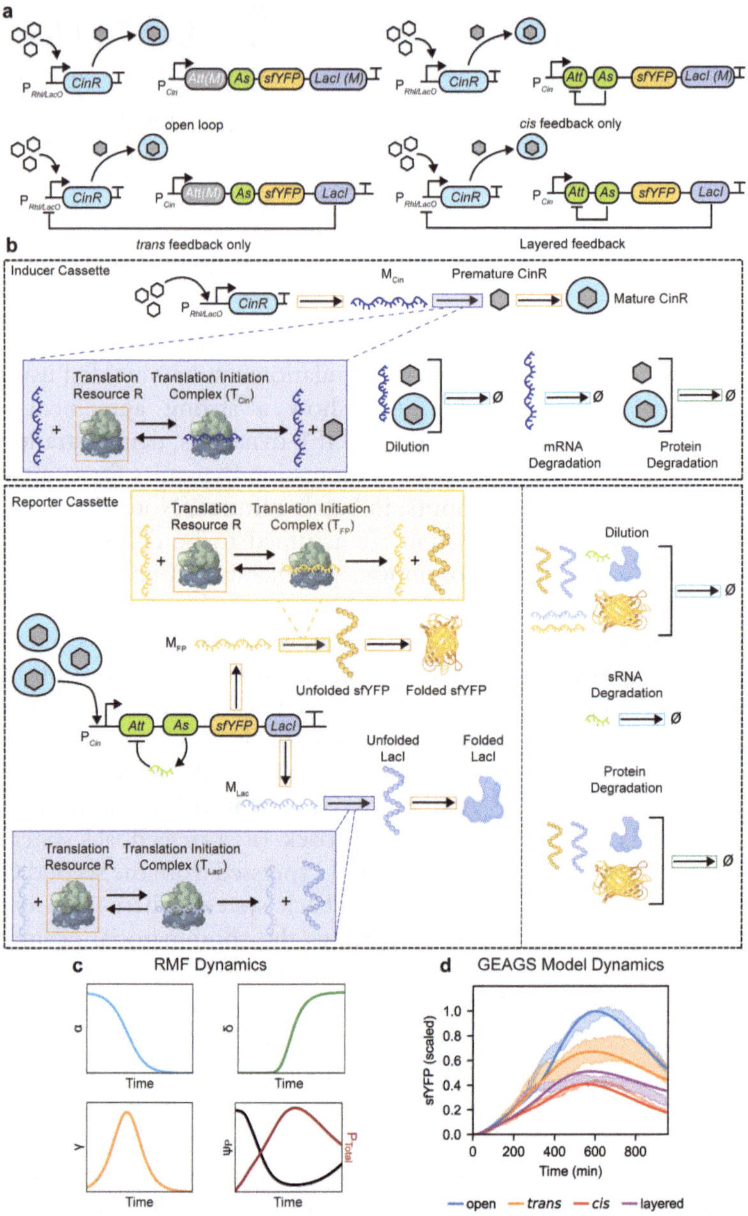

Fig. 7 Applying GEAGS framework to a layered feedback circuit. (a) Schematic of the open loop and the three feedback circuits studied by Hu et al. [12]. (b) Detailed gene expression steps for each construct and the associated species dynamics. Reactions rates and molecular species modified by the RMFs are highlighted with colored boxes, corresponding to the RMFs defined in (c). (c) Temporal profiles of the three RMFs. (d) Comparison of model simulations and experimental data. Solid lines denote model simulations, while the shaded regions represent the experimental mean ± standard deviation across replicates. (Data adapted from Hu et al. [12])

3. The transcription rate of *CinR* mRNA (M_{ind}) is $\beta_A \cdot \left(\frac{x}{K_x+x}\right) \cdot f_{trans}$, where x is the inducer concentration, K_x is the activation coefficient, and f_{trans} is the trans feedback term, where $f_{trans} = 1$ in the absence of trans feedback and $f_{trans} = \frac{K_p}{K_p+P_m}$ in the presence of trans feedback. Here, K_p is the repression constant for mature LacI (P_m).
4. The mRNA (M_{ind}) degradation, translation of premature CinR (P_{ind}), maturation of CinR to form the mature complex (C_{ind}), and protein degradation are modeled as outlined in Subheadings 2.1.
5. The ODEs for the inducer cassette can be written as:

$$\frac{dM_{ind}}{dt} = \beta_A \cdot \left(\frac{x}{K_x + x}\right) \cdot f_{trans} - d_m \cdot M_{ind} \tag{40}$$

$$\frac{dP_{ind}}{dt} = k_{tp} \cdot M_{ind} - d \cdot P_{ind} - K_r \cdot P_{ind} \tag{41}$$

$$\frac{dC_{ind}}{dt} = K_r \cdot P_{ind} - d \cdot C_{ind} \tag{42}$$

6. The reporter cassette consists of two genes *sfYFP* and *LacI* along with an antisense RNA (R_m) [19].
7. The transcription rate of the reporter cassettes' mRNA (M_G) is $\beta_B \cdot \left(\frac{C_{ind}}{K_{ind}+C_{ind}} + l_0\right) \cdot f_{cis}$, where C_{ind} is the product from the inducer cassette, K_{ind} is the activation constant, l_0 represents the coefficient of leaky expression, and f_{cis} is the cis feedback term, where $f_{cis} = 1$ in the absence of cis feedback and $f_{cis} = \frac{K_R}{K_R+R_m}$ in the presence of cis feedback. Here, K_R is the repression constant for the folded antisense RNA (R_m).
8. The transcription rate of the unfolded antisense RNA (R) is $\beta_B \cdot \left(\frac{C_{ind}}{K_{ind}+C_{ind}} + l_0\right) \cdot f_{cis}$. Then R folds into mature antisense RNA R_m following a first-order rate, with a rate constant k_{fold_R}. Both R and R_m degrade at a first-order rate of d_r.
9. The mRNA (M_G) degradation, translation of unfolded LacI and sfYFP (P, G), and maturation of P and G to form the folded LacI and sfYFP (P_m, G_m), and protein degradation are modeled as outlined in Subheadings 2.1.
10. The ODEs for the reporter cassette can be written as:

$$\frac{dM_G}{dt} = \beta_B \cdot (\frac{C_{ind}}{K_{ind} + C_{ind}} + l_0) \cdot f_{cis} - d_m \cdot M_G \tag{43}$$

$$\frac{dR}{dt} = \beta_B \cdot \left(\frac{C_{ind}}{K_{ind} + C_{ind}} + l_0\right) \cdot f_{cis} - d_r \cdot R - k_{fold_R} \cdot R \tag{44}$$

$$\frac{dR_m}{dt} = k_{fold_R} \cdot R - d_r \cdot R_m \tag{45}$$

$$\frac{dP}{dt} = k_{tr} \cdot M_G - d_{pr} \cdot P_{lac} - k_{fold_r} \cdot P \tag{46}$$

$$\frac{dP_m}{dt} = k_{fold_r} \cdot P - d_{pr} \cdot P_m \tag{47}$$

$$\frac{dG}{dt} = k_{tg} \cdot M_G - d \cdot G - k_{fold_G} \cdot G \tag{48}$$

$$\frac{dG_m}{dt} = k_{fold_G} \cdot G - d \cdot G_m \tag{49}$$

2.5.2 Constructing the GEAGS Model

Since the circuit two GECs that compete for translational resources, modeling growth-dependent resource allocation is essential to accurately capture this competition. In the GEAGS model, we therefore introduce a growth-dependent coarse-grained translational resource, *Rsrc*, and split-open the first-order translation rates as outlined in Subheading 2.4. To construct the GEAGS model, follow the steps:

1. Include the logistic growth equation (Eq. 5) in the system of ODEs.
2. Decompose the first-order translation rate k_{tp} into a CRN (Eqs. 50 and 51) where the mRNA (M_{ind}) binds with the freely available translation resource- *Rsrc* ($R_{rsrc_{free}}$) to form the translation initiation complex (TI_{cin}) (Fig 7b violet shaded box in the inducer cassette panel) with forward reaction rate constant $k_{tlib_{cin}}$ and reverse rate constant $k_{tliu_{cin}}$. TI_{cin} then undergoes translation elongation at a first-order rate of k_{tp} to form free M_{ind} and P_{ind}.

$$M_{ind} + R_{free} \underset{k_{tliu_{cin}}}{\overset{k_{tlib_{cin}}}{\rightleftarrows}} TI_{cin} \tag{50}$$

$$TI_{cin} \xrightarrow{k_{tp}} M_{ind} + R_{free} + P_{ind} \tag{51}$$

3. Similarly, decompose k_{tr}, and k_{tg} into CRNs where the reporter mRNA, (M_G) binds with $R_{rsrc_{free}}$ to form the translation initiation complex for LacI (TI_{lac}) and sfYFP (TI_{fp}) with forward reaction rate constants $k_{tlib_{lac}}$ and $k_{tlib_{fp}}$ and reverse rate constant $k_{tliu_{lac}}$ and $k_{tliu_{fp}}$ respectively (Eqs. 52, 53, 54, and 55). TI_{lac} and TI_{fp} then undergo translation elongation at a first-order rate of k_{tr} and k_{tg}, to form free M_G and corresponding proteins P and G (as shown in yellow and violet shaded boxes in Fig. 7b reporter cassette panel).

$$M_G + R_{free} \underset{k_{tli\,u_{lac}}}{\overset{k_{tli\,b_{lac}}}{\rightleftarrows}} TI_{lac} \tag{52}$$

$$M_G + R_{free} \underset{k_{tli\,u_{FP}}}{\overset{k_{tli\,b_{FP}}}{\rightleftarrows}} TI_{FP} \tag{53}$$

$$TI_{lac} \xrightarrow{k_{tr}} M_G + R_{free} + P \tag{54}$$

$$TI_{FP} \xrightarrow{k_{tg}} M_G + R_{free} + G \tag{55}$$

4. Since the mRNA (M_{ind} and M_G), sRNA (R and R_m), and proteins (P_{ind}, C_{ind}, P, P_m, G, and G_m) are diluted by cell growth, add a dilution rate d_{dil} to their respective degradation terms.
5. Apply α to all dilution rates (d_{dil}). Additionally, apply α to the mRNA and sRNA degradation rate (d_m, d_r) to capture the increase in mRNA stability as the culture enters the stationary phase [15].
6. Apply γ to the transcription rates (β_A, β_B) and the translation resource availability ($Rsrc_{Total}$). These rates peak during the mid-log phase, due to the increased abundance of resource.
7. Modify the protein degradation rates (d, d_{pr}) by $\delta \cdot (1 - \psi_p)$, where $(1 - \psi_p)$ captures the increase in degradation rate caused by molecular crowding [9, 10], and δ captures the increase in the endogenous proteolysis as the cells enter stationary phase [16].
8. Modify the protein folding rates (k_{fold_r} and k_{fold_G}) by $\gamma \cdot \psi_p + b_{fold}$, where ψ_p captures the reduction in protein maturation rate caused by molecular crowding [9], and γ captures the growth-dependence in the maturation rate [20].
9. The dynamics of the inducer cassette species after applying the RMFs (parameters modified by RMFs are in bold) can be written as:

$$\frac{dM_{ind}}{dt} = \boldsymbol{\beta_A} \cdot \left(\frac{x}{K_x + x}\right) \cdot f_{trans} - (\boldsymbol{d_m} + \boldsymbol{d_{dil}}) \cdot M_{ind} - r_{intn_{cin}} + r_{elng_{cin}} \tag{56}$$

$$\frac{dTI_{cin}}{dt} = r_{intn_{cin}} - r_{elng_{cin}} \tag{57}$$

$$\frac{dP_{ind}}{dt} = r_{elng_{cin}} - (\boldsymbol{d} + \boldsymbol{d_{dil}}) \cdot P_{ind} - K_r \cdot P_{ind} \tag{58}$$

$$\frac{dC_{ind}}{dt} = K_r \cdot P_{ind} - (\boldsymbol{d} + \boldsymbol{d}_{dil}) \cdot C_{ind} \tag{59}$$

$$r_{intn_{cin}} = k_{tlib_{cin}} \cdot Rsrc_{free} \cdot M_{ind} - k_{tliu_{cin}} \cdot TI_{cin} \tag{60}$$

$$r_{elng_{cin}} = TI_{cin} \cdot k_{tp} \tag{61}$$

10. The dynamics of the reporter cassette species after applying the RMFs (parameters modified by RMFs are in bold) can be written as:

$$\frac{dM_G}{dt} = \boldsymbol{\beta_B} \cdot \left(\frac{C_{ind}}{K_{ind} + C_{ind}} + l_0 \right) \cdot f_{cis} - (\boldsymbol{d_m} + \boldsymbol{d}_{dil}) \cdot M_G - (r_{intn_{lac}} + r_{intn_{FP}}) + (r_{elng_{lac}} + r_{elng_{FP}}) \tag{62}$$

$$\frac{dR}{dt} = \boldsymbol{\beta_B} \cdot \left(\frac{C_{ind}}{K_{ind} + C_{ind}} + l_0 \right) \cdot f_{cis} - (\boldsymbol{d_r} + \boldsymbol{d}_{dil}) \cdot R - k_{fold_R} \cdot R \tag{63}$$

$$\frac{dR_m}{dt} = k_{fold_R} \cdot R - (\boldsymbol{d_r} + \boldsymbol{d}_{dil}) \cdot R_m \tag{64}$$

$$\frac{dTI_{lac}}{dt} = r_{intn_{lac}} - r_{elng_{FP}} \tag{65}$$

$$\frac{dTI_{FP}}{dt} = r_{intn_{FP}} - r_{elng_{FP}} \tag{66}$$

$$\frac{dP}{dt} = r_{elng_{lac}} - (\boldsymbol{d_{pr}} + \boldsymbol{d}_{dil}) \cdot P - \boldsymbol{k_{fold_{lac}}} \cdot P \tag{67}$$

$$\frac{dP_m}{dt} = \boldsymbol{k_{fold_{lac}}} \cdot P_m - (\boldsymbol{d_{pr}} + \boldsymbol{d}_{dil}) \cdot P_m \tag{68}$$

$$\frac{dG}{dt} = r_{elng_{FP}} - (\boldsymbol{d} + \boldsymbol{d}_{dil}) \cdot G - \boldsymbol{k_{fold_{FP}}} \cdot G \tag{69}$$

$$\frac{dG_m}{dt} = \boldsymbol{k_{fold_{FP}}} \cdot G - (\boldsymbol{d} + \boldsymbol{d}_{dil}) \cdot G_m \tag{70}$$

$$r_{intn_{lac}} = k_{tlib_{lac}} \cdot Rsrc_{free} \cdot M_G - k_{tliu_{lac}} \cdot TI_{lac} \tag{71}$$

$$r_{elng_{lac}} = TI_{lac} \cdot k_{tr} \tag{72}$$

$$r_{intn_{FP}} = k_{tlib_{FP}} \cdot Rsrc_{free} \cdot M_G - k_{tliu_{sFP}} \cdot TI_{FP} \tag{73}$$

$$r_{elng_{FP}} = TI_{FP} \cdot k_{tg} \tag{74}$$

11. The availability of free resource $Rsrc_{free}$ can be written as a function of growth and intermediate species as:

$$Rsrc_{free} = Rsrc_{max} \cdot \gamma - (TI_{cin} + TI_{lac} + TI_{FP}) \tag{75}$$

The simulation of the GEAGS model (Fig. 7d) captures the experimentally observed dynamics, demonstrating the extensibility of the GEAGS framework to complex circuitry.

References

1. Klumpp S, Zhang Z, Hwa T (2009) Growth rate-dependent global effects on gene expression in bacteria. Cell 139(7):1366–1375. https://doi.org/10.1016/j.cell.2009.12.001
2. Poole W, Pandey A, Shur A, Tuza ZA, Murray RM (2022) BioCRNpyler: compiling chemical reaction networks from biomolecular parts in diverse contexts. PLoS Comput Biol 18(4): e1009987. https://doi.org/10.1371/journal.pcbi.1009987
3. Pandey A, Poole W, Swaminathan A, Hsiao V, Murray RM (2023) Fast and flexible simulation and parameter estimation for synthetic biology using bioscrape. J Open Source Softw 8(83). https://doi.org/10.21105/joss.05057
4. Alon U (2019) An introduction to systems biology: design principles of biological circuits. Chapman and Hall/CRC, Boca Raton
5. Verhulst P-F (1838) Notice sur la loi que la population suit dans son accroissement. Corresp Math Phys 10:113–129
6. Fujikawa H, Kai A, Morozumi S (2004) A new logistic model for growth at constant and dynamic temperatures. Food Microbiol 21(5): 501–509. https://doi.org/10.1016/j.fm.2004.01.007
7. Namboothiri HR, Pandey A, Hu CY (2025) Resolving emergent transient oscillations in gene circuits with a growth-coupled model. Sci Adv 12(6), eadz2310. https://doi.org/10.1126/sciadv.adz2310
8. Jefferys BR, Kelley LA, Sternberg MJ (2010) Protein folding requires crowd control in a simulated cell. J Mol Biol 397(5):1329–1338. https://doi.org/10.1016/j.jmb.2010.01.074
9. Mukherjee S, Waegele MM, Chowdhury P, Guo L, Gai F (2009) Effect of macromolecular crowding on protein folding dynamics at the secondary structure level. J Mol Biol 393(1): 227–236. https://doi.org/10.1016/j.jmb.2009.08.016
10. Goldberg AL (2003) Protein degradation and protection against misfolded or damaged proteins. Nature 426(6968):895–899. https://doi.org/10.1038/nature02263
11. Balleza E, Kim JM, Cluzel P (2018) Systematic characterization of maturation time of fluorescent proteins in living cells. Nat Methods 15(1):47–51. https://doi.org/10.1038/nmeth.4509
12. Hu CY, Murray RM (2022) Layered feedback control overcomes performance trade-off in synthetic biomolecular networks. Nat Commun 13(1):5393. https://doi.org/10.1038/s41467-022-33058-6
13. Taniguchi Y, Choi PJ, Li GW, Chen H, Babu M, Hearn J, Emili A, Sunney Xie X (2010) Quantifying E. coli proteome and transcriptome with single-molecule sensitivity in single cells. Am Assoc Adv Sci 329. https://doi.org/10.1126/SCIENCE.1188308/SUPPL_FILE/TANIGUCHI.SOM.PDF
14. Maurizi MR (1992) Proteases and protein degradation in Escherichia coli. Experientia 48(2): 178–201. https://doi.org/10.1007/BF01923511
15. Esquerre T, Laguerre S, Turlan C, Carpousis AJ, Girbal L, Cocaign-Bousquet M (2014) Dual role of transcription and transcript stability in the regulation of gene expression in Escherichia coli cells cultured on glucose at different growth rates. Nucleic Acids Res 42(4):2460–2472. https://doi.org/10.1093/nar/gkt1150
16. Shaikh AS, Tang YJ, Mukhopadhyay A, Martin HG, Gin J, Benke PI, Keasling JD (2010) Study of stationary phase metabolism via isotopomer analysis of amino acids from an isolated protein. Biotechnol Prog 26(1): 52–56. https://doi.org/10.1002/btpr.325
17. Schmidl SR, Sheth RU, Wu A, Tabor JJ (2014) Refactoring and optimization of light-switchable Escherichia coli two-component systems. ACS Synth Biol 3(11):820–831. https://doi.org/10.1021/sb500273n
18. Namboothiri HR, Pochana K, Jaiswal B, Emami A, Hu CY (2025) Closed-loop optogenetic control in a microplate reader. bioRxiv. https://doi.org/10.1101/2025.09.22.677797
19. Hu CY, Takahashi MK, Zhang Y, Lucks JB (2018) Engineering a functional small RNA negative autoregulation network with model-guided design. ACS Synth Biol 7(6): 1507–1518. https://doi.org/10.1021/acssynbio.7b00440
20. Hebisch E, Knebel J, Landsberg J, Frey E, Leisner M (2013) High variation of fluorescence protein maturation times in closely related Escherichia coli strains. PLoS One 8(10): e75991. https://doi.org/10.1371/journal.pone.0075991

Chapter 19

Coupled Perturbations of Gene Circuit Dynamics by Resource Competition and Growth Dilution

Abdelrahaman Youssef, Rong Zhang, and Xiao-Jun Tian

Abstract

Resource competition and growth-mediated dilution are major sources of context dependence in synthetic gene circuits. However, their coupled effects on circuit behavior remain insufficiently dissected. This chapter presents a step-by-step experimental workflow for constructing and characterizing inhibitory genetic cascades in *E. coli* to examine whether and how circuit dynamics depend on these two contextual factors. The protocol details the design and assembly of inhibitory circuits with tunable promoter strength, ribosome binding site strength, and plasmid copy number. Methods are provided for quantifying circuit dynamics using plate-reader fluorescence assays under varying inducer doses, with and without preinduction, and across different growth conditions. Finally, data processing and analysis procedures are outlined to characterize module interdependences and expression dynamics. Collectively, this protocol provides a practical framework for investigating how shared cellular resources and growth processes shape gene circuit behavior and serves as a guide for studying multiple context-dependent effects in synthetic biology.

Key words Inhibitory genetic cascades, Resource competition, Growth-mediated dilution, Host-circuit interactions, Gene expression dynamics

1 Introduction

Synthetic biology aims to design and assemble genetic circuits, engineered sequences of DNA that regulate gene expression in predictable ways. A central design principle is modularity, in which complex networks are built from smaller, well-characterized parts or modules [1, 2]. In theory, modules that perform distinct tasks can be interconnected, much like components in an electronic circuit, to generate higher-order system behavior. In practice, however, modules that function reliably in isolation often exhibit unexpected dynamics when combined. Major contributing factors include hidden interactions between the genetic circuit and host-cell growth, as well as indirect coupling between modules, whereby modules alter host-cell physiology or compete for shared cellular

Xiao-Jun Tian (ed.), *Synthetic Gene Circuits: Methods and Protocols*, Methods in Molecular Biology, vol. 3041,
https://doi.org/10.1007/978-1-0716-5304-3_19,

resources, leading to unintended interactions and context-dependent behavior [3–7].

Resource competition arises from the finite availability of transcriptional and translational resources, such as RNA polymerase, ribosomes, and nucleotides, which must be shared among concurrently active genes [8, 9]. When multiple modules are expressed simultaneously, they compete for these limited resources. Excessive demand by one module can deprive neighboring modules of resources, thereby altering their expression and perturbing the dynamics of the entire circuit [10–16]. Previous studies have demonstrated that resource competition can give rise to unexpected circuit behaviors. For example, two genes that are not directly connected can become linearly interdependent through shared resource usage [10]. An activation cascade may exhibit a biphasic (rise–then–fall) dose–response rather than the expected monotonic increase when an upstream module depletes resources required by downstream components [11]. And in networks with multiple bistable switches, winner-takes-all dynamics can emerge when one module monopolizes shared resources and prevents others from activating [12].

In addition to resource competition, growth-mediated dilution has been shown to significantly influence the behavior of synthetic gene circuits [17–25]. For example, we previously demonstrated that self-activation circuits can lose hysteresis under rapid growth conditions, as transcription factors are diluted by cell division and fail to reach or maintain the concentrations required for activation [17]. Importantly, this effect depends on circuit topology: self-activation circuits are particularly sensitive to growth-mediated dilution, whereas toggle switches remain robust and preserve hysteresis across growth conditions. However, the influence of growth-mediated dilution of cellular components on resource competition has not been systematically explored.

Recently, we constructed and analyzed a library of inhibitory genetic cascades to investigate how resource competition and growth-mediated dilution shape circuit behavior [26]. By varying promoter strength, ribosome binding site strength, and plasmid backbone, we systematically tuned the resource "load" imposed by each module. In many cases, the downstream module unexpectedly consumed a disproportionate share of the shared transcriptional and translational machinery, thereby suppressing expression of the upstream regulator. This resulted in a downstream-high/upstream-low expression state that cannot be explained by the intended regulatory logic alone. We further observed that cellular growth modulates these patterns: following dilution into fresh rich medium, the downstream-high/upstream-low state became unstable, whereas it persisted under minimal growth conditions. Together, these findings indicate that resource competition and growth jointly generate effective feedback mechanisms that must be accounted for in the design of robust genetic circuits.

Here, we present a detailed experimental workflow for the design, construction, and quantitative analysis of inhibitory genetic cascades to study resource competition and growth-mediated effects in bacterial gene networks. The protocol provides step-by-step instructions for circuit assembly, induction experiments, and plate-reader–based fluorescence measurements, along with practical guidance on plasmid construction, media preparation, and pre-induction versus non-preinduction conditions. By consolidating these methods, this chapter aims to make the study of resource competition and growth dilution accessible to early-stage researchers and to provide a practical framework for examining how hidden cellular interactions reshape genetic circuit behavior and impact the design of robust synthetic systems.

2 Materials

2.1 Strains

(a) For plasmid construction and propagation, *E. coli* strain DH5α is used. This strain is commonly chosen for DNA cloning because it is easy to transform and maintains plasmids stably.

(b) A single transformed DH5α colony is picked and grown overnight in a 15 mL conical tube containing 5 mL of culture medium at 37 °C with shaking at 250 rpm. The overnight culture provides enough cells and plasmid for downstream isolation.

(c) For testing purposes, all circuit induction and measurement experiments are performed in a derivative of *E. coli* K-12, strain MG1655ΔlacIΔaraCBAD.

2.2 Media and Chemicals

(a) Starter cultures of DH5α and MG1655ΔlacIΔaraCBAD are grown in Luria–Bertani (LB) broth (Sigma-Aldrich).

(b) To ensure that only cells carrying the desired plasmid are maintained, the LB medium is supplemented with an antibiotic corresponding to the resistance marker on each plasmid backbone: 25 μg/mL chloramphenicol, 50 μg/mL kanamycin, or 100 μg/mL ampicillin.

(c) For induction experiments, MG1655ΔlacIΔaraCBAD cells harboring the genetic circuits are cultured in 96-well plates or 15 mLs conical tubes. Gene expression is induced using anhydrotetracycline hydrochloride (aTc; Abcam), a small molecule that activates tetracycline-responsive promoters (pTet).

(d) A 1 mg/mL stock solution of aTc is prepared in double-distilled water (ddH_2O) and diluted to the desired working concentrations (50~400 ng/mL) immediately before use.

2.3 Equipment and Software

(a) Both multi-channel and single-channel micropipettes are used to transfer and mix liquids during circuit setup and reagent preparation.

(b) 5 mL and 15 mL conical tubes provided sterile, aerated containers for overnight cultivation of *E. coli* cultures.

(c) Plasmid extraction is performed with commercial kits from Qiagen or Zymo Research. PCR, ligation, and restriction digestion steps are carried out using enzymes and reaction mixes from New England Biolabs and Thermo Fisher Scientific.

(d) For fluorescence and absorbance assays, cells are cultured in transparent, opaque-bottom 96-well plates.

(e) Data acquisition is controlled using the plate reader's integrated software (Gen5; The Synergy H1's manufacturer-supplied software), and MATLAB R2024a is used to organize, analyze, and visualize the resulting fluorescence and absorbance data.

3 Methods

3.1 Inhibitory Genetic Cascades: Design and Rationale

(a) A series of two-module genetic circuits are designed. Circuit variants differ in either their M1 promoter strength, M2 RBS strength, or their plasmid backbone copy number. All circuit variants are described in (Fig. 1a).

(b) Module 1 (M1) encodes TetR fused to red fluorescent protein (RFP) and is expressed from a constitutive promoter, meaning it is always expressed at a steady level.

(c) Module 2 (M2) encodes green fluorescent protein (GFP) and is placed under the control of the pTet promoter, which can be repressed by TetR.

(d) In this design, M1 functions as a repressor–reporter module: TetR-RFP produced by M1 binds to the pTet promoter and blocks GFP expression from M2. The small-molecule inducer anhydrotetracycline (aTc) can relieve this repression by blocking TetR from binding to pTet, thereby tuning the strength of inhibition and allowing GFP expression (Fig. 1b, c).

(e) In an ideal case, this system would behave as a straightforward repression circuit, where M1 directly inhibits M2 through transcriptional control.

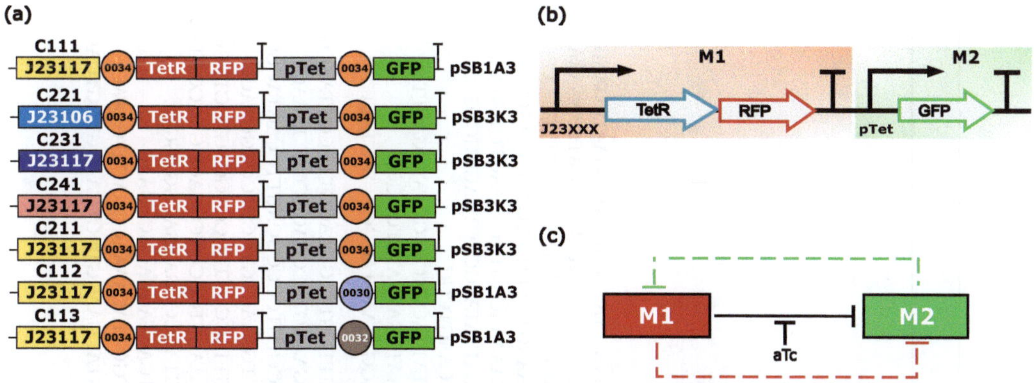

Fig. 1 Genetic circuit variants, architecture, and circuit topology. (**a**) Detailed schematics of the inhibitory cascade circuits showing the regulatory elements and genetic components used in each variant, including promoter type, RBS, reporter genes, and plasmid backbone. (**b**) Schematic design of the inhibitory cascade circuit. (**c**) The circuit topology showing both direct transcriptional regulation (solid lines) and potential resource competition between modules (dashed lines)

(f) However, cellular environments introduce additional layers of interaction, because RFP is strongly expressed from M1, it can also indirectly suppress M2 by competing for shared transcriptional and translational resources (Fig. 1c).

(g) Similarly, when GFP production is high, M2 can limit M1 activity by drawing from the same resource pool.

(h) Together, these effects create a bidirectional, resource-dependent interaction that overlays the intended genetic design.

3.2 Circuits Construction; Digestion and Ligation Cycles

(a) Assemble genetic circuits using the plasmid backbones pSB1A3 (high copy) or pSB3K3 (medium copy), depending on experimental context. Circuit components are listed in Table 1 (*see* **Note 1**).

(b) Digest all DNA fragments using the BioBrick restriction enzyme system (EcoRI, XbaI, SpeI, PstI; Thermo Fisher Scientific) (*see* **Note 2**).

(c) Prepare a 50 μL restriction digestion reaction by combining 5 μL of 10x Digestion Buffer, 2.5 μL of the first restriction enzyme, 2.5 μL of the second restriction enzyme, and 400–2000 ng of DNA template.

(d) Add ddH_2O to bring the final reaction volume to 50 μL. Mix gently by pipetting to ensure homogeneity of the reaction, then incubate the mixture at 37 °C for 30 minutes.

(e) Following digestion, purify the digestion products using either PCR cleanup (GenElute PCR Cleanup Kit, Sigma-Aldrich) or gel extraction (GelElute Gel Extraction Kit, Sigma-Aldrich) (*see* **Note 2**).

Table 1
Genetic parts used in circuit construction

Part	Description	Sequence/Reference number
J23117	Constitutive promoter	TTGACAGCTAGCTCAGTCCTAGGGATTGTGCTAGC
J23105	Constitutive promoter	TTTACGGCTAGCTCAGTCCTAGGTACTATGCTAGC
J23107	Constitutive promoter	TTTACGGCTAGCTCAGCCCTAGGTATTATGCTAGC
J23106	Constitutive promoter	TTTACGGCTAGCTCAGTCCTAGGTATAGTGCTAGC
B0034	Strong RBS	AAAGAGGAGAAA
B0030	Medium-strength RBS	ATTAAAGAGGAGAAA
B0032	Weak RBS	TCACACAGGAAAG
TetR	Tetracycline repressor	ATGTCCAGATTAGATAAAAGTAAAGTGATTAACAGCGCATTAGAGCTGCTTAATGAGGTCGGAATCGAAGG TTTAACAACCCGTAAACTCGCCCAGAAGCTAGGTGTAGAGCAGCCTACATTGTATTGGCATGTAAAAAA TAAGCGGGCTTTGCTCGACGCCTTAGCCATTGAGATGTTAGATAGGCACCATACTCACTTTTGCCC TTTAGAAGGGGAAAGCTGGCAAGATTTTTTACGTAATAACGCTAAAAGTTTTAGATGTGCTTTACTAAG TCATCGCGATGGAGCAAAAGTACATTTAGGTACACGGCCTACAGAAAAACAGTATGAAACTCTCGAAAA TCAATTAGCCTTTTTATGCCAACAAGGTTTTTCACTAGAGAATGCATTATATGCACTCAGCGCTG TGGGGCATTTTACTTTAGGTTGCGTATTGGAAGATCAAGAGCATCAAGTCGC TAAAGAAGAAAGGGAAACACCTACTACTGATAGTATGCCGCCATTATTACGACAAGCTATCGAATTA TTTGATCACCAAGGTGCAGAGCCAGCCTTCTTATTCGGCCTTGAATTGATCATATGCGGA TTAGAAAAACAACTTAAATGTGAAAGTGGGTCCGCTGCAAACGACGAAAACTACGCTTTAGTAGCT
RFP	Red fluorescent protein	ATGGCTTCCTCCGAAGACGTTATCAAAGAGTTCATGCGTTTCAAAGTTCGTATGGAAGGTTCCGTTAACGG TCACGAGTTCGAAATCGAAGGTGAAGGTGAAGGTCGTCCGTACGAAGGTACCCAGACCGCTAAAC TGAAAGTTACCAAAGGTGGTCCGCTGCCGTTCGCTTGGGACATCCTGTCCCCGCAGTTCCAGTACGG TTCCAAAGCTTACGTTAAACACCCGGCTGACATCCCGGACTACCTGAAACTGTCCTTCCCGGAAGG TTTCAAATGGGAACGTGTTATGAACTTCGAAGACGGTGGTGTTGTTACCGTTACCCAGGACTCCTCCC TGCAAGACGGTGAGTTCATCTACAAAGTTAAACTGCGTGGTACCAACTTCCCGTCCGACGGTCCGGTTA TGCAGAAAAAAACCATGGGTTGGGAAGCTTCCACCGAACGTATGTACCCGGAAGACGGTGCTC TGAAAGGTGAAATCAAAATGCGTCTGAAACTGAAAGACGGTGGTCACTACGACGCTGAAG TTAAAACCACCTACATGGCTAAAAAACCGGTTCAGCTGCCGGGTGCTTACAAAACCGACATCAAAC TGGACATCACCTCCCACAACGAAGACTACACCATCGTTGAACAGTACGAACGTGCTGAAGGTCGTCAC TCCACCGGTGCTTAA

pTet	Inducible promoter	TCCCTATCAGTGATAGAGATTGACATCCCTATCAGTGATAGAGATACTGAGCAC
GFP	Green fluorescent protein	ATGCGTAAAGGAGAAGAACTTTTCACTGGAGTTGTCCCAATTCTTGTTGAATTAGATGGTGATGTTAA TGGGCACAAATTTTCTGTCAGTGGAGAGGGTGAAGGTGATGCAACATACGGAAAACTTACCCTTAAA TTTATTTGCACTACTGGAAAACTACCTGTTCCATGGCCAACACTTGTCACTACTTTCGGTTATGGTG TTCAATGCTTTGCGAGATACCCAGATCATATGAAACAGCATGACTTTTTCAAGAGTGCCATGCCCGAAGG TTATGTACAGGAAAGAACTATATTTTTCAAAGATGACGGGAACTACAAGACACGTGCTGAAGTCAAG TTTGAAGGTGATACCCTTGTTAATAGAATCGAGTTAAAAGGTATTGATTTTAAAGAAGATGGAAACATTC TTGGACACAAATTGGAATACAACTATAACTCACACAATGTATACATCATGGCAGACAAACAAAAGAA TGGAATCAAAGTTAACTTCAAAATTAGACACAACATTGAAGATGGAAGCGTTCAACTAGCAGACCATTA TCAACAAAATACTCCAATTGGCGATGGCCCTGTCCTTTTACCAGACAACCATTACCTGTCCACACAATC TGCCCTTTCGAAAGATCCCAACGAAAAGAGAGACCACATGGTCCTTCTTGAGTTTGTAACAGCTGC TGGGATTACACATGGCATGGATGAACTATACAAATAATAA
B0015	Double terminator	CCAGGCATCAAATAAAACGAAAGGCTCAGTCGAAAGACTGGGCCTTTCGTTTTATCTGTTGTTTGTCGG TGAACGCTCTCTACTAGAGTCACACTGGCTCACCTTCGGGTGGGCCTTTCTGCGTTTATA
pSB1A3	High-copy-number backbone	pSB1A3
pSB3K3	Medium-copy-number backbone	pSB3K3

Adapted from our previous work [26]

(f) Ligate purified fragments using T4 DNA ligase (New England Biolabs) according to the ligation protocol listed below.

(g) Prepare a 15 μL ligation reaction by combining 1.5 μL of 10x Ligation Buffer, 1.5 μL of T4 DNA ligase, 4 μL of plasmid backbone, 4 μL of DNA fragment 1, and 4 μL of DNA fragment 2.

(h) Add ddH_2O to bring the final reaction volume to 15 μL. Mix gently by pipetting, then incubate the ligation mixture at 25 °C for 1 hour.

(i) Transform the ligation product into chemically competent *E. coli* DH5α cells by mixing 10 μL of the ligation mixture with 50 μL of competent cells (*see* **Notes 3** and **4**).

(j) Incubate the mixture on ice for 5 minutes, then heat-shock at 42 °C for 90 seconds. Immediately return the reaction to ice for 3 minutes, then add 200 μL of ZymoBroth supplemented with 0.4% glucose. (*see* **Notes 5** and **6**).

(k) Incubate the cells at 37 °C for at least 1 hour with shaking. Plate the entire transformation mixture onto LB agar supplemented with the appropriate antibiotic and incubate overnight at 37 °C (*see* **Note** 7).

3.3 Circuits Confirmation by Colony PCR and Plasmid Digestion

(a) Identify positive clones by colony PCR using circuit-specific primers.

(b) Confirm plasmid identity by diagnostic digestion with EcoRI and PstI.

(c) Validate all constructs by whole-plasmid sequencing (*see* **Note 8**).

(d) Culture verified clones overnight in LB broth supplemented with the appropriate antibiotic for plasmid maintenance.

3.4 Circuit Inductions to Characterize the Resource Competition

3.4.1 Inducers Stock and Working Solutions Preparation

(a) Prepare a 1 mg/mL stock of anhydrotetracycline (aTc) in ddH_2O (*see* **Note 9**).

(b) Prepare working stocks of 50, 100, 200, 300, and 400 μg/mL by diluting the stock solution (*see* **Note 9**).

(c) Dilute each working solution 1000-fold into LB medium (with appropriate antibiotic) to obtain final inducer concentrations of 50, 100, 200, 300, and 400 ng/mL.

3.4.2 Induction Protocol

Induction protocol to characterize the direct relationship between the upstream (M1) and downstream (M2) modules and to identify how resource competition shapes circuit dynamics.

(a) Pick three single colonies of *E. coli* K-12 MG1655ΔlacIΔaraCBAD carrying the circuit and inoculate into 5 mL LB with the appropriate antibiotic (50 μg/mL kanamycin for LBK or 100 μg/mL ampicillin for LBA).

(b) Grow starter cultures overnight at 37 °C, shaking at 250 rpm in 15-mL tubes.

(c) Prepare induction media by adding aTc to fresh LB (LBK or LBA) to final concentrations of 50–400 ng/mL.

(d) Dispense 200 μL of induction media into each well of a sterile 96-well plate.

(e) Dilute overnight cultures 40-fold into the wells containing induction media. Include at least triplicate wells for each condition.

(f) Include triplicate blank wells (medium + antibiotic only) to allow background subtraction of OD and fluorescence.

(g) Incubate at 37 °C with orbital shaking (e.g., 807 cpm) and record OD_{600}, GFP, and RFP every 20 min for 18 h, using the excitation/emission settings described in Subheading 3.6.

3.5 Effect of Growth-Mediated Dilution on Circuit Dynamics

To describe how growth-mediated dilution and medium-dependent growth conditions (100% LB, M9 + 5% LB, and M9) reshape the dynamics and expression states of the inhibitory cascade.

3.5.1 Media Preparation

(a) Prepare 100% LB medium and supplement with the appropriate antibiotic (50 μg/mL kanamycin for LBK or 100 μg/mL ampicillin for LBA).

(b) Prepare M9 minimal medium according to the manufacturer's instructions and add the appropriate antibiotic.

(c) Prepare M9 + 5% LB by mixing 1x M9 with 100% LB to obtain a final composition of 5% (v/v) LB in M9. Supplement with the appropriate antibiotic.

3.5.2 Growth-Mediated Dilution Assay

(a) Cells are first grown under saturating aTc dose before induction (400 ng/mL), to establish a defined "downstream-high / upstream-low" starting state.

(b) Re-inoculate overnight cultures into 5 mL LB containing 400 ng/mL aTc plus the appropriate antibiotic and grow overnight at 37 °C, shaking at 250 rpm.

(c) Harvest cells by centrifugation at 2500× *g* for 5 min, then wash twice with 1x PBS to remove residual aTc (*see* **Note 10**).

(d) After washing, prepare induction media and a 96-well plate exactly as described for the non-preinduction protocol.

(e) Dilute overnight cultures (**step 3.5.2.a**) 40-fold into the wells containing induction media.

(f) Incubate at 37 °C with orbital shaking (e.g., 807 cpm) and record OD_{600}, GFP, and RFP every 20 min for 18 h, using the excitation/emission settings described in Subheading 3.6.

(g) Repeat steps d-h for the other two media compositions (M9 + 5% LB and M9) following the same procedure.

3.6 Average Fluorescence Measurement by Plate Reader

(a) Measure fluorescence and optical density using a plate reader (Synergy H1 Hybrid Reader, BioTek).
(b) Set excitation/emission to 485/515 nm for GFP and 546/607 nm for RFP.
(c) Adjust the gains for both GFP and RFP fluorescence channels. Record GFP, RFP, and OD_{600} values every 20 minutes for 18 hours at 37 °C with orbital shaking (807 cpm).
(d) Export the blanked plate-reader data (OD_{600}, GFP, RFP) for each experiment as comma-separated values (.csv) or Excel (.xlsx) files, with time points and well identifiers.

3.7 Data Analysis

3.7.1 Data Import and Organization

(a) Open MATLAB and navigate to the folder containing the data files. Create variables corresponding to Time vector, OD_{600} readings, GFP and RFP readings, Inducer Concentrations (aTc).
(b) Arrange data into matrices organized by condition x replicate x time to facilitate averaging and plotting.
(c) Save the created variables as a data file by the Matlab function "save [Data File Name]".
(d) Call the data file using the function "Load [Data file name]".

3.7.2 Data Normalization (see Note 11)

(a) For Non-preinduction data, normalize fluorescence by OD600 to approximate per-cell fluorescence.
(b) For pre-induced and growth-mediated dilution data, normalize fluorescence by OD600, then normalize the outcome to the steady-state expression levels at 400 ng/mL aTc.

3.7.3 Averaging Replicates and Generating Plots

(a) Compute the mean and standard deviation across biological replicates.
(b) Generate dose-response curves by plotting steady-state GFP and/or RFP vs. inducer concentration.
(c) Plot steady states GFP vs RFP.
(d) Plot time-course dynamics for each condition.
(e) Overlay data from different circuit variants or different induction states, when applicable, to directly compare the effects of circuit parameters and growth-mediated dilution on circuit dynamics.
(f) When comparing the dose-response curves of different circuit variants, examine the steady-state of GFP/OD and RFP/OD as functions of inducer concentration. For example, Fig. 2a, b shows three distinct inhibitory cascade behaviors across

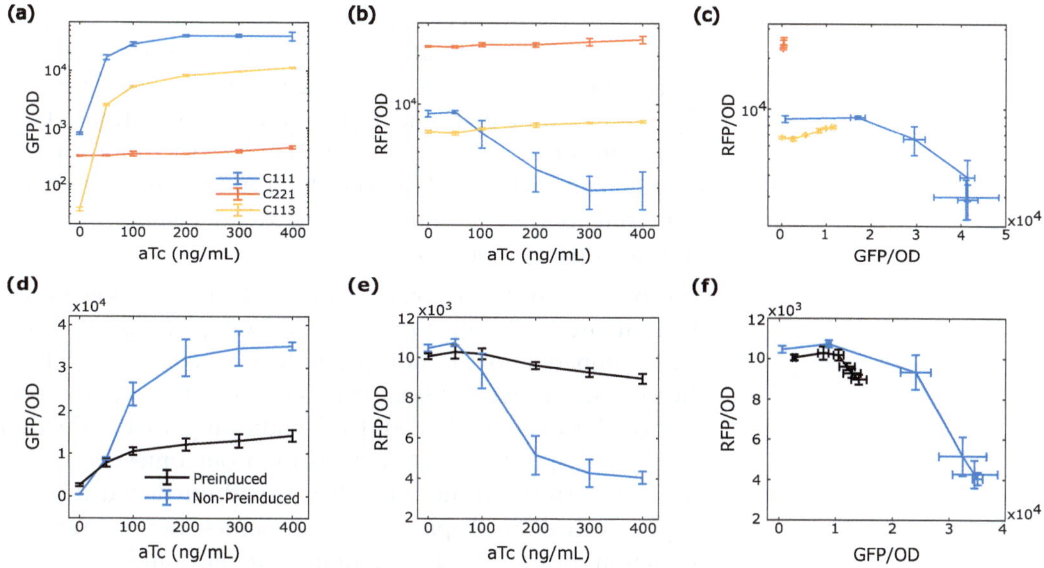

Fig. 2 Steady-state analysis of inhibitory cascade behavior with and without pre-induction. (**a–c**) Steady-state analysis of non-pre-induction experiments. (**a**) GFP expression and (**b**) RFP expression for three circuit variants (C111—main topology, C221—stronger M1 promoter, and C113—weaker M2 RBS), revealing three distinct steady-state behaviors across topologies. (**c**) GFP versus RFP steady-state expression for the three circuit variants, showing how tuning circuit parameters can reshape the circuit outcomes. (**d–f**) Steady-state analysis of pre-induction experiments (black) for the C111 topology, directly compared to nonpreinduction condition (blue) (**d**) GFP steady-state expression and (**e**) RFP steady-state expression. (**f**) GFP versus RFP steady-state expression for the C111 topology. Error bars represent standard deviation (n = 3). (Adapted from our previous work [26])

different circuit topologies. In C111, GFP increases with aTc while RFP decreases, indicating strong resource competition between modules. In C221, GFP remains low and insensitive to aTc, whereas RFP stays high with upstream module dominance. In C113, GFP follows the C111 trend at reduced levels due to a weaker M2 RBS, while RFP remains largely unchanged across inducer concentrations, indicating weak module coupling and minimal downstream influence on upstream expression.

(g) Next, plot GFP/OD versus RFP/OD to directly assess the relationship between the gene expression levels in the two modules. For example, Fig. 2c shows a clear negative correlation in C111 in favor of M2, while C221 shows a nonresponse state with M1 dominance over M2, and C113 displays largely uncoupled expression patterns with a smaller range of GFP expression. These plots illustrate how circuit parameter choices could lead to distinct expression outcomes.

(h) To evaluate the impact of growth-mediated dilution, compare steady-state GFP/OD and RFP/OD under nonpreinduced (blue) and preinduced (black) conditions. For example, Fig. 2d,e shows that under preinduction conditions, GFP reaches lower steady-state levels with reduced activation across the aTc range, while RFP remains high and relatively constant, exhibiting only a slight downward trend compared to the nonpreinduction condition.

(i) Finally, plot GFP/OD versus RFP/OD to assess how preinduction alters module coupling. For example, Fig. 2f shows that, compared to the non-pre-induction condition (blue), the pre-induction condition (black) modifies the negative correlation between GFP and RFP, indicating that induction history reshapes steady-state expression outcomes. A weaker negative correlation indicates that growth-mediated dilution plays a key role in shaping circuit dynamics, can override the induction history, and determines circuit states reset and retention outcomes.

4 Notes

The following notes highlight important technical considerations and troubleshooting tips that can improve reproducibility and data consistency across experiments. Attention to these details is essential for maintaining circuit integrity and obtaining interpretable results when studying resource competition effects in inhibitory genetic cascades.

1. **Plasmid Backbones and Copy Number.** Choosing between pSB1A3 (high copy) and pSB3K3 (medium copy) can significantly affect expression load. Circuits carried on pSB1A3 amplify resource competition effects, whereas pSB3K3 provides a more balanced output that minimizes metabolic stress. Select the backbone according to the desired sensitivity to resource dynamics.
2. **Restriction–Ligation Workflow.** Ensure complete digestion of both backbone and insert fragments to prevent religation. Incomplete digestion is a common source of false-positive colonies. For larger fragments (>2 kb), gel purification yields cleaner ligation products than PCR cleanup.
3. **Transformation Efficiency.** Ensure competent *E. coli* DH5α cells remain on ice throughout the preparation period prior to heat shock. Any warming significantly reduces transformation efficiency.

4. **Competent Cells Quality.** Use freshly thawed, high-efficiency competent DH5α cells. Avoid repeated freeze–thaw cycles, as these markedly reduce transformation competency.
5. **Heat Shock Timing.** Avoid exceeding 90 seconds during the 42 °C heat-shock step. Longer exposure can compromise cell viability and decrease recovery.
6. **Post-Shock Recovery Medium.** Using ZymoBroth supplemented with 0.4% glucose enhances recovery by reducing basal promoter activity, which is especially important for constructs expressing potentially burdensome or toxic components.
7. **Troubleshooting Failed Transformation.** Include a positive control transformation (e.g., a plasmid of known efficiency) when troubleshooting. Absence of colonies in the control indicates compromised competent cells rather than issues with the ligation reaction.
8. **Sequencing Verification.** Confirm all constructs by full plasmid sequencing before proceeding to induction experiments. Even minor sequence deviations, especially within promoter or RBS regions, can significantly alter expression strength and circuit dynamics.
9. **Induction Consistency.** Prepare fresh aTc working solutions immediately before use, as the compound is light-sensitive and can degrade over time. Store stock solutions at −20 °C, protected from light, and avoid multiple freeze–thaw cycles.
10. **Preinduction Procedure.** Ensure cells are thoroughly washed after preinduction to remove residual aTc. Even trace inducer carryover can distort initial conditions and downstream dose-response patterns.
11. **Normalization of Fluorescence Data.** Normalize fluorescence readings by optical density (OD_{600}) to correct for differences in cell density, then for pre-induction experiments, further normalize by the maximum expression value of each fluorophore to compare across conditions.

Acknowledgments

This work was supported by a grant from the US National Institutes of Health (R35GM142896 to X.-J.T.)

Competing Interests The authors declare no competing interests.

References

1. Slusarczyk AL, Lin A, Weiss R (2012) Foundations for the design and implementation of synthetic genetic circuits. Nat Rev Genet 13: 406–420. https://doi.org/10.1038/nrg3227
2. Brophy JAN, Voigt CA (2014) Principles of genetic circuit design. Nat Methods 11:508–520. https://doi.org/10.1038/nmeth.2926
3. Cardinale S, Arkin AP (2012) Contextualizing context for synthetic biology—identifying causes of failure of synthetic biological systems. Biotechnol J 7:856–866. https://doi.org/10.1002/biot.201200085
4. Boo ALC, Ellis T, Stan G-B (2019) Host-aware synthetic biology. Curr Opin Syst Biol 14:66–72. https://doi.org/10.1016/j.coisb.2019.03.001
5. Borkowski O, Ceroni F, Stan G-B, Ellis T (2016) Overloaded and stressed: whole-cell considerations for bacterial synthetic biology. Curr Opin Microbiol 33:123–130. https://doi.org/10.1016/j.mib.2016.07.009
6. Şimşek E, Yao Y, Lee D, You LC (2022) Toward predictive engineering of gene circuits. Trends Biotechnol 41:760–768. https://doi.org/10.1016/j.tibtech.2022.11.001
7. Stone A, Youssef A, Rijal S, Zhang R, Tian X-J (2024) Context-dependent redesign of robust synthetic gene circuits. Trends Biotechnol 42: 895–909. https://doi.org/10.1016/j.tibtech.2024.01.003
8. Ceroni F, Algar R, Stan G-B, Stan G-B, Stan G-B, Ellis T, Ellis T, Ellis T (2015) Quantifying cellular capacity identifies gene expression designs with reduced burden. Nat Methods 12:415–418. https://doi.org/10.1038/nmeth.3339
9. Carbonell-Ballestero M, Garcia-Ramallo E, Montañez R, Rodriguez-Caso C, Macía J (2016) Dealing with the genetic load in bacterial synthetic biology circuits: convergences with the ohm's law. Nucleic Acids Res 44: 496–507. https://doi.org/10.1093/nar/gkv1280
10. Gyorgy A, Jiménez JI, Yazbek J, Huang H-H, Chung H, Weiss R, Del Vecchio D (2015) Isocost lines describe the cellular economy of genetic circuits. Biophys J 109:639–646. https://doi.org/10.1016/j.bpj.2015.06.034
11. Qian Y, Huang HH, Jimenez JI, Del Vecchio D (2017) Resource competition shapes the response of genetic circuits. ACS Synth Biol 6:1263–1272. https://doi.org/10.1021/acssynbio.6b00361
12. Zhang R, Goetz H, Melendez-Alvarez J, Li J, Ding T, Wang X, Tian X-J (2021) Winner-takes-all resource competition redirects cascading cell fate transitions. Nat Commun 12:853. https://doi.org/10.1038/s41467-021-21125-3
13. Di Blasi R, Pisani M, Tedeschi F, Marbiah MM, Polizzi K, Furini SMN, Siciliano V, Ceroni F (2023) Resource-aware construct design in mammalian cells. Nat Commun 14:3576. https://doi.org/10.1038/s41467-023-39252-4
14. Roberto DB, Mara P, Tedeschi F, Marbiah M, Polizzi KM, Furini S, Siciliano V, Ceroni F (2023) Resource-aware construct design in mammalian cells. Nat Commun 14:3576. https://doi.org/10.6084/M9.FIGSHARE.22659211
15. Goetz H, Stone A, Zhang R, Lai Y, Tian X (2022) Double-edged role of resource competition in gene expression noise and control. Adv Genet 3:2100050. https://doi.org/10.1002/ggn2.202100050
16. Goetz H, Zhang R, Wang X, Tian X-J (2025) Resource competition-driven bistability and stochastic switching amplify gene expression noise. PLoS Comput Biol 21:e1012931. https://doi.org/10.1371/journal.pcbi.1012931
17. Zhang R, Li J, Melendez-Alvarez J, Chen XW, Sochor P, Goetz H, Zhang Q, Ding T, Wang X, Tian X-J (2020) Topology-dependent interference of synthetic gene circuit function by growth feedback. Nat Chem Biol 16:695–701. https://doi.org/10.1038/s41589-020-0509-x
18. Tan CEM, Marguet P, You LC (2009) Emergent bistability by a growth-modulating positive feedback circuit. Nat Chem Biol 5:842–848. https://doi.org/10.1038/nchembio.218
19. Stone A, Rijal S, Zhang R, Tian X-J (2024) Enhancing circuit stability under growth feedback with supplementary repressive regulation. Nucleic Acids Res 9:1512–1521. https://doi.org/10.1093/nar/gkad1233
20. Zhang R, Yang WF, Zhang RX, Godar A, Rijal S, Youssef A, Nielsen DR, Zheng WW, Tian X-J (2025) Phase separation to buffer growth-mediated dilution in synthetic circuits. Cell 188:1–15. https://doi.org/10.1016/j.cell.2025.10.017
21. Liao C, Blanchard AE, Lu T (2017) An integrative circuit–host modelling framework for predicting synthetic gene network behaviours. Nat Microbiol 2:1658–1666. https://doi.org/10.1038/s41564-017-0022-5

22. Melendez-Alvarez JR, Tian X-J (2022) Emergence of qualitative states in synthetic circuits driven by ultrasensitive growth feedback. PLoS Comput Biol 18:e1010518. https://doi.org/10.1371/journal.pcbi.1010518
23. Melendez-Alvarez J, He CH, Zhang R, Kuang Y, Tian X-J (2021) Emergent damped oscillation induced by nutrient-modulating growth feedback. ACS Synth Biol 10:1227–1236. https://doi.org/10.1021/acssynbio.1c00041
24. Melendez-Alvarez JR, Zhang R, Tian X-J (2023) Growth feedback confers cooperativity in resource-competing synthetic gene circuits. Chaos Solitons Fractals 173:113713. https://doi.org/10.1016/j.chaos.2023.113713
25. Julou T, Gervais T, de Groot D, van Nimwegen E (2025) Growth rate controls the sensitivity of gene regulatory circuits. Sci Adv 11:eadu9279. https://doi.org/10.1126/sciadv.adu9279
26. Youssef A, Rijal S, Zhang R, Tian X-J (2026) Resource competition and growth dilution modulate synthetic Gene Cascade dynamics. ACS Synth Biol. 15:181–190. https://doi.org/10.1021/acssynbio.5c00606

Chapter 20

Dynamic Constrained Allocation Flux Balance Analysis (dCAFBA)

Huili Yuan, Yang Bai, and Xiongfei Fu

Abstract

The survival of bacteria in dynamic environments depends on their ability to coordinate metabolism and gene expression. This interplay is fundamental: metabolism provides the energy and building blocks for protein synthesis, while the proteome, in turn, catalyzes and constrains metabolic fluxes. Here, we present a practical guide to dynamic Constrained Allocation Flux Balance Analysis (dCAFBA), a computational framework that integrates proteome allocation constraints with metabolic models to simulate this coordination. We provide step-by-step instructions for model construction, simulation, and analysis. The utility of dCAFBA is demonstrated through case studies of bacterial adaptation to carbon and amino acid shifts.

Key words Proteome allocation, Dynamic flux balance analysis (dFBA), Constraint-based modeling, Bacterial growth laws, Metabolic flux dynamics, Changing environment

1 Introduction

In their natural habits, bacteria inhabit dynamic environments where nutrient availability can shift rapidly. To survive, they must swiftly adjust their gene expression and metabolic fluxes to meet new demands for growth and protein synthesis. Understanding these adaptive mechanisms is crucial for deciphering microbial survival strategies and rational design of genetic circuits. This chapter provides a practical guide to using the dynamic Constrained Allocation Flux Balance Analysis (dCAFBA) model to simulate the kinetics of growth and metabolic fluxes during these fluctuating environments [1]. The chapter is organized as follows: Subheading 1 revisits the foundational concepts of flux balance analysis (FBA), bacterial growth laws, and the development of the dCAFBA model. Subheading 2 introduces the materials required for implementation. A step-by-step guide is presented in Subheading 3, followed by an illustrative application in Subheading 4.

Xiao-Jun Tian (ed.), *Synthetic Gene Circuits: Methods and Protocols*, Methods in Molecular Biology, vol. 3041, https://doi.org/10.1007/978-1-0716-5304-3_20,

1.1 Fundamentals of FBA

Flux Balance Analysis (FBA) is a computational method that predicts metabolic fluxes—the rates at which metabolites are converted in each chemical reaction inside a cell [2]. Its foundation is the principle of mass conservation for metabolites and reaction fluxes. FBA requires a genome-scale metabolic model (GSMM), typically an underdetermined system of linear equations, as input. By defining an objective function (e.g., biomass maximization), FBA identifies an optimal flux distribution that satisfies mass balance constraints. The output is a vector of fluxes for each reaction. For an extensive primer on FBA, see Orth et al. [2]. This section introduces the basic concepts.

1.1.1 Genome-Scale Metabolic Model

A genome-scale metabolic model (GSMM) is a computational reconstruction of the complete set of known metabolic functions in a specific organism or cell type. It catalogs biochemical reactions, their associated enzymes, and the genes encoding those enzymes [3, 4]. GSMM reconstruction is a multi-stage process, detailed in protocols involving over 100 steps [5]. The first stage typically involves generating a draft model automatically from genome annotation. This draft then undergoes rigorous manual curation to correct errors, fill gaps, and ensure network consistency [5]. To convert this reconstruction into a functional mathematical model, key steps include defining the biomass composition and specifying available nutrients. The model is validated by testing its ability to produce biomass precursors and comparing its predictions with experimental data. Validated models are often saved in the Systems Biology Markup Language (SBML) format [6] for compatibility with tools like the COBRA Toolbox [7].

A simplified metabolic network is illustrated in Fig. 1a. Metabolites within these networks are classified as internal or external. External metabolites (e.g., nutrients, byproducts) can be exchanged with the environment, while internal metabolites are metabolic intermediates.

1.1.2 Stoichiometric Matrix

Mathematically, a GSMM is represented by a stoichiometric matrix, $S_{m \times n}$ (Fig. 1b), which is central to constraint-based modeling. In this matrix, each row corresponds to a metabolite (m), and each column corresponds to a reaction (n). Each element, $\boldsymbol{S}(i, j)$, represents the stoichiometric coefficient of metabolite i in reaction j [8]. A negative coefficient indicates the metabolite is a reactant (consumed), while a positive coefficient indicates it is a product (formed). The matrix defines the connectivity of the metabolic network and is the primary constraint for determining feasible flux distributions.

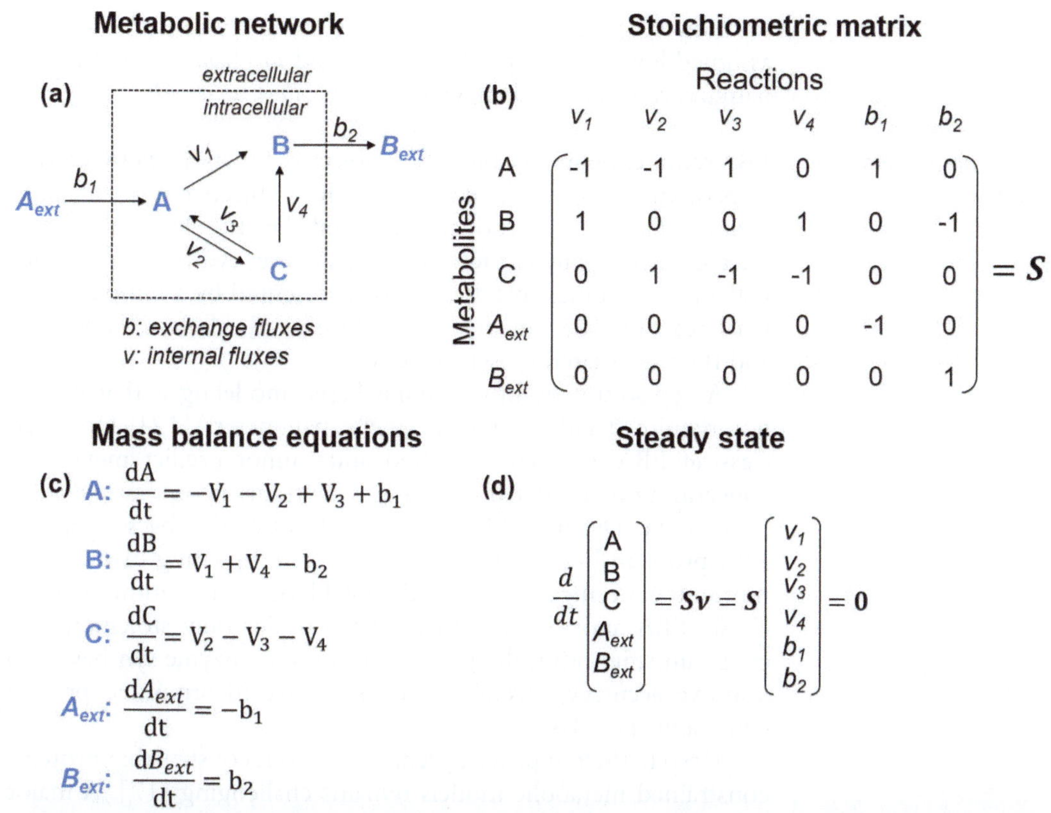

Fig. 1 Example of the stoichiometric matrix for a simple metabolic network. (**a**) The running example metabolic network, consisting of six reactions, two external metabolites (A_{ext}, B_{ext}) that act as source or sink, and three internal metabolites (**a**, **b** and **c**). (**b**) The corresponding stoichiometric matrix. (**c**) The mass balance equations. (**d**) The steady-state assumption

1.1.3 Steady-State Assumption

A key principle of FBA is the steady-state assumption, which posits that the concentration of internal metabolites remains constant over time. This implies that for any internal metabolite, its total production rate must equal its total consumption rate (Fig. 1d). This assumption is justified because intracellular metabolic dynamics are typically much faster than extracellular environmental changes or cellular growth [9].

The mass conservation constraint at steady state is formulated using the stoichiometric matrix. For instance, the mass balance for metabolite A in Fig. 1a is given by: $\frac{dA}{dt} = -\nu_1 - \nu_2 + \nu_3 + b_1$. Applying this to all metabolites (Fig. 1c) yields the system-wide equation:

$$S\boldsymbol{\nu} = 0 \tag{1}$$

where S is the stoichiometric matrix and $\boldsymbol{\nu}$ is a vector of metabolic fluxes.

Since metabolic networks often have more reactions (columns) than metabolites (rows) [10], the system is underdetermined, leading to infinitely many flux solutions that satisfy Eq. 2. To reduce the solution space, additional constraints are applied. Each flux ν_i is assigned lower and upper bounds (lb and ub) based on thermodynamic reversibility and enzyme capacity: $lb \ll \nu_i \ll ub$.

1.1.4 The Objective Function

FBA requires defining a biological objective function relevant to the physiological context. The most common objective is the maximization of biomass production, based on the hypothesis that microbes, especially in the exponential phase, evolve to grow optimally [11]. Biomass production is represented by a lumped "biomass reaction" that incorporates all essential cellular components and their associated biosynthetic costs.

A key advantage of constraint-based modeling is that it does not require detailed enzyme kinetic parameters [12]. However, classical FBA is a static method and cannot predict metabolite concentrations or dynamic changes without incorporating additional data. Dynamic FBA (dFBA) addresses this by solving the FBA problem iteratively at each time step, updating extracellular nutrient concentrations accordingly [13]. A limitation of both classic FBA and dFBA is that they consider only stoichiometric constraints, ignoring the proteomic costs of enzyme synthesis. To improve accuracy, several frameworks have incorporated protein constraints [14–16].

Despite their improved predictive power, constructing protein-constrained metabolic models remains challenging [17]. A major hurdle is the reliance on enzymatic parameters like turnover numbers (e.g., k_{cat}), which are often unavailable at the genome-scale or measured under non-physiological conditions. An alternative approach is to integrate phenomenological growth laws into FBA, bypassing the need for extensive kinetic data [18, 19]. The following section provides an introduction to growth laws and proteome allocation theory, which form the phenomenological foundation for the dCAFBA model.

1.2 Growth Laws and Proteome Allocation

Bacterial growth has been studied for decades. Monod's seminal work in the 1940s established a relationship between growth rate and extracellular nutrient concentration [20]. Subsequent research revealed robust correlations between cellular composition and physiological state, forming the basis of empirical growth laws [21]. A key insight was that faster-growing cells have higher concentrations of macromolecules like RNA and protein, indicating a positive correlation between cell size and growth rate.

1.2.1 Steady-State Growth Laws

Growth laws formally linking gene expression to growth rate were established by Hwa and colleagues to describe steady-state exponential growth [22]. In balanced exponential growth, all cellular components (e.g., proteins, RNA, and others) increase at the same rate. If M_X denotes the total abundance of a cellular component in a standard culture volume, its abundance M_X increases as:

$\frac{dM_X}{dt} = \lambda M_X$, where λ is the growth rate. This rate is constant under stable conditions but varies with nutrient quality or other environmental factors.

A key observation is that the ribosomal protein content is strongly and linearly correlated with the growth rate [21, 23, 24]. To sustain the protein synthesis demands of growth, cells must modulate ribosomal abundance. For *E. coli*, this empirical relationship is described by:

$$\phi_R = \phi_{R,0} + \frac{\lambda}{\gamma} \quad (2)$$

Here, ϕ_R is the mass fraction of the proteome allocated to ribosomal proteins. The intercept, $\phi_{R,\,0}$, represents the basal ribosomal fraction at zero growth rate. The constant γ is related to the maximal translation rate of the ribosomes [25].

Proteome allocation is a central concept in bacterial growth laws, linking gene expression to metabolism [22, 26, 27]. The total proteome is a finite resource partitioned among functional sectors:

- C-sector (ϕ_C): Catabolic proteins for carbon uptake and transport.
- E-sector (ϕ_E): Metabolic enzymes for biosynthesis.
- R-sector (ϕ_R): Ribosomal and translational proteins.
- Q-sector (ϕ_Q): Housekeeping proteins with constant expression across conditions.

The fraction of each sector $\phi_i = \frac{M_i}{M_P}$, where M_i is the mass of that sector *i*, and M_P is the total protein mass. These fractions sum to unity: $\phi_C + \phi_E + \phi_R + \phi_Q = 1$. This partition creates a trade-off: increasing allocation to one sector necessitates a decrease in others, allowing efficient resource reallocation in response to environmental changes [27].

1.2.2 From Steady-State to Dynamic Environments

While steady-state growth laws describe physiology in stable environments, microbes in nature face frequent nutrient fluctuations. Studies of bacterial behavior during nutrient shifts [28–30] have been synthesized in the Flux-Controlled Regulation (FCR) model. A key advance is the Flux-Controlled Regulation (FCR) model introduced by Erickson et al. [29]. This model extends proteome allocation theory to dynamic conditions. It proposes that during a fluctuation, the *expression rate* of each protein sector, $\chi(\sigma)$, changes rapidly in response to metabolic signals (e.g., cAMP, ppGpp). In contrast, the actual *allocation* (ϕ) of each sector evolves more slowly, governed by protein synthesis and dilution by growth:

$$\frac{d\phi_i}{dt} = v_R(t)\chi_i(\sigma) - \lambda\phi_i(t) \quad (3)$$

where $v_R(t)$ is the protein synthesis rate (unit in per mass) and λ is the growth rate. The regulation function $\chi(\sigma)$ is derived from steady-state proteomic data.

The translational activity $\sigma(t)$ is defined as $\sigma(t) \equiv \frac{v_R(t)}{\phi_R(t)}$. When amino acid production is coordinated, the protein synthesis rate $v_R(t)$ is proportional to the total amino acid synthesis rate $(v_{aa}(t))$: $v_R(t) \propto v_{aa}(t)$. The dynamics of translational activity can be derived by its full derivative:

$$\frac{d\sigma(t)}{dt} = d\left(\frac{\beta v_{aa}(t)}{\phi_R(t)}\right) = \beta \frac{1}{\phi_R(t)}\left(\frac{dv_{aa}}{dt} - v_{aa}(t)\sigma(t)(\chi_R - \phi_R)\right) \quad (4)$$

During amino acid depletion, where biosynthesis is activated sequentially, the dynamics of $\sigma(t)$ must be represented with stepwise functions, as detailed in the case study on amino acid downshift.

1.3 Integrating Proteome Allocation with FBA: The dCAFBA Model

The FCR model successfully captures proteome dynamics during nutrient shifts without modeling complex regulatory networks. However, it does not predict metabolic flux dynamics. To bridge this gap, we developed the dynamic Constrained Allocation FBA (dCAFBA) model [1], which integrates the FCR framework with metabolic models (Fig. 2).

In dCAFBA, environmental changes trigger a redistribution of metabolic fluxes to synthesize amino acids and other metabolites.

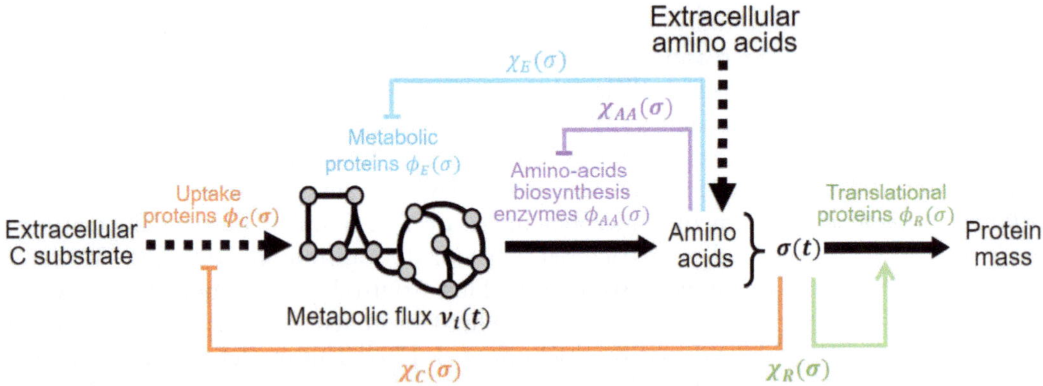

Fig. 2 A schematic illustrating the integration of protein synthesis within the *E. coli* metabolic network, incorporating regulation functions of protein synthesis and flux constraints. This integration allows for the prediction of protein allocation and metabolic fluxes. Extracellular carbon substrates are taken in by uptake proteins ϕ_C in the carbon influx v_C and are converted into central precursors and amino acids by metabolic proteins ϕ_E. These amino acids, which affect the ribosome's translational activity $\sigma(t)$, are further consumed by the ribosome for synthesizing uptake (ϕ_C), metabolic (ϕ_E), translational (ϕ_R) proteins, AAB enzymes (ϕ_{AAB}) and others. The regulatory effects are implemented through regulation functions $\chi_C(\sigma)$, $\chi_E(\sigma)$, $\chi_R(\sigma)$, and $\chi_{AA}(\sigma)$ which determine the amounts of uptake, metabolic, ribosomal proteins, and AAB enzymes, respectively

These metabolites act as precursors for protein synthesis, influencing the protein synthesis rate (v_R) and translational activity (σ). In turn, σ governs the allocation of the proteome among functional sectors (C, R, Q, E, and a dedicated AAB-sector for amino acid biosynthesis) [1, 29]. Changes in protein allocation (ϕ) impose time-varying constraints on metabolic fluxes; for example, the maximum uptake rate is constrained by ϕ_C, and metabolic reaction fluxes are constrained by ϕ_E.

Given that metabolic equilibration (~seconds) [31] is much faster than proteome reallocation (tens of minutes), we use FBA to compute the metabolic flux distribution at each time point [2]. The dynamically calculated protein allocations are used to set time-varying constraints on reaction fluxes based on their type. The remainder of this chapter provides a step-by-step guide to constructing and simulating a dCAFBA model.

2 Materials

To build and run a dCAFBA model, you will need the following components.

2.1 Software

1. MATLAB: dCAFBA is implemented in the MATLAB environment.
2. COBRA toolbox: This toolbox is required for handling and simulating GSMMs. It provides functions to load, modify, and simulate metabolic networks. Installation instructions are available at https://github.com/opencobra/cobratoolbox (*see* **Note 1**).
3. Solver: dCAFBA requires a linear programming (LP) solver to perform the optimizations. It is compatible with GLPK (a free solver included in the COBRA toolbox) and Gurobi (a commercial solver, but free for academic use).
4. CAFBA package: Since dCAFBA model is a dynamic extension of the CAFBA framework, its implementation relies on core functions from the CAFBA package provided by Mori et al. [18].

2.2 Prepared Metabolic Models

dCAFBA has been successfully tested with the *E. coli i* JR904 [32] and *i*ML1515 [33] metabolic models. You can adapt it to other models by modifying the *addProteinGroupsToModel()* and *setWeights()* functions from the CAFBA package. *E. coli* GSMMs can be downloaded from databases such as BiGG models (http://bigg.ucsd.edu/) [34] and EcoCyc (http://ecocyc.org/) [35].

2.3 Data Requirements

To parameterize the proteome allocation constraint in dCAFBA, you need experimental data of growth rate λ and proteome data ϕ. These data are used to determine key modelling parameters like protein efficiencies (w_C, w_E), and the specific regulation functions (h_1, h_2) (*see* **Note 3**).

3 Methods

This section provides a step-by-step guide to implementing the dCAFBA simulation for *E. coli* using the *i*JR904 model as an example. We will go through the entire procedure, from setting up the computational environment to analyzing the results. The scripts for the simulations described here are available at https://github.com/BaiYangBqdq/dCAFBA. This process involves six main steps: (1) initialize the modelling environment, (2) prepare the metabolic model, (3) define the initial conditions, (4) set up the proteome allocation constraints, (5) formulate and solve the optimization problem, and (6) analyze the results.

3.1 Initialize the Modeling Environment

The dCAFBA model is implemented using MATLAB. The first step is to configure your environment by initializing the COBRA toolbox and selecting a linear programming solver.

To perform these in MATLAB, you would enter the following commands:

```
% Initialize the Cobra toolbox
>> initCobraToobox;
% Select GLPK as the solver
>> changeCobraSolver('glpk','LP');
```

3.2 Prepare the Metabolic Model

Once your simulation environment is ready, the next step is to prepare a metabolic model for use with dCAFBA. This preparation involves dividing the metabolic reactions into four groups (C-sector, E-sector, R-sector, and Q-sector) and assigning a protein cost to each sector (*see* **Note 2**). This cost represents the protein investment required to carry a unit of flux. This is accomplished by using two key functions [18]:

- *addProteinGroupsToModel()*: This function categorizes reactions into the four sectors, creating a new field called '*protGroup*'.
- *setWeights()*: This function assigns a specific protein cost, or weight, to each reaction, which is stored in the new '*w*' field.

A detailed description of the function *addProteinGroupsToModel()* can be found in Algorithm 1.

Algorithm 1: Grouping Metabolic Reactions into Proteome Sectors

Input: Metabolic model, a carbon source *(optC)*

Output: Updated model with a '*.protGroup*' field containing four reaction groups

Steps:

1. If *optC* is not provided, set *optC* = '*glc*'(glucose).
2. Define E-sector reactions: set *phiEopt* = '*subSys*'.
3. Identify biomass reaction index: *Biomass* = *find(model.c)*.
4. If model lacks '*protGroup*' field, initialize *protGroup* as a structure with 4 entries (C, E, R, Q sectors). For each sector, define *name* (sector identifier), *phi0* (offset of protein allocation), *rxns* (list of reactions in the sector). If *protGroup* exists, only update the C-sector reactions using *optC.*
5. Define C-sector reactions (*phiC*):
   - Find exchange reaction(s) corresponding to the specified carbon source (*optC),* such as '*EX_glc_e_*' for glucose.
   - Assign these reactions to the C-sector (*see* **Note 2**).
6. Define E-sector reactions (*phiE*):
   - Identify reactions that are not in predefined subsystems, such as transport, exchange, or alternate carbon metabolism.
   - Remove any reactions that overlaps with C-sector from this group,
7. Define R-sector reactions (*phiR*):
   - Assign biomass reaction to R-sector.
   - Remove overlaps with C- and E-sector reactions,
8. Define Q-sector reactions (*phiQ and others*):
   - Include all remaining reactions that are not in C, E, and E-sectors.
   - Remove overlaps with C-, E-, and R-sector reactions.
9. Update model by assigning *protGroup* structure to *model.protGroup.*

3.3 Define Initial Conditions

The dCAFBA model is a dynamic simulation, indicating it tracks changes over time. To start the simulation, we must first establish the initial state of the cell growth. This includes:

- Initial growth rate (λ): Set this to the experimentally measured pre-shift growth rate.
- Initial proteome fractions and other parameters: Calculate the initial fractions for ribosomal protein ϕ_R based on the steady-state growth law: $\phi_R = \phi_{R,0} + \frac{\lambda}{\gamma}$. Calculate the initial value for ϕ_C based on $\phi_C = 1 - \frac{\sigma}{\lambda_C}\phi_R$. Compute the initial value for

translational activity σ using $\sigma = \frac{\lambda}{\phi_R}$. Set the values of regulation functions (χ) the same value as their corresponding protein fraction (ϕ). $\phi_{R,\ 0}$, γ and λ_C are the constants from steady-state measurements.

- Growth environment: Configure the exchange reactions of the metabolic model to reflect the available nutrients in the pre-shift medium. For example, in a glucose minimal medium, set the upper bound of the glucose uptake flux to a non-zero value and all other carbon source uptake fluxes to zero.

3.4 Set Proteome Allocation Constraints

The central feature of dCAFBA is linking metabolic fluxes to the available proteome. This is done by imposing upper bounds on metabolic fluxes, which are proportional to the mass fraction of the proteins that catalyze them. The two primary constraints are:

- Carbon uptake flux (ν_C): The influx of the carbon substrate is limited by the amount of C-sector proteins via $\nu_C \leq \frac{\phi_C(t)}{w_C}$.
- Total metabolic flux ($\sum|v_i(t)|$): The sum of all the absolute metabolic fluxes (v_i) is constrained by the E-sector enzymes via $\sum_i |v_i(t)| \leq \frac{\phi_E(t)}{w_E}$.

Here, w_C and w_E characterize the cost of carbon uptake (ϕ_C) and metabolic proteins (ϕ_E) needed to carry one unit flux, respectively. These are determined by fitting the model to steady-state experimental data.

3.5 Formulate and Solve the Optimization Problem

dCAFBA model is a hybrid approach that combines FBA with ordinary differential equations (ODEs). The simulation proceeds through an iterative loop for each time step (*see* **Note 4**).

- Constrain the model: Use the current proteome fractions (ϕ_C, ϕ_E, etc.) to set the upper bounds on the corresponding fluxes as described in **step 4**. The associated parameters are described in Table 1.
- Solve FBA: With the constraints in place, solve the FBA problem to find the optimal flux distribution (v_i) and the corresponding biomass production rate (λ). The optimization problem is formulated as:

$$\text{Maximize} : \lambda$$

$$\text{Subject to} \sum_i S_{\mu i} v_i = 0 \forall \mu \ \text{Mass balance}$$

$$l_i \ll v_i \ll \mu_i \forall i \ \text{Flux bounds}$$

$$\sum_i w_E |v_i| \leq 1 - \varnothing_C - \varnothing_R - \varnothing_Q \tag{5}$$

Table 1
Model variables used in this study

Variables	Related equation	Description
σ	$\sigma(t) \equiv \frac{\lambda(t)}{\phi_R(t)}$	Translational activity
χ_R	$\chi_R(t) = \frac{\phi_{Rb,0}}{1-\sigma/\gamma}$	Regulation function of ribosomal protein
χ_C	$\chi_C(t) = \left(1 - \frac{\sigma}{\lambda_C}\chi_R\right)\phi_{Cmax}$	Regulation function of catabolic protein allocation
$\chi_{A,\ tot}$	$\chi_{A,tot}(t) = \phi_{A,tot}^{max} - \alpha_A \sigma \frac{\phi_{Rb,0}}{1-\sigma/\gamma}$	Regulation function of total AAB enzymes
$\chi_{A,\ j}$	$\chi_{A,\ j}(t) = \eta_j(t)\chi_{A,\ tot}(t)$	Regulation function of individual AAB enzyme group
η_j	$\eta_j(t) = \begin{cases} 0 \; j > \tilde{j}(t) \\ \phi^*_{A,j}/\sum_{m=1}^{\tilde{j}(t)}\phi^*_{A,m} \; j \le \tilde{j}(t) \end{cases}$	Fraction of protein synthesis for total AAB enzyme allocated to group *j*
$\tilde{j}(t)$		The number of AAB enzyme groups being synthesized at time t
q_j	$q_j(t) \equiv \frac{\phi_{A,j}(t)}{\phi^*_{A,j}}$	Abundance of AAB enzyme group *j* relative to its post-shift steady-state
σ	$\frac{d\sigma(t)}{dt} = \beta\frac{1}{\phi_R(t)}\left(\frac{dv_{aa}}{dt} - v_{aa}(t)\sigma(t)(\chi_C - \chi_R)\right)$	Dynamics of σ used in carbon shift or titrating PHB gene expression
ϕ_R	$\frac{d\phi_R}{dt} = \sigma\phi_R(\chi_R - \phi_R)$	Mass fraction of ribosomal protein over total protein
ϕ_C	$\frac{d\phi_{C1,2}}{dt} = \sigma\phi_R(b_{1,2}\chi_C - \phi_C)$	Mass fraction of catabolic protein over total protein
$\phi_{A,\ tot}$	$\phi_{A,tot} = \sum_j \phi_{A,j}$	Mass fraction of the total AAB enzymes over total protein
$\phi_{A,\ j}$	$\frac{d\phi_{A,j}}{dt} = \sigma\phi_R\left(\chi_{A,j} - \phi_{A,j}\right)$	Mass fraction of individual AAB enzyme group
ϕ_E	$\phi_E(t) = 1 - \phi_C(t) - \phi_R(t) - \phi_Q$	Mass fraction of metabolic protein over total protein
ν_{aa}		Protein synthesis flux
ν_C		Carbon uptake rate
ν_E		The total metabolic flux
ν_{ji}		Flux carried by each AAB biosynthetic reaction
λ		Growth rate

In this table, we summarized all the variables used in the model. Here the subscript ***i*** represents a protein group. The subscript ***j*** only represents amino acid biosynthetic (AAB) enzyme groups. The asterisk represents the value in the post-shift steady-state

Table 2
Description of parameters

Symbol	Description
γ	Inverse slope of growth laws
λ_C	Strain specific constant of growth laws
$\phi_{R,0}$	a growth rate-independent offset of R sector
ϕ_Q	Fraction of house-keeping protein
ϕ_C^{max}	Maximum of proteome attainable to C, E and R sectors
β	Conversion factor between moles and gram
$h_{1,2}$	Carbon-specific allocation
ω_E	Enzyme cost per unit metabolic influx
ω_C	Enzyme cost per unit carbon influx
$\phi_{A,tot}^{max}$	Maximum of AAB enzyme at a given condition
α_A	Relation parameter between AAB enzyme and growth rate

$$l \ll v_C \ll \frac{\varnothing_C}{w_C} \tag{6}$$

$$v_{Aj,i} \leq \frac{\phi_{Aj}}{\phi_{Aj}^*} \times \left| \nu_{Aji}^f \right| \tag{7}$$

where the sum of all absolute metabolic fluxes (ν_i) is constrained by: $\sum_i |v_i(t)| \leq \frac{\phi_E(t)}{w_E}$ (Eq. 4). Equation 6 is the flux boundary set on carbon influx during the carbon shift. Upon the AA downshift, we incorporate Eq. 7 to FBA as additional constraints on amino acid biosynthetic reactions (see case studies).

- Calculate protein synthesis rate: From the optimal flux vector v_i, we calculate the total amino acid synthesis flux (v_{aa}), which serves as a proxy for the total protein synthesis rate (v_R).
- Update proteome fractions: Use the calculated protein synthesis rate and the regulatory functions χ to solve the system of ODEs (listed in Table 2) for one time step, then update the proteome fractions (ϕ) for the next iteration. This is solved using a simple forward Euler method.

This iterative framework enables us to simulate the dynamic cross-feedback between metabolic fluxes and proteome reallocation over time. A detailed description of this step can be found in Algorithm 2.

Algorithm 2: Methods for Solving the dCAFBA Problem

Load model, experimental data and model parameters ($\gamma, \lambda_C, \phi_{R,0}, w_C, w_E, \phi_Q, dt, t_{shift}$).

Initialize all arrays (e.g., $\sigma, \chi_C, \chi_R, \phi_C, \phi_R, \lambda, \nu_{aa}, \nu_C, \nu$) with the length of simulation time (t).

Set initial conditions and define the growth medium.

% main loop

For each time step i from 2 to length (t):

% pre-shift dynamics

If $t(i) < t_{shift}$ (pre-shift dynamics):

% update proteome allocation (ϕ) using the forward Euler method

$\phi(i) = \phi(i-1) + dt \cdot (\sigma(i-1) \cdot \phi(i-1)(\chi(i-1) - \phi(i-1))$

% update the E-sector proteome fraction ϕ_E:

$$\phi_E(i) = 1 - \phi_C(i) - \phi_R(i) - \phi_Q$$

% constrain the reaction fluxes with the newly updated protein fractions via Eq. 5, Eq. 6, Eq. 7, Eq. 8, and Eq. 9 accordingly

% solve FBA problem using *CAFBA_OptimizeCbModel_glpk()* to find the optimal growth rate λ and flux vector ν

$\{\lambda(i), \quad \nu_R(i), \quad \nu(i)\} =$ *CAFBA_OptimizeCbModel_glpk* (model, ϕ_C, ϕ_R, ϕ_Q)

% update the translational activity σ and regulatory functions for the next time

$$\sigma(i) = \sigma(i-1) + dt \cdot \left(\beta \cdot \frac{1}{\phi_R(i-1)} \cdot \left(\frac{\nu_{aa}(i) - \nu_{aa}(i-1)}{dt} - \nu_{aa}(i-1) \cdot \sigma(i-1) \cdot (\chi_R(i-1) - \phi_R(i-1)\right)\right)$$

$$\chi_R(i) = \phi_{R,0} / \left(1 - \frac{\sigma(i)}{\gamma}\right)$$

If $t(i) = t_{shift}$ (at the instant of shift)

% change the model substrates accordingly, update translational activity σ by

$$\sigma(i) = \nu_R(i) / \phi_R(i)$$

% update the regulatory function χ, protein allocation ϕ and fluxes ν as done in the pre-shift stage

If $t(i) \geq t_{shift}$ (post-shift dynamics)

% update the values of all variables using the same procedure as the pre-shift stage until a new steady-state is reached.

End for

3.6 Analyze the Results

After the simulation completes, the final step is to visualize and interpret the results. The model outputs time-course data for the growth rate (λ), proteome allocation (ϕ), and individual metabolic fluxes (ν_i). Visualizing the simulated metabolic fluxes is particularly useful for understanding how the cell adjusts its metabolism in response to a nutrient shift. Tools like Escher [36] provide a great framework for contextualizing data about metabolic pathways and models. Finally, comparing model predictions with experimental data is crucial for validation.

4 Case Studies

The core dCAFBA simulation described above is general, but specific scenarios require model adjustments. Here, we demonstrate its application to two common scenarios: a carbon downshift and an amino acids (AA) downshift. While dCAFBA can also simulate a carbon upshift and an AA upshift, we will focus on these more complex downshift cases for this chapter.

4.1 Carbon Downshift

We simulated a carbon downshift where cells are transited from a medium containing two carbon sources (pyruvate and glycerol) to a medium with only one carbon source (pyruvate). When two carbon substrates are co-utilized, the C-sector is split into two sub-sectors: ϕ_{C1} and ϕ_{C2}, each corresponding to the uptake proteins for one substrate. The allocation of protein synthesis flux to each sub-sector (ϕ_{C1}, ϕ_{C2}) is controlled by specific regulation functions $h_{1,\ 2}$ that are introduced by a step function based on the presence or absence of the cognate carbon substrate. This parameter $h_{1,\ 2}$ denotes the partition of a carbon to ensure the predicted growth rate and protein fractions matched experimental data. We calculate the protein fractions for each carbon substrate during the shift by $\frac{d\phi_{C1,2}}{dt} = \sigma\phi_R\left(h_{1,2}\chi_C - \phi_{C1,2}\right)$. The metabolic enzymes ϕ_E can be calculated based on $\phi_E = 1 - \phi_R - (\phi_{C1} + \phi_{C2}) - \phi_Q$. With the updated protein fractions ϕ_E, and $\phi_{C1,\ 2}$, we can bound the reaction fluxes based on Eq. 4 and 6, respectively, and proceed with the dCAFBA simulation in Algorithm 2. The parameter values and constants used in the simulation of carbon downshift are shown in Table 3.

Upon depletion of glycerol, the model captures an abrupt drop in the growth rate, followed by a recovery to a new, lower steady state (Fig. 3a). During the transitions, cells change their gene expression patterns in response to changes in their growth conditions. A key insight from the model is the prediction of a transient overshoot in growth rate during the transition, which arises from a limitation switch from carbon uptake proteins to metabolic enzymes [1].

The changes in metabolic fluxes are distinct (Fig. 3b). Before the shift, fluxes through both the pyruvate ('PYRt2r') and glycerol ('GLYK') uptake reactions are active. After the shift, the glycerol uptake flux ('GLYK') drops to zero. However, the flux of 'G3PD2'

Table 3
The parameters and constants used in the simulation of carbon shifts using *E. coli* iJR904 metabolic model

	Substarte1	Substarte2	λ_1 (h^{-1})	λ_{12} (h^{-1})	w_{C1} gh/mmol	w_{C2}	h_1	h_2	ϕ_{C1}^{max}	ϕ_{C2}^{max}	w_E gh/mmol	γ (h^{-1})	λ_C (h^{-1})	ϕ_Q	$\phi_{R,0}$	β (g/mmol)
Down-shift	Pyruvate	Glycerol	0.68	0.96	5.05×10^{-3}	3.07×10^{-3}	0.6	1	0.47	0.41	8.3×10^{-4}	11.02	1.17	0.45	0.049	0.1968

Value of w_E is obtained from Mori et al. [18]
Values of γ, λ_C, ϕ_Q and $\phi_{R,0}$ were obtained from Erickson et al. [29]

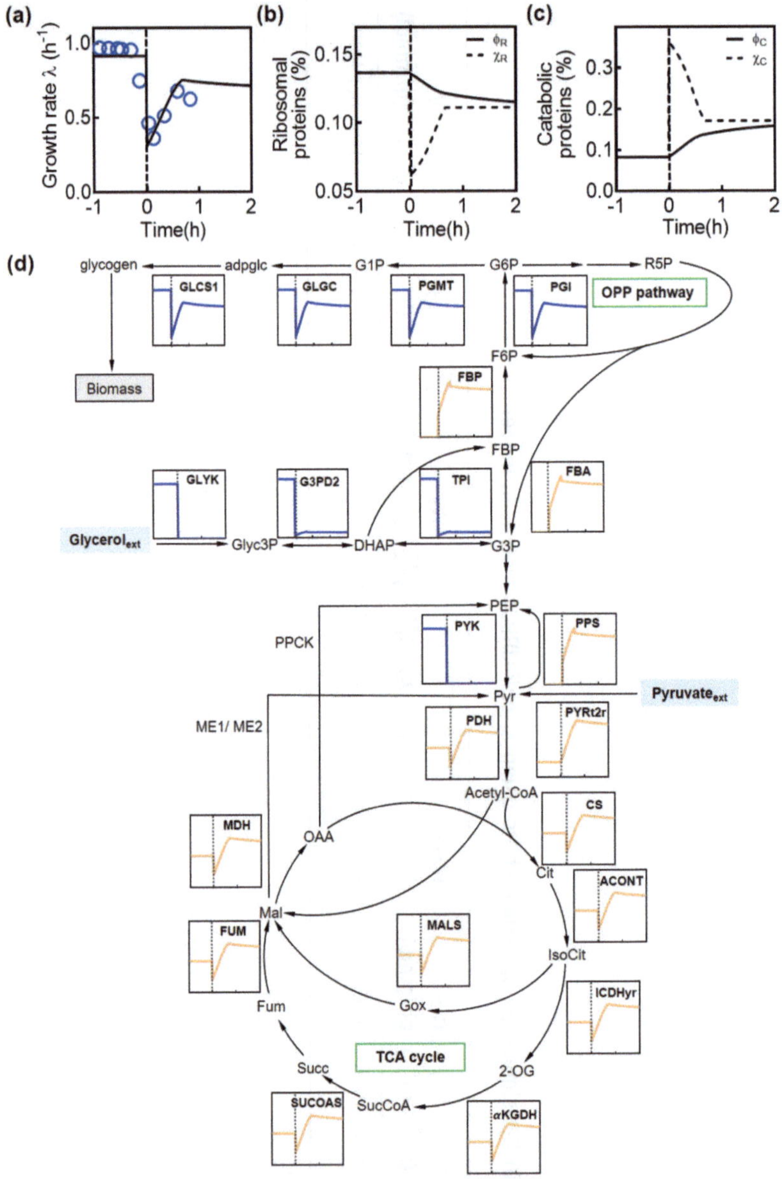

Fig. 3 Model illustration of carbon downshift and simulated results. (**a**), Dynamics of growth rate $\lambda(t)$, regulatory functions (χ_C and χ_R), protein fractions (ϕ_R and ϕ_C) during a carbon downshift from pyruvate and glycerol to pyruvate. Model predictions (solid lines) are compared with experimental data (blue circles). (**b**) The flux dynamics of selected reactions in the glycolysis, gluconeogenesis, and TCA pathways during the carbon downshift. The x-axis represents time (in hours), and the y-axis represents reaction flux (mmol/gDW/h). The vertical grey line at 0 hour indicates the shifting time, separating the pre-shift (left) from the post-shift. **Abbreviations: Metabolites**: *$Glycerol_{ext}$* extracellular glycerol, *G1P* glucose-1-phosphate, *G6P* glucose-6-phosphate, *F6P* fructose-6-phosphate, *FBP* fructose-1,6-biphosphate, *G3P* glyceraldehyde-3-phosphate, *DHAP* dihydroxyacetone phosphate, *PEP* phosphoenol pyruvate, *Pyr* pyruvate, *OAA* oxalacetic acid, *Cit* citrate, *IsoCit* threo-isocitrate, *Succ* succinate, *$Succ_{ext}$* extracellular succinate, *$Pyruvate_{ext}$* extracellular pyruvate, *SucCoA* succinyl-CoA, *Fum* fumarate, *Mal* malate, *2-OG* 2-oxoglutarate, *adpglc* ADP-glucose, *Gly3P* glycerol-3-phosphate

Enzymes: *PGI* glucose-6-phosphate isomerase, *PFK* phosphofructokinase, *FBP* fructose bisphosphatase, *FBA* fructose bisphosphate aldolase, *TPI* triose phosphate isomerase, *PPS* phosphoenolpyruvate synthase, *PYK*

remained nonzero in the post-shift as 'G3PD2' operated in reverse direction, producing glycerol-3-phosphate (Glyc3P) from dihydroxyacetone phosphate (DHAP). In the post-shift, the pyruvate uptake reaction remained active, exhibiting two distinct transition stages. Initially, the pyruvate uptake flux increased rapidly, accompanied by an increasing in the expression of its uptake proteins. In the second stage, the pyruvate influx gradually decreased until a final steady state was achieved. Similarly, the flux dynamics of other reactions, such as pyruvate dehydrogenase (PDH) and citrate synthase, also exhibited an overshoot phenomenon, which has been extensively discussed in our recent work [1].

4.2 Amino Acid Downshift

We also simulated an amino acid (AA) downshift, where cells transition from a rich medium (glycerol supplemented with amino acids) to a minimal medium (glycerol only).

4.2.1 Model Adjustments

For this scenario, we define a new proteome sub-sector: AA sector ($\phi_{A,\ tot}$), which contains all amino acid biosynthesis (AAB) enzymes and is a subset of the E-sector. To model the sequential activation of different AAB pathways, we incorporated an as-needed allocation strategy that is protein synthesis is preferentially allocated to the enzymes for the most growth-limiting amino acid first [37]. The protein synthesis flux directed to total AAB enzymes $\chi_{A,\ tot}$ is further sub-allocated to the enzymes of individual pathways (for example, alanine, leucine, indicated by protein fractions ϕ_{A1}, ϕ_{A2}, etc.) by the pathway-specific allocation function $\chi_{A,\ j}$ (Fig. 4). In our model, the total AAB enzymes are classified into 14 groups based on operon structure and 89 reactions (*see* Table 4).

Similar to the carbon shifts, protein synthesis flux allocated to ribosomal proteins (ϕ_R) and total AAB enzymes ($\phi_{A,\ tot}$) is controlled by the regulatory functions χ_R and $\chi_{A,\ tot}$, respectively. These functions are mathematically expressed as $\chi_R(\sigma) = \frac{\phi_{R,0}}{1-\sigma/\gamma}$ and $\chi_{A,tot}(\sigma) = \phi_{A,tot}^{max} - \alpha_A \sigma \frac{\phi_{R,0}}{1-\sigma/\gamma}$, where α_A, γ, $\phi_{A,tot}^{max}$, $\phi_{R,\ 0}$ are constants determined from pre- and post-shift steady-state measurements (Table 5). The translational activity $\sigma(t)$ is defined by the protein synthesis rate $v_R(t)$ and ribosome allocation $\phi_R(t)$: $\sigma(t) \equiv \frac{v_R(t)}{\phi_R(t)}$.

The synthesis flux directed to total AAB enzymes $\chi_{A,\ tot}$ is further sub-allocated to the enzymes of individual pathways by pathway-specific allocation function $\chi_{A,\ j}$: $\chi_{A,\ j}(t) \equiv \eta_j(t)\chi_{A,\ tot}$.

Fig. 3 (continued) pyruvate kinase, *CS* citrate synthase, *ACONT* aconitase, *ICDHyr* isocitrate dehydrogenase, *AKGDH* 2-Oxogluterate dehydrogenase, *SUCOAS* succinyl CoA synthetase, *SUCD1i* succinate dehydrogenase, *FUM* fumarase, *MDH* malate dehydrogenase, *PPCK* phosphoenolpyruvate carboxykinase, *ME1/ME2* malic enzyme, *PGMT* Phosphoglucomutase, *GLGC* glucose-1-phosphate adenylyltransferase, *GLCS1* glycogen synthase, *G3PD2* glycerol-3-phosphate dehydrogenase

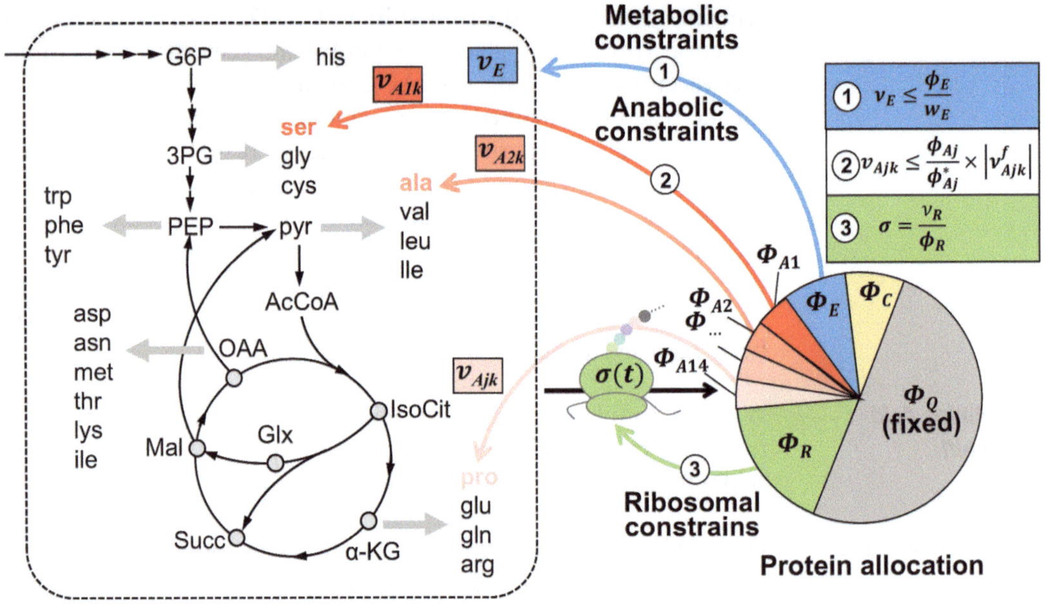

Fig. 4 Model illustration of AA down-shift and its simulated results. The taken-in substrates are converted to amino acids through a series of metabolic reactions, the fluxes of which ν_E are constrained by metabolic proteins ϕ_E. The pool of the amino acids reflects the ribosomal translational activity $\sigma(t)$, which controls the allocation of protein synthesis flux to catabolic proteins (ϕ_C), AAB enzymes (ϕ_{Aj}), metabolic proteins (ϕ_E), ribosomal proteins (ϕ_R) and housekeeping proteins (ϕ_Q). The reaction flux of each AAB pathway ν_{Aji} is constrained by its protein abundance level (ϕ_{Aj}) relative to the final post-shift level (ϕ^*_{Aj}) as well as its flux under the steady state in the post-shift $\left(\nu^f_{Aji}\right)$. The dynamics of the model is set by the translational activity $\sigma(t)$, the metabolic fluxes are assumed to be in quasi-steady state

Referring to Wu et al. [37], the group-specific relative allocation functions $\eta_j(t)$ are described as

$$\eta_j(t) = \begin{cases} 0 \ j > \tilde{j}(t) \\ \phi^*_{A,j} / \sum_{m=1}^{\tilde{j}(t)} \phi^*_{A,m} \ j \leq \tilde{j}(t) \end{cases} \tag{8}$$

where $\tilde{j}(t)$ is the number of enzyme groups being synthesized at time t. Groups with $j \leq \tilde{j}(t)$ are growth-limiting, while those with $j > \tilde{j}(t)$ are not Whether an enzyme group is growth-limiting is determined by its present enzyme abundance ϕ_{Aj} relative to the final abundance $\phi^*_{Aj,}$ denoted as $q_j(t) = \frac{\phi_{A,j}}{\phi^*_{A,j}}$. At any time t, the fractional enzyme reserve $q_j(t)$ divides the AAB groups into two classes: non-limiting groups receive no allocation ($\eta_j(t) = 0$), and limiting groups with the lowest $q_j(t)$ are allocated in proportion to their final steady state abundance $\phi^*_{A,j}$.

With the regulatory functions χ_R, $\chi_{A,\ tot}$ and $\chi_{A,\ j}$, we linearize this system into the following forward Euler scheme to update the protein fractions at each time step. With the updated ϕ_E, $\phi_{A,\ tot}$ and $\phi_{A,\ j}$, we can set the bounds on the corresponding reaction flux by Eqs. 5 and 7. Then, we compute the flux distribution v_i by

Table 4
The AAB enzyme groups used in the simulation of AA down-shift

AAB enzyme group	Genes	Reactions ID in *i*JR904
Arg group	*argA, argB, argC, argD, argE, argF, argG, argH, argI*	ACGS, ACGK, AGPR, ACOTA, ACODA, NACODA, ARGSS, ARGSL, OCBT
Aro group	*aroA, aroB, aroC, aroD, aroE, aroF, aroG, aroH, aroK, aroL*	PSCVT, DHQS, CHORS, DHQD, SHK3Dr, DDPA, SHKK
Cys group	*cysC, cysD, cysE, cysH, cysI, cysJ, cysK, cysM, cysN*	ADSK, SADT2, SERAT, PAPSR, SULR, CYSS
Glt group	*gdhA, gltB, gltD*	GLUDy, GLUSy
His group	*hisA, hisB, hisC, hisD, hisF, hisG, hisH, hisI*	PRMICIi, IGPDH, HISTP, HSTPT, HISTD, IG3PS, ATPPRT, PARTPP, PRAMPC
Ilv group	*ilvA, ilvB, ilvC, ilvD, ilvE, ilvG_1, ilvH, ilvI, ilvM, ilvN*	THRD_L, ACHBS, ACLS, KARA1i, KARA2i, DHAD1, DHAD2, ILETA, VALTA, ACHBS, ACLS
Leu group	*leuA, leuB, leuC, leuD*	IPPS, IMPD, OMCDC, IPPMIa, IPPMIb
Lys group	*Asd, dapA, dapB, dapD, dapE, dapF, lysA, lysC*	ASAD, DHDPS, DHDPRy, THDPS, SDPDS, DAPE, DAPDC
Met group	*metA, metB, metC, metE, metH, metL*	HSST, SHSL1, CYSTL, METS, ASPK, HSDy
Other group	*alaA, alaC, asnA, asnB, aspC, glnA, glyA, proA, proB, tyrB, proC*	ALATA_L, ASNS2, ASNS1, ASPTA, GLNS, THRAr, GHMT2, G5SD, GLU5K, P5CR
Phetyr group	*pheA, tyrA, tyrB*	CHORM, PPNDH, PHETA1, LEUTAi, TYRTA
Ser group	*serA, serB, serC*	PGCD, PSP_L, PSERT
Thr group	*thrA, thrB, thrC*	HSK, THRS
Trp group	*trpA, trpB, trpC, trpD, trpE*	TRPS1, TRPS2, TRPS3, PRAIi, IGPS, ANS, ANPRT

optimizing the biomass synthesis flux. The translational activity σ is then calculated to drive further protein reallocation kinetics, forming an iterative loop. Since the carbon source remains constant during the AA shift, the carbon uptake proteins ϕ_C is fixed at 3%. The housekeeping protein ϕ_Q is set to 4.5% based on a previous study.

4.3 Simulated Results

The model enables us to predict the dynamics of protein allocation and flux redistribution during AA downshift. During the transition, the predicted dynamics of growth rate $\lambda(t)$ and proteome fractions

Table 5
Experimental data and the constants of the substrates used in the simulation of AA downshift with the *E. coli* iJR904 metabolic model

	Pre-shift Glycerol+18AA[a]	Post-shift Glycerol[a]
$\phi_{A,\ met}$	0.60‰	32.2‰
$\phi_{A,\ arg}$	0.14‰	4.71‰
$\phi_{A,\ leu}$	0.25‰	5.85‰
$\phi_{A,\ trp}$	0.12‰	2.31‰
$\phi_{A,\ glt}$	0.72‰	7.23‰
$\phi_{A,\ ilv}$	1.60‰	13.1‰
$\phi_{A,\ his}$	0.74 ‰	4.15‰
$\phi_{A,\ ser}$	1.00‰	3.79‰
$\phi_{A,\ lys}$	2.63‰	6.32‰
$\phi_{A,\ thr}$	2.40‰	4.13‰
$\phi_{A,\ aro}$	1.97‰	2.97‰
$\phi_{A,\ phetyr}$	0.65‰	1.09‰
$\phi_{A,\ cys}$	8.66 3‰	9.88‰
$\phi_{A,\ other}$	11.1‰	11.5‰
$\phi_{A,tot}^{max}$	18.1%	Nd
α_A	0.11 h	Nd
$\phi_{R,\ 0}$	4.45%	
γ	8.35/h	
λ	1.40 h	0.68 h
ϕ_Q	0.45	0.45
w_E(gh/mmol)	2.0×10^{-5}	8.3×10^{-4}
ϕ_C	3%	3%

$\phi_{A,tot}^{max}$ and α_A were calculated based on pre-shift and post-shift growth rate and total AAB enzyme abundance as described in Wu et al. [37]
[a]Data were obtained from Wu et al. [37]

(ϕ_R and $\phi_{A,\ tot}$) were in good alignment with experimental results (Fig. 5a–c), indicating our model has good predictive power.

Regarding the metabolic fluxes during AA downshift, the model captures diverse and complex flux dynamics (Fig. 5d). Fluxes of certain reactions, such as those catalyzed by pyruvate dehydrogenase (PDH) and phosphoenolpyruvate carboxykinase (PPCK) dropped to zero immediately after the shift. In contrast, fluxes of other reactions, such as those catalyzed by phosphoenolpyruvate

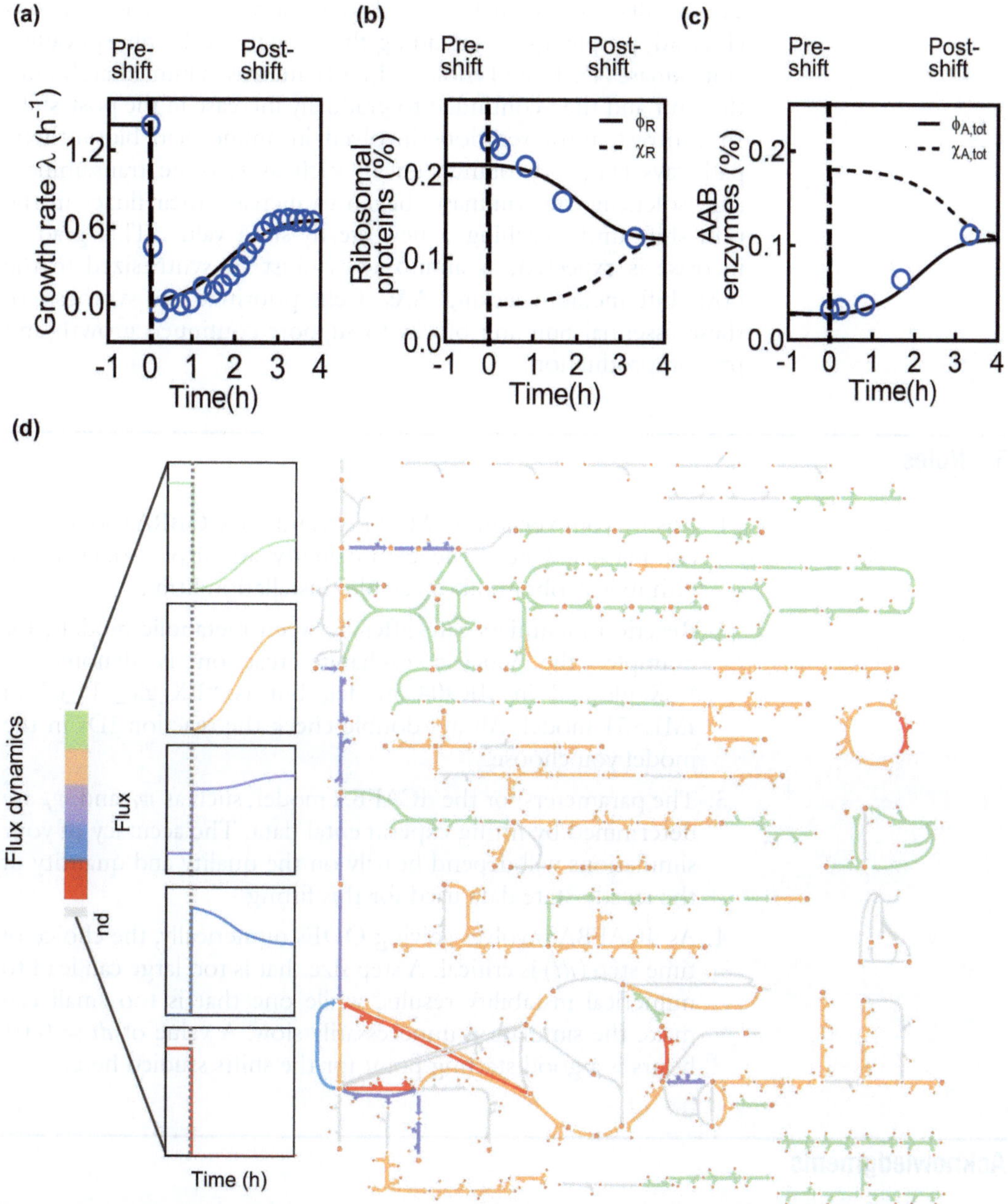

Fig. 5 Growth kinetics and metabolic flux reconfiguration during an amino acid upshift. (**a**) Dynamics of growth rate $\lambda(t)$, (**b**) protein fractions ϕ_{AA} and (**c**) ϕ_R in the conditions of transition from glycerol supplemented with amino acids to glycerol. (**d**) Illustration of changes in reaction fluxes during AA downshift. Colors (left) represent the flux dynamics of metabolic reactions. While dashed lines in the legend indicate the shift time, i.e., $t = 0$ h. The study condition here is from glycerol supplemented with 18AA to glycerol. This figure was created using Escher [36]

synthase (PPS), increased abruptly at the shift time and then gradually decreased. Based on the patterns of reaction fluxes, the reactions in the central metabolic pathway were categorized into five groups, each assigned a distinct color for visualization (Fig. 5d).

The results showed that fluxes through lower glycolytic reactions (Fig. 5d, purple lines), including those catalyzed by phosphoglycerate kinase (PGK) and enolase (ENO), increased immediately after the shift and then continued to gradually increase in the post-shift. In contrast, most reactions involved in amino acid biosynthetic pathways (Fig. 5d, orange lines), such as tyrosine transaminase and isoleucine transaminase, began to increase their fluxes in the post-shift until reaching a new steady-state value. This gradual increase is expected, as amino acids must be synthesized in the post-shift medium lacking AAs. Cells prioritize the synthesis of these essential building blocks to support continued growth and protein production.

5 Notes

1. Ensure your versions of MATLAB and the COBRA toolbox are compatible. Check the compatibility at: https://opencobra.github.io/cobratoolbox/stable/installation.html.
2. Reaction identifiers can differ between metabolic models, for example, the glucose exchange reaction is denoted as "EX_glc_e_" in *i*JR904 model, but is "EX_glc__D_e" in *i*ML551 model. Always double-check the reaction IDs in the model you choose.
3. The parameters for the dCAFBA model, such as w_C and w_E are determined by fitting experimental data. The accuracy of your simulations will depend heavily on the quality and quantity of the steady-state data used for this fitting.
4. As dCAFBA involves solving ODEs numerically, the choice of time step (dt) is critical. A step size that is too large can lead to numerical instability results, while one that is too small can make the simulation unnecessarily slow. A value of $dt = 0.01$ hours is a good starting point for the shifts studied here.

Acknowledgments

This work is partially supported by the National Key Research and Development Program of China (2024YFA0919600), Strategic Priority Research Program of the Chinese Academy of Sciences (XDB0480000), and NSFC (T2525031, 32571666).

References

1. Yuan H, Bai Y, Li X, Fu X (2024) Cross-regulation between proteome reallocation and metabolic flux redistribution governs bacterial growth transition kinetics. Metab Eng 82:60–68. https://doi.org/10.1016/j.ymben.2024.01.008
2. Orth JD, Thiele I, Palsson BO (2010) What is flux balance analysis? Nat Biotechnol 28(3):

245–248. https://doi.org/10.1038/nbt.1614

3. Yuan H, Cheung CY, Poolman MG, Hilbers PA, van Riel NA (2016) A genome-scale metabolic network reconstruction of tomato (Solanum lycopersicum L.) and its application to photorespiratory metabolism. Plant J 85(2):289–304. https://doi.org/10.1111/tpj.13075
4. Yuan H, Cheung CY, Hilbers PA, van Riel NA (2016) Flux balance analysis of plant metabolism: the effect of biomass composition and model structure on model predictions. Front Plant Sci 7:537. https://doi.org/10.3389/fpls.2016.00537
5. Thiele I, Palsson BO (2010) A protocol for generating a high-quality genome-scale metabolic reconstruction. Nat Protoc 5(1):93–121. https://doi.org/10.1038/nprot.2009.203
6. Hucka M, Bergmann FT, Chaouiya C, Drager A, Hoops S, Keating SM, Konig M, Novere NL, Myers CJ, Olivier BG, Sahle S, Schaff JC, Sheriff R, Smith LP, Waltemath D, Wilkinson DJ, Zhang F (2019) The systems biology markup language (SBML): language specification for level 3 version 2 Core release 2. J Integr Bioinform 16(2). https://doi.org/10.1515/jib-2019-0021
7. Heirendt L, Arreckx S, Pfau T, Mendoza SN, Richelle A, Heinken A, Haraldsdottir HS, Wachowiak J, Keating SM, Vlasov V, Magnusdottir S, Ng CY, Preciat G, Zagare A, Chan SHJ, Aurich MK, Clancy CM, Modamio J, Sauls JT, Noronha A, Bordbar A, Cousins B, El Assal DC, Valcarcel LV, Apaolaza I, Ghaderi S, Ahookhosh M, Ben Guebila M, Kostromins A, Sompairac N, Le HM, Ma D, Sun Y, Wang L, Yurkovich JT, Oliveira MAP, Vuong PT, El Assal LP, Kuperstein I, Zinovyev A, Hinton HS, Bryant WA, Aragon Artacho FJ, Planes FJ, Stalidzans E, Maass A, Vempala S, Hucka M, Saunders MA, Maranas CD, Lewis NE, Sauter T, Palsson BO, Thiele I, Fleming RMT (2019) Creation and analysis of biochemical constraint-based models using the COBRA Toolbox v.3.0. Nat Protoc 14(3):639–702. https://doi.org/10.1038/s41596-018-0098-2
8. Palsson BØ (2006) Systems biology: properties of reconstructed networks. Cambridge University Press
9. Kauffman KJ, Prakash P, Edwards JS (2003) Advances in flux balance analysis. Curr Opin Biotechnol 14(5):491–496. https://doi.org/10.1016/j.copbio.2003.08.001
10. Klamt S, Schuster S, Gilles ED (2002) Calculability analysis in underdetermined metabolic networks illustrated by a model of the central metabolism in purple nonsulfur bacteria. Biotechnol Bioeng 77(7):734–751. https://doi.org/10.1002/bit.10153
11. Feist AM, Palsson BO (2010) The biomass objective function. Curr Opin Microbiol 13(3):344–349. https://doi.org/10.1016/j.mib.2010.03.003
12. Smallbone K, Simeonidis E, Broomhead DS, Kell DB (2007) Something from nothing: bridging the gap between constraint-based and kinetic modelling. FEBS J 274(21):5576–5585. https://doi.org/10.1111/j.1742-4658.2007.06076.x
13. Mahadevan R, Edwards JS, Doyle FJ (2002) Dynamic flux balance analysis of diauxic growth in Escherichia coli. Biophys J 83(3):1331–1340
14. Salvy P, Hatzimanikatis V (2021) Emergence of diauxie as an optimal growth strategy under resource allocation constraints in cellular metabolism. Proc Natl Acad Sci USA 118(8). https://doi.org/10.1073/pnas.2013836118
15. Yang L, Ebrahim A, Lloyd CJ, Saunders MA, Palsson BO (2019) DynamicME: dynamic simulation and refinement of integrated models of metabolism and protein expression. BMC Syst Biol 13:2
16. Waldherr S, Oyarzun DA, Bockmayr A (2015) Dynamic optimization of metabolic networks coupled with gene expression. J Theor Biol 365:469–485. https://doi.org/10.1016/j.jtbi.2014.10.035
17. Heckmann D, Lloyd CJ, Mih N, Ha Y, Zielinski DC, Haiman ZB, Desouki AA, Lercher MJ, Palsson BO (2018) Machine learning applied to enzyme turnover numbers reveals protein structural correlates and improves metabolic models. Nat Commun 9(1):5252. https://doi.org/10.1038/s41467-018-07652-6
18. Mori M, Hwa T, Martin OC, De Martino A, Marinari E (2016) Constrained allocation flux balance analysis. PLoS Comput Biol 12(6)
19. Zeng H, Yang A (2019) Quantification of proteomic and metabolic burdens predicts growth retardation and overflow metabolism in recombinant Escherichia coli. Biotechnol Bioeng 116(6):1484–1495. https://doi.org/10.1002/bit.26943
20. Monod J (2012) The growth of bacterial cultures 139:606 Annu Rev Microbiol
21. Schaechter M, Maaloe O, Kjeldgaard NO (1958) Dependency on medium and temperature of cell size and chemical composition during balanced grown of salmonella typhimurium. J Gen Microbiol 19(3):

592–606. https://doi.org/10.1099/00221287-19-3-592

22. Scott M, Gunderson CW, Mateescu EM, Zhang Z, Hwa T (2010) Interdependence of cell growth and gene expression: origins and consequences. Science 330(6007): 1099–1102. https://doi.org/10.1126/science.1192588
23. Neidhardt FC, Magasanik B (1960) Studies on the role of ribonucleic acid in the growth of bacteria. Biochim Biophys Acta 42:99–116. https://doi.org/10.1016/0006-3002(60)90757-5
24. Bremer H, Dennis PP (2008) Modulation of chemical composition and other parameters of the cell at different exponential growth rates. EcoSal Plus 3(1). https://doi.org/10.1128/ecosal.5.2.3
25. Klumpp S, Scott M, Pedersen S, Hwa T (2013) Molecular crowding limits translation and cell growth. Proc Natl Acad Sci USA 110(42): 16754–16759. https://doi.org/10.1073/pnas.1310377110
26. You CH, Okano H, Hui S, Zhang ZG, Kim M, Gunderson CW, Wang YP, Lenz P, Yan DL, Hwa T (2013) Coordination of bacterial proteome with metabolism by cyclic AMP signalling. Nature 500(7462):301–306
27. Hui S, Silverman JM, Chen SS, Erickson DW, Basan M, Wang JL, Hwa T, Williamson JR (2015) Quantitative proteomic analysis reveals a simple strategy of global resource allocation in bacteria. Mol Syst Biol 11(2)
28. Basan M, Honda T, Christodoulou D, Horl M, Chang YF, Leoncini E, Mukherjee A, Okano H, Taylor BR, Silverman JM, Sanchez C, Williamson JR, Paulsson J, Hwa T, Sauer U (2020) A universal trade-off between growth and lag in fluctuating environments. Nature 584(7821):470–474. https://doi.org/10.1038/s41586-020-2505-4
29. Erickson DW, Schink SJ, Patsalo V, Williamson JR, Gerland U, Hwa T (2017) A global resource allocation strategy governs growth transition kinetics of Escherichia coli. Nature 551(7678):119–123. https://doi.org/10.1038/nature24299
30. Zhu M, Dai X (2023) Stringent response ensures the timely adaptation of bacterial growth to nutrient downshift. Nat Commun 14(1):467. https://doi.org/10.1038/s41467-023-36254-0
31. Bi S, Kargeti M, Colin R, Farke N, Link H, Sourjik V (2023) Dynamic fluctuations in a bacterial metabolic network. Nat Commun 14(1):2173. https://doi.org/10.1038/s41467-023-37957-0
32. Reed JL, Vo TD, Schilling CH, Palsson BO (2003) An expanded genome-scale model of Escherichia coli K-12 (iJR904 GSM/GPR). Genome Biol 4(9):R54
33. Monk JM, Lloyd CJ, Brunk E, Mih N, Sastry A, King Z, Takeuchi R, Nomura W, Zhang Z, Mori H, Feist AM, Palsson BO (2017) iML1515, a knowledgebase that computes Escherichia coli traits. Nat Biotechnol 35(10):904–908. https://doi.org/10.1038/nbt.3956
34. Norsigian CJ, Pusarla N, McConn JL, Yurkovich JT, Drager A, Palsson BO, King Z (2020) BiGG models 2020: multi-strain genome-scale models and expansion across the phylogenetic tree. Nucleic Acids Res 48(D1):D402–D406. https://doi.org/10.1093/nar/gkz1054
35. Karp PD, Paley S, Caspi R, Kothari A, Krummenacker M, Midford PE, Moore LR, Subhraveti P, Gama-Castro S, Tierrafria VH, Lara P, Muniz-Rascado L, Bonavides-Martinez C, Santos-Zavaleta A, Mackie A, Sun G, Ahn-Horst TA, Choi H, Juenemann R, Knudsen CNM, Covert MW, Collado-Vides J, Paulsen I (2025) The EcoCyc database (2025). EcoSal Plus:eesp00192024. https://doi.org/10.1128/ecosalplus.esp-0019-2024
36. King ZA, Drager A, Ebrahim A, Sonnenschein N, Lewis NE, Palsson BO (2015) Escher: a web application for building, sharing, and embedding data-rich visualizations of biological pathways. PLoS Comput Biol 11(8):e1004321. https://doi.org/10.1371/journal.pcbi.1004321
37. Wu C, Mori M, Abele M, Banaei-Esfahani A, Zhang Z, Okano H, Aebersold R, Ludwig C, Hwa T (2023) Enzyme expression kinetics by Escherichia coli during transition from rich to minimal media depends on proteome reserves. Nat Microbiol 8(2):347–359. https://doi.org/10.1038/s41564-022-01310-w

Chapter 21

Constructing an NF-κB Oscillator for Dual-Input Control in Yeast

Yuanxu Jiang, Yingying Fan, Zhibo Zhang, and Ping Wei

Abstract

Interactions between biological oscillators and environmental periodic signals are crucial for the correct regulation of cellular behaviors. Most synthetic biological oscillators only respond to one signal that initiates the oscillation, limiting the study of the interaction between the biological oscillator and multiple environmental cues. Here, we present a detailed protocol for constructing a synthetic oscillatory circuit that can respond to both α-factor and ethanol inputs. Based on the mammalian NF-κB negative feedback loop, this circuit produces sustained and tunable oscillation in the budding yeast and can be operated by dual periodic inputs.

Key words Synthetic oscillation, Entrainment, Coupled oscillator, Molecular cloning, Yeast transformation, Time-lapse microscopy, Microfluidics

1 Introduction

The ability to generate and sustain oscillations provides organisms with a crucial competitive advantage, allowing them to anticipate and adapt to predictable environmental changes, separate incompatible biochemical processes in time, and enhance the robustness of cellular decision-making through temporal filtering of noise. One particularly powerful approach to unraveling the complexity of natural oscillatory networks is by reconstructing core network motifs from well-characterized parts in living cells [1, 2]. However, despite these compelling proofs of concept, the reliable design and implementation of robust, predictable synthetic oscillators remain a significant challenge.

A core principle underpinning all oscillatory systems is a delicate balance between the synthesis and degradation rates of its core components. This parameter sensitivity often makes initial designs fragile and highly susceptible to cell-to-cell variability, environmental fluctuations, and genetic context effects [3, 4]. Furthermore,

Xiao-Jun Tian (ed.), *Synthetic Gene Circuits: Methods and Protocols*, Methods in Molecular Biology, vol. 3041,
https://doi.org/10.1007/978-1-0716-5304-3_21,

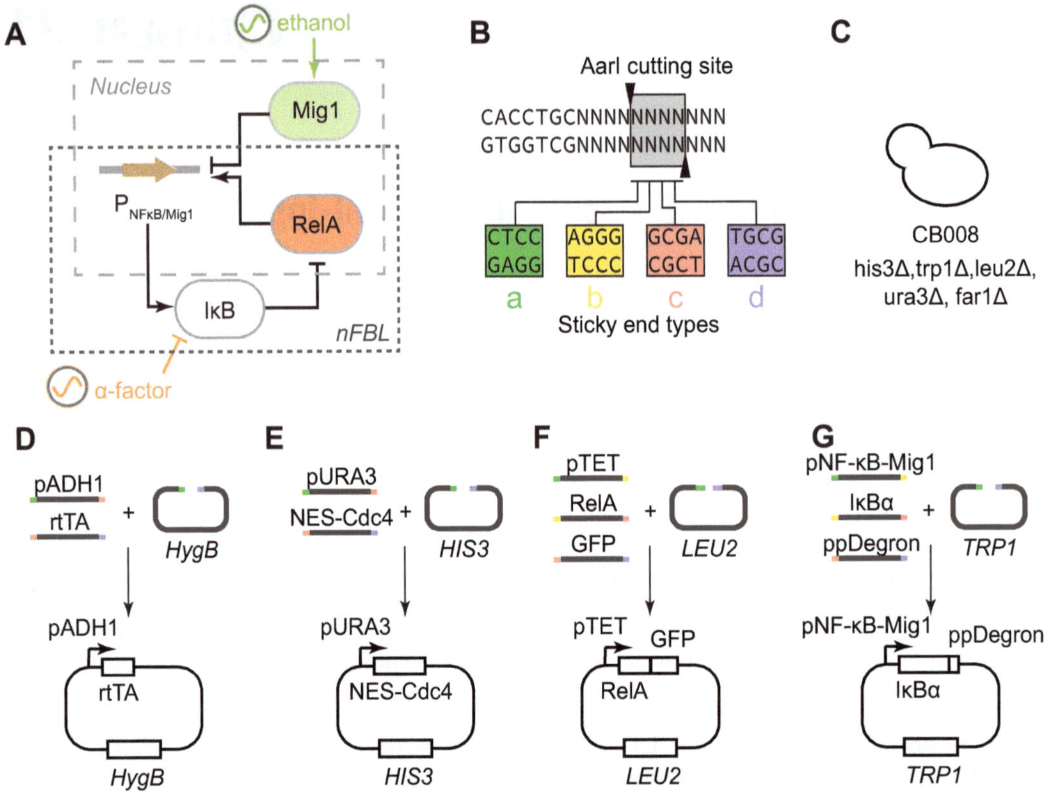

Fig. 1 Genetic cassettes used to construct the NF-κB oscillator. (**a**) Synthetic circuit of the yeast NF-κB oscillator. (**b**) Upper panel: Four types of sticky ends designed for AarI digestion. Lower panel: Schematic figure of the primer design. (**c**) Yeast chassis genomic background used for transformation. (**d–g**). Detailed cloning strategies for each recombinant plasmid

expanding the functionality of synthetic oscillators beyond simple, isolated loops to make them responsive to external cues represents an even greater frontier. Designing synthetic systems that can similarly process complex inputs to modulate oscillation characteristics (e.g., period, phase, or amplitude) is an enormous challenge. It requires not only a stable core oscillator but also the careful integration of input modules that must be orthogonally wired to avoid disruptive crosstalk while effectively modulating the oscillator's kinetics.

To address these challenges and provide a robust, tunable, and modifiable platform for studying biological rhythms, we have designed and implemented a novel synthetic NF-κB oscillator in the yeast *Saccharomyces cerevisiae* (Fig. 1a) [5, 6]. The system is designed to be highly modular, allowing for relatively straightforward replacement of components to alter its dynamic properties. Furthermore, we have engineered the negative feedback promoter within the oscillator network to be responsive to both α-factor and ethanol, enabling user-directed control over the oscillatory behavior.

This protocol is intended as a comprehensive guide for researchers seeking to implement this synthetic oscillator in their own laboratories. We provide detailed, step-by-step instructions for the construction and transformation of the genetic modules, along with best practices for culturing conditions and quantitative time-series measurement of oscillations using live-cell microscopy or flow cytometry. We also include guidance on critical validation steps and troubleshooting common pitfalls. By disseminating this detailed methodology, we aim to lower the barrier to entry for working with synthetic oscillators, providing a reliable tool that can be adapted and built upon to explore the dynamics of biological timekeeping, cellular computation, and the engineering of complex rhythmic behaviors in eukaryotic cells.

2 Materials

2.1 Biological Parts, Plasmid Backbone, and Yeast Chassis Strain

1. Plasmids or other kinds of DNA templates that respectively contain the yeast promoter pADH1/pURA3/pTET07, the EGFP gene, the yeast E3 ligase Cdc4 (N-terminally tagged with a nuclear export signal peptide), the rtTA transcription factor, and the human RelA and IκBα gene should be constructed, extracted, or synthesized before the experiment. An engineered dual-responsive promoter, namely the NF-κB-Mig1-inducible promoter, consisting of four repeats of a 10 bp-κB binding site (GGGACTTTCCGCTGGGACTTTC CAGGGACTTTCCGCTGGGACTTTCC) followed by two repeats of Mig1 binding motif (AATAAAAATGCGGGGAA) upstream to the yeast CYC1 promoter, should also be synthesized ahead of the experiment [7].
2. Prepare yeast single-integrating plasmids with auxotrophic markers HIS3, TRP1, LEU2, and HYGB.
3. Prepare the standard *Saccharomyces cerevisiae* strain with the following genetic background: W303 MATa, far1Δ, pFUS1-GFP::MFA2, his3Δ,trp1Δ,leu2Δ,ura3Δ (Fig. 1c) [8].

2.2 PCR Primers

1. A total of four recombinant plasmids will be constructed to complete the oscillatory circuit. In this protocol, we recommend a standard Digestion-Ligation strategy with the Type II restriction enzyme AarI, enabling seamless cloning. Therefore, primers with corresponding flanking sequences, recognition sites, and paired cutting sites should be designed and synthesized ahead of the experiment. Here, we list all the essential information of the primer pairs used in the molecular cloning process (Table 1, Fig. 1b, d–g).

Table 1
Primer information for molecular cloning

Recombinant vector 1 Selective Marker: *HygB*		
Primer name	*Template*	*AarI cut site type*
Primer 1–1 F Primer 1–1 R	pADH1	a (paired with the vector) b
Primer 1–2 F Primer 1–2 R	rtTA	b d (paired with the vector)
Recombinant vector 2 Selective Marker: *HIS3*		
Primer name	*Template*	*AarI cut site type*
Primer 2–1 F Primer 2–1 R	pURA3	a (paired with the vector) c
Primer 2–2 F Primer 2–2 R	NES-Cdc4	c d (paired with the vector)
Recombinant vector 3 Selective Marker: *LEU2*		
Primer name	*Template*	*AarI cut site type*
Primer 3–1 F Primer 3–1 R	pTet	a (paired with the vector) b
Primer 3–2 F Primer 3–2 R	RelA	b c
Primer 3–3 F Primer 3–3 R	GFP	c d (paired with the vector)
Recombinant vector 4 Selective Marker: *TRP1*		
Primer name	*Template*	*AarI cut site type*
Primer 4–1 F Primer 4–1 R	pNF-κB-Mig1	a (paired with the vector) b
Primer 4–2 F Primer 4–2 R	IκBα	b c
Primer 4–3 F Primer 4–3 R	ppDegron	c d (paired with the vector)

2.3 Yeast Transformation

1. Prepare one tube A and tube B for each transformation: Tube A contains 480 uL 50%(m/v) PEG3350, 60 uL 10xTE (pH 7.5). Vortex to make sure the reagents mix thoroughly (*see* **Note 1**) Tube B contains 3 uL ssDNA and 10 uL reaction mix. The latter contains 2ug of the plasmid for transformation, 1ul 10x Cutsmart buffer, and 0.5ul restriction enzyme PmeI. Incubate the mixture at 37 °C for 30 min. The ssDNA should be boiled at 100 °C for 5 min and immediately kept on ice before use.

2.4 Time-Lapse Microscopy of the Yeast Oscillator

1. CellASIC Y04C microfluidic plates and CellASIC ONIX microfluidic platform for the microfluidic control of environmental inputs (*see* **Note 2**).
2. Yeast α-factor stock, resolved in ddH_2O and diluted to 1 mM.
3. SD complete media supplemented with 5 μg/mL doxycycline.

3 Methods

3.1 Plasmid Construction

1. Prepare 50 μL PCR mix in a PCR tube with standard PCR reagents to amplify each DNA fragment from the corresponding template. Run PCR to amplify the target promoters, pADH1, pURA3, pTET07, and pNF-κB-Mig1, and the target genes, rtTA, NES-Cdc4, RelA, EGFP, IκBα, and ppDegron, all flanked with the corresponding cutting sites.
2. Load the PCR products onto a 1–2% agarose gel and run DNA electrophoresis.
3. Cut out the gel band and place them into microcentrifuge tubes.
4. Use TIANGEN Universal DNA Purification Kit to purify the DNA fragments (including PCR products and linearized vectors).
5. Resuspend the DNA in 30 ~ 50 μL ddH2O in a microcentrifuge tube. Use the NanoDrop Spectrophotometer to measure the concentration of the purified DNA.
6. Use AarI (or other corresponding restriction enzymes in your own plasmid backbone) to linearize the backbone plasmids and the PCR insert plasmids mentioned above (Fig. 1b). Set up a microcentrifuge tube and add 100 μL of the restriction enzyme digestion reaction according to Table 2, mix and incubate for 3 h at 37 °C. Load the restriction enzyme digestion mixture onto a 1–2% agarose gel. Run DNA electrophoresis and isolate the purified linearized vectors and PCR inserts.
7. Use T4 DNA ligase to ligate the PCR inserts with the corresponding linearized cloning vector: Ligate pADH1, rtTA with linearized cloning vector (HYGB marker) to get HYGB-pADH1-rtTA. Ligate pURA3, NES-Cdc4 with linearized cloning vector (HIS3 marker) to get HIS3-pURA3-NES-Cdc4. Ligate pTET07, RelA, EGFP with linearized cloning vector (LEU2 marker) to get LEU-pTET07-RelA-EGFP. Ligate pNF-κB-Mig1, IκBα, ppDegron with linearized cloning vector (TRP1 marker) to get TRP-pNF-κB-Mig1-IκB-α-ppDegron (Fig. 1d–g, *see* **Note 3**). Detailed reagents can be seen in Table 3. Incubate the reaction mixture at room temperature for 30 min.

Table 2
Reagents of the restriction enzyme AarI digestion

Reagent	Amount
Plasmid/fragment (0.5–1 μg/μL)	5 μl
10 x AarI buffer	10 μl
50 x oligonucleotide (0.025 mM)	2 μl
AarI	2.2 μl
Nuclease-free water	Total to 100 μl

Table 3
Reagents of the ligation reaction

Reagent	Amount
Vector	0.3 μl
Fragment 1	0.5 μl
Fragment 2	0.5 μl
Fragment 3 (if applicable)	0.5 μl
10 x T4 DNA ligase buffer	1 μl
T4 DNA ligase	0.5 μl
ddH_2O	Total to 10 μl

8. Transform competent *Escherichia coli* (*E. coli*) with the ligation reaction mixture. Isolate the positive colonies in which PCR fragments are correctly inserted by performing colony PCR and sequencing. Inoculate into 10 mL LB liquid media. Shake the culture in the incubator at 250 rpm at 37 °C for 12 ~ 16 h.
9. Use the QIAprep Spin Miniprep Kit to extract the plasmid from the culture. Elute the plasmid in a smaller volume of elution buffer (e.g., 50 μL) in a microcentrifuge tube from each 10 mL culture. Use the NanoDrop Spectrophotometer to measure the concentration of the extracted plasmids.

3.2 Yeast Transformation

1. Pick single colony of the original chassis yeast strain (background strain for transformation), inoculate into 5 mL YPD, and incubate it overnight at 30 °C.
2. Dilute the culture with 1:20–1:50 ratio into 3 mL YPD medium. Incubate it for 3–5 h to until it reaches the mid-log phase (OD600 ≈ 0.6). Centrifuge the culture at 2000 rpm for 2 min to harvest the yeast cells.

3. Resuspend the cells with 1 mL 1xTE (pH 7.5) and centrifuge it at 2000 rpm for 2 min. Repeat this step twice.
4. Resuspend the cells with 100 uL 1xTE (pH 7.5).
5. Add 100 uL yeast cells to tube B, then transfer the mix to tube A and vortex.
6. Add 60 uL of 1 M LiOAc to tube A and vortex. Incubate the mixture at 42 °C for 30 min. Centrifuge the mixture at 2000 rpm for 2 min to harvest the cells (*see* **Note 4**).
7. Resuspend the cells with 500 uL 1xTE (pH 7.5) and centrifuge at 2000 rpm for 2 min. Repeat this step.
8. Resuspend the cells with 80 uL 1xTE (pH 7.5) and plate them on selective media. Incubate at 30 °C for 1–2 days until the single colonies become clear (*see* **Note 5**).
9. Transform this newly acquired single colony as described in **steps 1–8**, introducing the remaining three plasmids one at a time. The resulting yeast strain will contain all four genetic elements and be capable of performing oscillatory dynamics under constant α-factor input.

3.3 Measurement of the RelA Dynamics

1. Pick a single colony and inoculate into 5 mL SD-dox medium (SD with 5 μg/mL doxycycline) at 30 °C, 50 rpm overnight.
2. The next morning, dilute the yeast culture with a ratio of 1:30–1:100 to 3 mL new SD-dox medium. Incubate the culture at 30 °C, 50 rpm for 3–5 h until it reaches OD600 ≈ 0.5.
3. Add 200 μL yeast culture to the cell loading wells of the Y04C microfluidic plate. Add 100 μL corresponding stimuli (10 μM α-factor/2.5%) ethanol to the medium loading wells. The Y04C microfluidic plate contains four rows and six independently controlled panels of medium-loading wells, enabling simultaneous monitoring of the oscillatory dynamics under four input conditions. The first two panels of wells are normally used for system stabilizing, only SD-dox media shall be added.
4. To verify that the engineered yeast strain can respond to and be entrained by both signals, we recommend the following setting of the medium loading wells: **Condition 1**, continuous 10 μM α-factor, **Condition 2**, oscillatory 10 μM α-factor, **Condition 3**, oscillatory 2.5% ethanol, **Condition 4**, dual-oscillatory signals with tunable phase.
5. Place the plate under the Nikon NIS microscope and load the cells into the chamber with 0.3 psi for 120 s, followed by 4 psi for 0.5 s. Repeat the latter step for 5–10 times until the density reaches about 30 cells per field under a 100x oil immersion lens (Fig. 2A, *see* **Note 6**).

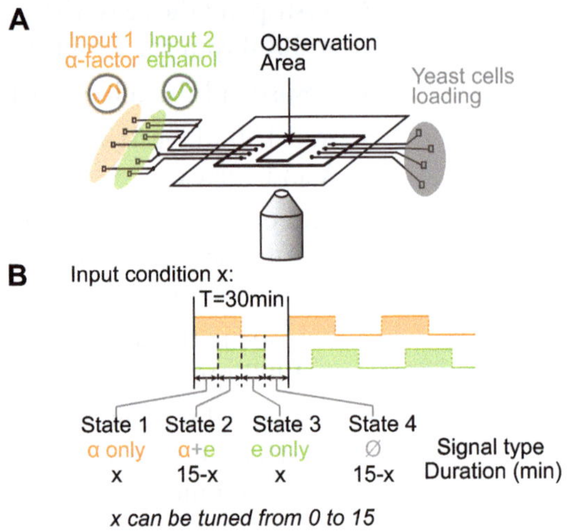

Fig. 2 Setting of the microfluidic program. (**a**) Schematic figure of the setting of the microfluidic device. (**b**) Schematic figure showing the four input states of the dual-oscillatory condition

6. Sparsely select the fields for long-time observation, with no fewer than 5 selected fields per condition. Set the time interval between every two capturing points to be 3–5 min (*see* **Note** 7).

7. Set the microfluidic control program. The control program is divided into 2 stages. In Stage I, panels 1 and 2 are simultaneously activated for 30 min, aiming to stabilize the system. In Stage II, inputs with various dynamics are added. According to the theory of coupled oscillators, synchronization (i.e., 1:1 entrainment) happens when the periods of external and internal oscillations are close. Therefore, since the engineered yeast strain has a free-oscillating period of roughly 30 min, we recommend setting the period of oscillatory inputs to also be 30 min, which means the present and absent time of the oscillatory inputs to be 15 min in a single period. For the dual-oscillatory signal condition (Condition 4), there are four states of input types in one period, i.e., only α-factor present, only ethanol present, both present, and both absent. If the time lag between α-factor and ethanol is set to x (min), then the period would start with the state that only α-factor present, lasting for x (min), then moves to the both present state, lasting for 15-x (min), thirdly to the only ethanol present state, lasting for x (min), and finally to the both absent state, lasting for 15-x (min), thus completing a whole oscillation period. We recommend the total observation time of oscillatory behavior to be 4 hours, which corresponds to 4 cycles for Stage II. The flow rate is fixed to 2 psi throughout the experiment. (Fig 2B).

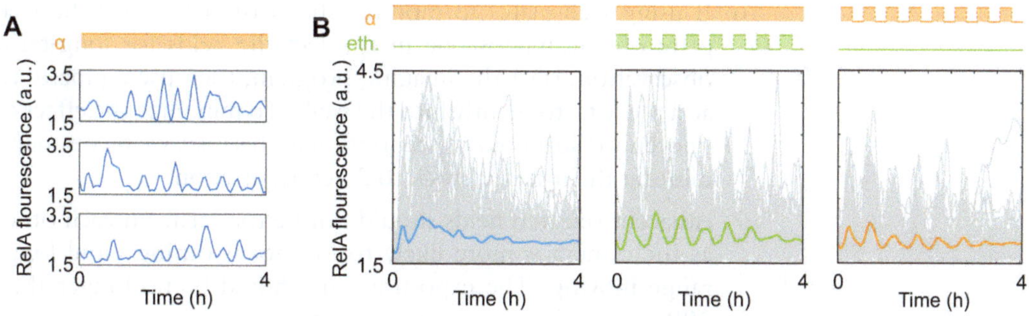

Fig. 3 The single-cell oscillatory dynamics of the synthetic oscillator. (**a**) Single cell RelA nuclear fluorescence dynamics under the condition of continuous α-factor. (**b**) Populational dynamics under various conditions. Left panel: continuous α-factor, middle panel: continuous α-factor with periodic ethanol (Period = 30 min), right panel: periodicα-factor (Period = 30 min)

8. Simultaneously run the microfluidic device and the NIS-element to acquire time-lapse images. The total imaging time should be no longer than 6 h to avoid overgrowth and phototoxicity.
9. Use ImageJ to subtract the background of the fluorescence photos and run Cell segmentation programs to extract the oscillatory single-cell dynamics (Fig. 3a,b).

4 Notes

1. We found it better to prepare this fresh.
2. If microfluidic devices are not available, 96-well or 384-well plates with flat bottoms can also be used to monitor the dynamics. We add 50uL/100uL cell culture to the 96/384-well plate, respectively. After adding the cells, the plate should be centrifuged at 2000 rpm for 2 min to make sure the yeast cells are loaded to the bottom of the well.
3. In order to characterize RelA and IκBα protein expression in yeast, Flag tag can be added to their N-terminal to enable the detection through immunoblotting. The flag tag is not necessary for the proper functioning of the oscillator.
4. Yeast cells in the mixture can be difficult to separate as the liquid is sticky due to the existence of PEG3350. If no clear deposit is observed at the bottom of the tube, extend the centrifuge time to make sure that the cells have been adequately centrifuged.
5. For the antibiotic selective marker, plate the cells on YPD for one night to recover, and replicate them on drug plates the next morning, then grow for another 2 days.

6. If using well-plates to monitor the yeast cells, place the well plate on the microscope first, select the fields for long-term observation, start the imaging program, then use a pipette to add the α-factor gently into the wells. To minimize the effect of unevenly distributed concentration in this case, we recommend diluting the α-factor stock to 2x concentration.
7. Adjacent selected fields should not be too far from each other as the camera is more likely to lose focus during rapid long-range moving. The exposure time should be no longer than 200 ms to minimize phototoxicity.

Acknowledgments

This work was supported partly by the National Key R&D Program of China (2024YFA0919500, 2023YFF1204500), the National Science Foundation of China (31622022, 31470819), and the Strategic Priority Research Program of the Chinese Academy of Sciences (XDA0510000, XDB0480000).

References

1. Elowitz MB, Leibler S (2000) A synthetic oscillatory network of transcriptional regulators. Nature 403(6767):335–338. https://doi.org/10.1038/35002125
2. Gardner TS, Cantor CR, Collins JJ (2000) Construction of a genetic toggle switch in Escherichia coli. Nature 403(6767):339–342. https://doi.org/10.1038/35002131
3. Pomerening JR, Kim SY, Ferrell JE Jr (2005) Systems-level dissection of the cell-cycle oscillator: bypassing positive feedback produces damped oscillations. Cell 122(4):565–578. https://doi.org/10.1016/j.cell.2005.06.016
4. Novák B, Tyson JJ (2008) Design principles of biochemical oscillators. Nat Rev Mol Cell Biol 9(12):981–991. https://doi.org/10.1038/nrm2530
5. Zhang ZB, Wang QY, Ke YX, Liu SY, Ju JQ, Lim WA, Tang C, Wei P (2017) Design of Tunable Oscillatory Dynamics in a synthetic NF-κB signaling circuit. Cell Syst 5(5):460–470.e5. https://doi.org/10.1016/j.cels.2017.09.016
6. Heltberg MS, Jiang Y, Fan Y, Zhang Z, Nordentoft MS, Lin W, Qian L, Ouyang Q, Jensen MH, Wei P (2023) Coupled oscillator cooperativity as a control mechanism in chronobiology. Cell Syst 14(5):382–391.e5. https://doi.org/10.1016/j.cels.2023.04.001
7. Gordley RM, Williams RE, Bashor CJ, Toettcher JE, Yan S, Lim WA (2016) Engineering dynamical control of cell fate switching using synthetic phospho-regulons. Proc Natl Acad Sci USA 113(47):13528–13533. https://doi.org/10.1073/pnas.1610973113
8. Bashor CJ, Helman NC, Yan S, Lim WA (2008) Using engineered scaffold interactions to reshape MAP kinase pathway signaling dynamics. Science 319(5869):1539–1543. https://doi.org/10.1126/science.1151153

Index

Xiao-Jun Tian (ed.), *Synthetic Gene Circuits: Methods and Protocols*, Methods in Molecular Biology, vol. 3041, https://doi.org/10.1007/978-1-0716-5304-3,